Biotic Recovery from Mass Extinction Events

Geological Society Special Publications
Series Editor A. J. FLEET

GEOLOGICAL SOCIETY SPECIAL PUBLICATION NO. 102

Biotic Recovery from Mass Extinction Events

EDITED BY

M. B. HART
Department of Geological Sciences
University of Plymouth
UK

1996
Published by
The Geological Society
London

THE GEOLOGICAL SOCIETY

The Society was founded in 1807 as The Geological Society of London and is the oldest geological society in the world. It received its Royal Charter in 1825 for the purpose of 'investigating the mineral structure of the Earth'. The Society is Britain's national society for geology with a membership of around 8000. It has countrywide coverage and approximately 1000 members reside overseas. The Society is responsible for all aspects of the geological sciences including professional matters. The Society has its own publishing house, which produces the Society's international journals, books and maps, and which acts as the European distributor for publications of the American Association of Petroleum Geologists, SEPM and the Geological Society of America.

Fellowship is open to those holding a recognized honours degree in geology or cognate subject and who have at least two years' relevant postgraduate experience, or who have not less than six years' relevant experience in geology or a cognate subject. A Fellow who has not less than five years' relevant postgraduate experience in the practice of geology may apply for validation and, subject to approval, may be able to use the designatory letters C Geol (Chartered Geologist).

Further information about the Society is available from the Membership Manager, The Geological Society, Burlington House, Piccadilly, London W1V 0JU, UK. The Society is a Registered Charity, No. 210161.

Published by The Geological Society from:
The Geological Society Publishing House
Unit 7, Brassmill Enterprise Centre
Brassmill Lane
Bath BA1 3JN
UK
(*Orders:* Tel. 01225 445046
           Fax 01225 442836)

First published 1996

The publishers make no representation, express or implied, with regard to the accuracy of the information contained in this book and cannot accept any legal responsibility for any errors or omissions that may be made.

© The Geological Society 1996. All rights reserved. No reproduction, copy or transmission of this publication may be made without written permission. No paragraph of this publication may be reproduced, copied or transmitted save with the provisions of the Copyright Licensing Agency, 90 Tottenham Court Road, London W1P 9HE. Users registered with the Copyright Clearance Center, 27 Congress Street, Salem, MA 01970, USA: the item-fee code for this publication is 0305-8719/96/$7.00.

British Library Cataloguing in Publication Data
A catalogue record for this book is available from the British Library.

ISBN 1-897799-45-4

Typeset by Bath Typesetting, Bath, UK

Printed by The Alden Press, Osney Mead, Oxford, UK

Distributors

USA
 AAPG Bookstore
 PO Box 979
 Tulsa
 OK 74101-0979
 USA
 (*Orders:* Tel. (918) 584-2555
            Fax (918) 584-2652)

Australia
 Australian Mineral Foundation
 63 Conyngham Street
 Glenside
 South Australia 5065
 Australia
 (*Orders:* Tel. (08) 379-0444
            Fax (08) 379-4634)

India
 Affiliated East-West Press PVT Ltd
 G-1/16 Ansari Road
 New Delhi 110 002
 India
 (*Orders:* Tel. (11) 327-9113
            Fax (11) 326-0538)

Japan
 Kanda Book Trading Co.
 Tanikawa Building
 3-2 Kanda Surugadai
 Chiyoda-Ku
 Tokyo 101
 Japan
 (*Orders:* Tel. (03) 3255-3497
            Fax (03) 3255-3495)

Contents

Preface   vii

Acknowledgements   viii

General

BOTTJER, D. J., SCHUBERT, J. K. & DROSER, M. L. Comparative evolutionary palaeoecology: assessing the changing ecology of the past   1

KAUFFMAN, E. G. & HARRIES, P. J. The importance for crisis progenitors in recovery from mass extinction   15

HARRIES, P. J., KAUFFMAN, E. G. & HANSEN, T. A. Models for biotic survival following mass extinction   41

KRASSILOV, V. A. Recovery as a function of community structure   61

JARZEMBOWSKI, E. A. & ROSS, A. J. Insect origination and extinction in the Phanerozoic   65

Palaeozoic

ZHURAVLEV, A. Yu. Reef ecosystem recovery after the Early Cambrian extinction   79

SWAIN, F. M. Ostracode speciation following Middle Ordovician extinction events, north central United States   97

ARMSTRONG, H. A. Biotic recovery after mass extinction: the role of climate and ocean-state in the post-glacial (Late Ordovician–Early Silurian) recovery of the conodonts   105

BERRY, W. B. N. Recovery of post-Late Ordovician extinction graptolites: a western North American perspective   119

KALJO, D. Diachronous recovery patterns in Early Silurian corals, graptolites and acritarchs   127

ČEJCHAN, P. & HLADIL, J. Searching for extinction/recovery gradients: the Frasnian–Famennian interval, Mokrá Section, Moravia, central Europe   135

HOUSE, M. R. Juvenile goniatite survival strategies following Devonian extinction events   163

KOSSOVAYA, O. L. The mid-Carboniferous rugose coral recovery   187

DIMICHELE, W. A. & PHILLIPS, T. L. Climate change, plant extinctions and vegetational recovery during the Middle–Late Pennsylvanian Transition: the Case of tropical peat-forming environments in North America   201

Mesozoic

ERWIN, D. H. & HUA-ZHANG, P. Recoveries and radiations: gastropods after the Permo-Triassic mass extinction   223

HALLAM, A. Recovery of the marine fauna in Europe after the end-Triassic and Early Toarcian mass extinctions   231

CONTENTS

TEWARI, A., HART, M. B. & WATKINSON, M. P. Foraminiferal recovery after the mid-Cretaceous oceanic anoxic events (OAEs) in the Cauvery Basin, southeast India — 237

PERYT, D. & LAMOLDA, M. Benthonic foraminiferal mass extinction and survival assemblages from the Cenomanian–Turonian Boundary Event in the Menoyo Section, northern Spain — 245

TUR, N. A. Planktonic foraminifera recovery from the Cenomanian–Turonian mass extinction event, northeastern Caucasus — 259

HART, M. B. Recovery of the food chain after the Late Cenomanian extinction event — 265

FITZPATRICK, M. E. J. Recovery of Turonian dinoflagellate cyst assemblages from the effects of the oceanic anoxic event at the end of the Cenomanian in southern England — 279

YAZYKOVA, E. A. Post-crisis recovery of Campanian desmoceratacean ammonites from Sakhalin, far east Russia — 299

KOCH, C. F. Latest Cretaceous mollusc species 'fabric' of the US Atlantic and Gulf Coastal Plain: a baseline for measuring biotic recovery — 309

Cenozoic

KOUTSOUKOS, E. A. M. Phenotypic experiments into new pelagic niches in early Danian planktonic Foraminifera: aftermath of the K/T boundary event — 319

BUGROVA, E. M. Recovery of North Caucasus foraminiferal assemblages after the pre-Danian extinctions — 337

SPEIJER, R. P. & VAN DER ZWAAN, G. J. Extinction and survivorship of southern Tethyan benthic foraminifera across the Cretaceous/Palaeogene boundary — 343

KELLEY, P. H. & HANSEN, T. A. Recovery of the naticid gastropod predator-prey system from the Cretaceous–Tertiary and Eocene–Oligocene extinctions — 373

Index — 387

Preface

The majority of the papers in this volume are those presented at the project meeting held at the University of Plymouth, 3–11 September 1994. The remainder of the papers were accepted for the conference (and publication) although, in the event, the authors were unable to attend in person.

I.G.C.P. 335, **Biotic Recovery from Mass Extinction Events**, is a successor to the highly successful I.G.C.P. 216 **Global Bioevents**. It was initiated in 1993 and the co-leaders are Dr Doug Erwin (Smithsonian Institution, Washington D.C., USA) and Dr Erle Kauffman (University of Colorado, Boulder, Colorado, USA). In following on the work of I.G.C.P. 216 it was accepted that many of the extinction events considered in some detail during that work would be revisited, but on this occasion concentrating on how the fauna and flora recovered from the actual 'event'. It was always recognized that there would, inevitably, be a need to debate further the mechanisms and causes of extinction as the nature of the event is key to the mode of recovery we see preserved in the geological record. As in all branches of science there is a complex terminology, much of which is not well known or used in an inconsistent way. Many of those at the Plymouth meeting requested that one of the roles of the project could be the development of a standardized, or at least properly defined, terminology. While most participants admitted to a knowledge of *Lazarus* taxa, many were unsure of the exact nature of *Refugia* and the majority were somewhat baffled by the identification of *Elvis* taxa. It was readily agreed that the development of a glossary would be an aid to communication as the project proceeds. It was also agreed that meetings be held on a regular basis throughout the duration of the project either as piggy-back sessions at other meetings or as special conferences in their own right. This volume, which will appear part-way through the life of the project, represents a collection of papers that present the current state of knowledge on floral/faunal recovery following some of the global bioevents recognized and discussed during I.G.C.P. 216. A wide range of taxonomic groups are considered in this collection of papers, including those from both terrestrial and marine ecosystems. Following a number of more theoretical papers the remainder of the contributions are arranged in stratigraphical order.

M. B. Hart, Plymouth, UK
May 1995

Acknowledgements

As the Local Secretary involved in the organization of the project meeting at the University of Plymouth I would like to thank Nicola Lobb and her assistants in the way in which the administration of the meeting was organized. This allowed me the time to concentrate on the scientific content of the meeting. All the speakers made a tremendous contribution to the success of the meeting with the overwhelming majority delivering their manuscripts on time. This enabled me to set in train the reviewing process quite quickly and I wish to thank all those involved for the excellent advice they gave to the contributors. This feedback has ensured that this volume will make a significant contribution to our knowledge on global bioevents. I would also like to thank Professor Michael House (University of Southampton) and Dr Meriel FitzPatrick (University of Plymouth) who, at the very last minute, kindly stepped in to lead the Dorset field excursion when personal circumstances prevented me from carrying out the original programme.

Comparative evolutionary palaeoecology: assessing the changing ecology of the past

DAVID J. BOTTJER,[1] JENNIFER K. SCHUBERT[2] & MARY L. DROSER[3]

[1]*Department of Earth Sciences, University of Southern California, Los Angeles, CA 90089, USA*
[2]*Department of Geological Sciences, University of Miami, Miami, FL 33124, USA*
[3]*Department of Earth Sciences, University of California, Riverside, CA 92521, USA*

Abstract: Various palaeoecological trends have been identified in the Phanerozoic, each focusing on different aspects of the fossil record. Patterns that have been described include histories of tiering, palaeocommunity species richness, and guild occupation in evolutionary faunas, as well as onshore–offshore trends in origination, expansion and retreat. Patterns of change through time have also been documented from biosedimentological features (ichnofabrics, microbial structures, shell beds). Such trends can be compared and contrasted to yield unique insights into understanding the changing ecology of the past, and in particular may be helpful in evaluating the relative degree of ecological degradation caused by a mass extinction. This comparative approach can also shed light on a variety of fundamental palaeobiological problems, for example, why no new body plans (phyla) have evolved since the early Phanerozoic. Causes of this phenomenon are thought to be either: (1) ecospace was not sufficiently open after the early Phanerozoic for survival of new body plans; or (2) accumulating developmental constraints after the early Phanerozoic have prevented the evolution of new body plans. Because the Permian–Triassic mass extinction was the most devastating biotic crisis of the Phanerozoic, one might expect new body plans to appear if ecospace were the primary limiting factor and opened sufficiently by this mass extinction. Although previous studies have shown that ecospace availability in the Cambrian and Early Triassic was indeed different, this comparative approach indicates that ecological conditions in the Early Triassic were most like those of the Late Cambrian/Early Ordovician. Thus, if ecospace availability has constrained the survival of new body plans, then ecospace has always been sufficiently filled after the Cambrian explosion to inhibit their evolution.

Evolutionary palaeoecology examines the interplay of evolution and ecology on a geological time scale through the filter of the fossil and stratigraphic record. Although a variety of approaches has been utilized in previous studies concerning evolutionary palaeoecology, perhaps the most significant achievement of this research has been the recognition of long-term palaeoecological trends over 10^6 to 10^9 years. Such trends include changes in Phanerozoic benthic marine palaeocommunity species richness, tiering and evolutionary fauna guild occupation as well as onshore–offshore patterns and changes through time of biosedimentological features (ichnofabrics, microbial structures, shell beds). These trends can be compared and contrasted among themselves and with other biotic trends (such as Phanerozoic familial diversity, e.g. Sepkoski 1992) to yield unique insight in understanding the changing ecology of the past. In particular, comparative evolutionary palaeoecology can be an especially effective approach for deciphering the ecological context of mass extinction aftermaths.

Phanerozoic palaeoecological trends

Palaeoecology originally developed as a methodology for palaeoenvironmental reconstruction (e.g. Bottjer *et al.* 1995). Widespread application of ecological approaches in palaeontological data collection through the 1960s and 1970s (e.g. Valentine 1973) led to the development of a suitably large database to facilitate syntheses of long-term ecological trends through some or all of the Phanerozoic. Perhaps the earliest Phanerozoic trend to be documented was that of palaeocommunity species richness (Bambach 1977). Application of ecological approaches towards an understanding of the Precambrian fossil records has also been fruitful, with perhaps the most convincing Precambrian palaeoecological trend being that of long-term stromatolite decline, which began in the late Proterozoic (e.g.

From Hart, M. B. (ed.), 1996, *Biotic Recovery from Mass Extinction Events*, Geological Society Special Publication No. 102, pp. 1–13

(a)

MEDIAN NUMBER OF SPECIES	HIGH STRESS ENVIRONMENTS	VARIABLE NEARSHORE ENVIRONMENTS	OPEN MARINE ENVIRONMENTS
CENOZOIC	8.5	39	61.5
MESOZOIC	7.5	17	25
UPPER PALEOZOIC	8	16	30
MIDDLE PALEOZOIC	9.5	19.5	30.5
LOWER PALEOZOIC	7	12.5	19

(b)

CAMBRIAN FAUNA

PELAGIC

SUSPENSION	HERBIVORE	CARNIVORE
2		

EPIFAUNA

	SUSPENSION	DEPOSIT	HERBIVORE	CARNIVORE
MOBILE		3	2	
ATTACHED LOW	2			
ATTACHED ERECT	1			
RECLINING	1			

INFAUNA

	SUSPENSION	DEPOSIT	CARNIVORE
SHALLOW PASSIVE			
SHALLOW ACTIVE	1	2	1
DEEP PASSIVE		*(stippled)*	
DEEP ACTIVE			

(c)

MIDDLE AND UPPER PALEOZOIC FAUNA

PELAGIC

SUSPENSION	HERBIVORE	CARNIVORE
3		4

EPIFAUNA

	SUSPENSION	DEPOSIT	HERBIVORE	CARNIVORE
MOBILE	1	4	5	5
ATTACHED LOW	7			
ATTACHED ERECT	7			
RECLINING	5			

INFAUNA

	SUSPENSION	DEPOSIT	CARNIVORE
SHALLOW PASSIVE	2		
SHALLOW DEEP	2	4	2
DEEP		*(stippled)*	
DEEP ACTIVE		1	

(d)

MESOZOIC - CENOZOIC FAUNA

PELAGIC

SUSPENSION	HERBIVORE	CARNIVORE
3	2	5

EPIFAUNA

	SUSPENSION	DEPOSIT	HERBIVORE	CARNIVORE
MOBILE	2	2	5	5
ATTACHED LOW	7			
ATTACHED ERECT	6			
RECLINING	4			

INFAUNA

	SUSPENSION	DEPOSIT	CARNIVORE
SHALLOW	3	1	1
SHALLOW ACTIVE	3	4	3
DEEP PASSIVE	1	*(stippled)*	
DEEP ACTIVE	3	2	1

Fig. 1. Trends in Phanerozoic benthic marine palaeo-community species richness and evolutionary fauna guild occupation. (**a**) Median number of species in palaeocommunities for various Phanerozoic times and environments, modified from Bambach (1977). (**b**) General adaptive strategies that typify the more diverse classes of the Cambrian evolutionary fauna. (**c**) General adaptive strategies that typify the more diverse classes of the Palaeozoic evolutionary fauna. (**d**) General adaptive strategies that typify the more diverse classes of the Mesozoic–Cenozoic (Modern) evolutionary fauna. For b, c, d the number of more diverse classes which show each adaptive strategy is indicated in the boxes, with empty boxes indicating no classes present at that time and stippled boxes indicating not biologically practical adaptive strategies; modified from Bambach (1983), identity of summed classes can be found in Bambach (1983).

Garrett 1970; Awramik 1971, 1990, 1991; Walter & Heys 1985). To illustrate the variety of ways in which palaeoecological trends are documented, several examples are described below. A detailed summation of some of these as well as a variety of other palaeoecological trends can be found in Bambach (1986), Vermeij (1987) and Sepkoski *et al.* (1991).

Palaeocommunity species richness

Bambach (1977) synthesized data on species richness from 366 Phanerozoic palaeocommunities occurring in a variety of level-bottom marine benthic palaeoenvironments. From this he determined the median number of species/palaeocommunity for five Phanerozoic time intervals. Throughout the Phanerozoic the median number of species always increases from marginal marine high stress environments to offshore open marine environments (Fig. 1a). Both variable nearshore and open marine environments show an increase in the median number of species from the lower to the middle Palaeozoic, but these values then stay relatively the same until the Cenozoic, when the median number of species/palaeocommunity for each of these environments more than doubles corresponding Mesozoic values (Fig. 1a). The median number of species/palaeocommunity stays about the same throughout the Phanerozoic in high stress marginal marine environments (Fig. 1a). This analysis was the first to show that within-habitat species richness has increased through the Phanerozoic.

Evolutionary fauna guild occupation

Bambach (1983) synthesized the adaptive strategies of major classes comprising Sepkoski's (1981) three great Phanerozoic evolutionary faunas (Fig. 1b–d), to document changes in guild occupation and hence ecospace utilization through time. The ecospace parameters considered by Bambach (1983) for a broad definition of guilds are mode of life and feeding type (Fig. 1b–d). Cambrian faunas utilized relatively little ecospace, occupying mostly epifaunal guilds (Fig. 1b). The Palaeozoic fauna shows increased occupation of epifaunal modes of life, with diversification also into infaunal and pelagic guilds (Fig. 1c). The Mesozoic–Cenozoic (Modern) fauna is characterized by increased diversification into infaunal guilds, as well as an increase in carnivores, when compared with the preceding faunas (Fig. 1d). Bambach (1983) concluded from this analysis that each succeeding evolutionary fauna was characterized by

exploitation of more ecospace than was typical of the preceding fauna.

Tiering

Ausich & Bottjer (1982) defined tiering as the distribution of benthic organisms within the space above and below the seafloor. In ecological studies the term 'stratification' is used, but because this term has other meanings in geology, tiering has received widespread acceptance in analyses of palaeocommunities (e.g. Watkins 1991) as well as trace fossils (e.g. Droser & Bottjer 1993). Ausich & Bottjer (1982) proposed a Phanerozoic history for tiering of suspension feeders in level-bottom settings of one broad palaeoenvironment, that of shallow subtidal shelves and epicontinental seas. This history has been modified in subsequent publications as additional data on tiering have accumulated (Bottjer & Ausich 1986; Ausich & Bottjer 1991) (Fig. 2a). From this analysis a four-phase tiering history can be recognized. The initial phase, during the Cambrian, had very low levels of both infaunal and epifaunal tiering (Fig. 2a). Phase two, from the Ordovician through the Permian, is characterized by complex epifaunal tiering as well as, towards the end of the Palaeozoic, increases in infaunal tiering (Fig. 2a). The mid-Triassic through Jurassic constitutes phase three, and was a time where the potential existed for maximum infaunal and epifaunal tiering (Fig. 2a). Phase four (Cretaceous to present) has been a time dominated by infaunal tiering (Fig. 2a). These changes in tiering structure are interpreted as due to a variety of processes, such as increased ecospace utilization, the effects of mass extinctions, and increased biotic interactions (Bottjer & Ausich 1986; Ausich & Bottjer 1991).

Onshore–offshore patterns

Shifts in habitat occupation through time, in an onshore to offshore direction, of benthic marine invertebrates were first documented by Sepkoski & Sheehan (1983) for the Cambrian/Ordovician and Jablonski & Bottjer (1983) for the post-Palaeozoic. Such onshore–offshore trends have since been documented for assemblages of taxa throughout the Palaeozoic (e.g. Sepkoski & Miller 1985), individual clades of body fossils (e.g. Bottjer & Jablonski 1988; Jablonski & Bottjer 1991; Sepkoski; 1992) (Fig. 2b), as well as trace fossil genera (Bottjer *et al.* 1988). Onshore origination in nearshore or inner-shelf settings can typically be followed by two other environmental patterns, expansion and retreat

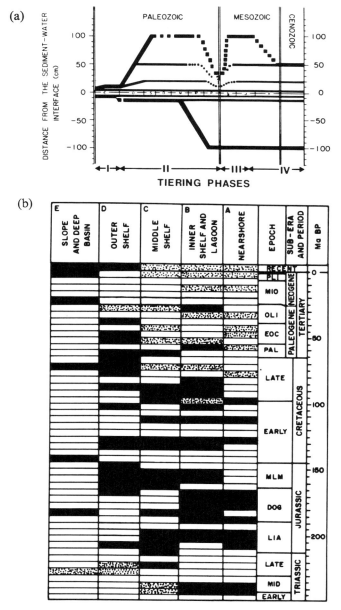

Fig. 2. Tiering and onshore–offshore patterns. (**a**) Four-phase history of Phanerozoic tiering in soft substrata suspension-feeding palaeocommunities. The heaviest lines indicate the maximum infaunal and epifaunal tiering levels at any time. Other lines represent tier subdivisions. Solid lines represent data and dotted lines are inferred. From Ausich & Bottjer (1991). (**b**) Time–environment diagram of isocrinid crinoid presence–absence. Black boxes indicate presence of isocrinids in an assemblage of that age and environment; blank boxes indicate no data yet found for that age and environment; stippled boxes indicate presence of one or more taphonomic control taxa but absence of isocrinids in an assemblage of that age and environment. Taphonomic control taxa and palaeoenvironments are defined in Bottjer & Jablonski (1988); references for each data point are in Bottjer & Jablonski (1988); from Bottjer & Jablonski (1988). Note that Mesozoic decline in epifaunal tiering shows general correspondence with onshore–offshore retreat of isocrinid crinoids.

(Bottjer & Jablonski 1988). Expansion is defined as origination onshore with later movement offshore, while retaining representatives onshore. Retreat is movement offshore with loss of onshore representatives. As an example, the pattern documented for isocrinid crinoids shows expansion in the Triassic and Jurassic, and retreat in the Late Cretaceous and Tertiary (Fig. 2b). A variety of mechanisms that might account for these long-term onshore–offshore trends has been proposed (e.g. Bottjer & Jablonski 1988; Sepkoski 1991); they do not correspond with any large-scale physical trends, such as sea-level changes, and no consensus on which biological processes lead to this intriguing pattern currently exists.

Ichnofabrics

Sedimentary rock fabric resulting from bioturbation has been termed ichnofabric (Ekdale & Bromley 1983). Ichnofabric includes discrete identifiable trace fossils as well as mottled bedding, all produced through the activities of organisms forming surface traces and trails and infaunal burrows (and borings). One approach to studying ichnofabric emphasizes determination of the extent or amount of bioturbation recorded in a sedimentary rock. The ichnofabric index method is a semi-quantitative approach for estimating the extent of bioturbation recorded in a sedimentary rock (Droser & Bottjer 1986) (Fig. 3a). Using this method, ichnofabric index data were recorded from Cambrian and Ordovician carbonate strata of the western United States, in order to determine the nature of the metazoan radiation into the infaunal habitat (Droser & Bottjer 1988, 1989). From this data an average ichnofabric index for carbonate inner shelf paleoenvironments was computed (Droser & Bottjer 1989) (Fig. 3a). This average ichnofabric index trend shows a large increase in bioturbation within the Lower Cambrian and a second major increase between the Middle and Upper Ordovician (Fig. 3a). Droser & Bottjer (1989) interpreted this trend in average ichnofabric index as indicating, for inner-shelf palaeoenvironments, increased utilization of infaunal ecospace during the early Palaeozoic.

In the same way that sedimentary facies occur in strata of different ages and from different geographical regions, ichnofabrics, which are types of biogenic sedimentary fabrics, also recur (e.g. Droser 1991; Droser & Bottjer 1993). Because ichnofabrics are produced by biological processes, the temporal or stratigraphic distribution of a given ichnofabric is controlled to some extent by evolution. The nearshore terrigenous

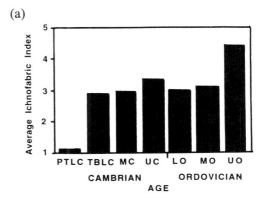

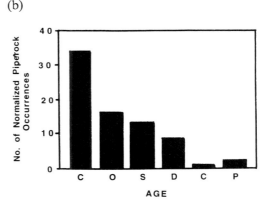

Fig. 3. Trends in ichnofabrics. (a) Average ichnofabric index for carbonate inner shelf palaeoenvironments for Cambrian and Ordovician strata of the Great Basin in western North America. PTLC, pre-trilobite Lower Cambrian; TBLC, trilobite-bearing Lower Cambrian. Modified from Droser & Bottjer (1993). (b) Distribution of Skolithos piperock ichnofabric through the Palaeozoic. Histogram represents the number of piperock occurrences as a function of age, normalized to correct for differences in map area and duration such that the area and duration of the Cambrian is set equal to 1. Data source listed in Droser (1991); modifed from Droser (1991).

clastic setting is characterized, particularly in the Palaeozoic, by the presence of the trace fossil Skolithos, which forms as a vertical tube of variable length. Dense accumulations of Skolithos produce a characteristic ichnofabric, commonly termed 'piperock' (e.g. Droser 1991). Droser (1991) has compiled occurrences of piperock throughout the Palaeozoic, which are defined as densely packed Skolithos which show an ichnofabric index of 3 or greater (Fig. 3b). This analysis shows that Skolithos piperock occurs most abundantly in the Cambrian, and then decreases in occurrence through the remainder of the Palaeozoic (Fig. 3b). Droser (1991) has attributed this post-Cambrian decline in occurrence of piperock to two possible causes: (1) the consequences of the Ordovician faunal radiation, when bioturbation increased in other environments and, thus, may reflect the appearance of a new and better adapted fauna in the nearshore terrigenous clastic setting; and (2) a systematic decrease in (preserved) shallow marine sandstones in the post-Cambrian Palaeozoic.

Microbial structures

Stromatolites first appeared in the Archaean, became increasingly abundant and diverse in the Early Proterozoic, and by the Late Proterozoic had attained their maximum abundance and diversity (Awramik 1971, 1990, 1991) (Fig. 4a). However, during the latest Proterozoic, stromatolites declined precipitously in diversity and abundance (Fig. 4a), most likely owing to effects of the early diversification of metazoans (e.g. Garrett 1970; Awramik 1971, 1990, 1991), such as increased grazing and sediment disturbance. Stromatolites from normal marine palaeoenvironments were still somewhat common in the Cambrian and Early Ordovician, but from the Middle Ordovician through the remainder of the Phanerozoic, stromatolites typically formed in environments characterized by hypersalinity or hyposalinity and strong currents or wave action, which typically reduced activity of epifaunal, grazing and/or burrowing animals (Awramik 1990). This second phase of stromatolite decline is attributed to the effects in marine environments of the Ordovician radiation of metazoans (Awramik 1990, 1991), including increased space competition for substrates favourable for colonization, and accelerated generation and deposition of carbonate sediment (in the form of skeletal debris and silt- and sand-sized bioclasts and pellets) that would bury microbial mats (e.g. Pratt 1982).

Shell beds

Shell concentrations or shell beds (any relatively dense accumulation of biomineralized remains with various amounts of sedimentary matrix, irrespective of taxonomic composition and degree of post-mortem modification) are a conspicuous and significant part of the Phanerozoic stratigraphic record (Kidwell 1990, 1991; Kidwell & Bosence 1991). However, few studies have examined temporal patterns and trends in shell beds. Kidwell (1990) predicted that patterns of shell accumulations should vary over Phanerozoic time because of changes in the diversity and environmental distribution of shell-producers and of organisms that interact with skeletal hardparts. Two different modes of shell concentrations (archaic v. modern mode) were recognized by Kidwell (1990) using features such as physical dimension, taxonomic composition and taphonomic attributes of the shell beds. The archaic mode of shell beds is represented in Palaeozoic and Triassic strata and characterized by relatively thin concentrations dominated by brachiopods and other epifaunal and semi-infaunal organisms (excepting crinoidal calcarenites). In contrast, the modern mode is primarily in Cretaceous and Cenozoic strata and characterized by thin pavements plus full three-dimensional bioclastic concentrations dominated by molluscs and other epifaunal and fully infaunal organisms (Kidwell 1990). Li & Droser (1992) further recognized distinct differences between Cambrian and Ordovician shell beds. These differences reflect the dominance of the Cambrian and Palaeozoic faunas respectively. Cambrian shell beds are dominated by trilobites and occur as thin pavements whereas the archaic mode described by Kidwell (1990) first really occurs in the stratigraphic record in the Ordovician with the appearance of the Ordovician fauna and the diversification of the articulate brachiopods. Thus, changes in the temporal and environmental distribution of shell beds can be directly linked to the types and ecologies of the organisms living at the time. Future work (e.g. Kidwell 1993) on shell beds will help refine Kidwell's (1990) model and indeed, it may be possible to recognize mass extinction events and subsequent radiations in the shell bed record.

Comparative evolutionary palaeoecology

These palaeoecological trends can be compared and contrasted with each other to yield unique insights into understanding the changing ecology of the past. Such trends can also be compared with other biotic trends, such as the history of marine familial diversity developed by Sepkoski (1981, 1992). For example, Ausich & Bottjer (1985) postulated that there was a correlation between the amount of characteristic tiering present (Fig. 2a) and global marine familial diversity (as determined by Sepkoski 1981). For suspension-feeders, this relationship holds true for the Palaeozoic, with the pattern of low epifaunal tiering in the Cambrian and rising epifaunal tiering in the Ordovician, stabilizing through the remainder of the Palaeozoic, almost mirroring changes in marine familial diversity (Ausich & Bottjer 1985). However, amount of tiering and overall familial diversity are decoupled after the Palaeozoic (Ausich & Bottjer 1985). Ausich & Bottjer (1985) postulated that this decoupling of diversity from tiering, particularly in the Cenozoic, may have been due to increased biogeographical provinciality during that time, which could have been the primary mechanism driving Cenozoic diversity increase. It is also possible that post-Palaeozoic diversity increase, not driven by increases in tiering, could have been at least partially influenced by the increase in mobile carnivores during the Mesozoic–Cenozoic (Bambach 1983) (Fig. 1b–d), which would be less likely to live in a tiered structure.

Most previous applications of comparative evolutionary palaeoecology have focused on the early metazoan radiation and increases in palaeoecological complexity (e.g. Ausich & Bottjer 1985; Allison & Briggs 1993). An examination of the effects of mass extinctions on the features documented in long-term palaeoecological trends, in order to understand better the causes and consequences of mass extinctions, is also an effective application of comparative evolutionary palaeoecology. Palaeoecological data can be collected from fossils and strata of the mass extinction aftermath. These can be used to gauge the type and degree of ecological degradation caused by the mass extinction, by determining if aftermath ecological indicators resemble those of any earlier time periods documented in the long-term trends. Results from this comparative analysis might indicate that all ecological parameters show degradation to a single earlier time interval, or to several different time intervals.

Permian–Triassic mass extinction aftermath

The Permian–Triassic mass extinction is the greatest of all Phanerozoic mass extinctions (e.g. Raup 1979; Sepkoski 1992). Recent studies indicate that the aftermath of this mass extinction lasted through the Early Triassic (e.g. Hallam 1991, 1995; Schubert 1993; Schubert & Bottjer 1995), a time interval approximately 4 Ma long (Harland *et al.* 1990). A variety of palaeoecological data has been collected from Lower Triassic marine strata, which can be used to assess the ecological degradation caused by the mass extinction.

From an extensive literature search Schubert & Bottjer (1992*a*) compiled reported occurrences of stromatolites from strata of Silurian and younger ages deposited in normal-marine level-bottom palaeoenvironments. Very few records of stromatolites in normal-marine level-bottom settings were forthcoming, with the greatest

number occurring in the Early Triassic. This compilation has been expanded with newly published data, and still shows a similar pattern but with an even greater number of Early Triassic occurrences (Fig. 4b). Although the number of occurrences of normal marine stromatolites documented from the literature is small, their relative prominence in the Early Triassic (Fig. 4b) is suggestive of a real phenomenon. It appears that the abundance of stromatolites in the Early Triassic, although by no means as widespread as in the latest Proterozoic (Fig. 4a, b), may more closely approximate that of the Late Cambrian/Early Ordovician (Fig. 4a). This increase of normal-marine level-bottom stromatolites in the Early Triassic may be because metazoan-imposed barriers to the nearshore normal-marine environments previously dominated by stromatolites were removed, so that opportunities for stromatolites to form in such settings were increased (Schubert & Bottjer 1992*a*).

It has long been known that Early Triassic macroinvertebrate assemblages characteristically have a low species richness (e.g. Hallam 1991; Erwin 1993). Only one study of Early Triassic palaeocommunity structure has been done, from Lower Triassic (Nammalian, Spathian) carbonate strata in western North America, deposited in shelfal settings, where 8 palaeocommunities were identified (Schubert 1993; Schubert & Bottjer 1995). The number of samples which define these palaeocommunities is low, so that a statistical comparison of species richness is not possible. However, the average number of species in collections defining these palaeocommunities is 13 (Schubert 1993; Schubert & Bottjer 1995). Using Bambach's (1977) environmental classification (Fig. 1a), these Early Triassic palaeo-communities would have been living in either variable nearshore environments or open marine environments. Comparison with Bambach's (1977) raw data and determination of median palaeocommunity species richness (Fig. 1a), as well as with data from particular Permian and later Triassic palaeocommunities (Schubert & Bottjer 1995), indicates that species richness of these Early Triassic palaeocommunities is lower than typical upper Palaeozoic or other Mesozoic palaeocommunities from similar environments, and is most compatible with species richness from lower Palaeozoic palaeocommunities.

Ecospace was to some extent emptied by the Permian–Triassic mass extinction (Erwin *et al.* 1987). Of the possible guilds in Bambach's (1993) ecospace utilization analysis (Fig. 2b–d), only 5 are occupied in the Early Triassic

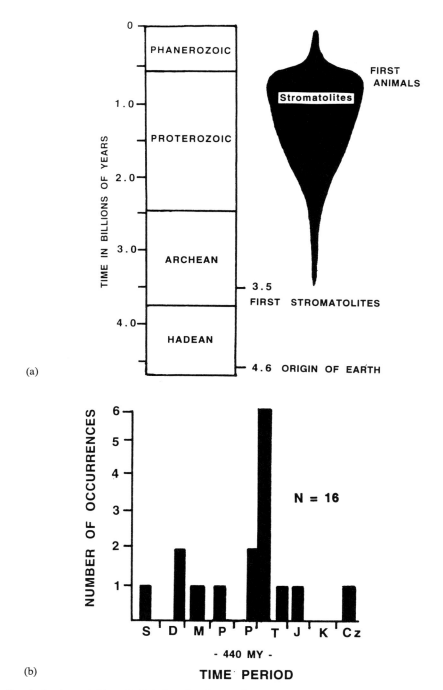

Fig. 4. Trends in stromatolites. (a) Relative abundance of stromatolites plotted against time; modified from Awramik (1991). (b) Histogram of normal-marine level-bottom stromatolites (left to right) in Silurian, Late Devonian, Mississippian, Pennsylvanian, Late Permian, Early Triassic, Late Triassic, Jurassic and Cenozoic (K, Cretaceous; Cz, Cenozoic). Data sources listed in Schubert & Bottjer (1992a) with the addition of an Early Triassic occurrence (Wignall & Hallam 1992) and a Cenozoic occurrence (Braga et al. 1995); modified from Schubert & Bottjer (1992a).

palaeocommunities documented from western North America (Schubert 1993; Schubert & Bottjer 1995). Comparison with Bambach's (1983) analysis of ecospace utilization by the three great evolutionary faunas shows that these Early Triassic palaeocommunities (although not representing a worldwide database as is Bambach's (1983)) are more similar to the Cambrian fauna, which occupies 9 of the ecospace categories, than the Palaeozoic fauna (14 ecospace categories) or the Mesozoic–Cenozoic (Modern) fauna (20 ecospace categories) (Fig. 2b–d). Erwin *et al.* (1987), from available Triassic data, calculated that at least 15 ecospace categories were occupied during this time period. However, by combining data from the Early Triassic with those from the Middle and Late Triassic, this analysis (Erwin *et al.* 1987), while useful for understanding the Mesozoic radiation, is very likely an overestimation of ecospace occupation during the Early Triassic aftermath.

This reduction in ecospace utilization is also shown in Early Triassic patterns of tiering. Early Triassic palaeocommunities, not only in western North America but worldwide, had very low levels of epifaunal tiering (Schubert & Bottjer 1992b, 1995), with only the 0 to +5 cm tier present in the earliest Early Triassic (Griesbachian, Nammalian) and the 0 to +5 cm and +5 to +20 cm tiers present in the later Early Triassic (Spathian). Analysis of bioturbation in lower Triassic strata of western North America (Bottjer *et al.* 1993; Schubert 1993; Schubert & Bottjer 1995) and the southern Alps (Italy) indicates that only the 0 to −6 cm and −6 to −12 cm tiers are present, and that this pattern may be worldwide. When compared with characteristic Phanerozoic tiering patterns (e.g. Bottjer & Ausich 1986; Ausich & Bottjer 1991), these reduced levels of tiering are most like those exhibited by Late Cambrian or Early Ordovician faunas (Fig. 2a).

Systematic measurement of ichnofabric indices has yet to be done for Lower Triassic strata. Preliminary observations indicate that all ichnofabric indices, from 1 (no preserved bioturbation) to 5 (completely bioturbated), can be found in Lower Triassic carbonate strata deposited in shelf palaeoenvironments of western North America (Schubert 1993; Schubert & Bottjer 1995) and the southern Alps (Italy). However, because bioturbation in these rocks is either horizontal or relatively shallow, storm beds with relatively low amounts of bioturbation are commonly preserved (Schubert 1993), so that the preserved amount of bioturbation appears to be less like that of the Upper Ordovician and more like that of the trilobite-bearing Lower Cambrian through Middle Ordovician strata studied by Droser & Bottjer (1988, 1989) (Fig. 3a).

Nearshore sandstones characterized by *Skolithos* piperock ichnofabric (e.g. Droser & Bottjer 1990) are rare in the post-Palaeozoic (Droser 1991). Nevertheless, the only Lower Triassic nearshore sandstone that we are aware of, the Bjorne Formation in the Canadian Arctic, exhibits a *Skolithos* piperock ichnofabric (Devaney 1991). Thus, ichnofabric type in this setting is a holdover from the Palaeozoic. A similar Palaeozoic holdover pattern is found with shell beds (Kidwell 1990).

This analysis indicates that ecospace occupation was diminished by the Permian–Triassic mass extinction to levels characteristic of the Late Cambrian/Early Ordovician. The evidence which provides the most robust support for this conclusion is from stromatolite occurrence and tiering structure. The other lines of evidence – palaeocommunity species richness, ecospace utilization, ichnofabrics and shell beds – are in need of additional development but show every indication of being compatible with this interpretation. That ecospace utilization was altered by the Permian–Triassic mass extinction is not a novel conclusion (e.g. Erwin *et al.* 1987), but demonstration that the Early Triassic had patterns of ecospace utilization similar to an earlier time in the Phanerozoic has implications for our understanding of constraints on origination of higher taxa.

Why no new phyla after the Cambrian explosion?

One of the most striking patterns of the fossil record is that origination of 10 to 11 phyla with skeletonized representatives is confined to the Vendian–Cambrian (e.g. Erwin *et al.* 1987; Valentine 1995), a phenomenon that is commonly called the Cambrian explosion. This interval of phylum-level origination is basically over by the end of the Middle Cambrian, with only the Bryozoa appearing later in the Early Ordovician (although this Early Ordovician skeletonized bryozoan could have been preceded by non-skeletonized members) (e.g. Erwin *et al.* 1987; Valentine 1995). The two leading mechanisms for the restriction of phylum-level origination to the Cambrian attribute this pattern to either of what have come to be termed the genome and ecospace hypotheses (e.g. Valentine 1973, 1980, 1986, 1992, 1995; Erwin *et al.* 1987; Valentine & Erwin 1987; Jablonski & Bottjer 1990; Erwin 1994).

The ecospace hypothesis holds that the more

open ecospace of the Vendian–Cambrian presented an ecological setting of greater opportunity for divergent morphologies than that of later times (e.g. Bambach 1983) (Fig. 1b–d). As these phyla became established and diversified in different ecological roles, and as ecospace filled, they would competitively exclude members of other clades with innovations that might lead them to similar or overlapping modes of life (e.g. Valentine 1973, 1980, 1995; Erwin *et al.* 1987; Jablonski & Bottjer 1990; Erwin 1994). A central tenet of the ecological hypothesis is that evolutionary innovation, particularly at the phylum-level, has occurred generally at constant rates since the appearance of the first metazoans (e.g. Jablonski & Bottjer 1990). Thus, by the Late Cambrian, occupation of marine ecospace would have risen to the level where no innovations at the phylum-level would be successful.

The genome hypothesis as originally developed maintained that major alterations in early metazoan development were more easily attainable because their genomes were less highly canalized and had fewer among-gene interactions (e.g. Valentine 1986, 1992; Valentine & Erwin 1987). Later metazoans with more canalized genomes would have faced increasing developmental constraints that would not have allowed the generation of major innovations at the phylum level (e.g. Valentine 1986, 1992; Jablonski & Bottjer 1990). Origin of new phyla would thus have been temporally restricted to the early phase of metazoan evolution. Therefore, the major difference between the two hypotheses is that the ecospace hypothesis predicts a constant rate of phylum-level innovation, while the genome hypothesis predicts a rate of appearance for such innovations that rapidly declined to zero.

A traditional test of these two hypotheses has been to search for other times in the Phanerozoic where ecospace occupation might be reduced to levels of the Vendian/Early–Middle Cambrian. The most appropriate time would be during a mass extinction aftermath. If ecospace during such a time was reduced to the levels prevalent during the Cambrian explosion, and yet no new phyla appeared, then this could be interpreted as support for the genome hypothesis.

Because no mass extinction has been larger, the Permian–Triassic mass extinction aftermath has been the logical place to conduct such a test (e.g. Erwin *et al.* 1987; Valentine 1992). Using a similar but simpler approach to the comparative evolutionary palaeoecology approach outlined herein, Erwin *et al.* (1987) evaluated ecospace utilization for the entire Triassic and compared this with Bambach's (1983) results on ecospace utilization for the remainder of the Phanerozoic. Because 15 of the 20 possible marine adaptive types defined by Bambach (1983) were present in the Triassic, actually a slight increase over the Palaeozoic fauna (Fig. 1c), Erwin *et al.* (1987) concluded that survivors of the mass extinction represented a broad variety of body plans thinly scattered among adaptive zones. Such a pattern or ecospace utilization could have effectively served to inhibit the occurrence of phylum-level innovations (Erwin *et al.* 1987).

However, an even more revealing test would be to determine ecospace utilization during the Early Triassic, the genuine aftermath interval of this mass extinction (e.g. Hallam 1991, 1995; Schubert 1993) and very likely the time of least ecospace utilization during the post-Middle Cambrian. The analysis of ecospace utilization for this time period outlined herein indicates that Early Triassic ecospace was occupied at levels similar to those found in the Late Cambrian/Early Ordovician and thus was not degraded to levels prevalent during the Cambrian explosion. These results, in conjunction with those of Erwin *et al.* (1987), do not lead to conclusions as to the primacy of the ecospace or genome hypotheses. Instead they support the conclusion that if ecospace availability has indeed constrained the survival of new phylum-level innovations, then ecospace has always been sufficiently filled after the Cambrian explosion to inhibit their evolution. As an intriguing postscript, Valentine (1995) has recently concluded, through an analysis of the *Hox/HOM* homeobox genes, that the current understanding of the molecular basis of development provides no support for the genome hypothesis, and that the ecospace hypothesis is more consistent as an explanation for the concentrated origin of phyla during the Cambrian explosion.

Conclusions

The comparative evolutionary palaeoecology approach can produce unique insights into the evolutionary and ecological history of life, and the processes which have driven this history. This approach can grow through further refinement of the long-term palaeoecological trends that have already been identified, as well as continued development of new palaeoecological trends. In particular, application of comparative evolutionary palaeoecology promises to be especially useful in allowing a better understanding of the ecological processes operating during and after mass extinctions.

Such an approach can also help to characterize systematically differences between the several mass extinctions and their aftermaths. For example, before the Frasnian–Famennian mass extinction glass sponges lived only in deeper water regions, but during the mass extinction aftermath they are found in shallow water palaeoenvironments, where they underwent a burst of diversification (e.g. McGhee 1996). This reversed onshore–offshore pattern is considered to be a key piece of evidence for unravelling the cause of the Frasnian–Famennian mass extinction (e.g. McGhee 1996). That no such similar patterns are found in the Early Triassic is indicative of somewhat different ecological conditions for the Permian–Triassic mass extinction and its aftermath.

Ultimately, the study of mass extinction aftermaths is about ecosystem modification and even collapse, with subsequent recovery. Evolutionary processes operate upon an ecological template, so that the palaeoecological study of mass extinction aftermaths and rebounds is critical to understanding the evolutionary pathways which life has taken. This comparative evolutionary palaeoecology study has concluded that although in the Phanerozoic ecosystems have undergone massive perturbations, such as that caused by Permian–Triassic mass extinction processes, they have never experienced degradation to the ecological conditions present during the Cambrian explosion. Thus, although mass extinctions are exceptional features in the Phanerozoic history of life, the Earth's ecosystems show a remarkable resilience to breakdown.

DJB thanks the University of Southern California Faculty Research and Innovation Fund, the National Geographic Society, and the US National Science Foundation (EAR-90-04547) for funding support. JKS thanks the Paleontological Society, the Geological Society of America, the American Association of Petroleum Geologists, Sigma Xi, the Theodore Roosevelt Memorial fund of the American Museum of Natural History, and the Department of Geological Sciences, University of Southern California, for funding support. MLD acknowledges the Petroleum Research Fund, administered by the American Chemical Society, for support, as well as the US National Science Foundation (EAR-92-19731), the National Geographic Society and the White Mountain Research Station. We are grateful to those who have been generous with advice and encouragement during the course of this study, particularly Douglas H. Erwin, Paul B. Wignall, A. Hallam and A. G. Fischer, as well as to Simon Conway Morris for a critical review of the manuscript.

This paper is a contribution to IGCP 335 and IGCP 366.

References

ALLISON, P. A. & BRIGGS, D. E. G. 1993. Exceptional fossil record: Distribution of soft-tissue preservation through the Phanerozoic. *Geology*, **21**, 527–530.

AUSICH, W. I. & BOTTJER, D. J. 1982. Tiering in suspension-feeding communities on soft substrata throughout the Phanerozoic. *Science*, **216**, 173–174.

——— & ——— 1985. Phanerozoic tiering in suspension-feeding communities on soft substrata: Implications for diversity. *In:* VALENTINE, J. W. (ed.) *Phanerozoic Diversity Patterns.* Princeton University Press, 255–274.

——— & ——— 1991. History of tiering among suspension-feeders in the benthic marine ecosystem. *Journal of Geological Education*, **39**, 313–319.

AWRAMIK, S. M. 1971. Precambrian columnar stromatolite diversity: reflection of metazoan appearance. *Science*, **174**, 825–827.

——— 1990. Stromatolites. *In:* BRIGGS, D. E. G. & CROWTHER, P. R. (eds) *Palaeobiology – A Synthesis.* Blackwell, Oxford, 336–341.

——— 1991. Archaean and proterozoic stromatolites. *In:* RIDING, R. (ed.) *Calcareous Algae and Stromatolites.* Springer, Berlin, 289–304.

BAMBACH, R. K. 1977. Species richness in marine benthic habitats through the Phanerozoic. *Paleobiology*, **3**, 152–167.

——— 1983. Ecospace utilization and guilds in marine communities through the Phanerozoic. *In:* TEVESZ, M. J. S. & McCALL, P. L. (eds) *Biotic Interactions in Recent and Fossil Benthic Communities.* Plenum, New York, 719–746.

——— 1986. Phanerozoic marine communities. *In:* RAUP, D. M. & JABLONSKI, D. (eds) *Patterns and Processes in the History of Life.* Springer, Berlin, 407–428.

BOTTJER, D. J. & AUSICH, W. I. 1986. Phanerozoic development of tiering in soft substrata suspension-feeding communities. *Paleobiology*, **12**, 400–420.

——— & JABLONSKI, D. 1988. Paleoenvironmental patterns in the evolution of post-Paleozoic benthic marine invertebrates. *Palaios*, **3**, 540–560.

———, CAMPBELL, K. A., SCHUBERT, J. K. & DROSER, M. L. 1995. Palaeoecological models, non-uniformitarianism, and tracking the changing ecology of the past. *In:* BOSENCE, D. W. J. & ALLISON, P. A. (eds) *Marine Palaeoenvironmental Analysis from Fossils.* Geological Society, London, Special Publication, **83**, 7–26.

———, DROSER, M. L. & JABLONSKI, D. 1988. Palaeoenvironmental trends in the history of trace fossils. *Nature*, **333**, 252–255.

———, SCHUBERT, J. K. & DROSER, M. L. 1993. Bioturbation and the Permian–Triassic mass extinction. *Geological Society of America Abstracts with Programs*, **25**, A-155.

BRAGA, J. C., MARTIN, J. M. & RIDING, R. 1995. Controls on microbial dome fabric development along a carbonate–siliciclastic shelf-basin transect,

Miocene, SE Spain. *Palaios*, **10**, 347–361.

DEVANEY, J. R. 1991. Sedimentological highlights of the Lower Triassic Bjorne Formation, Ellesmere Island, Arctic Archipelago. *Current Research, Part B, Geological Survey of Canada*, Paper **91-1B**, 33–40.

DROSER, M. L. 1991. Ichnofabric of the Paleozoic *Skolithos* ichnofacies and the nature and distribution of *Skolithos* piperock. *Palaios*, **6**, 316–325.

—— & BOTTJER, D. J. 1986. A semiquantitative classification of ichnofabric. *Journal of Sedimentary Petrology*, **56**, 558–559.

—— & —— 1988. Trends in extent and depth of bioturbation in Cambrian carbonate marine environments, western United States. *Geology*, **16**, 233–236.

—— & —— 1989. Ordovician increase in extent and depth of bioturbation: Implications for understanding early Paleozoic ecospace utilization. *Geology*, **17**, 850–852.

—— & —— 1990. Ichnofabric of sandstones deposited in high-energy nearshore environments: Measurement and utilization. *Palaios*, **4**, 598–604.

—— & —— 1993. Trends and patterns of Phanerozoic ichnofabrics. *Annual Review of Earth and Planetary Sciences*, **21**, 205–225.

EKDALE, A. A. & BROMLEY, R. G. 1983. Trace fossils and ichnofabric in the Kjolby Gaard Marl, Upper Cretaceous, Denmark. *Bulletin of the Geological Society of Denmark*, **31**, 107–119.

ERWIN, D. H. 1993. *The Great Paleozoic Crisis: Life and Death in the Permian*. Columbia University Press, New York.

—— 1994. Early introduction of major morphological innovations. *Acta Palaeontologica Polonica*, **38**, 281–294.

——, VALENTINE, J. W. & SEPKOSKI, J. J. JR. 1987. A comparative study of diversification events: The early Paleozoic versus the Mesozoic. *Evolution*, **41**, 1177–1186.

GARRETT, P. 1970. Phanerozoic stromatolites: noncompetitive ecologic restriction by grazing and burrowing animals. *Science*, **169**, 171–173.

HALLAM, A. 1991. Why was there a delayed radiation after the end-Palaeozoic extinctions? *Historical Biology*, **5**, 257–262.

—— 1995. The earliest Triassic as an anoxic event, and its relationship to the end-Palaeozoic mass extinction. *Canadian Society of Petroleum Geologists Memoir*, **17**, 797–804.

HARLAND, W. B., ARMSTRONG, R. L., COX, A. V., CRAIG, L. E., SMITH, A. G. & SMITH, D. G. 1990. *A Geologic Time Scale 1989*. Cambridge University Press, Cambridge.

JABLONSKI, D. & BOTTJER, D. J. 1983. Soft-bottom epifaunal suspension-feeding assemblages in the Late Cretaceous: Implications for the evolution of benthic paleocommunities. *In:* TEVESZ, M. J. S. & McCALL, P. L. (eds) *Biotic Interactions in Recent and Fossil Benthic Communities*. Plenum, New York, 747–812.

—— & —— 1990. The ecology of evolutionary innovation: The fossil record. *In:* NITECKI, M. H. (ed.) *Evolutionary Innovation*. University of Chicago Press, 253–288.

—— & —— 1991. Environmental patterns in the origins of higher taxa: The post-Paleozoic fossil record. *Science*, **252**, 1831–1833.

KIDWELL, S. M. 1990. Phanerozoic evolution of macroinvertebrate shell accumulations. *In:* MILLER, W. III (ed.) *Paleocommunity Temporal Dynamics: The Long Term Development of Multispecies Assemblages*. Paleontological Society, Special Publication, **5**, 305–327.

—— 1991. The stratigraphy of shell concentrations. *In:* ALLISON, P. A. & BRIGGS, D. E. G. (eds) *Taphonomy: Releasing the Data Locked in the Fossil Record*. Plenum, New York, 211–290.

—— 1993. Patterns of time-averaging in the shallow marine fossil record. *In:* KIDWELL, S. M. & BEHRENSMEYER, A. K. (eds) *Taphonomic Approaches to Time Resolution in Fossil Assemblages*. Paleontological Society, Short Courses in Paleontology, **6**, 275–300.

—— & BOSENCE, D. W. J. 1991. Taphonomy and time-averaging of marine shelly faunas. *In:* ALLISON, P. W. & BRIGGS, D. E. G. (eds) *Taphonomy: Releasing the Data Locked in the Fossil Record*. Plenum, New York, 211–290.

LI, X. & DROSER, M. L. 1992. The development of Early Paleozoic shell concentrations: Evidence from the Cambrian and Ordovician of the Great Basin. *Fifth North American Paleontological Convention Abstracts and Programs*, **183**.

McGHEE, G. R. JR. 1996. *The Late Devonian Mass Extinction: The Frasnian/Famennian Crisis*. Columbia University Press, New York.

PRATT, B. R. 1982. Stromatolite decline – A reconsideration. *Geology*, **10**, 512–515.

RAUP, D. M. 1979. Size of the Permian/Triassic bottleneck and its evolutionary implications. *Science*, **206**, 217–218.

SCHUBERT, J. K. 1993. *Rebound from the Permian–Triassic mass extinction event: Paleoecology of Lower Triassic carbonates in the western U.S.* PhD Thesis, University of Southern California.

—— & BOTTJER, D. J. 1992a. Early Triassic stromatolites as post-mass extinction disaster forms. *Geology*, **20**, 883–886.

—— & —— 1992b. Paleobiology of the oldest known articulate crinoid. *Lethaia*, **25**, 97–110.

—— & —— 1995. Aftermath of the Permian–Triassic mass extinction event: Palaeoecology of Lower Triassic carbonates in the western U.S. *Palaeogeography, Palaeoclimatology, Palaeoecology*, **116**, 1–39.

SEPKOSKI, J. J. JR. 1981. A factor analytic description of the Phanerozoic marine fossil record. *Paleobiology*, **7**, 36–53.

—— 1991. A model of onshore–offshore change in faunal diversity. *Paleobiology*, **17**, 58–77.

—— 1992. Phylogenetic and ecologic patterns in the Phanerozoic history of marine biodiversity. *In:* ELDREDGE, N. (ed.) *Systematics, Ecology, and the Biodiversity Crisis*. Columbia University Press, 77–100.

—— & MILLER, A. I. 1985. Evolutionary faunas and the distribution of Paleozoic marine communities in space and time. *In:* VALENTINE, J. W. (ed.) *Phanerozoic Diversity Patterns: Profiles in Macroevolution.* Princeton University Press, 153–190.

—— & SHEEHAN, P. M. 1983. Diversification, faunal change, and community replacement during the Ordovician radiation. *In:* TEVESZ, M. J. S. & McCALL, P. L. (eds) *Biotic Interactions in Recent and Fossil Benthic Communities.* Plenum, New York, 673–717.

——, BAMBACH, R. K. & DROSER, M. L. 1991. Secular changes in Phanerozoic event bedding and the biological overprint. *In:* EINSELE, G., RICKEN, W. & SEILACHER, A. (eds) *Cycles and Events in Stratigraphy.* Springer, Berlin, 298–312.

VALENTINE, J. W. 1973. *Evolutionary Paleoecology of the Marine Biosphere.* Prentice-Hall, Englewood Cliffs, New Jersey.

—— 1980. Determinants of diversity in higher taxonomic categories. *Paleobiology,* **6**, 444–450.

—— 1986. Fossil record of the origin of Baupläne and its implications. *In:* RAUP, D. M. & JABLONSKI, D. (eds) *Patterns and Processes in the History of Life.* Springer, Berlin, 209–222.

—— 1992. The macroevolution of phyla. *In:* LIPPS, J. H. & SIGNOR, P. W. (eds) *Origin and Early Evolution of the Metazoa.* Plenum, New York, 525–553.

—— 1995. Why no new phyla after the Cambrian? Genome and ecospace hypotheses revisited. *Palaios,* **10**, 192–196

—— & ERWIN, D. H. 1987. Interpreting great developmental experiments: The fossil record. *In:* RAFF, R. A. & RAFF, E. C. (eds) *Development as an Evolutionary Process.* Alan R. Liss, New York, 71–107.

VERMEIJ, G. J. 1987. *Evolution and Escalation: An Ecological History of Life.* Princeton University Press.

WALTER, M. R. & HEYS, G. R. 1985. Links between the rise of the Metazoa and the decline of stromatolites. *Precambrian Research,* **29**, 149–174.

WATKINS, R. 1991. Guild structure and tiering in a high-diversity Silurian community, Milwaukee County, Wisconsin. *Palaios,* **6**, 465–478.

WIGNALL, P. B. & HALLAM, A. 1992. Anoxia as a cause of the Permian/Triassic extinction: Facies evidence from northern Italy and the western United States. *Palaeogeography, Palaeoclimatology, Palaeoecology,* **92**, 21–46.

The importance of crisis progenitors in recovery from mass extinction

ERLE G. KAUFFMAN,[1] & PETER J. HARRIES[2]

[1]*Department of Geological Sciences, University of Colorado, Boulder, CO 80309-0250 USA*
[2]*Department of Geology, University of South Florida, Tampa, FL 33620-5200, USA*

Abstract: Progenitor taxa are defined as species or lineages which arise, commonly through punctuated or macroevolutionary processes, during the main phases of a mass extinction interval, and which then survive to seed the evolution of dominant groups during ensuing radiation and ecosystem recovery. Their success in surviving the severe environmental perturbations commonly associated with mass extinctions and their immediate aftermath lies in the fact that they are initially adapted in their evolution to these dynamically changing environments. This differentiates them from other surviving clades of ecological generalists, opportunists, disaster taxa, taxa with specialized survival mechanisms, etc., all of which may have a long pre-extinction evolutionary history. Progenitor taxa characterize those ecosystems which are most severely affected by mass extinction processes (perturbations and feedback loops), e.g. those of tropical to warm temperate climate zones. Progenitor taxa are rarer in those ecosystems with relatively minor response to environmental perturbations of mass extinction intervals (deep sea and more poleward areas), where many established pre-extinction lineages survive the extinction event(s) with little change. In several published records of 'explosive radiation' among new lineages following mass extinctions, high-resolution stratigraphic sampling has shown that many of these 'new' recovery taxa actually had their origins as small, relatively rare progenitor taxa during the preceding mass extinction intervals. Examples from Cretaceous mass extinction intervals are presented (Cenomanian–Turonian, Cretaceous–Tertiary).

The documentation and interpretation of Phanerozoic mass extinctions (Sepkoski 1993) has become a focal point for interdisciplinary research in the Earth sciences and allied fields (e.g. papers in Silver & Schultz 1982; Nitecki 1984; Elliot 1986; Walliser 1986; Donovan 1989; Lamolda *et al.* 1988; Kauffman & Walliser 1990; Sharpton & Ward 1990; and references therein). Sufficient high-resolution (cm- to dm-scale) stratigraphic, geochemical and palaeobiological data now exist internationally to synthesize general theory and models concerning mass extinction. Whereas there is a consensus that the nature, rate, and magnitude of environmental perturbations, and biogeographical/ecological constraints of mass extinction intervals may strongly affect subsequent recovery, few data exist on survival and recovery patterns, especially during the first 1–2 Ma after these biotic crises. This lack of comparable data before and after mass extinction episodes has led to a series of assumptions regarding the extinction–survival–recovery process that are only now beginning to be challenged as extensive high-resolution data from post-extinction intervals appear, largely under the auspices of IGCP Project 335 – Biotic Recoveries From Mass Extinction.

Three widely held and interrelated hypotheses concerning mass extinction–survival–recovery intervals are especially challenged by a new generation of high-resolution data from post extinction intervals: (a) the concept that mass extinctions are abrupt global events, even catastrophes within a single year, caused by terrestrial or extraterrestrial forces of such magnitude that they are directly responsible for mass death and widespread extinction among genetically and ecologically diverse life forms; (b) as a consequence of (a), that mass extinction is relatively non-selective among coeval taxa and communities, and does not exhibit an ecological or genetic gradient in time or space (Raup & Jablonski 1993); this infers that characteristics which enhance survivorship during background conditions are not effective during mass extinctions (Jablonski 1986*a*, *b*) and that survivors will mainly be represented by ecological generalists, taxa within refugia (e.g. Lazarus Taxa), and 'lucky individuals,' reflecting contingency (Gould 1989); and (c) based on the assumption of limited numbers and types of surviving clades, that evolutionary radiations following mass extinction, especially during short-term recovery intervals, will be extraordinarily rapid, or 'explosive' in nature, and predominantly reflect punctuated and macroevolutionary processes.

These are not unreasonable assumptions

From Hart, M. B. (ed.), 1996, *Biotic Recovery from Mass Extinction Events*,
Geological Society Special Publication No. 102, pp. 15–39

based on the stratigraphic resolution of available data a decade or more ago for all mass extinctions (and still for many events), in which (a) bed-by-bed occurrences v. non-occurrences of taxa were not differentiated, and taxa range data were plotted as solid lines between the first and last appearance datums; (b) published biostratigraphic ranges were commonly extended beyond actual occurrences to formation, substage, or even stage boundaries; (c) data were derived only from one or a few well-studied sections, reflecting local rather than broad biogeographical patterns of extinction; and (d) stratigraphic data on extinction–survival–recovery patterns were compiled from multiple stratigraphic levels within broad, prescribed time intervals (e.g. substage- or stage-level resolution; Raup & Sepkoski 1982, 1984; Sepkoski 1993), each of which was then represented by a single point on a graph. Consequently, major changes in extinction and origination levels appeared as sharp peaks or troughs in the data.

Are any mass extinctions catastrophic? A significant number of Phanerozoic mass extinctions have now been studied in detail over extended intervals of 1–4 Ma bracketing the extinction 'boundary', integrating cm- to dm-scale physical, geochemical and palaeobiological data sets. All of these high-resolution analyses have shown minor origination and either a stepwise pattern of species- and genus-level extinction, or a rapidly graded pattern of extinction with subtle steps in the observed data, spread over one or more Ma. These include the end-Ordovician (O–S) (e.g. Brenchley 1989; Armstrong 1994, 1996), Frasnian–Famennian (F–F) (Late Devonian; e.g. McGhee 1982, 1988, 1989; Sandberg et al. 1988; Walliser et al. 1989; Schindler 1990a, b), mid-Carboniferous (CARB), Permo-Triassic (P-T) (e.g. Teichert 1990; Erwin 1993), end-Triassic to earliest Jurassic (T–J) (e.g. Hallam 1981), Cenomanian–Turonian (C–T) (middle Cretaceous; e.g. Jefferies 1961; Hart & Bigg 1981; Kauffman 1984, 1988, 1995; Eicher & Diner 1985; Leckie 1985; Jarvis et al. 1988; Elder 1985, 1987a, b, 1989; Harries, 1993b; Harries & Kauffman 1990), Cretaceous–Tertiary (K–T) (e.g. Hansen 1988; Hansen et al. 1987, 1993; Kauffman 1988; Keller 1988a, b; 1989a, b; Ward 1988; Ward et al. 1991), and Eocene–Oligocene (E–O) (e.g. Keller 1986; Kauffman 1988) mass extinction intervals at specific sites or basins. Detailed, interregionally correlated data showing the same relationships have been compiled for the F–F, P-T, C–T, and K–T intervals. These steps or rapidly graded extinction intervals do not resemble normal background patterns that pre- and post-date

them (although such events do occur episodically among taxa with high turnover rates, e.g. ammonites), and they may involve groups which show high extinction rates only during the mass extinction interval. Because most of these stepwise extinctions are based on observed data, they still need to be statistically tested for the Signor–Lipps Effect (Signor & Lipps 1982); preliminary tests on Cenomanian–Turonian data by the authors and Charles Marshall (in manuscript) using the Marshall (1990, 1991) and newly derived equations suggest that many of the observed C–T extinction steps can also be recognized in predictive range data. Neither the observed nor the statistically predicted, high-resolution biostratigraphic data describe a catastrophic pattern for mass extinctions.

Detailed sedimentological and geochemical data, where available, suggest that possible causal mechanisms for stepwise extinction, extraordinary environmental perturbations that appear to have exceeded the adaptive ranges of genetically and ecologically diverse taxa, may occur at numerous levels within the mass extinction interval. Supportive data include numerous, high-frequency, large scale fluctuations in trace elements, stable isotopes, and/or organic carbon (e.g. Fig. 4; and Orth et al. 1993 for C–T boundary), major disturbance events in the sedimentary record, and even evidence for one to several, temporally clustered extraterrestrial impacts (Alvarez & Muller 1984; Hut et al. 1987; Kauffman 1995). Whereas a single large catastrophe might indeed change the rules for survival, and in turn demand an explosive radiation to account for the relatively rapid restructuring of many ecosystems (excluding the tropics), the observed high-resolution data suggest instead that most mass extinctions are multicausal, predominantly the result of repeated environmental perturbations with complex feedback loops, and spread out over a million or more years. This would clearly affect patterns of survival and recovery.

Individual extinction events within a stepwise or rapidly graded mass extinction interval might not be as severe in terms of environmental stress or ecosystem shock as would be a single large catastrophe. Thus, the cumulative effect of stepwise extinction events with intervals of partial environmental normalization between steps might allow a greater number and diversity of taxa to survive, numerous survival mechanisms to be employed, and even evolutionary innovation and adaptation to prevailing extinction-related conditions, creating Progenitor Taxa (subsequently discussed). Clearly, the greater the number and diversity of survivors, the less we will have to call on punctuated and macroevolutionary processes to account for the rapid recovery of global ecosystems after mass extinction.

What survival mechanisms are effective during mass extinctions? Jablonski (1986b) proposed that normal survival mechanisms which might give taxa longevity during background conditions would be largely ineffective during the extraordinary perturbations of a mass extinction interval, and that survival would be mainly among rare ecological generalists (eurytopic taxa) and those occupying effective refugia. Gould (1989) further suggested that survival was a matter of chance, not pre-adaptation, and that survival levels were contingent upon the numbers and diversity of taxa within pre-extinction biotas. Both hypotheses predict low levels of survival across mass extinction intervals, and that rapid subsequent radiation would result from punctuated and macroevolutionary processes rooted in a few surviving clades. Yet, in Jablonski's (1986a, b) hypothesis, most of the assumed survivors of mass extinctions belong to clades characterized by very slow evolutionary rates and long taxa durations. This contradicts the assumption that they form the rootstocks for explosive radiation during ecosystem recovery. These patterns would certainly be more predictable in catastrophic extinction theory than in mass extinctions characterized by stepwise or graded patterns. But inasmuch as stepwise and graded extinction patterns seem to dominate observed high-resolution data sets, it is logical also to re-evaluate the nature of survivors of mass extinction from equivalent data.

Harries & Kauffman (1990), Harries (1993a, b) and Harries et al. (1996) have surveyed the behavioural and ecological attributes of known survivors from various Phanerozoic mass extinctions, especially those for which high-resolution stratigraphic data are available. From this survey, 16 successful survival mechanisms have been identified, most of which are commonly utilized by organisms during background conditions (e.g. preadaptation, biogeographically widespread dispersion, opportunism, dormancy, eurytopy, and the potential for rapid evolutionary response). Not only is the diversity of survival mechanisms observed higher than would be predicted by present theory for mass extinctions, but also some of the strategies employed are highly specialized adaptations, and many surviving clades are not necessarily ecological generalists nor refugia taxa.

Harries (1993a, b) used very high-resolution stratigraphic data from the Western Interior Basin of North America to document the nature of survival and repopulation following the Cenomanian–Turonian (C–T) mass extinction, and found that species-level survival was high in most larger taxonomic groups. Survivors used diverse ecological and biological strategies, and were not only ecological generalists and opportunists. If this is broadly the case for survival from mass extinctions, it affects our views on the rate and magnitude of evolutionary innovation necessary to account for the rapid recovery of global ecosystems after mass extinction.

Is mass extinction–survival–recovery ecologically and biogeographically graded? Numerous authors (e.g. Kauffman 1979, 1984, 1988; Kauffman & Fagerstrom 1993; Copper 1994a, b) have proposed that mass extinction events have a more severe effect on ecosystems that are composed of diverse, equitable, biologically regulated communities, especially tropical reefs, than they do on those composed of less diverse, inequitable, physico-chemically regulated communities (e.g. shallow, cold-temperate marine communities, and more poleward ecosystems); this, in turn, suggests a latitudinal gradient to mass extinction. The key factors in this hypothesis relate to the smaller and more specialized adaptive ranges among stenotopic organisms that comprise equitable, time-stable, highly structured, tropical–subtropical communities with narrow environmental thresholds, compared to those found in environmentally more dynamic environments where broader adaptive ranges and simpler but more stable community structure are characteristic. This concept was recently challenged by Raup & Jablonski (1993) for Maastrichtian bivalves across the K–T boundary; they found no significant difference in extinction rate between tropical and more temperate faunas. But because the reef-building rudistid bivalves (113 Maastrichtian species representing 58 genera in the Caribbean alone) were eliminated from their analysis, the great majority of data points were within the warm to mild temperate Cretaceous climate zones, and substage-level data within the Maastrichtian were not differentiated, the results of this analysis must be viewed with caution. Two related proposals must therefore be tested: (1) does extinction follow an ecologically regulated latitudinal gradient? and (2) is there a temporal pattern to mass extinction reflecting varying sensitivity of different taxa and communities to the environmental perturbations of mass extinction intervals?

Kauffman & Fagerstrom (1993) summarized the Phanerozoic history of reef diversity and found that mass extinction events were primarily responsible for major depletion in reef diversity, that reef ecosystems collapsed very early in the extinction intervals, before the main levels of temperate taxonomic loss, and that recovery of reef ecosystems took from 2 to 10 Ma, compared to < 1–2 Ma for temperate ecosystems. This long recovery time is probably related to the evolutionary time required to redevelop the complex species interactions, symbioses, and food webs characteristic of highly structured tropical communities today. Kauffman (1988) and Elder (1985, 1989) further noted that more subtropical to warm temperate molluscs disappeared first, and more mild to cool temperate and cosmopolitan molluscs disappeared last, in the observed stepwise extinction data across the Cenomanian–Turonian (C–T) mass extinction boundary in the Western Interior of North America; Birkelund & Hakansson (1982), Ward (1988) and Ward et al. (1991) noted similar patterns among late Cretaceous ammonites in Spain and Denmark respectively. Keller (1989a) noted a latitudinal gradient in the timing and magnitude of planktic foraminifer extinctions across the K–T boundary. These high-resolution observed data support the idea that there is at least a broad, temporal, ecological gradient to mass extinctions, including the Cretaceous–Tertiary 'catastrophe.' Early high-resolution studies of survival and recovery intervals suggest a similar temporal gradient to recovery, with more poleward, temperate and deeper water ecosystems recovering earlier than those of tropical–subtropical habitats. If these relationships can be shown to persist in additional survival and recovery data, the survival probability of certain clades can be more clearly defined, and the diachronous nature of ecosystem recovery in geographically different parts of the world better understood.

Raup & Jablonski's (1993) findings further seem to conflict with observed fossil data (including bivalves) for the Antarctic (Zinsmeister & Macellari 1988; Elliot et al. 1994) and northern Alaska (Marincovitch 1993), reflecting the historical Danian faunal problem, where the K–T extinction is difficult to define clearly because of the low levels of lineage and clade extinction, and the great similarity of latest Cretaceous and earliest Palaeocene biotas. Contrast this to the Late Maastrichtian tropical carbonate platforms, where the bivalve-rich reef ecosystem collapses 1.5–3 Ma below the K–T boundary (Johnson & Kauffman 1995), there is a punctuated loss of tropical to sub-tropical molluscan diversity throughout the Late Maastrichtian (Sohl pers. comm. 1990), and the last recorded faunas are paucispecific small mollusc–echinoderm–algal associations and oyster biostromes (Johnson & Kauffman 1995). Hansen (1988) and Hansen et al. (1987, 1993) have further documented a loss of over 60% of bivalve species in a series of accelerated extinction levels, or steps, through the Late Maastrichtian of the subtropical to warm-temperate Texas Gulf Coastal Plain. These geographically scattered high-resolution data sets, which finely divide the stratigraphic record of extinction within the Maastrichtian, generally describe major geographical differences in the magnitude of Late Cretaceous extinction.

There seems to be a strong geographical and ecological influence on the magnitude of mass extinction, and thus on survivorship in different ecosystems, in both time and space. This predicts a greater number and diversity of survivors in temperate as opposed to tropical environments, as well as more rapid recovery of ecosystems, and greater similarity of community structure before and after mass extinction, in those geographic areas where the rates and magnitude of extinction were significantly less. These observations suggest that a broad spectrum of ecologically and genetically distinct taxa and clades may survive mass extinction events to seed ensuing radiation, and that in general these will be more diverse in temperate, mid-latitude to poleward habitats than in more tropical-subtropical, equatorial habitats. The predictions of this complex picture of survival affect the rates and patterns of ensuing ecosystem recovery, which may not be due so much to explosive radiation involving macroevolutionary processes as it is to the complex template of survival in different areas.

Common groups of survivors

Harries & Kauffman (1990), Harries (1993a, b), and Harries et al. (1996) have progressively developed data for biological and ecological adaptations that allow diverse taxa/clades to survive mass extinctions and other biological crises. These observations have given rise to a new model for survival and recovery from mass extinctions (Fig. 1) in which the successful survival mechanisms are grouped into a series of characteristic stratigraphic patterns. The various names applied to these patterns are rooted in modern evolutionary theory and ecological observations, but they also have temporal (stratigraphic) and spatial (palaeogeographical) connotations when applied to palaeobiological data sets. These integrated concepts, as used herein, are not readily available in the literature; we therefore define our usage of these terms below. In general, these terms apply to

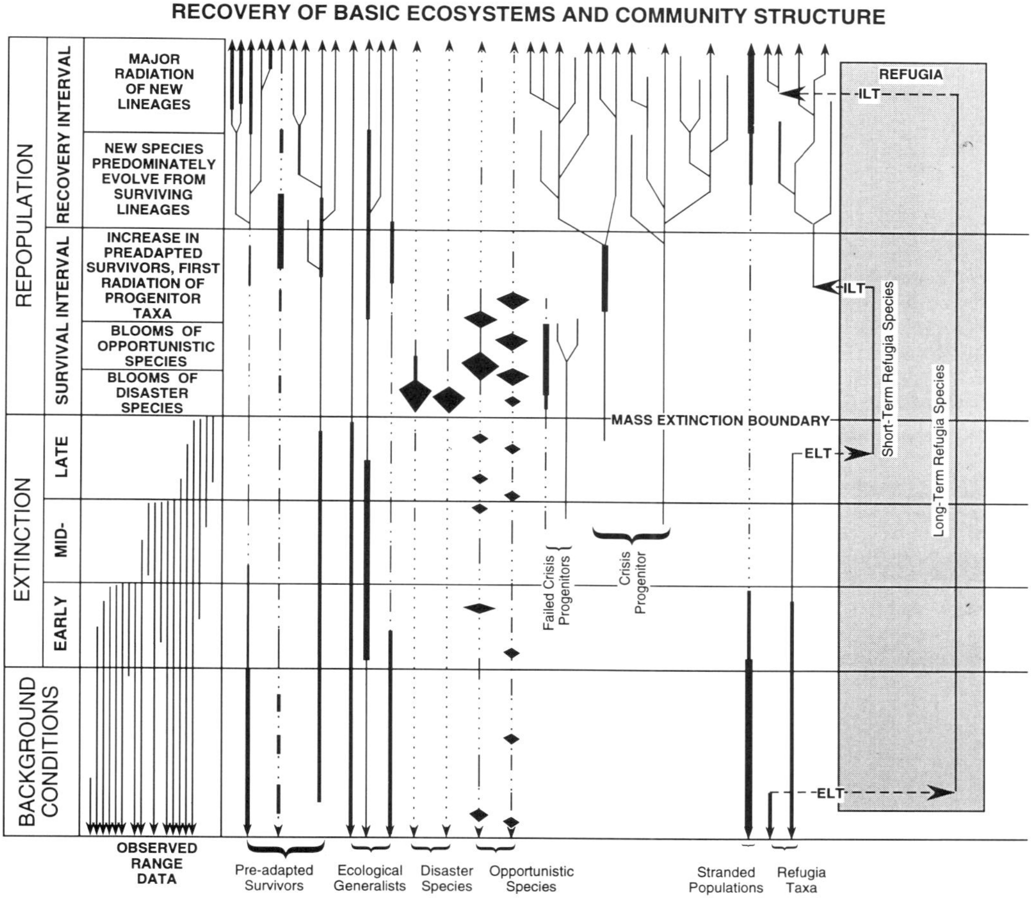

IMPORTANCE OF CRISIS PROGENITORS

Fig. 1. Generalized model showing typical stratigraphic patterns of occurrence for taxa or clades which survive mass extinction intervals, and which therefore comprise the rootstocks for subsequent radiation and basic restructuring of ecosystems within the survival and recovery intervals. The model is constructed from observed patterns of high-resolution biostratigraphic and palaeoecological data, derived from several Phanerozoic mass extinction–recovery intervals. See manuscript for definition of terms. ELT, emigrant Lazarus Taxa (into refugia?); ILT, immigrant Lazarus Taxa (from refugia?). Varying widths of solid lines depict relative robustness of species populations (not to scale); exceptional population blooms among disaster species and opportunistic taxa are represented by black diamonds. Intervals of non-occurrence of a taxon within its biostratigraphic range are represented by lines of dots. Arrows to right represent direction of emigration (>) or immigration (<) of refugia taxa. Modified from Kauffman & Erwin (1995).

specific lineages or species that survive mass extinction intervals to seed ensuing radiations and ecosystem recovery. Confident assignment of a species or lineage to one of these categories can only be done after high-resolution stratigraphic analysis of pre- and post-extinction fossil occurrences, and studies of their autecology. Care must also be taken not to assign an entire taxonomic group or clade to a survival category based on the attributes of a single taxon.

Ecological generalists (Eurytopic Taxa) have broad adaptive ranges, large niche sizes, and can tolerate large-scale changes in one or several environmental factors such as salinity (euryhaline taxa), temperature (eurythermal taxa), food (euryphagic taxa), and/or habitat selection (Fig. 1). Ecological generalists are rare to common, but non-dominant elements of relatively stable, diverse, biologically regulated communities (i.e. those primarily constrained by competition, symbioses, complex food webs, etc.). Eurytopic species may become co-dominant to dominant in biologically stressed versions of these communities, and commonly dominate physico-chemically regulated, inequitable communities. Many ecological generalists exhibit relatively primitive morphological adaptations within their lineages, and may represent ancestral stocks within clades. Stratigraphically, ecological generalists occur consistently, but rarely in great numbers, in diverse facies and over broad palaeogeographical areas, and may be components of more than one community. They are common survivors of biological crises and may actually show a relative increase in abundance during and after these events. Species will most commonly have long stratigraphic ranges, slow evolutionary rates, and tend to evolve anagenetically.

Disaster species are taxa which have special adaptations (e.g. encystment, chemosymbiosis) to environmental factors which characteristically produce extremely high levels of biological stress in more equitable, stable communities, and/or which have been almost wholly excluded from primary habitats through biological competition, returning only when competitor taxa are decimated. Disaster species rarely occur as isolated individuals within stable communities (Fig. 1). In the stratigraphic record, disaster species are normally absent or show very rare temporal and spatial occurrences (commonly after severe perturbational events) within facies representing more stable background environments. They begin to appear consistently as secondary elements of depleted biotas during stratigraphic intervals bearing physical and chemical evidence of rapid environmental changes associated with late phases of regional to global biological crises. Disaster taxa undergo very short-term, large-scale population blooms immediately following these crises, early in the survival phase (Fig. 1), within a thin stratigraphic interval containing rare fossils ('dead zone' of Harries & Kauffman 1990). This is the point of lowest biological diversity in the extinction–recovery sequence. This suggests that they are r-strategists with high profundity and short life cycles. Shortly after these blooms, and early in the phase of environmental equilibrium, disaster species characteristically disappear or become very rare, and are commonly replaced by blooms of ecological opportunists. Many disaster species have long species durations and belong to primitive groups with an extensive history of inhabiting refugia. In the stratigraphic record, disaster species are normally absent or show very rare temporal and spatial occurrences (commonly after severe perturbational events) within facies representing more stable back-

ground environments.

Ecological opportunists (Fig. 1) commonly occupy minor ecological roles and normally occur as small populations in locally disturbed areas within relatively time-stable, high-diversity, biologically regulated communities characterized by a low opportunist : equilibrium species ratio. But these same opportunistic species rapidly undergo short-lived population blooms and dominate inequitable, physico-chemically regulated communities with high opportunist : equilibrium species ratios, during environmental perturbations resulting in biological stress, population depletion, and/or local disappearance of equilibrium species. Opportunists may also inhabit refugia at the margins of their preferred range. Opportunists commonly have long species ranges, small body size, and, because many of them are r-selected strategists (i.e. adapted for rapid population growth, including increased fecundity and accelerated maturation), they have large populations with short life histories. Stratigraphically, opportunistic taxa characteristically have temporally disjunct occurrences, with low population size (except locally) and small individuals during intervals characterized by relatively long-term environmental stability (thick stratigraphic expression of similar facies suites) and diverse, equitable fossil associations. Opportunist-dominated associations are short-lived, but continuously represented in the stratigraphic record, with at least some thin, widespread, intervals displaying high abundance peaks, strong numerical dominance, and an even greater number of size cohorts than normal (Levinton 1970). They commonly occur during intervals characterized by disrupted environments (numerous perturbational event beds, rapid facies changes) and a prevalence of low diversity, inequitable fossil associations.

Preadapted survivors (Fig. 1) are taxa whose normal biological and ecological adaptations for fluctuations in background conditions include survival mechanisms which are also successful, at least among small populations, during severe environmental perturbations associated with mass extinction intervals. Harries & Kauffman (1990), Harries (1993a, b), and Harries et al. (1996) define these survival 'strategies' and review known fossil evidence for them; other preadapted strategies will probably be found as high-resolution stratigraphic studies continue. Because these taxa may represent very different habitat groups and community associations, their stratigraphic records as survivors during mass extinction intervals will vary; Fig. 1 attempts to model some of this variation. In general, preadapted survivors show reduced abundance, dispersion, and stratigraphic levels of occurrence across mass extinction and recovery intervals, compared to their pre-extinction record, and begin to expand their population size, dispersion range and habitat utilization during the later phases of the survival interval and throughout the recovery interval.

Crisis progenitor taxa evolve, normally through punctuated and possibly macroevolutionary processes, during the highly stressed phases of a mass extinction interval (as differentiated from progenitors arising during normal background conditions). The concept was first introduced by Kauffman and Harries (as 'progenitor taxa', 1993) and has already been applied to other Phanerozoic mass extinctions (e.g. Koren 1994; papers herein). In stepwise and graded mass extinction intervals (Kauffman 1988), the first appearance of crisis progenitor species commonly lies between major extinction steps in the observed data, reflecting partial normalization of environments. But observed data also suggest that some species arise during times of severe environmental stresses in association with extinction steps. Crisis progenitors are initially adapted to perturbed environmental conditions of the mass extinction interval, readily survive this interval, and are among the first groups to seed subsequent radiation into unoccupied ecospace during the survival and recovery intervals. They initially occur as rare fossils in small, temporally and spatially disjunct populations throughout the late extinction and earliest survival intervals, but rapidly increase in abundance, size, facies distribution, and biogeographical dispersion throughout the late survival and recovery intervals (Fig. 1). Radiation of basic stocks may begin within the late extinction interval, but normally occurs within physicochemically regulated, inequitable communities during the survival and early recovery intervals (Fig. 1). Crisis progenitor taxa are commonly replaced, leading to extinction, and/or competitively displaced to refugia or secondary habitats by new taxa/clades arising through later radiations to found more diverse, stable, equitable communities signalling ecosystem recovery. This replacement/displacement may be very rapid, especially in temperate to polar ecosystems where post-extinction environmental amelioration is fast. However, the process of replacement may be very slow where extinction-related stressed environments are prolonged (e.g. the Early Triassic), where ecosystem shock has been severe as a result of mass extinction (e.g. tropical to subtropical communities, like reefs), and/or where community reorganization takes millions of years.

Stranded populations are successful, commonly widely dispersed species whose populations are greatly reduced for short time periods during mass extinction and early recovery intervals (Fig. 1). They then expand their populations to fill their original niche, or a modified niche, during post-extinction normalization of environments. Long-term isolation of such populations may result in speciation or extinction. Stratigraphically, these species will have a robust fossil record before and during early phases of environmental change leading to mass extinction, but their abundance and palaeobiogeographical range become greatly restricted, and their stratigraphic occurrences episodic in many areas during late phases of the mass extinction and early survival intervals. They subsequently display population expansion to near normal proportions by the end of the recovery interval. The stratigraphic record of stranded populations differs from that of long-term Lazarus Taxa; when sporadically found in the stratigraphic record between main intervals of occurrence, stranded populations occur within facies representing their primary habitat, whereas Lazarus Taxa are more likely to be found in different facies, representing the non-primary habitats in which they survive mass extinction intervals.

Refugia species occupy very small (e.g. reef cavities, artesian springs) to very large (e.g. the deep ocean) protected habitats (a) which have been permanently isolated or left behind by long-term shifts in Earth environments (e.g. glacial retreat); (b) to which organisms have emigrated from their original habitat during times of severe environmental stress or competition; and/or (c) to which local populations have adapted, and ultimately become restricted, at the margins of their primary species range. Primary habitation of refugia by surviving species populations commonly results from environmental deterioration in, and/or competitive displacement from their original preferred habitat. Refugia effectively protect species from the severe environmental perturbations that lead to a significant decline in biomass, mass death and mass extinction of taxa in more exposed primary habitats. The character of these refugia may vary among mass extinction events, depending on the nature of forcing mechanisms and their feedback loops. Most refugia taxa retain their ability to reoccupy the primary habitats from which they have been displaced during mass extinction, as **Immigrant Survivors**, as those environments stabilize; refugia taxa may thus comprise **Lazarus Taxa**. Their ability to do this, however, declines as the evolutionary time in the refugium lengthens. Two types have been recognized in the biostratigraphic record of mass extinctions: Short- and Long-term Refugia species.

Short-term Refugia species are those taxa which were forced into refugia habitats they did not previously occupy, or which they sparsely occupied as small populations at the margins of their adaptive range, by highly stressful environmental conditions during regional biological crises or mass extinctions. As post-crisis environmental conditions normalize, these taxa rapidly return to occupy their primary habitats as short-term Lazarus Taxa (Fig. 1) or early immigrant survivors, usually without having undergone speciation. Stratigraphically, short-term refugia taxa have relatively continuous biostratigraphic ranges, though with diminishing abundance and dispersion, into early phases of the extinction interval, at which time they essentially disappear from the fossil record or have very scattered occurrences (possibly representing immigration episodes during short-term, less stressful, environmental intervals) during the major period of environmental decline. After a geologically short interval of non-occurrence, they then return abundantly as immigrant Lazarus Taxa within the survival and early recovery intervals.

Long-term Refugia species comprise those taxa which have been isolated within favourable microhabitats by permanent change in their preferred environments, and/or have been displaced from their primary habitat (usually through biological competition), long before the onset of stressed environments associated with biodiversity crises and mass extinction. Long-term refugia species may thus undergo species-level evolution and adapt to refugia habitats prior to rehabitation of their original niches. If the original displacing mechanism or competitor no longer exists following a crisis event, these long-term refugia taxa may return to their original or related niches either as the same species (in slowly evolving lineages) or as new, phylogenetically related species with similar habitat requirements. Conceivably, long-term refugia species could have evolved characteristics that would allow them to occupy a different primary habitat after mass extinction than that which they originally inhabited. Stratigraphically these long-term refugia taxa become rare and disappear from the fossil record well below the first perturbation associated with mass extinction intervals. After long stratigraphic intervals of non-occurrence they reappear as late immigrants, or long-term Lazarus Taxa (Fig. 1) during ecosystem recovery. Some work-ers (e.g. Vermeij 1986) have a more expanded view of refugia species, suggesting that the term also applies to species whose populations were greatly reduced and restricted to within a small portion of the original primary habitat by some environmental crisis, and which may subsequently reoccupy much of their primary ecospace as environmental conditions ameliorate. Because this example does not involve significant migration of the main surviving species population(s) to new or marginal habitats (i.e. refugia) that are more protected from perturbation, nor does it represent a permanent migration or destruction of the primary habitat, it does not fit the original concept of a refugium species (or population). We have termed these types of survivors, 'stranded populations' (previously discussed).

Patterns of recovery and radiation

A review of existing high-resolution Phanerozoic stratigraphic data spanning survival and recovery intervals after numerous mass extinctions is summarized as a series of stratigraphic patterns among survivors and newly evolved taxa in Fig. 1. These patterns define the characteristics of surviving species and clades, and their relative contribution to early post-extinction radiation leading to the restructuring of ecosystems decimated by the mass extinction process.

As would be predicted from their ecological and evolutionary records, many groups of survivors make only a limited contribution, as potential evolutionary rootstocks, to the major radiations that follow mass extinctions. Disaster and opportunistic taxa experience population blooms during the early to middle survival interval, but become rare, without significant diversification, during ensuing ecosystem recovery. These groups characteristically evolve slowly and have very long species ranges. Lineages of ecological generalists also show limited post-extinction diversification, reflecting their broad adaptive ranges and, in many cases, more primitive bauplans. Eurytopic taxa do exhibit major increases in population size, habitat range, and biogeographical dispersion, however, following an extinction interval. Long-term refugia taxa return late in the recovery interval, or even after re-establishment of basic ecosystems, and thus do not normally have the 'evolutionary time' to contribute greatly to the first major post-extinction radiations. They may play a significant role in later phases of community re-organization, however.

This leaves three major groups of survivors, as well as new evolutionary innovations, with the potential to make major contributions to the rapid or 'explosive' patterns of short-term, post-extinction radiation: (a) Pre-adapted Survivors, (b) Short-term Refugia Taxa, and (c) Crisis Progenitors. A tentative model of the relative importance of these groups in seeding post-extinction radiations is presented in Fig. 1. Because pre-adapted survivors and short-term refugia taxa already have specific habitat requirements and will probably reoccupy these habitats and undergo radiation primarily at their niche margins, their post-extinction radiation is predictable, and may not involve especially high levels of evolutionary innovation. Ecosystems based on these groups will restructure rapidly after mass extinction, and look very much like those that existed in these habitats prior to the extinction interval. For example, Danian shallow-water, north-temperate molluscan faunas (Heinberg 1979, 1992) and the bryozoan mounds of outer carbonate shelf facies in Denmark (Birkelund & Hakansson 1982), are very similar in taxonomic composition and the functional use of morphospace to those of the Late Maastrichtian. This was the basis of the so-called 'Danian Problem', i.e. whether or not the Danian was latest Cretaceous in the North-Temperate Realm.

Major evolutionary innovations of post-extinction radiation are therefore dependent on three processes: (a) long recovery duration (millions of years) for decimated ecosystems, e.g. tropical reefs, which commonly undergo extensive restructuring after mass extinctions (Kauffman & Fagerstrom 1993; Copper 1994a, b, and references therein); (b) large-scale, short-term, punctuated and macroevolutionary changes may occur within even generalized surviving lineages during the early phases of recovery; and (c) radiation from crisis progenitor taxa, which arose abruptly from existing but commonly dissimilar lineages during highly perturbed mass extinction intervals, and which were therefore well-adapted to early post-extinction environments, and had the initial potential to radiate into unoccupied ecospace. Process (a) is well documented; process (b) is implied from the rapid and innovative radiations following mass extinction, but has rarely been well documented with high-resolution data (e.g. Danian planktic foraminifers: Norris & Berggren 1992; Koutsoukos 1994); process (c), involving crisis progenitors, is a previously undocumented and poorly understood pattern of post-extinction diversification, but one that is commonly encountered during high-resolution stratigraphic analyses of mass extinction–survival–recovery intervals. Our research in this paper will focus on this group of survivors and describe a typical Cretaceous example.

The nature of crisis progenitor taxa

Crisis progenitors, as previously defined, arise abruptly as evolutionary innovations under environmentally stressed conditions of mass extinction intervals, are well adapted to these environments, and therefore survive the extinction to potentially seed the early recovery and repopulation of global biotas. In this sense, some crisis progenitors may have had broad adaptive ranges; others may have been more specifically adapted to highly stressed habitats (e.g. to low oxygen, low light, episodic cold or temporarily 'sterile' substrates). Their sparse, localized, early fossil record within mass extinction intervals may explain why so few crisis progenitors were described prior to high-resolution stratigraphic data collection across mass extinction–recovery intervals, and why their more abundant and diverse descendent species have commonly been regarded as the product of 'explosive' post-extinction radiation.

Species that arose during the extinction interval but which did not survive the final phases of mass extinction were unsuccessful evolutionary experiments, not progenitors. Progenitor species which arose during mass extinction intervals and which survived these crises, but which themselves went extinct without giving rise to new forms (or to rare species which themselves soon became extinct without descendants) during the survival and recovery intervals are termed 'failed crisis progenitors'. Crisis progenitors are those species which survived mass extinction and which were among the first groups to rapidly diversify, spreading widely into unoccupied ecospace during the survival and early recovery intervals, where they commonly dominate low diversity, low equitability communities. Crisis progenitors and their derived species are commonly replaced by newly derived taxa in more equitable, diverse communities during and after the late recovery interval. The major radiation of crisis progenitor taxa, with or without supplemental diversification among other types of survivors, may comprise the largest portion of the early post-extinction radiation. Thus, rather than representing some macroevolutionary process based in rare and generalized surviving lineages, the initial post-extinction burst of radiation seen in the fossil record may be rooted in well-adapted survivors, primarily in crisis progenitor, pre-adapted and short-term refugia (Lazarus) taxa.

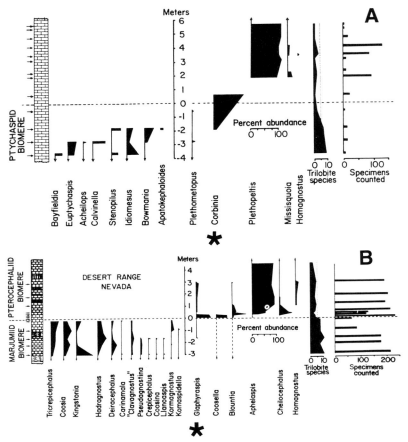

Fig. 2. Examples of possible crisis progenitor trilobite genera (asterisks) from Cambrian biomere (regional extinction) boundaries, discussed in text and by Palmer (1982, 1984). (**A**) The genus *Corbinia* (*) first occurs and initiates population expansion during the extinction interval at the top of the Ptychaspid Biomere, survives the biomere boundary extinction, reaches peak population expansion and gives rise to new species just above the boundary, and then is rapidly replaced and becomes extinct, with subsequent diversification of typical biomere assemblages. (**B**) The crisis progenitor genus *Glaphyraspis* (*) which originates during the extinction interval near the end of the Marjumid Biomere in Palmer's (1982) dataset, survives the Marjumid–Pterocephalid Biomere boundary crisis and then rapidly undergoes population expansion and speciation just above the boundary. It diminishes in dominance, and eventually disappears, during subsequent diversification of the typical Pterocephalid Biomere assemblage. Modified with permission from Palmer (1982).

Phanerozoic examples of crisis progenitor taxa

Although the high-resolution collecting programs essential to the documentation of the early evolutionary history of crisis progenitors have been applied to relatively few extinction–survival–recovery intervals, some outstanding examples of crisis progenitors (though not identified as such) have been documented throughout the Phanerozoic, lending support to the hypothesis that they are important components of post-extinction radiations.

Palmer (1982, 1984) proposed the term biomere for specific taxa sets of shallow-water Cambrian invertebrates, mainly trilobites, separated by short-term regional extinction events (biomere boundaries). Biomere boundaries have four common characteristics. (1) Abrupt extinction of most major genera and species of trilobites, associated with a significant decline in diversity (50–60%). In some cases the observed data show an apparent stepwise pattern of population decline and taxa loss (2–3 steps, including the biomere boundary; Fig. 2), with less abundant genera and species disappearing first, and the final extinction level among more abundant genera and species beginning a few centimetres below the biomere boundary (Fig. 2B; top Marjumid Biomere). However, this pattern may, in part, be an artifact of collecting or preservation. In cases where nearly all species are abundant, they collectively become extinct within a few centimetres of the biomere boundary, although some show decline in species abundance beginning a metre or less below the boundary (Fig. 2B). (2) Biomere boundary survivors are few and may be rare genera with a long evolutionary history, those with offshore, cool-water habitats, or morphologically generalized species within lineages. Many reappear as long-term Lazarus taxa just above the biomere boundary where they undergo a short interval of species-level diversification. (3) A few genera and species first appear high in the biomere, during an interval where observed data show apparent initiation of population decline and/or extinction below the biomere boundary. These newly derived species survive the boundary, abruptly increase in abundance, and in some cases give rise to new species to comprise the first step of the ensuing radiation. These have the typical stratigraphic pattern of crisis progenitors (if they give rise to new post-biomere taxa) or failed crisis progenitors (if they do not speciate). (4) The short-lived expansion of crisis progenitors at the base of the younger biomere is followed, and in a few cases accompanied, by the major interval of post-extinction diversification among newly evolved and immigrant species.

Figure 2 shows these characteristics for selected sections of two Cambrian biomere boundaries (modified from Palmer 1982). (1) The Marjumid–Pterocephalid boundary, in which apparent loss of taxa and population decline in the observed data starts between 1 and 2 m below the boundary and the final extinction lies within centimetres of the boundary; the crisis progenitor genus *Glaphyraspis* apparently originates near the beginning of the apparent extinction interval in most sections, increases in abundance and speciates just above the boundary, and then becomes a secondary element of the biota during the ensuing radiation of new generic lineages. The longer-ranging *Coosella* and *Blountia* also seem to show crisis progenitor patterns, and give rise to two new species in the base of the *Aphelaspis* zone (Palmer pers. comm. 1994), but it is now known that their first appearances are below strata recording the first environmental changes leading up to the biomere boundary extinction in some sections. (2) The top–Ptychaspid Biomere boundary, in which extinction is again apparently stepwise,

initiates just after a major population expansion in both common and rare species, with the last major step nearly 2 m below the boundary. The crisis progenitor genus *Corbinia* arises during the latest phases of the trilobite extinction, increases in abundance through the last phase of the extinction interval (top 1.8 m of the Ptychaspid Biomere), and undergoes its major population expansion, and ultimate extinction, within the first 0.5 m above the biomere boundary, well below the major expansion and diversification of new trilobite genera like *Plethopetis* and *Missisquoia*.

At the end-Ordovician mass extinction, Koren & Karpinsky (1992), Koren (1994 & pers. comm.) and Berry (1994, 1996) noted the rise of the (crisis) progenitor lineage of monograptid graptolites late in the most severe part of the O–S extinction interval, and survival and early radiation of monograptids in the survival interval prior to diversification of more characteristic Silurian graptolite clades. Armstrong (1994, 1996) noted that the origins of crisis progenitors among conodonts occurred in deep water during the Late Ordovician, just prior to end-Ordovician glaciation. These crisis progenitors then migrated into shallow water during glaciation and rise of the oceanic thermocline, where they survived the extinction boundary and were stranded by development of a mid-water oxygen minimum zone during return to greenhouse conditions. They subsequently diversified rapidly in shallow water facies during the survival interval to seed ensuing Silurian conodont radiation.

Many authors have noted that successful Famennian and Carboniferous groups of brachiopods, molluscs, conodonts and echinoderms evolved from apparent crisis progenitor stocks that arose during the late part of the Frasnian (Late Devonian) extinction crisis. For example, the origination of shallow water conodont lineages within the genus *Icriodus* in the Late Frasnian mass extinction interval, during the upper Kellwasser oxygen depletion event, and its subsequent radiation in the basal Famennian (Schindler 1990a, b), follows the evolutionary pattern of crisis progenitors. After decades of uncertainty about the origins of many Carboniferous echinoderm clades based on a meagre fossil record, the exciting new discovery of a possible early Famennian refugium in China, preserving diverse, previously unknown surviving Frasnian echinoderm lineages, including progenitor taxa, and ancestral taxa to Famennian–Carboniferous radiations (Maples *et al.* 1994) greatly clarifies these phylogenetic relationships.

The extensive high-resolution biostratigraphic

data collected across the Cretaceous–Tertiary boundary in many parts of the world have provided many examples of crisis progenitor taxa. For example, the rediversification of the planktonic foraminifers in the Danian had its roots in surviving crisis progenitors like *Guembelitria trifolia* (Keller 1988a, b, 1989a, b; Keller & Barrera 1990). Hansen (1988) and Hansen *et al*. (1987, 1993) have shown among the Cretaceous molluscan species that survived the K–T boundary in the Gulf Coastal Plain (mainly Brazos River area), that newly appearing latest Cretaceous species like *Vetericardiella webbervillensis*, *Nuculana corbetensis?*, *Lucina* sp. (new) aff *L. chatfieldana*, and possibly *Striarca webbervillensis* gave rise to successful Danian lineages. The species of *Vetericardiella* and *Striarca* show major population expansion just above the K–T boundary, a crisis progenitor pattern. Kauffman & Hansen (in prep.) have noted the abrupt appearance of large crisis progenitor bivalves like *Venericardia*, *Cerastoderma* and possibly large *Cucullaea* in the latest Maastrichtian of the Parras Basin of Mexico, just below the final ammonite occurrence; these are all typical dominant Palaeocene bivalves on the Gulf and Atlantic Coastal Plain, and were previously unknown in the latest Cretaceous.

These are but a few obvious examples of crisis progenitor taxa in well-studied mass extinction–survival–recovery intervals. As high-resolution stratigraphic analysis of more of these intervals is completed and the evolutionary history of more groups becomes better known, it is very likely that crisis progenitor taxa will appear more commonly in the data, and that their impact on the ensuing history of radiation and recovery of ecosystems can be more clearly evaluated. In the following section, we review the history of discovery and subsequent interpretation of a typical middle Cretaceous crisis progenitor lineage, the inoceramid bivalve *Mytiloides*.

Mytiloides: The evolutionary history of a crisis progenitor bivalve genus

Mytiloides (Family Inoceramidae) was one of the most common and characteristic Lower Turonian and Upper Turonian–Lower Coniacian epifaunal to semi-infaunal marine bivalves in outer shelf and basinal facies worldwide (Fig. 3). It numerically dominated temperate, low diversity, physico-chemically regulated, inequitable benthic communities, especially in oxygen-restricted facies. *Mytiloides* also spread into shallow-water carbonate and siliciclastic facies,

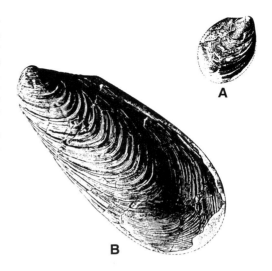

Fig. 3. Typical examples (×0.5) of the crisis progenitor bivalve *Mytiloides*. (**A**) A specimen of *Mytiloides* sp. showing small size and fine ornamentation characteristic of earliest known North American *Mytiloides*, e.g. *Mytiloides* sp. cf *M*. '*latus*' of Sageman (1993), from the highest part of the middle Hartland Shale Member of Sageman (1993), Late Cenomanian *Metoicoceras mosbyense* biozone, and *M*. sp. cf. *M. kattini* of Elder (1991). These earliest forms (n. sp.) of the crisis progenitor genus *Mytiloides* characteristically have very small, subovate, thin mytiliform shells with weak growth lines, but lacking rugae. (**B**) A typical Early Turonian *Mytiloides* (*M*. '*labiatus*' *sensu* Woods, 1911, pl. L, fig. 1: non Schlotheim, 1813), which evolved during the peak of the radiation of the most successful lineage (*M. hattini*–*M. mytiloides*–*M*. '*labiatus*' *sensu* Woods, 1911, lineage), following the Cenomanian–Turonian mass extinction. Note moderate size, thin, strongly mytiliform shells with regularly alternating rugae and raised growth lines, features which characterize most Turonian *Mytiloides*.

and even into palaeotropical marine environments, as secondary to rare elements of more diverse, equitable communities. Species have cosmopolitan to inter-regional dispersion, reflecting their apparently long-lived planktotrophic larvae and broad adaptive ranges. Yet species and subspecies have short biostratigraphic ranges, reflecting rapid evolutionary rates, which seems an enigma considering their broad distribution. The genus has been most recently redefined by Kauffman & Powell (1979) and by Harries *et al*. (1995). Figure 3B shows the typical growth form and size of an adult valve.

Species of *Mytiloides* rapidly diversified and strongly dominated Early Turonian basinal marine assemblages, appearing in considerable abundance shortly after the Cenomanian–Turonian (C–T) boundary extinction (MX5, Fig. 4) (Ernst *et al.* 1983; Kauffman 1984, 1988; Elder 1985, 1987*a, b*, 1989, 1991; Harries and Kauffman 1990; Harries 1993*a, b*). The rapid appearance of abundant *Mytiloides* in the basal Turonian has not only implied an explosive radiation of a newly derived genus following the C–T extinction but has also led past authors (e.g. Cobban & Reeside 1952; Kauffman 1975; Kennedy *et al.* 1987; Kaplan 1992) to use it as a biostratigraphic marker genus for the base of the Turonian, suggesting that the C–T boundary always lies at or below the first appearance of *Mytiloides*. These interpretations are challenged as a result of cm- to dm-scale sampling across the C–T mass extinction boundary interval in Europe and North America.

A preliminary phylogeny of known taxa belonging to dominant Turonian lineages of *Mytiloides* has been constructed from international literature and personal observations (Fig. 5). This compilation shows the rapid and large-scale nature of the Early Turonian radiation of the group, beginning just above extinction level MX5 (Fig. 4) at the Cenomanian–Turonian boundary. This apparently 'explosive' radiation comprises the basis for using *Mytiloides* as a basal Turonian biostratigraphic marker. But the phylogenetic compilation also reveals the presence of rare ancestral lineages of *Mytiloides* which arise and even begin to diversify in the latest Cenomanian, during the mass extinction interval. *Mytiloides* thus appears to be a crisis progenitor genus rather than one that arises through macroevolutionary processes in the basal Turonian. Supportive data for this interpretation from the Western Interior Basin of North America are as follows.

Sageman (1991), in his high-resolution stratigraphic and palaeobiological analysis of the Late Cenomanian Hartland Shale Member, Greenhorn Formation, recorded what appears to be the oldest known *Mytiloides* (*M.* sp. cf. *M. 'latus'*), a very small (2–5 cm), flat, thin-shelled, subovate, nearly smooth form with fine, evenly spaced growth lines which is most closely similar in character, but is not necessarily related to Lower Turonian *M. latus* (*sensu* Woods 1911, p. 283, fig. 38; p. 285, fig. 41). A superficial resemblance to early growth stages of *Inoceramus tenuis* was noted, but does not necessarily suggest a phylogenetic relationship, and no clear ancestor to *Mytiloides* is known. This earliest *Mytiloides* is represented by only 16 specimens at four stratigraphic levels (one level contains 13 specimens), among several hundred bivalves collected in a section near Pueblo, Colorado. This is the most basinal of seven sections studied

by Sageman (1991) and the only one yielding *Mytiloides*. These specimens occur in the upper part of Sageman's middle Hartland Shale unit, 9–10.2 m below the C–T boundary (93.4 Ma). This and subsequent age interpretations are based on the time scale of Obradovich (1993), as calibrated for biozones by Kauffman *et al.* (1993). Approximately the same stratigraphic interval at Eldorado Springs, Colorado, has yielded specimens that may represent an early *Mytiloides*, but are still under study. The Pueblo Cenomanian *Mytiloides* are associated with a moderately diverse (N = 18 taxa), *Inoceramus*-dominated fauna of the middle *Metoicoceras mosbyense–Inoceramus ginterensis* biozone (= *Dunveganoceras albertense* biozone to the north; Kauffman *et al.* 1993, fig. 4). The diversity and equitability of this fauna at first suggest that the earliest *Mytiloides* may have arisen not as a crisis progenitor, but under favourable marine conditions some 400 Ka below the first step of C–T mass extinction in north-temperate biotas (see Elder 1985, 1989; Kauffman 1988, 1995). However, early extinction in Caribbean reef ecosystems was already under way at this time (Johnson & Kauffman 1990), and the fauna associated with the first known occurrence of *Mytiloides* is habitat-specific. Associated strata are well laminated and contain 2.19–3.05 %/wt TOC. Suspension-feeding epibenthic bivalves, especially low-oxygen tolerant Inoceramidae and small *Phelopteria* and ostreids that commonly attach to inoceramid shell islands (Kauffman 1982), almost wholly dominate the assemblage; both nektonic ammonites and infaunal detritovores (only *Planolites* burrows) are rare (Sageman 1991). This biota and associated geochemistry suggest significant oxygen restriction within the sediment and up to the sediment–water interface, where the redoxcline was probably situated; the water column was probably dysoxic. These are conditions producing biological stress among normal marine biotas. The first *Mytiloides*-bearing strata predate the first $\delta^{13}C$, $\delta^{18}O$ and trace-element excursions associated with the earliest temperate-zone extinction events in the C–T mass extinction interval by about 390–400 Ka (Kauffman 1995). The abrupt rise in TOC values at this level, however, signals the onset of the upper Hartland Regional Oxygen Depletion Event (RODE). This event, though only basinal in scale, was environmentally similar to and immediately predated OAE II, spanning the C–T boundary. Thus, the initial *Mytiloides* stock evolved under moderately stressful marine conditions, but gave rise to two additional lineages during subsequent, even more severe Late Cenomanian environmental

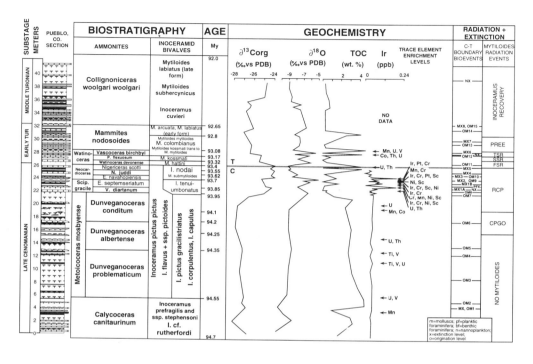

Fig. 4. Integrated, high-resolution, physical, geochemical, chronological and biostratigraphical data across the Late Cenomanian–Early Turonian mass extinction interval at Pueblo, Colorado, the proposed C–T boundary stratotype (Kennedy & Cobban 1991); modified from Kauffman (1995). Time scale adapted from Obradovich (1993), Kauffman et al. (1993), and Kauffman (1995). The origin and first major radiation of *Mytiloides* spans this interval; phases of *Mytiloides* evolution (see text) are coded in the right column, as follows: CPGO, crisis progenitor genus origination; RCPL, initial radiation of crisis progenitor lineages during late phases of the C–T mass extinction; FSR, first species-level radiation of *Mytiloides* in the early survival interval; SSR, second species-level radiation of *Mytiloides* in the late survival and early recovery intervals; TSR, third species-level radiation of *Mytiloides* in the recovery interval; PREE, post-radiation evolution and extinction of *Mytiloides* spp. during initial restructuring of marine ecosystems. Note the coincidence or the origin and early radiation of *Mytiloides* lineages with low oxygen levels (evidenced by elevated TOC values) and/or with large-scale geochemical perturbations, defining conditions of biological stress characteristic of the C–T extinction interval.

changes associated with the mass extinction interval.

Kauffman & Powell (1977, pl. 6, figs 7, 10) noted the second oldest occurrence of *Mytiloides* (as *M.* sp. aff. *M. submytiloides* (Seitz)) near the base of the late Late Cenomanian upper Hartland Shale unit of Sageman (1991), Cimarron County, Oklahoma. This form first occurs about 5.6 m below the Cenomanian–Turonian boundary in calcareous, hemipelagic, basin-platform transition facies of the *Dunveganoceras conditum–Inoceramus pictus neocaledonicus* biozone of Sageman (1991) and Kauffman et al. (1993) (= upper one-third of the southern *Metoicoceras mosbyense* biozone; Fig. 4, herein). Additional individuals were collected in similar facies of the overlying latest Cenomanian *Sciponoceras gracile* biozone, and in the basal bed of the Turonian *Watinoceras devonense* biozone, lower Bridge Creek Limestone Member, Greenhorn Formation (Kauffman et al. 1979). Only four, very small (3–4 cm length), poorly preserved internal molds of *Mytiloides* sp. aff. *M. submytiloides* were recovered among several hundred bivalve specimens during high-resolution collecting. All are very thin mytiliform shells with weakly and irregularly developed rugae and sparse intervening growth lines typical of the genus; they are most closely allied to a primitive German species previously unknown in North

America. This new lineage, which is dissimilar in shape and ornamentation to the older *Mytiloides* sp. cf. *M. latus* discovered by Sageman (1991), abruptly appears without obvious ancestry, but goes on to give rise to one major Early Turonian lineage of *Mytiloides* (*M. labiatus*, *sensu* Mantell, lineage: see Woods 1911, pl. L, figs 1, 5).

Subsequently, Elder's (1985, 1987, 1991) regional high-resolution analysis of the Late Cenomanian mass extinction interval in the Western Interior of North America turned up rare, small specimens of yet a third type of *Mytiloides* (*Mytiloides* sp. aff. *M. hattini* Elder), which first appears 220 Ka below the C–T boundary in the basal part of the *Neocardioceras juddi* biozone, and again in the uppermost Cenomanian *Nigericeras scotti* biozone (Fig. 4), in both cases associated with basinal hemipelagic limestone–shale bedding rhythms of the lower Bridge Creek Limestone Member. This lineage is characterized by broad, flat, thin mytiliform shells, covered with finely and evenly developed raised growth lines and very weak rugae, or none at all, and may have been derived from the somewhat older *Mytiloides* noted by Sageman (1991), with which it shows broad similarities; it is very dissimilar to the coeval *Mytiloides submytiloides* and *M.* sp. cf. *M. 'latus'* lineages. Elder (1991) named the first descendant species of this Late Cenomanian lineage, *M. hattini* and noted that its species range was restricted to the lower half of the Early Turonian, although possibly extending down into the latest few centimetres of the Cenomanian. Elder stated (1991, pp. 239–240, fig. 1) that the "first specimens unquestionably assignable to *M. hattini* are found in a shale bed at, or within 10 cm above the top of limestone marker bed PBC-14 in the central Western Interior basin . . ."; based on his Fig. 1 (1991) this would be of latest Cenomanian age. The ancestral members of the crisis progenitor *M. hattini* lineage arose during the most severe phases of environmental perturbations associated with peak development of OAE II and major steps of the C–T mass extinction interval (Fig. 4).

Thus, three discrete lineages of Late Cenomanian crisis progenitor *Mytiloides* are known from a sparse fossil record in basinal hemipelagic facies of the Western Interior Basin of North America. Based on the calibrated Late Cenomanian time scale of Kauffman (1995), *Mytiloides* first appears about 850 Ka below the C–T boundary and undergoes a first radiation into three basic surviving lineages between 700 Ka and 220 Ka below the C–T boundary. The early origin and first division of lineages within *Mytiloides* are associated with the onset and/or

peak development of environmental changes characterizing the Hartland Regional Oxygen Depletion Event and initiation of the Late Cenomanian mass extinction interval in the Caribbean Tethys. Whereas various lineages of Inoceramidae (e.g. *Inoceramus pictus* and subspp.; *I.* sp. aff. *I. tenuistriatus* and related forms) are abundant in beds containing the first *Mytiloides*, none are morphologically very similar or obviously ancestral. The thicker, more convex shells of *Inoceramus* spp., their thicker, more coarsely pitted ligamental plate, and major differences in musculature clearly separate this genus from Late Cenomanian and younger *Mytiloides* (Kauffman and Powell, 1979). This, in turn, suggests punctuated derivation of *Mytiloides*, though probably from some Late Cenomanian inoceramid stock, possibly that of *Inoceramus tenuis* (see Woods 1911, figs 31, 32) or *I. pictus bohemicus* (see Tröger 1967, pl. 3, figs 10, 11). Whereas the detailed fossil record of these evolutionary events is best developed in the Western Interior Basin of North America, there remains the possibility that derivation of *Mytiloides*, a cosmopolitan genus, occurred elsewhere in the world.

In the latest phases of Late Cenomanian mass extinction, during which the geochemical record (Fig. 4) suggests rapid, large-scale shifts in marine chemistry and the ocean-climate system, peak development of the global positive $\delta^{13}C$ excursion and of OAE II, fossils become rare at the top of the *Neocardioceras juddi* biozone and are very rare just above the C–T boundary in the so-called 'dead zone' of Harries & Kauffman (1990). Yet surviving *Mytiloides* are scattered through this interval and belong to all three of the morphologically discrete lineages described above. Early radiation during the final and environmentally most severe phases of the Late Cenomanian mass extinction interval represents an unusual pattern for crisis progenitors, insofar as they are known, but it strongly supports the contention that this group of survivors arises, adapts and diversifies under severe environmental stresses characteristic of extinction and at least early survival intervals, and is thus well situated to seed early radiation during biosphere recovery.

The major radiation of *Mytiloides* begins about 20 Ka above the base of the Turonian (Figs 4, 5), in the basal *Watinoceras devonense* ammonite biozone, where species populations of *M. hattini* (Elder 1991) increase significantly and become relatively common in the fossil record, especially on the eastern side of the Western Interior Seaway. *Mytiloides submytiloides* (*sensu* Seitz 1934; Kauffman & Powell 1977), and a

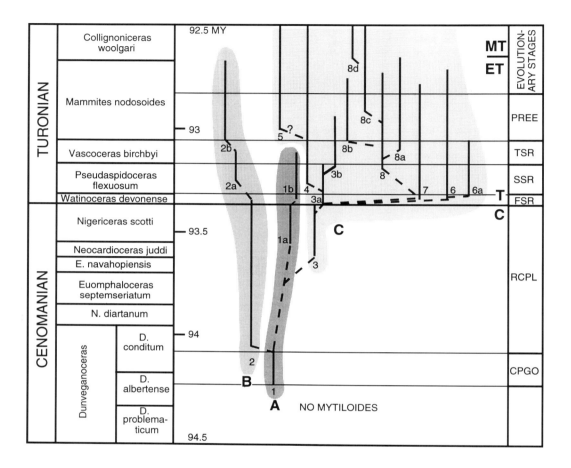

Fig. 5. A proposed phylogeny of Cenomanian–Turonian, crisis progenitor *Mytiloides* species and lineages, based on the North American record. Standard North American ammonite biostratigraphy, keyed to radiometrically based ages (Obradovich 1993), interpolated by Kauffman *et al.* (1993), shown on left. Evolutionary phases of Late Cenomanian–Early Turonian *Mytiloides* coded to right and explained in text and Fig. 4 description. The evolutionary records of the major lineages of crisis progenitor *Mytiloides* are shown by the three shaded areas. Area A (middle shaded area) comprises the rootstocks of *Mytiloides*, as represented by small, subovate, nearly smooth *Mytiloides* sp. cf. *M. 'latus'* (1), by a more mytiliform, smoother, small species (1a), and by larger, nearly smooth, mytiliform species (1b), all rare and undescribed. Area B (left shaded area) includes the strongly mytiliform, coarsely and irregularly rugate *M.* sp. aff. *M. submytiloides* (2), early transitional forms to typical *M. labiatus* (Schlotheim) (2a), and typical *M. labiatus* s.s. (see Seitz 1934, fig. 9a) (2b). Area C (right shaded area) includes the major radiating lineages of *Mytiloides* in the Early Turonian: the *M. hattini–M. mytiloides* lineages, including the crisis progenitor lineage of ancestral Late Cenomanian *M.* sp. aff. *M. hattini* (see Elder 1991) (3), *M. hattini* (3a), and *M. hattini* n. subsp. with weak regular rugae, transitional to *M. mytiloides* (3b), and the three lineages derived from it; (a) the *M. latus* s.s. lineage (4), which may eventually have given rise to the genus *Sergipia*; (b) the *M. kossmati* (6), *M. kossmati elongata* (6a) lineage; and (c) the *M. colombianus* (7)–*M. mytiloides mytiloides* (8)–*M. 'labiatus'* (*sensu* Woods, 1911, pl. L, fig. 1) (8a)–*M. mytiloides arcuata* (8b)–*M. subhercynicus* (8c)–*M. hercynicus* (8d) lineage. Cenomanian (C)–Turonian (T), and Early Turonian (ET)–Middle Turonian (MT) boundaries indicated to right of diagram.

large, nearly smooth new species of *Mytiloides* apparently descended from the first crisis progenitor *Mytiloides* sp. cf. *M. latus* (Fig. 5), but with a more mytiliform shell, occur sparsely with these *M. hattini*. Higher in the *W. devonense* zone, about 80 Ka above the C–T boundary and before the first major Turonian radiations of other biotas, the surviving *M. hattini* lineage initiates its major post-extinction radiation within the C–T survival interval (Fig. 5), rapidly giving rise to a group of interrelated species commonly known as the *M. kossmati–M. mytiloides* lineage, and including the origins of *M. kossmati kossmati* and *M. kossmati elongata*, transitional forms between *M. kossmati* and *M. mytiloides*, and *M. colombianus*. At the same time, the ancestral *Mytiloides* sp. cf. *M. latus* and *M. submytiloides* lineages remain rare and conservative, with populations changing only gradually in size and details of ornamentation through the 200 Ka interval of the *W. devonense* and basal *Pseudaspidoceras flexuosum* ammonite biozones. This first Turonian diversification of *Mytiloides* is episodic due to continuing large-scale fluctuations in post C–T extinction environments (Fig. 4).

A second wave of *Mytiloides* radiation occurs between 93.2 and 93.3 Ma (Figs 4, 5) with continued rapid diversification of the *M. kossmati–M. mytiloides* lineage, as marked by the derivation of *M. mytiloides mytiloides* and *M. 'labiatus' sensu* Mantell (see Woods 1911, pl. L, figs 1, 5), and a second division of the *M. hattini* root lineage into two discrete forms – a mytiloid-shaped derivative of *M. hattini* (n. subsp.) with weak rugae (see morphs illustrated by Elder 1991, figs 3–5, 9, 18) and a more ovate-shaped group characterized by *M. latus* (*sensu* Woods 1911, fig. 41) and its subspecies. A last major interval of Early Turonian radiation takes place within the *Vascoceras birchbyi* ammonite biozone (Fig. 5) between 93.08 and 93.2 Ma, and this is limited to the final diversification of the *M. mytiloides* lineage with the origination of *M. mytiloides arcuata* (*sensu* Seitz 1934, p. 439, pl. 37, fig. 4; text figs 3d–f) and the first appearance of *M. labiatus* s.s. (Schlotheim) (see Seitz 1934, pl. 9, fig. a; pl. 38, fig. 1), possibly derived from surviving descendents of *M. submytiloides* (Fig. 5).

Thus, during the last 850 Ka of the Cenomanian and the first 50 Ka of the Turonian, the genus *Mytiloides* underwent virtually all of its major radiation, well before the radiation of most coeval groups, and before re-establishment of normal marine ecosystems near the end of the Early Turonian. During virtually the entire 1.4 Ma-long radiation of *Mytiloides*, origination

levels far exceeded extinction of taxa (11 2 species and subspecies), and all major species groups became established. One lineage, that of *M. hattini* and its descendent branches *M. hattini* n. subsp. and *M. kossmati–M. mytiloides* dominated the radiation (9 of 11 originations) and seems to have exhibited high rates of evolution. A second lineage (*M. submytiloides*) showed conservative radiation (2 new taxa) and apparently gradualistic evolutionary patterns. The third, and most primitive, probably ancestral clade, *Mytiloides* smooth sp. cf. *M. latus*, which gave rise to all Late Cenomanian crisis progenitor *Mytiloides* lineages, survived the extinction following that radiation, but died out without further radiation near the end of the *Vascoceras birchbyi* biozone (Fig. 5), 320 Ka above the C–T boundary. Turonian survivors of this ancestral clade thus comprise a failed crisis progenitor lineage (Fig. 1).

The origination and subsequent radiation of *Mytiloides* can be divided into five intervals (Fig. 5), which can then be sequentially analysed for their evolutionary and ecological history, as follows: (oldest) CPGO (crisis progenitor genus origination), RCPL (radiation of crisis progenitor species/lineages during the late extinction phase), FSR (first species-level radiation within the survival interval), SSR (second species-level radiation; early recovery interval), and TSR (third species-level radiation; Late recovery interval). Following this radiation, a final interval of post-radiation evolution and extinction (PREE) of crisis progenitor *Mytiloides* occurred midway through the late Early Turonian *Mammites nodosoides* biozone after initial recovery of marine ecosystems, near peak Early Turonian eustatic highstand (Fig. 4). In this youngest interval, populations of *M. mytiloides* and subsequently *M. 'labiatus'* (*sensu* Mantell; see Woods 1911, pl. L, figs 1, 5) became dominant, but only one new species (*M. subhercynicus*) and possibly one genus (*Sergipia*) newly arose and remained common into the early Middle Turonian. Just above this, in the upper *Mammites nodosoides* zone, four species of *Mytiloides* died out and there were no originations. This signals the almost total demise of crisis progenitor lineages of *Mytiloides*, with only two new and uncommon species arising over the next 2–3 Ma in North America, and their replacement by newly evolved lineages of *Inoceramus* s.s., in the Middle and Late Turonian. The genus *Inoceramus*, therefore, seems to have been a long-term refugium survivor, and did not return in abundance until after more stable Middle Turonian environments and communities had been re-established. Rare species of

32 E. G. KAUFFMAN & P. J. HARRIES

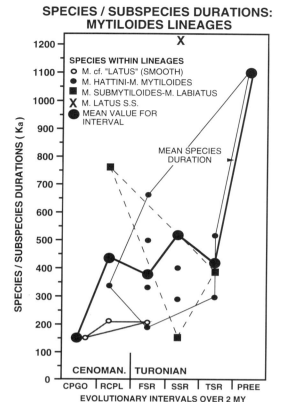

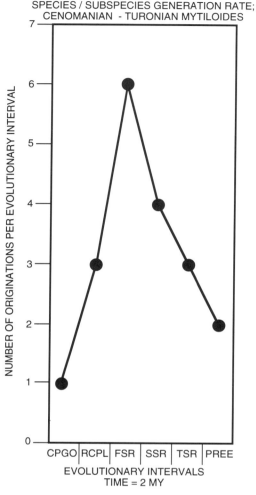

Fig. 6. Plot of individual species/subspecies durations (vertical axis, in Ka) per evolutionary interval (horizontal axis) for taxa within crisis progenitor *Mytiloides* lineages across the Cenomanian–Turonian boundary interval. Large black dots and heavy lines represent mean values for all species per evolutionary interval. See explanation of Fig. 4 for letter codes of evolutionary intervals (bottom), and upper left of this figure for key to lineage affiliations and plotting symbols. Data from each of the three main crisis progenitor lineages are enclosed in polygons. Note that the mean durations for species and subspecies initially appear to be short, suggesting rapid evolutionary rates, during the late mass extinction (CPGO, RCPL) and early survival (FSR) intervals of the C–T biotic crisis (150–450 Ka/taxon range). Mean taxa durations increase slightly during the late survival and early recovery intervals (SSR, TSR) to 420–525 Ka/taxon range, and climb abruptly to 1.05 Ma/taxon range in the late recovery interval characterized by basic restructuring of ecosystems (PREE) in late Early to early Middle Turonian time. The highest rates of evolution (shortest taxa ranges) seem to be initially among the core *Mytiloides* sp. cf. *M. 'latus'* lineage during late stages of mass extinction, although the fossil record of these taxa is still sparse and this may partially be an artifact of sampling; evolutionary rates among the dominant *M. hattini–M. mytiloides* lineages are also very high during the survival and early recovery intervals, and based on a robust fossil record.

Fig. 7. Species/subspecies generation rates per evolutionary interval (horizontal axis; see Fig. 4 for key to symbols), for crisis progenitor *Mytiloides* across the Cenomanian–Turonian boundary interval. Note that the evolutionary rate of *Mytiloides*, measured in this manner, increases rapidly through the late extinction interval (CPGO, RCPL), peaks during the early survival interval (time of major radiation; FSR), and declines through the Early and Middle Turonian. The Cenomanian–Turonian boundary is between intervals RCPL and FSR.

Mytiloides are known through the Middle and Late Turonian; others apparently persisted in refugia as Lazarus Taxa. These survivors seeded a second major radiation of species in the latest Turonian–earliest Coniacian boundary interval (Kauffman 1975; Kauffman *et al*. 1993), during which environmental conditions were very similar to, but less severe than those characterizing the Cenomanian–Turonian boundary interval. This remarkable case of iterative evolution in *Mytiloides*, with many species very similar in form to those of the Early Turonian, is currently being documented by the authors.

Evolutionary rates and habitat utilization of crisis progenitor *Mytiloides*

Rapid evolutionary rates and habitat diversification into recently vacated ecospace are predictable for surviving and newly evolved taxa during initial post-mass extinction radiations. But documentation and testing of these hypothetical relationships requires a very high-resolution biostratigraphic record and a refined chronology such as that available for the Cenomanian–Turonian boundary interval in North America (e.g. Kauffman 1988; 1995; Kauffman *et al*. 1993), Europe (e.g. Seibertz 1979; Ernst *et al*. 1983, 1984; Gale *et al*. 1993), and northern South America (e.g. Villamil 1994). *Mytiloides* data from the Western Interior of North America, integrated with a high-resolution biostratigraphy (Kauffman *et al*. 1993), radiometric time scale (Obradovich 1993; as interpolated by Kauffman *et al*. 1993), and a Milankovitch time scale (Kauffman 1995) (Fig. 4) are used here to document the evolutionary dynamics and patterns of habitat utilization during the late phases of the Late Cenomanian mass extinction, and the subsequent survival and recovery intervals during the Early Turonian.

Evolutionary rates for species and subspecies of *Mytiloides* have been calculated utilizing the methodology applied by Kauffman (1978) to assess patterns of evolution among different habitat and trophic groups of Western Interior Cretaceous bivalves. Thus, evolutionary rates during the 1.4 Ma-long *Mytiloides* radiation and ensuing 700 Ka-long Middle Turonian demise of these crisis progenitors have been measured (a) by calculating observed and mean species/subspecies durations for taxa arising during each of the six radiation intervals defined herein and illustrated on Figs 5–7; and (b) by calculating the number of new taxa arising in each of these intervals during the radiation of *Mytiloides*. The results of these analyses are shown in Figs 6 and 7.

The analysis of taxa durations (Fig. 6) shows, on average, short mean species/subspecies ranges (150, 440 Ka, respectively) for those crisis progenitor taxa arising during late mass extinction intervals (CPGO, RCPL; range 150–760 Ka), and their descendents in the first post-extinction radiation interval (FSR; mean 376 Ka; range 170–660 Ka). Species durations lengthen somewhat (means 523; 407 Ka), and become more variable (range 160–1240 Ka) during the second and third radiation intervals (SSR, TSR), and then rise significantly to 1.1 Ma/taxon (excluding the new genus *Sergipia*; 1.0 Ma mean) during the post-radiation (PREE) phase of evolution (Fig. 6). An analysis of the number of new taxa arising during each of these intervals (Fig. 7) shows a more symmetrical pattern, with the curve increasing from one to three in the late extinction interval, peaking at 6 taxa during the first phase of post-extinction radiation (with initial amelioration of extinction-related environmental stress), and falling to four, then to three and then two new taxa during the last three evolutionary intervals of crisis progenitor *Mytiloides*.

Thus, species/subspecies durations for newly evolved crisis progenitors are initially short during late extinction and earliest survival intervals, lengthen slightly during late survival and early recovery intervals, and increase markedly, eventually to 1.1 Ma/taxon during the middle and late phases of the radiation (late recovery interval). The generation rate of new taxa is initially low during late extinction, peaks during the early survival interval, and diminishes again during the recovery interval, when *Mytiloides* is displaced in its primary habitat by newly evolving lineages of *Inoceramus*. Further, two different lineages of *Mytiloides* (*M.* sp. cf. *M. latus: M. hattini–M. mytiloides* lineages) show the same general trends, but different rates in both analyses (Figs 6, 7). The most speciose radiating lineages (*M.* sp. aff. *M. hattini* to *M. kossmati–M. mytiloides* lineages) generally show the highest rates of origination as well as the shortest mean and absolute species ranges during the latest extinction and survival intervals. The more conservative trunk lineage of *M.* sp. cf. *M. latus* shows lower levels of origination and, after its earliest occurrence, somewhat longer species durations than many coeval *Mytiloides*, before its earliest Turonian extinction. The *Mytiloides submytiloides* lineage is the most conservative and long-lived, with moderate to low values for evolutionary rates in both analyses. The fabric of this major radiation in *Mytiloides* is thus complex, distinct for different lineages, and at different times, during the

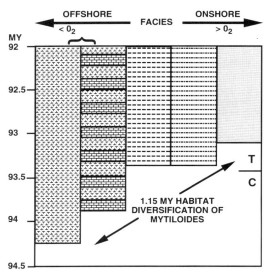

Fig. 8. Facies spread and habitat diversification among crisis progenitor *Mytiloides* in the Western Interior Seaway of North America during 1.15 Ma of Late Cenomanian (C)–Early Turonian (T) time. Initial Cenomanian occurrences and radiation of *Mytiloides* are in basinal, low-oxygen, calcareous shale and shaly chalk facies (left column), and subsequently in even deeper water facies of cyclic limestone–shale bedding rhythms representing Milankovitch Climate Cycle deposits (second column from left). The bracket at the top of these left two facies columns indicates that both are offshore, basinal deposits, with the calcareous shale/shaly chalk representing the most oxygen-depleted facies. Shoreward spread of *Mytiloides* into middle to inner platform/shelf siliciclastic shale and mudstone facies (middle and second from right columns) did not occur until the earliest Turonian, and into shoreface sandstone/sandy mudstone facies (right column) until the mid-Early Turonian, 1.15 Ma after the deeper water origins of *Mytiloides* crisis progenitor lineages. The base of each lithologic column denotes the time of first appearance of *Mytiloides* in this facies in the Western Interior Basin of North America; time scale calibrated from Obradovich (1993).

survival and recovery intervals, and rapid ('explosive') only for one surviving crisis progenitor lineage (*M. hattini–M. kossmati–M. mytiloides*).

High-resolution facies analysis for *Mytiloides*-bearing C–T boundary strata has been completed for over 100 stratigraphic sections spread throughout the Western Interior Basin of North America. These can be precisely correlated using the integrated radiometric-event/cycle chronostratigraphic–biostratigraphic chronology developed for the basin (Kauffman *et al.* 1993;

Kauffman 1995) (Fig. 4). This, in turn, allows the habitat diversification of *Mytiloides* to be progressively documented from the time of its origin, through the various phases of its radiation, to its general demise in the early Middle Turonian. Figure 8 shows the results of this analysis and demonstrates the following. The origin and differentiation of the three major lineages of *Mytiloides* during 850 Ka of the Late Cenomanian extinction interval took place in basinal facies represented by finely laminated calcareous shale, chalky shale and, somewhat later, cyclically bedded units of hemipelagic limestone (or chalk or calcarenite) and calcareous shale/marl within the axial basin of the Western Interior Seaway (see Kauffman 1985). By 80 Ka above the C–T boundary, during the initial radiation, the *M. kossmati–M. mytiloides* and *M. hattini* lineages had spread shoreward to fine-grained shelfal siliciclastic facies, and by 320 Ka above the C–T boundary, the *M. mytiloides* and *M. labiatus* lineages had spread to proximal offshore and shoreface facies. The ancestral small, smooth *Mytiloides* lineage (*M.* sp. cf. *M. latus*) and the *M. submytiloides* lineage remained in basinal facies as small populations until extinction of the former lineage in mid-Early Turonian time. However, rapid dispersal of *M. labiatus* s.s. (possibly derived from *M. submytiloides*) into all facies occurred in the late Early and early Middle Turonian. At the end of the ecological dominance of *Mytiloides* in the late Early and early Middle Turonian, surviving *Mytiloides* (*M. subhercynicus–M. hercynicus* lineage) generally retreated back to outer-shelf and basinal facies as species of *Inoceramus* s.s. radiated into both basinal and shallow-water environments.

Conclusions

Integrated high-resolution sedimentological, chemostratigraphic and biostratigraphic analyses of several Phanerozoic mass extinction–survival–recovery intervals have shown that they are characterized by numerous short-term, stepwise and/or narrowly graded extinction events separated by short intervals reflecting a partial return to background conditions, collectively spanning 1–3 Ma on average. These events are associated with, and in some cases completely enveloped by, environmentally dynamic, highly perturbed intervals in which the rates and magnitude of changes in ocean-climate systems far exceed background conditions. These intervals may or may not contain extraordinary or catastrophic events (e.g. bolide impacts). The precise correlation of individual extinction intervals with perturbational events and severe environmental feedback loops caused by them, as well as factors such as oxygen depletion, trace-element advection, and sharp temperature changes in various mass extinction intervals, suggests that they are multicausal in nature, and may be forced by both terrestrial and extra-terrestrial phenomena.

Whereas large-scale catastrophic extinction events may substantially change the rules for survival during mass extinction, a series of stepwise to graded extinction events separated by intervals of lower stress environments – the observed pattern of high-resolution analysis – leaves open the possibility that at least some normal survival mechanisms, evolved during background conditions, will successfully operate through these biotic crises, and that surviving clades may be relatively diverse. A survey of surviving Phanerozoic clades/lineages discovered in high-resolution palaeontological analysis of certain late extinction and early survival intervals has yielded a far greater number and diversity of survivors, most of which are characterized by 'normal' survival mechanisms, than would be expected if mass extinction did represent a unique macroevolutionary regime (*sensu* Jablonski 1986*b*), and survival was limited to a few ecological generalists and refugia taxa. As an alternative to 'explosive' radiation and widespread evolutionary convergence, this could explain, at least for temperate and poleward ecosystems, the relatively rapid recovery of basic community structure following many mass extinctions (at 1 Ma or less), in many cases based largely on the same lineages and/or adaptive plans as found in similar pre-extinction communities.

The observed stepwise to graded pattern of many mass extinctions, with brief periods of background patterns and partial normalization of environments, also creates a situation whereby some lineages might rapidly be able to evolve successful adaptations to deteriorating environmental conditions broadly forcing mass extinction events (keep-up evolution). Successful adaptation of evolving lineages to extinction-related environments would, in turn, enhance their survival potential and position them for early radiation during the ensuing survival interval when environmentally stressful conditions still existed but had begun to ameliorate. These new extinction-adapted taxa or lineages comprise crisis progenitors, and they are characterized by (a) rapid to punctuated evolution of new character suites during the mass extinction interval; (b) successful survival of these lineages through the final and most severe phases of extinction and associated environmental perturbations; and (c) rapid early radiation into at least partially stressed but unoccupied ecospace early in post-extinction history (i.e. during the survival interval). Crisis progenitor-dominated communities are usually low in diversity and equitability. Depending upon the duration of post-extinction environmental stress, crisis progenitor radiation can also continue for a relatively long time interval (> 1 Ma) into the recovery phase of repopulation. But normally, crisis progenitor radiation ends before full recovery of basic ecosystems, when progenitors are apparently replaced in the same habitats and even driven into refugia by more advanced taxa comprising more equitable, diverse, biologically regulated communities. The bivalve genus *Mytiloides* of the Cenomanian–Turonian (Cretaceous) mass extinction interval provides an excellent example of the evolutionary and extinction history of a crisis progenitor lineage.

These high-resolution stratigraphic analyses of several mass extinction–survival–recovery intervals suggest that mass extinction events may not be as different in process as proposed by Jablonski (1986*b*) and others, except during major catastrophes (e.g. large impacts). Instead, these data suggest that most mass extinctions may be only different in the rates and magnitude of environmental change (i.e. a dominance of perturbational or large scale, short-term events), allowing an ecologically and genetically diverse spectrum of taxa to survive and to seed ensuing ecosystem recovery, including lineages employing normal survival mechanisms evolved during background conditions. These data also reveal the persistence and importance of crisis progenitor species, which evolved during stressed environments characteristic of extinction events, and were thus already adapted to early post-extinction conditions and well positioned to seed the earliest, major, post-extinction radiations. Collectively, these survivor clades and lineages may account for much of the so-called 'explosive' nature of post-extinction radiation. If so, rapid recovery of basic communities from mass extinction would not require extraordinary post-extinction macroevolutionary regimes to explain rapid restructuring of ecosystems recorded in at least temperate to polar biotas after many mass extinctions.

References

ALVAREZ, W. & MULLER, R. A. 1984. Evidence from crater ages for periodic impacts on Earth. *Nature*, **308**, 718–720.

ARMSTRONG, H. A. 1994. Conodonts as a proxy for

biotic recovery following the Late Ordovician glaciation. *IGCP 335: Biotic Recovery from Mass Extinction Events. Plymouth, UK, Abstracts.*
—— 1996. Biotic recovery after mass extinction – the role of climate and ocean-state in the post glacial (Late Ordovician–Early Silurian) recovery of the conodonts. *This volume.*
BERRY, W. B. N. 1994. Early Llandovery graptolite biotic recovery: A western North American perspective. *IGCP 335: Biotic Recovery from Mass Extinction Events. Plymouth, UK, Abstracts.*
—— 1996. Recovery of post-late Ordovician extinction graptolites: a Western North American perspective. *This volume.*
BIRKELUND, T. & HAKANSSON, E. 1982. The terminal Cretaceous extinction in Boreal shelf seas – A multicausal event. *In:* SILVER, L. T. & SCHULTZ, P. H. (eds) *Geological Implications of impacts of large Asteroid and Comets on the Earth.* Geological Society of America, Special Paper **190**, 373–384.
BRENCHLEY, P. J. 1989. The Late Ordovician extinction. *In:* DONOVAN, S. K. (ed.) *Mass Extinctions: Processes and Evidence.* Ferdinand Enke Verlag, Stuttgart, 104–132.
COBBAN, W. A. & REESIDE, J. B. JR. 1952. *Correlation of the Cretaceous formations of the Western Interior of the United States.* American Association of Petroleum Geologists, **63**, 1011–1044.
COPPER, P. 1994a. Reefs under stress: the fossil record. Courier Forschungsinstitut *Senckenberg,* **172**, 87–94.
—— 1994b. Ancient reef ecosystem expansion and collapse. *Coral Reefs,* **13**, 3–11.
DONOVAN, S. K. (ed.) 1989. *Mass Extinctions: Processes and Evidence.* Columbia University Press, New York.
EICHER, D. L. & DINER, R. 1985. Foraminifera as indicators of water mass in the Cretaceous Greenhorn Sea, Western Interior. *In:* PRATT, L. M., KAUFFMAN, E. G. & ZELT, F. B. (eds) *Fine-grained Deposits and Biofacies of the Cretaceous Western Interior Seaway: Evidence of Cyclic Sedimentary Processes.* Society of Economic Paleontologists and Mineralogists, 1985 Midyear Meeting, Golden, CO, Fieldtrip Guidebook **4**, 60–71.
ELDER, W. P. 1985. Biotic patterns across the Cenomanian–Turonian extinction boundary near Pueblo, Colorado. *In:* PRATT, L. M., KAUFFMAN, E. G. & ZELT, F. B. (eds) *Fine-grained Deposits and Biofacies of the Cretaceous Western Interior Seaway: Evidence of Cyclic Sedimentary Processes.* Society of Economic Paleontologists and Mineralogists, 1985 Midyear Meeting, Golden, CO, Fieldtrip Guidebook **4**, 157–169.
—— 1987a. *The Cenomanian–Turonian (Cretaceous) stage boundary extinctions in the Western Interior of the United States.* PhD Thesis, University of Colorado, Boulder.
—— 1987b. The paleoecology of the Cenomanian–Turonian (Cretaceous) stage boundary extinction at Black Mesa, Arizona. *Palaios,* **2**, 24–40.
—— 1989. Molluscan extinction patterns across the Cenomanian–Turonian stage boundary in the Western Interior of the United States. *Paleobiology,* **15**, 299–320.
—— 1991. *Mytiloides hattini* n. sp.: A guide fossil for the base of the Turonian in the Western Interior of North America. *Journal of Paleontology,* **65**, 234–241.
ELLIOT, D. H., ASKIN, R. A., KYTE, F. T. & ZINSMEISTER, W. J. 1994. Iridium and dinocysts at the Cretaceous–Tertiary boundary on Seymour Island, Antarctica: implications for the K–T event. *Geology,* **22**, 675–678.
ELLIOT, D. K. 1986. *Dynamics of Extinction.* Wiley, New York.
ERNST, G., SCHMID, F. & SEIBERTZ, E. 1983. Event-Stratigraphie im Cenoman und Turon von NW-Deutschland. *Zitteliana,* **10**, 531–554.
——, WOOD, C. J. & HILBRECHT, H. 1984. The Cenomanian–Turonian boundary problems in NW-Germany with comments on the north–south correlation to the Regensburg area. *Bulletin of the Geological Society of Denmark,* **33**, 103–113.
ERWIN, D. H. 1993. *The Great Paleozoic Crisis. Life and Death in the Permian.* Columbia University Press, New York.
GALE, A. S., JENKYNS, H. C., KENNEDY, W. J. & CORFIELD, R. M. 1993. Chemostratigraphy versus biostratigraphy: Data from around the Cenomanian-Turonian boundary. *Journal of the Geological Society, London,* **150**, 29–32.
GOULD, S. J. 1989. *Wonderful Life: The Burgess Shale and the Nature of History.* Norton, New York.
HALLAM, A. 1981. The end-Triassic bivalve extinction event. *Palaeogeography, Palaeoclimatology, Palaeoecology,* **45**, 1–44.
HANSEN, T. A. 1988. Early Tertiary radiation of marine molluscs and the long-term effects of the Cretaceous–Tertiary extinction. *Paleobiology,* **14**, 37–51.
—— & UPSHAW, B. 1990. Aftermath of the Cretaceous–Tertiary extinction: Rate and nature of the Early Paleocene molluscan rebound. *In:* KAUFFMAN, E. G. & WALLISER, O. H. (eds) *Extinction Events in Earth History* Springer, Berlin, Lecture Notes in Earth History, **30**, 401–409.
——, FARRAND, R. B., MONTGOMERY, H. A., BILLMAN, H. G. & BLECHSCHMIDT, G. 1987. Sedimentation and extinction patterns across the Cretaceous–Tertiary boundary interval in east Texas. *Cretaceous Research,* **8**, 229–252.
——, FARREL, B. R. & UPSHAW, B. 1993a. The first 2 million years after the Cretaceous–Tertiary boundary in east Texas: rate and paleoecology. *Paleobiology,* **19**, 251–265.
——, UPSHAW, B., KAUFFMAN, E. G. & GOSE, W. 1993b. Patterns of molluscan extinction and recovery across the Cretaceous–Tertiary boundary in east Texas; report on new outcrops. *Cretaceous Research,* **14**, 685–706.
HARRIES, P. J. 1993a. *Patterns of repopulation following the Cenomanian–Turonian (Upper Cretaceous) mass extinction.* PhD Thesis, University of Colorado, Boulder.

—— 1993b. Dynamics of survival following the Cenomanian–Turonian (Upper Cretaceous) mass extinction event. *Cretaceous Research*, **14**, 563–583.

—— & KAUFFMAN, E. G. 1990. Patterns of survival and recovery following the Cenomanian–Turonian (Late Cretaceous) mass extinction in the Western Interior Basin, United States. *In:* KAUFFMAN, E. G. & WALLISER, O. H. (eds) *Extinction Events in Earth History.* Springer, Berlin, Lecture Notes in Earth History, **30**, 277–298.

——, ——, CRAMPTON, J. *ET AL.* 1995. Systematic concepts, morphology, and biostratigraphy of Late Cenomanian–Early Turonian Inoceramidae. *Proceedings of the 4th International Cretaceous Symposium*, Hamburg, Germany.

——, —— & HANSEN, T. A. 1996. Models of biotic survival following mass extinction. *This volume.*

HART, M. B. & BIGG, P. J. 1981. Anoxic events in the late Cretaceous chalk seas of North-West Europe. *In:* NEALE, J. W. & BRAISER, M. D. (eds) *Microfossils from Recent and Fossil Shelf Seas.* British Micropalaeontological Society, 177–185.

HEINBERG, C. 1979. Bivalves from the latest Maastrichtian of Stevns Klint and their stratigraphic affinities. *In:* BIRKELUND, T. & BROMLEY, R. G. (eds) *Cretaceous–Tertiary Boundary Events I. The Maastrichtian and Danian of Denmark.* University of Copenhagen, Copenhagen, 58–64.

—— 1992. Ecology of the lastest Maastrichtian bivalve fauna – Denmark. *V. International Conference on Bio-Events: Phanerozoic Global Bio-Events and Event-Stratigraphy.* Göttingen, Germany, 49.

HUT, P., ALVAREZ, W., ELDER, W. P., HANSEN, T., KAUFFMAN, E. G., KELLER, G., SHOEMAKER, E. M. & WEISSMAN, P. 1987. Comet showers as a cause of mass extinctions. *Nature*, **329**, 118–126.

JABLONSKI, D. 1986a. Evolutionary consequences of mass extinctions. *In:* RAUP, D. M. & JABLONSKI, D. (eds) *Patterns and Processes in the History of Life.* Springer, Berlin, Report of the Dahlem Conference Workshop on Patterns of Change in Earth Evolution, 313–329.

—— 1986b. Background and mass extinctions: The alternation of macroevolutionary regimes. *Science*, **231**, 129–133.

JARVIS, I., CARSON, G. A., COOPER, M. K. E., HART, M. B., LEARY, P. N., TOCHER, B. A., HORNE, D. & ROSENFELD, A. 1988. Microfossil assemblages and the Cenomanian–Turonian (Late Cretaceous) Oceanic Anoxic Event. *Cretaceous Research*, **9**, 3–103.

JEFFERIES, R. P. S. 1961. The paleoecology of the *Actinocamax plenus* Subzone (lowest Turonian) in the Anglo-Paris Basin. *Palaeontology*, **4**, 609–647.

JOHNSON, C. C. & KAUFFMAN, E. G. 1990. Originations, radiations and extinctions of Cretaceous rudisitid bivalve species in the Caribbean Province. *In:* KAUFFMAN, E. G. & WALLISER, O. H. (eds) *Extinction Events in Earth History.* Springer, Berlin, Lecture Notes in Earth History, **30**, 305–324.

—— & —— 1995. Maastrichtian extinction patterns of Caribbean Province rudistids. *In:* MACLEOD, N. & KELLER, G. (eds) *Biotic and Environmental Effects of the Cretaceous–Tertiary Boundary Event.* Norton, New York, in press.

KAPLAN, U. 1992. Die Oberkriede-Aufschlüsse im Raum Lengerich/Westfalen. *Geologie und Paläontologie Westfalen*, **21**, 7–37.

KAUFFMAN, E. G. 1975. Dispersal and biostratigraphic potential of Cretaceous benthonic Bivalvia in the Western Interior. *In:* CALDWELL, W. G. E. (ed.) *The Cretaceous System in the Western Interior of North America.* Geological Association of Canada, Special Paper **13**, 163–194.

—— 1977. Illustrated guide to biostratigraphically important Cretaceous macrofossils, Western Interior Basin, U.S.A. *In:* KAUFFMAN, E. G. (ed.) *Cretaceous Facies, Faunas, and Paleoenvironments across the Western Interior Basin, Field Guide.* Mountain Geologist, **13**, 225–274.

—— 1978. Evolutionary rates and patterns among Cretaceous Bivalvia. *Transactions of the Royal Society of London*, **B 284**, 277–304.

—— 1979. The ecology and biogeography of the Cretaceous–Tertiary extinction event. *In:* BIRKELUND, T. & CHRISTENSEN, W. K. (eds) *Cretaceous–Tertiary Boundary Events symposium, II.* Proceedings. University of Copenhagen, Copenhagen, 29–37.

—— 1982. The community structure of 'shell islands' on oxygen-depleted substrates in Mesozoic dark shales and laminated carbonates. *In:* EINSELE, G. & SEILACHER, A. (eds) *Cyclic And Event Stratigraphy.* Springer, Berlin, 502–503.

—— 1984. The fabric of Cretaceous marine extinctions. *In:* BERGGREN, W. A. & VAN COUVERING, J. A. (eds) *Catastrophes in Earth History.* Princeton University Press, Princeton, 151–246.

—— 1985. Cretaceous evolution of the Western Interior Basin of the United States. *In:* PRATT, L. M., KAUFFMAN, E. G. & ZELT, F. M. (eds) *Fine-Grained Deposits and Biofacies of the Cretaceous Western Interior Seaway: Evidence of Cyclic Sedimentary Processes.* Society of Economic Paleontologists and Mineralogists, Midyear Meeting, Golden, CO, Field Trip Guidebook, **4**, iv–xiii.

—— 1988. The dynamics of marine stepwise mass extinction. *In:* LAMOLDA, M. A., KAUFFMAN, E. G. & WALLISER, O. H. (eds) *Paleontology and Evolution: Extinction Events.* Revista Española de Paleontología, No. Extraordinario, 57–71

—— 1995. Global change leading to biodiversity crisis in a greenhouse world: The Cenomanian–Turonian (Cretaceous) mass extinction. *In:* STANLEY, S. M., KNOLL, A. H. & KENNETT, J. (eds) *The Effects of Past Global Change on Life.* Washington, DC, National Academy Press, 47–71.

—— & ERWIN, D. H. 1995. Surviving mass extinctions. *Geotimes*, **14**, 3, 14–17.

—— & FAGERSTROM, J. A. 1993. The Phanerozoic evolution of reef diversity. *In:* RICKLEFS, R. E. & SCHLUTER, D. (eds) *Species Diversity in Ecological Communities.* Chicago University Press, 315–329.

—— & HARRIES, P. J., 1993. A model for survival and recovery after mass extinction. *Global Boundary Events, Conference Abstracts*. Warsaw, Polish Geological Institute, 27.

—— & POWELL, J. D. 1977. Paleontology. *In: Stratigraphic, paleontologic and paleoenvironmental analysis of the Upper Cretaceous rocks of Cimarron County, Northwestern Oklahoma*. Geological Society of America Memoir, **149**, 47–114.

—— & WALLISER, O. H. (eds) 1990. *Extinction Events in Earth History*. Springer, Berlin, **30**, 432.

——, HATTIN, D. E. & POWELL, J. D. 1977. *Stratigraphic, paleontologic and paleoenvironmental analysis of the Upper Cretaceous rocks of Cimarron County, Northwestern Oklahoma*. Geological Society of America Memoir, **149**.

——, SAGEMAN, B. B., KIRKLAND, J. I., ELDER, W. P., HARRIES, P. J. & VILLAMIL, T. 1993. Molluscan biostratigraphy of the Cretaceous Western Interior Basin, North America. *In:* CALDWELL, W. G. E. & KAUFFMAN, E. G. (eds) *Evolution of the Western Interior Basin*. Geological Association of Canada, Special Paper, **39**, 397–434.

KELLER, G. 1986. Stepwise mass extinctions and impact events: Late Eocene and early Oligocene. *Marine Micropaleontology*, **10**, 267–293.

—— 1988*a*. Extinction, survivorship and evolution of planktic foraminifera across the Cretaceous/Tertiary boundary at El Kef, Tunisia. *Marine Micropaleontology*, **13**, 239–263.

—— 1988*b*. Biotic turnover in benthic foraminifera across the Cretaceous/Tertiary boundary at El Kef, Tunisia. *Palaeogeography, Palaeoecology, Palaeoclimatology*, **66**, 153–171.

—— 1989*a*. Extended period of extinction across the Cretaceous/Tertiary boundary in planktonic foraminifers of continental-shelf sections, implications of impact and volcanism theories. *Bulletin of the Geological Society of America*, **101**, 1408–1419.

—— 1989*b*. Extended Cretaceous/Tertiary Boundary Extinctions and Delayed Population Change in Planktonic Foraminifera from Brazos River, Texas. *Paleoceanography*, **4**, 287–332.

—— & BARRERA, E. 1990. The Cretaceous/Tertiary boundary impact hypothesis and the paleontological record. *In:* SHARPTON, V. L. & WARD, P. D. (eds) *Global Catastrophes in Earth History: An Interdisciplinary Conference on Impacts, Volcanism, and Mass Mortality*. Geological Society of America, Special Paper, **247**, 563–575.

KENNEDY, W. J. & COBBAN, W. A. 1991. Stratigraphy and interregional correlation of the Cenomanian–Turonian transition in the Western Interior of the United States near Pueblo, Colorado, a potential boundary stratotype for the base of the Turonian Stage. *Newsletter of Stratigraphy*, **24**, 1–33.

——, WRIGHT, C. W. & HANCOCK, J. M. 1987. Basal Turonian ammonites from West Texas. *Palaeontology*, **30**, 27–74.

KOREN, T. N. 1994. Post-glacial adaptive radiation of the Early Silurian multispinose biserial graptolites. *IGCP 335: Biotic Recovery from Mass Extinction Events. Plymouth, UK. Abstracts*, **23**.

—— & KARPINSKY, A. P. 1992. Extinctions and recoveries of graptolite faunas in Late Wenlock through Ludlow. *V. International Conference on Bio-Events: Phanerozoic Global Bio-Events and Event-Stratigraphy. Göttingen, Germany*, 68.

KOUTSOUKOS, E. A. M. 1994. Phenotypic experiments into new pelagic niches in Early Danian planktonic foraminifera. *IGCP 335: Biotic Recovery from Mass Extinction Events. Plymouth, UK. Abstracts*, 25.

LAMOLDA, M. A., KAUFFMAN, E. G. & WALLISER, O. H. (eds) 1988. *Paleontology and Evolution: Extinction Events*. Revista Sociedad Española de Paleontología, Numero Extraordinaro.

LECKIE, R. M. 1985. Foraminifera of the Cenomanian–Turonian boundary interval, Greenhorn Formation, Rock Canyon Anticline, Pueblo, Colorado. *In:* PRATT, L. M., KAUFFMAN, E. G. & ZELT, F. B. (eds) *Fine-grained Deposits and Biofacies of the Cretaceous Western Interior Seaway: Evidence of Cyclic Sedimentary Processes*. Society of Economic Paleontologists and Mineralogists, 1985 Midyear Meeting, Golden, CO, Field Trip Guidebook No. 4, 139–149.

LEVINTON, J. S. 1970. The paleontological significance of opportunistic species. *Lethaia*, **3**, 69–73.

McGHEE, G. R. JR. 1982. The Frasnian–Fammenian extinction event: A preliminary analysis of Appalachian marine ecosystems. *In:* SILVER, L. T. & SCHULTZ, P. H. (eds) *Geological Implications of large Asteroid and Comets on the Earth*. Geological Society of America, Special Paper, **190**, 491–500.

—— 1988. Evolutionary Dynamics of the Frasnian–Famenian Extinction Event. *In:* McMILLAN, N. J., EMBRY, A. F. & GLASS, D. J. (eds) *Devonian of the World*. Canadian Society of Petroleum Geologists, **3**, 23–28.

—— 1989. The Frasnian–Fammenian extinction event. *In:* DONOVAN, S. K. (ed.) *Mass Extinctions: Processes and Evidence*. Ferdinand-Enke, Stuttgart, 133–151.

MAPLES, C. G., WATERS, J. A., LANE, N. G., HONG-FEI, H., MARCUS, S. A. & JIN-XING, W. 1994. Famennian echinoderm recovery from Late Devonian mass extinction events: evidence from the People's Republic Of China. *IGCP 335: Biotic Recovery From Mass Extinction, Plymouth, UK, Abstracts*, 28

MARINCOVITCH, L. 1993. Danian mollusks from Prine Cree Formation, northern Alaska, and implications for Arctic Ocean paleogeography. *Journal of Paleontology [supplement], Memoir*, **67**, 31–35.

MARSHALL, C. R. 1990. Confidence intervals on stratigraphic ranges. *Paleobiology*, **16**, 1–10.

—— 1991. Estimation of taxonomic ranges from the fossil record. *In:* GILINSKY, N. L. & SIGNOR, P. W. (eds) *Analytical Paleobiology*. Paleontological Society, Short Courses in Paleontology No. **4**, 19–38.

NITECKI, M. H. 1984. *Extinctions*. University of Chicago Press.

NORRIS, R. D. & BERGGREN, W. A. 1992. Recovery from mass extinction in pelagic biotas. *Fifth North*

American Paleontological Convention. Chicago, USA, 223.

OBRADOVICH, J. 1993. A Cretaceous time scale. *In:* CALDWELL, W. G. E. & KAUFFMAN, E. G. (eds) *Evolution of the Western Interior Basin.* Geological Association of Canada, Special Paper, **39**, 379–396.

ORTH, C. J., ATTREP, M. JR, QUINTANA, L. R., ELDER, W. P., KAUFFMAN, E. G., DINER, R. & VILLAMIL, T. 1993. Elemental abundance anomalies in the late Cenomanian extinction interval: A search for the source(s). *Earth and Planetary Science Letters,* **117**, 189–204.

PALMER, A. R. 1982. Biomere boundaries: a possible test for extraterrestrial perturbation of the biosphere. *In:* SILVER, L. T. & SCHULTZ, P. H. (eds) *Geological Implications of Impacts of Large Asteroids and Comets on Earth.* Geological Society of America Special Paper, **190**, 469–476.

—— 1984. The biomere problem: Evolution of an idea. *Journal of Paleontology,* **58**, 599–611.

RAUP, D. M. & JABLONSKI, D. 1993. Geography of end-Cretaceous marine bivalve extinctions. *Science,* 971–973.

—— & SEPKOSKI, J. J. JR 1982. Mass extinctions in the marine fossil record. *Science,* **215**, 1501–1503.

—— & —— 1984. Periodicity of extinctions in the geological past. *Proceedings of the National Academy of Sciences, USA,* **81**, 801–805.

SAGEMAN, B. B. 1991. *High-resolution event stratigraphy, carbon geochemistry, and paleobiology of the Upper Cenomanian Hartland Shale Member (Cretaceous), Greenhorn Formation, Western Interior, U.S.* PhD Thesis, University of Colorado.

SANDBERG, C. A., ZIEGLER, W., DREESEN, R. & BUTLER, J. L. 1988. Late Frasnian mass extinction: Conodont event stratigraphy, global changes, and possible causes. *Courier Forschungsinstitut Senckenberg,* **102**, 263–307.

SCHINDLER, E. 1990*a.* The Late Frasnian (Upper Devonian) Kellwasser crisis. *In:* KAUFFMAN, E. G. & WALLISER, O. H. (eds) *Extinction Events in Earth History.* Springer, Berlin, Lecture Notes in Earth Sciences, **30**, 151–159.

—— 1990*b.* Die Kellwasser-Krise (hohe Frasne-Stufe, Ober-Devon). *Göttinger Arbeiten zur Geologie und Paläontologie,* **46**, 115.

SEIBERTZ, E. 1979. Stratigraphisch-fazielle Entwicklung des Turon im südöstlichen Münsterland (Oberkriede, NW-Deutschland). *Newsletters of Stratigraphy,* **8**, 3–60.

SEITZ, O. 1934. Die Variabilität des Inoceramus labiatus v. Schlotheim. *Jahrbuch der Preußischen Geologischen Landesanstalt zu Berlin,* **55**, 429–474.

SEPKOSKI, J. J. JR 1993. Ten years in the library: New data confirm paleontological patterns. *Paleobiology,* **146**, 43–51.

SHARPTON, V. L. & WARD, P. D. (eds) 1990. *Global Catastrophes in Earth History: An Interdisciplinary Conference on Impacts, Volcanism, and Mass* Mortality. Geological Society of America, Special Paper, **247**

SIGNOR, P. W. III & LIPPS, J. H. 1982. Sampling bias, gradual extinction patterns and catastrophes in the fossil record. *In:* SILVER, L. T. & SCHULTZ, P. H. (eds) *Geological Implications of impacts of large Asteroid and Comets on the Earth.* Geological Society of America, Special Paper, **190**, 291–303.

SILVER, L. T. & SCHULTZ, P. H. 1982. *Geological Implications of impacts of large Asteroid and Comets on the Earth.* Geological Society of America, Special Paper, **190**.

TEICHERT, C. 1990. The Permian–Triassic boundary revisited. *In:* KAUFFMAN, E. G. & WALLISER, O. H. (eds) *Extinction Events in Earth History.* Springer, Berlin, Lecture Notes in Earth History, **30**, 199–238.

TRÖGER, K.-A. 1967. Zur Paläontologie, Biostratigraphie und faziellen Ausbildung der unteren Oberkreide (Cenoman bis Turon). Teil I. Paläontologie und Biostratigraphie der Inoceramen des Cenomans bis Turons Mitteleuropas. *Abhandlungen des Staatlichen Museums für Mineralogie und Geologie zu Dresden,* **12**, 13–207.

VERMEIJ, G. 1986. Survival during biotic crises: the properties and evolutionary significance of refuges. *In:* ELLIOT, D. K. (ed.) *Dynamics of Extinction.* Wiley, New York, 231–246.

VILLAMIL, T. 1994. Chronology, relative sea level history, and a new sequence stratigraphic model for basinal Albian to Santonian facies, Colombia. PhD Thesis, University of Colorado.

WALLISER, O. H. (ed.) 1986. *Global Bio-Events.* Springer, Heidelberg, **8**, 442.

——, GROOS-UFFENORDE, H., SCHINDLER, E & ZIEGLER, W. 1989. On the Upper Kellwasser Horizon (Boundary Frasnian/Famennian). *Courier. Forschungs.-institut Senckenberg,* **110**, 247–255.

WARD, P. D. 1988. Maastrichtian ammonite and inoceramid ranges from Bay of Biscay. *In:* LAMOLDA, M. A., KAUFFMAN, E. G. & WALLISER, O. H. (eds) *Paleontology and Evolution: Extinction Events.* Revista Española de Paleontología, No. Extraordinario, 119–126.

——, KENNEDY, W. J., MACLEOD, K. G. & MOUNT, J. F. 1991. Ammonite and inoceramid bivalve extinction patterns in Cretaceous/Tertiary. *Geology,* **19**, 1181–1184.

WOODS, H. 1911. *A monograph of the Cretaceous Lamellibranchia of England: Volume 2, Part 7,* Inoceramus. Monograph Palaeontological Society, **2, 7**, 262–340.

ZINSMEISTER, W. J. & MACELLARI, C. E. 1988. Bivalvia (Mollusca) from Seymour Island, Antarctic Peninsula. *In:* FELDMANN, R. M. & WOODBURNE, M. O. (eds) *Geology and Paleontology of Seymour Island.* Geological Society of America Memoir **169**, 253–284.

Models for biotic survival following mass extinction

PETER J. HARRIES,[1] ERLE G. KAUFFMAN[2] & THOR A. HANSEN[3]

[1] *Department of Geology, University of South Florida, 4202 E. Fowler Ave, SCA 203, Tampa, FL 33620-5200, USA*

[2] *Department of Geological Sciences, CB-250, University of Colorado, Boulder, CO 80309-0250, USA*

[3] *Department of Geology, Western Washington University, Bellingham, WA 98225, USA*

Abstract: Mass extinction intervals are characterized by three dynamic processes: extinction, survival, and recovery. It has been assumed that the taxa surviving a mass extinction are composed predominantly of eurytopic groups and opportunistic/disaster species. However, high-resolution stratigraphic and palaeontological analyses of several mass extinction intervals show that the repopulation of the global ecosystem takes place among ecologically and genetically diverse and complex taxa and occurs far too rapidly to be solely attributed to rapid radiation from a few ecological generalists. We suggest a number of potential survival mechanisms or strategies *(sensu* Fryxell 1983) which have evolved in diverse taxa and which could have allowed them to survive mass extinction intervals. These mechanisms consist of: rapid evolution, preadaptation, neoteny/progenesis, protected and/or unperturbed habitat, refugia species, disaster species, opportunism, broad adaptive ranges, persistent trophic resources, widespread and rapid dispersion, dormancy, bacterial-chemosymbioses, skeletonization requirements, reproductive mechanisms, larval characteristics and chance. Because of the wide variety of potential survival mechanisms, the range of survivors may be far higher than previously hypothesized. This would account, in part, for the diversity and evolutionary state of Lazarus taxa and for the rapid re-establishment of some complex ecosystems following many mass extinction intervals, without calling on "explosive" radiation from generalist/opportunist stocks following a mass extinction interval.

The Phanerozoic record of marine invertebrates is punctuated by numerous, geologically short-term intervals (generally < 3 Ma) during which biotic diversity and abundance declined significantly (< 40% at the familial level and < 63% at the generic level; Raup & Sepkoski 1986). The evolutionary and ecological patterns of extinction during these crises may be significantly different from normal background patterns (Jablonski 1986a), and they are followed abruptly by major evolutionary radiations (or revolutions *sensu* Wiedmann 1973) which change the character of the global biota. These intervals are termed mass extinctions and have been recognized by palaeontologists for over a century (Cuvier 1812; Newell 1967).

Yet mass extinction is only one part of a three-stage, geologically dynamic biological process and is closely linked to subsequent survival and recovery intervals or, collectively, repopulation (see Harries & Kauffman 1990). Jablonski (1986a) pointed out that mass extinction events may be one of the most important factors in evolution – reducing diversity significantly and opening ecospace for the rapid evolution of new

forms. Episodes of repopulation may record some of the most rapid, large-scale biotic changes in life history. Yet, the mechanisms of survival and the nature of repopulation during and following mass extinctions have been poorly studied.

This paper is focused on adaptive traits that may aid in survival through the exceptional stresses of a mass extinction interval. Although most survival mechanisms discussed herein are common to living taxa and have some geological expression, their extensive testing across well-studied mass-extinction intervals needs to be performed in order to validate or discard them. This will also have implications as to their effectiveness relative to magnitude, duration, and causation among Phanerozoic mass extinctions. We expect to find differences in the relative effectiveness of various survival strategies associated with different mass extinction events.

It has been proposed that marine taxa which survive mass extinction events have one or more of the following characteristics: (1) broad adaptive range or eurytopy; many of these taxa are trophic generalists as well (e.g. Sheehan &

From Hart, M. B. (ed.), 1996, Biotic Recovery from Mass Extinction Events,
Geological Society Special Publication No. 102, pp. 41–60.

Hansen 1986); (2) opportunistic taxa adapted to highly stressed environments (see Levinton 1970) which characterize intervals of mass extinction but are normally found in low population numbers within most robust ecosystems (e.g. Palmer 1988); and (3) taxa living primarily or commonly in protected niches or refuges (Vermeij 1986) including the deep ocean. These models predict that survivors of mass extinction should normally have long species ranges, low diversity during the survival interval, usually large, widely dispersed populations prior to the extinction interval(s) (although the opposite may be true for opportunists and refugia species), broad adaptive ranges and thus broad environmental and biogeographic distributions prior to the mass extinction event where not excluded through competition. Their domination of newly available ecospace immediately following mass extinction events should occur rapidly and should give rise through punctuated evolution to diversifying successor species with progressively more restricted niches. In addition, Lazarus taxa (Flessa & Jablonski 1983) – species that survive the extinction event(s) but disappear from the record for an interval spanning a portion or the entire mass extinction event – should reappear abruptly during and following the flood of ecological generalists. In this scenario of survival and recovery, a short interval with little or no obvious biota would be followed by the rapid colonization of the ecospace with a few, usually small-sized (indicative of stress?) eurytopic and opportunistic taxa. These would, in turn, be succeeded by the reappearance of Lazarus taxa and the rapid radiation of new, more specialized taxa derived from these more generalized stocks (e.g. Surlyk & Johansen 1984).

However, a number of observations of recovery intervals following Cretaceous and Tertiary mass extinctions suggest a more complex picture (Kauffman 1984, 1988a; Keller 1986, 1988a,b, 1989; Hansen et al. 1987; Hansen 1988; Hansen & Upshaw 1990; Harries & Kauffman 1990; Harries 1993a). In many well-studied recovery intervals, especially in temperate settings (1) Lazarus taxa which return after mass extinction events are ecologically and evolutionarily more diverse and specialized than previously hypothesized; (2) the first marine recovery biotas are far more diverse, and arise much too rapidly (100–500 ka) to be accounted for by radiation from a few generalized, eurytopic stocks; (3) deeper water and/or poleward, diverse marine assemblages are much less affected than their Tropical counterparts by mass extinction (Kauffman 1979, 1984, 1988; Wilde & Berry 1986, unless

oxygen depletion or other chemical 'poisoning' events are operative. This is due to their broad adaptive ranges and colder temperature tolerance and suggests that a robust Temperate gene pool may remain following mass extinction to provide genetically and ecologically advanced stocks for rapid recolonization of those environments/ecospace (i.e. shallow water and more equatorward) which were most affected by the perturbations associated with mass extinction events (but see discussion Raup & Jablonski 1993 and in Kauffman & Harries, this volume).

If these factors hold true, we would expect diverse survival mechanisms, many of which evolved to cope with normal background conditions operated during the perturbations of mass extinction. This resulted in high levels of ecological/genetic specialization among a certain portion of the survivors. This would explain not only the rapid diversification and apparent rapid emplacement of new taxa shortly after some mass extinction events, and especially individual extinction steps, but also the apparent survival of complex ecosystems across mass extinction boundaries (e.g. the highly diverse bryozoan-brachiopod mound associations found in the Maastrichtian and Danian intervals of the Danish Chalk; Birkelund & Hakansson 1982; Surlyk & Johansen 1984, and Danian deepwater reefs). Although some of the taxa comprising such communities/ecosystems pass through the boundaries unaffected, even newly appearing taxa perform very similar trophic and ecological functions as their extinct predecessors or ancestors (Hansen, 1988). It is quite possible that these new taxa evolved from extinct forms rapidly enough to prevent destruction of complex ecosystems. Thus, we believe that the mechanisms of survival are diverse, and these ideas are at least partially corroborated by the initial data collected at highly refined levels of resolution from the Cenomanian–Turonian (C–T), Cretaceous–Tertiary (K–T) and Eocene–Oligocene (E–O) mass extinction boundaries and their survival–recovery intervals.

Survival mechanisms

There are a number of adaptive traits by which some populations of genetically and ecologically diverse species appear to survive the rapid, large-scale environmental and ecological perturbations commonly associated with mass extinction intervals (e.g. Boucot 1981, 1988; Harries et al. 1987, 1988), and for which there are known examples. Many of these are common traits operative during background conditions. Although each of these mechanisms can function

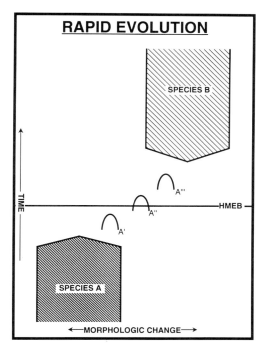

Fig. 1. Shows the relative temporal patterns (vertical dimension) and hypothetical subspecies evolved in the rapid evolution of species B from species A. HMEB, hypothetical mass extinction boundary. Its position depends on the type of mass extinction (ranging from catastrophic to graded (see text) and may occur at any point and/or over a longer temporal interval. In the latter case, the boundary is placed at the last significant extinction episode.

independently to enhance survival, there is also the probability that two or more of these survival mechanisms may function in concert. Several of these mechanisms are closely related, if not functionally or genetically linked, in the sense that if one is in operation, the other, by necessity, is as well. In addition, one adaptive characteristic may fit several categories.

The principal adaptive traits or survival mechanisms can be divided into six primary groups, although there is a certain degree of overlap between them. The main divisions are: evolutionary, habitat, trophic/life habit, population dynamics, physiological, and reproductive. These groups encompass a number of specific 'mechanisms' which are as follows:

Evolutionary

(i) Rapid evolution. Lineages capable of rapid evolutionary rates (macromutation, punctuation, allopatric speciation, etc.) that keep pace with the rate of environmental and ecological changes forcing mass extinction have a higher survival potential (Fig. 1). These rates might be markedly accelerated compared to those of marine taxa evolving under more normal environmental conditions. The inoceramid bivalves at the C–T boundary show a rapid rate of speciation (1 sp./0.2–0.4 Ma; Kauffman 1979; Elder 1987a, b, 1989; Harries & Kauffman 1990; Harries 1993b; Kauffman & Harries 1996), and this may indeed reflect the rapid spread of adaptive traits which allowed this group to pass through the event. In addition, Perch-Nielson (1988) documented the rapid evolution of horseshoe-shaped nannofossils, apparently adapted to hypersaline waters, from rod-shaped forms during Mesozoic and Cenozoic salinity crises.

(ii) Preadaption. The presence of certain previously selected, morphological, behavioural, and/or physiological traits may allow a species to survive at a higher rate during increasing stress associated with mass extinction intervals, to which these traits are strongly adaptive (Fig. 2). A similar effect may be gained when more polymorphic populations of a species develop unique genetic, physiological, behavioural, and/ or morphological traits which are in a sense 'preadapted' for the stressed conditions occurring during mass extinction intervals (i.e. they are better adapted to living in environments that are not optimum for the species as a whole). However, this does not imply that a taxon has the ability to predetermine which adaptations may be advantageous to survival. Kitchell *et al.* (1986), in attempting to explain the resilience to extinction in diatoms across the K–T mass extinction interval in contrast to other planktic groups, suggested that their inherent ability to form resting spores during periods of low nutrient supply, insufficient light, and/or potential lethal dosages of light may have allowed them to survive better the perturbations associated with mass extinctions as well. This adaptation evolved independently of any mass extinction pressures, but was critical to their survival during normal environmental fluctuations in polar regions.

(iii) Paedomorphosis. If a taxon can respond to large-scale perturbations by regulating the rate of ontogenetic development, so that the most resilient life stage is rapidly attained and maintained for longer periods of time, this may improve their chances for survival (Fig. 3). Kennedy & Cobban (1990) have documented a case of progenesis in the ammonite *Rhamphidoceras saxatilis* Kennedy & Cobban from the lowest Turonian, and MacLoed & Kitchell

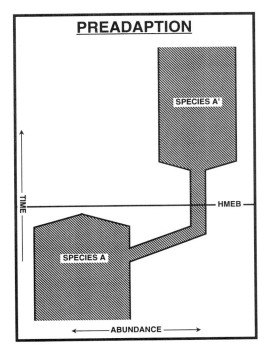

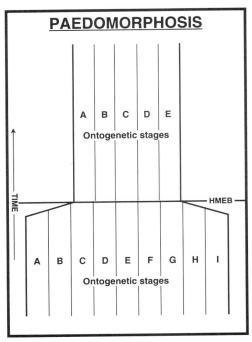

Fig. 2. Shows the relative temporal (vertical dimension) and abundance pattern (horizontal dimension) for a species with a previously evolved trait or behavioural characteristic which allows it to survive through a mass extinction interval. In this case, the survivors are derived from a stock located at the fringes of species A's biogeographical range or niche, possibly reflecting certain populations which may already be adapted to higher levels of stress than the main pre-extinction population(s).

Fig. 3. Shows the temporal distribution and varying hypothetical ontogenetic stages for a neotenic species. The diagram implies that in the survival/recovery intervals, the species is more resilient to environmental perturbations during earlier ontogenetic stages and hence these stages become dominant.

(1988) described progenesis in the Miocene foraminifer *Subbotina linaperta* during a stratigraphic interval which contains microspherules (indicative of impact), although there is no associated mass extinction event.

Habitat

(i) Protected or buffered habitats and/or regions. Taxa which normally occupy habitats in populations/communities that are not as greatly perturbed by the forcing mechanisms of mass extinction (e.g. deep, level-bottom communities, more polar regions) have high survival potential, although these may not be the most diverse of survival biotas. They differ from refugia species in that they remain in their primary habitats throughout their stratigraphic range (i.e. throughout their temporal ranges there is little change in their biogeographical distribution)

although dispersion patterns may become more disjunct. There are many such examples such as the persistence of robust Cretaceous biotas into the Palaeocene of the Alaskan North Slope (cool temperate; Brouwers 1988) and of temperate Cretaceous flora (Fleming & Nichols 1990) reflect this strategy. In addition, Elder (1989) made a case for infaunal habitats potentially allowing for a higher rate of survivorship among C–T boundary bivalves.

(ii) Primary or temporary refugia species. Those taxa which can migrate to secondary habitats not profoundly affected by the environmental mechanisms forcing mass extinction (Vermeij 1986). Most refugia taxa represent species (or their descendants) that have been driven into more limited habitats by competitive displacement from normal marine settings (Fig. 4). Refugia habitats can be small (e.g. reef caves) or geographically widespread (e.g. deep-sea or high latitude regions). Survival of at least small populations in these restricted habitats may

allow refugia taxa to expand their ranges rapidly to previous and even greater levels following the elimination of competitive species and/or the return of favourable environmental conditions. The longer lineages have been in refugia and thus evolved into endemic refugia species, the longer it will take them to re-radiate into prime habitats occupied by their ancestors, if at all. Refugia species usually occur in low diversity, small, scattered populations, although robust populations are known such as brachiopods in reef caves.

There are three possible roles for refugia species in survival and recovery intervals following mass extinction.

(a) Long-term refugia taxa consist of lineages or clades whose ancestral species were forced through competitive displacement or deterioration of their preferred habitat into refugia habitats, to which they have subsequently adapted and even diversified through niche-partitioning long before the onset of mass extinction events. In particular, biotas with affinities to dominant Palaeozoic shallow-water communities occupy such refugia today. Brachiopods and certain groups of 'stromatoporoids' (coralline sponges) are common in reef caves (Jackson *et al.* 1971; Hartman & Goreau 1977; Jackson & Winston 1982); brachiopods, glass sponges, and stalked crinoids are most abundant today in deep-water refugia. Thus, long-term adaptation to refugia is a viable survival strategy; reinvasion of decimated shallow-water environments by possibly deep-water refugia stocks following Cambrian biomere events is an example of this strategy (Palmer 1988).

(b) Some taxa with initially wide habitat ranges were decimated in their primary habitats during mass extinction, leaving only (or mostly) the migrated, refugia populations as survivors. Because of the short residence time in refugia, rapid migration back into, and habitation of, previously occupied environments is possible once conditions ameliorate to the extent that temporary refugia species can repopulate their favoured environments. Stanley (1988) has shown that accreted terrains found in north-western North America, formerly representing Triassic island arc systems located somewhere in the South Pacific, contained forms of corals, algae, and sponges that were thought to have gone extinct in the Permian, but flourished in the Middle Triassic as members of various reef guilds.

(c) It is possible that certain characteristic refugia environments spread during mass extinction intervals. An example of this may be the advection of somewhat toxic deep-water to the surface and onto epicontinental shelves allowing the survival and spread of adapted species (Wilde *et al.* 1990). The presence of anoxia has been correlated with several mass extinction events and has been documented in examples such as the spread of the genus *Mytiloides* at the C–T boundary (see Kauffman & Harries, this volume).

Trophic/life habit

(i) Disaster species (Fig. 5) are taxa specifically adapted to the stressed environmental conditions associated with mass extinction intervals (as opposed to broader ranging opportunists, see below), or which have specific behavioural patterns that become prevalent during these stressed intervals. These taxa develop relatively large populations ('blooms') during the early survival interval. The population increase of these taxa is extremely short-lived, as they are quickly replaced by opportunists and other survivors early in the repopulation. The thoracospheres and braarudospheres most commonly found just above the impact level in K–T sections (Percival & Fischer 1977, Thierstein 1981, Perch-Nielson 1988) may represent this strategy. They are generally absent or found in low abundances prior to the extinction level, but at and just above the boundary they proliferate dramatically. Another example of this is *Discinisca* sp., an inarticulate orbiculoid brachiopod, which is normally very rare, but which became more abundant and briefly dominated level-bottom communities almost exclusively in the 'dead zone' of a number of C–T boundary sections in the Western Interior of North America (WINA; Elder 1987a, Harries & Kauffman 1990); similar communities have also been found at the E–O boundary. In addition, the benthic foraminifers *Gavelinella berthelini* (Keller) (synonymous with *G. dakotensis* of the WINA) and *Lingulogavelinella globosa* (Brotzen) are found in great abundances directly at the C–T mass extinction boundary in the United Kingdom. These populations are reduced quite dramatically as a more robust ecosystem is re-established (Jarvis *et al.* 1988).

(ii) Opportunists. These taxa are persistent, common members in pre-extinction environments, suppressed by competition from equilibrium species (Fig. 6). But owing primarily to their eurytopy and r-selection, opportunists are capable of prolific population expansion and rapid biogeographical dispersal into stressed environments (see Levinton 1970), including

(a)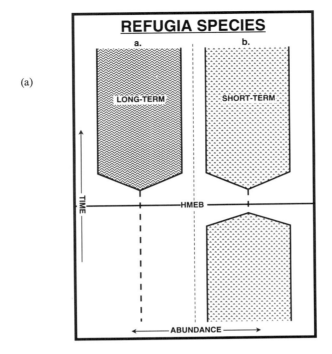

(b)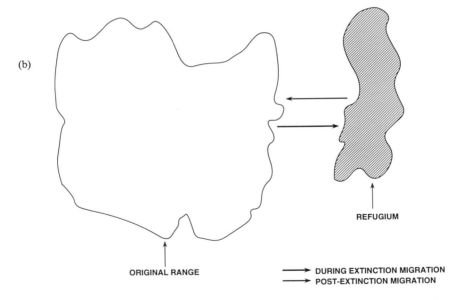

(Fig. 4. (a) Shows the relative temporal and abundance patterns of refugia species. Two possible patterns are depicted: (1) survival in a refugium prior to the extinction representing earlier displacement, followed by radiation into an expanded niche once its previous competitors (?) are removed by extinction; and (2) survival in a refugium for the duration of the extinction interval until conditions ameliorate. (b) Shows the biogeographical pattern depicting the active migration of a certain number of individuals to (dark arrow) and from (light arrow) a refugium. Note that the original range of the species does not encompass the refugium. This figure is not meant to imply that the post-extinction geographical distribution will be equivalent to the pre-extinction range in all cases.

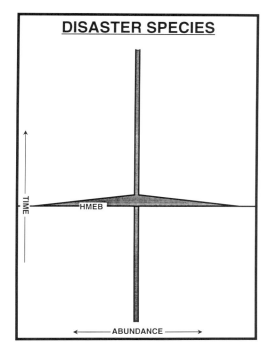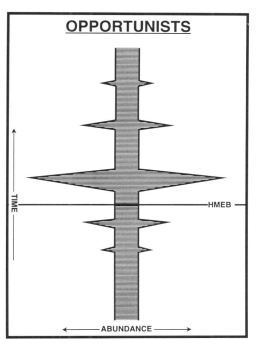

Fig. 5. Shows the relative temporal and abundance patterns for disaster species. Note that during 'background' conditions their abundance is extremely low, although they 'bloom' once the environmental conditions markedly decline and once there is a large amount of available ecospace for them to occupy.

Fig. 6. Shows the relative temporal and abundance patterns for an opportunistic species. Note the slight population 'blooms' above and below the boundary; these are indicative of background environmental perturbations to the ecosystem. The larger peak in the post-extinction interval implies more volatility in the ecological structure following a mass extinction event. The largest bloom occurs close to the extinction boundary as conditions begin to ameliorate.

those characterizing mass extinctions. These are commonly pioneer species and represent an early phase of recolonization of vacated ecospace. Palmer (1984) has shown that there are a few trilobites and brachiopods which pass through the various biomere boundaries at low abundances, but then completely dominate the open ecospace for a short period before being relegated to pre-extinction population levels by a diversifying recovery fauna. Fleming and Nichols (1990) noted a similar pattern among ferns at the K–T boundary (the 'fern spike'). Although they may be one, if not the, dominant biotic components of early post-extinction communities, there is little evidence to suggest that these were the primary stocks seeding post-extinction radiations.

(iii) Ecological generalists or eurytopes. Those groups (Fig. 7) with broad adaptive ranges and reproductive and survival limits which are not breached by the environmental perturbations causing mass extinction. However, their biogeographical dispersion may become more restricted and disjunct during mass extinction intervals. They may be regarded as generalists in terms of their broad habitat range, broad range of trophic resources (relative omnivores), and/or in their ability to withstand or adapt to wide-ranging environmental fluctuations (Sheehan & Hansen 1986). Most are characterized by slow evolutionary rates (Kauffman 1978) and many belong to relatively primitive, but persistent, lineages. 'Living fossils' (e.g. marsupial opossums, *Crassostrea*, *Lingula*, *Nautilus* and *Limulus*) are outstanding examples. Among resistant survivors of Cretaceous extinction events with long species/genus ranges, Kauffman (1978) identified diverse fresh, brackish, and intertidal molluscs, especially unionid, ostreid, mytilid and donacid bivalves and neritinid gastropods, as the most extinction resistant. More rapidly evolving marine inoceramid, generalized ostreid, pectinid and pteriid bivalves are

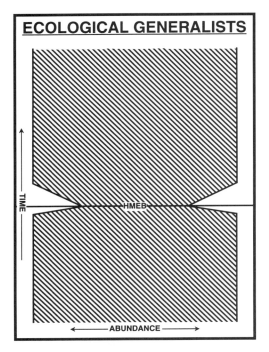

Fig. 7. Shows the relative temporal and abundance patterns of a eurytopic taxon across a mass extinction boundary.

also noted as having broad adaptive ranges and high survival potential. Keller (1988a, 1989) noted that generalized, cosmopolitan planktonic foraminifera survived the K–T mass extinction interval, whereas the more specialized forms became extinct.

(iv) Persistent trophic resource exploiters. Mass extinction events are commonly associated with disruption and/or partial collapse of the global food chain (Kauffman 1977; Hsü et al. 1982). Mass extinction should most severely affect trophic specialists dependent upon food resources that are sensitive to changes in temperature, chemistry, light, moisture, etc. (e.g. upper-water-column plankton), and give higher survival potential not only to trophic generalists but also to those groups whose trophic requirements, no matter how specialized, are time-predictable and not affected by the causal mechanisms for mass extinction (e.g. benthic organic detritus). The most common trophic class among long-lived marine survivors are those based around non-specific detritus feeding (e.g. protobrachiate bivalves; Levinton 1974; Kauffman 1978; Sheehan & Hansen 1986).

Other common survivors are generalized algal–fungal grazers (many gastropods; Sohl 1987), selective detritus feeders (Tellinidae, many Nuculanidae among bivalves; Kauffman 1977), and many deep-water microcarnivores (poromyacean bivalves). Many of these groups, as well as trophic generalists, dominated survival faunas of the C–T (Elder 1987a, b; Harries & Kauffman 1990; Roy et al. 1990) and K–T boundary intervals (Hansen 1988; Hansen & Upshaw 1990; Roy et al. 1990).

Population dynamics

(i) Widespread (continuous or discontinuous) adult biogeographical dispersion. Extensive biogeographic distribution by long-lived planktotrophic larvae or adult mobility, without loss of gene flow, may ensure that at least some marginal populations will survive the large-scale perturbances of mass extinction intervals (Fig. 8). The cause of survivorship may be that a population is located beyond the range of mass extinction processes or due to the broader adaptive ranges of marginal populations. This may be especially applicable to taxa which have a biogeographical range extending from the Tropics towards the more polar regions. Jablonski (1986a) indicated that a broad geographic range was the best trait for survival, although at the clade level. *Lenticulina rotulata* (Lamarck), a C–T benthic foraminifera, was widely distributed prior to the mass extinction event and, in addition, had developed a spectrum of morphologies (polymorphism), possibly attributable to ecophenotypic variation, in response to the varying conditions inhabited by widely dispersed local populations. Expansion of the oxygen-minimum zone during the C–T Bonarelli Oceanic Anoxic Event (OAE II) reduced the once broad range of this species, but owing to its wide original dispersal, the species survived in certain regions which were unaffected and the species survived (Jarvis et al. 1988). Surviving populations were the more primitive forms (i.e. the stock from which subspecies proliferated). Palaeozoic brachiopods which survived mass extinction episodes were primarily generalized cosmopolitan forms (Boucot 1990). The final surviving ammonites and inoceramids in the C–T 'dead zone' and early recovery were among the most cosmopolitan lineages with widespread dispersion resulting from adult mobility (ammonites) and exceptional planktic larval drift (inoceramids; Kauffman 1975; Elder 1987a, b, 1989, Harries and Kauffman 1990).

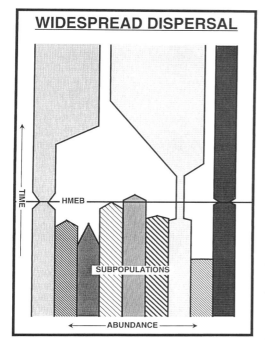

Fig. 8. Shows the relative temporal and abundance patterns of a species with widespread dispersal. This model implies that several of the subpopulations may become extinct in certain regions, whereas others may survive. These surviving subpopulations can then expand into the available ecospace following the extinction boundary.

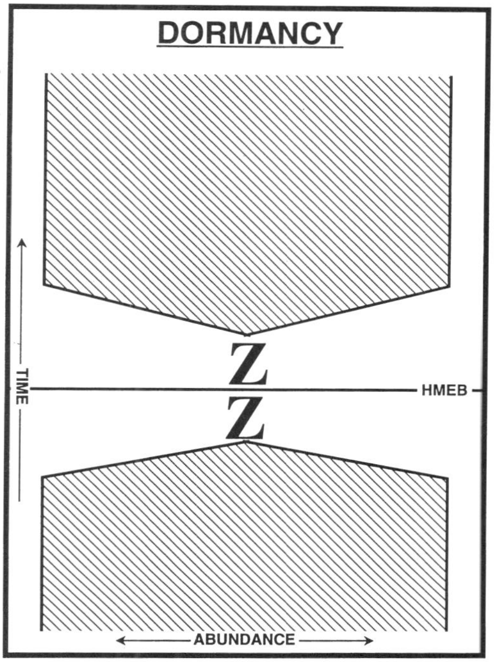

Fig. 9. Shows the relative temporal and abundance patterns for a species that is able to remain in a dormant state during the ecologically stressed intervals of a mass extinction event. This allows the species to repopulate available niche(s) once conditions ameliorate.

Physiological

(i) Dormancy. If a group, such as the algae (see papers in Fryxell 1983, Vermeij 1987) or the dinoflagellates, is able to encyst during the onset of harmful environmental stress, they may be able to retain encystment long enough to survive the mass extinction event. An example of this may be the proliferation of the algae *Renalcis* in the Frasnian–Fammenian reefal facies of the Canning Basin (Playford 1980). This group replaced the previously dominant stromatoporoids during this Late Devonian mass extinction. Other taxa, such as fresh-water sponges (e.g. *Spongilla*), can also encyst and survive for considerable periods of time. Many plants are able to produce seeds and spores which can maintain their vitality for considerable periods of time. Plants, in general, have higher survivorship than many terrestrial and marine animals across mass extinction intervals.

(ii) Bacterial chemosymbioses. Diverse marine invertebrate taxa associated with submarine thermal vents, H_2S and methane seeps, petroleum seeps, and H_2S-enriched anaerobic substrates are now known to have symbioses with sulphide- or methane-oxidizing bacteria which provide both a source of C_{org} and O_2 for vital physiological processes in otherwise inhospitable environments (Desbruyeres & Laubier 1983; Grassle 1983; Hessler & Smithey 1983; Cavanaugh 1985; Felbeck *et al.* 1985; Hecker 1985; Hessler *et al.* 1985; McLean 1985; Suess *et al.* 1985; Turncliffe *et al.* 1985). Kauffman (1988b) has suggested that this unique ecosystem may have been much more pervasive in the past under more typical Phanerozoic conditions characterized by warmer, more equable climates, slower deep-ocean circulation with pervasive oxygen restriction, and more tectonically and methanogenically produced seepage on the seafloor. Because the taxa associated with these environments were basically immune to the environmental perturbations which decimated more normal trophic and O_2-based physiological

groups during mass extinction, they had very high survivorship potential. Further, because many of the molluscan lineages associated with these springs and seeps have representatives (species) which are also common in normal marine environments (with or without symbiotic bacteria), it is probable that these survivor groups could have served as the rootstocks for radiation into available normal marine ecospace following mass extinction events. Some species could potentially have moved back and forth between the two life modes to escape environmental deterioration associated with global biotic crises. Most of the Bivalvia commonly associated with well-documented Cretaceous (Howe & Kauffman 1986; Howe 1987) and modern submarine springs (Desbruyeres & Laubier 1983; Grassle 1983; Hessler & Smithey 1983; Cavanaugh 1985; Felbeck *et al.* 1985; Hecker 1985; Hessler *et al.* 1985; McLean 1985; Suess *et al.* 1985; Turncliffe *et al.* 1985) have bacterial symbioses, and have a long history of survival across many mass extinction intervals (e.g. Lucinidae, Mytilidae, *Calyptogena,* Inoceramidae, Solemyacea, possibly Nuculidae and Nuculanidae; see also Seilacher 1990).

(iii) Skeletonization requirements. Major thermal and chemical changes in the world's oceans associated with mass extinction intervals are evident from stable isotope fluctuations of a rate and magnitude that well exceeds background levels (Hsü *et al.* 1982; Kauffman 1986, 1988*a*; Hut *et al.* 1987). These changes and possible large-scale fluctuations in the Calcium Compensation Depth (Worsely 1974; Arthur *et al.* 1987) could strongly affect the availability and physiological mineralization processes of shell-building (especially carbonate) materials. This, in turn, might favour, as survivor taxa, those groups with siliceous skeletons (diatoms, radiolarians, hexacxtinellids), organic-walled skeletons (dinoflagellates, worms, sponges, many arthropods), phosphatic skeletons (inarticulate brachiopods), chemically inert, external organic layers that are capable of protecting the shell (bivalve periostracum in deep-water taxa), agglutinated test-builders (certain foraminifera, many worms), and naked soft-bodied taxa (worms, coelenterates). Diatoms and dinoflagellates are dominant survivors among the plankton across the K–T boundary interval (Tappan 1979, Kitchell *et al.* 1986); bioturbation patterns attributable to various worms and chitinous arthropods pass through the C–T and K–T boundary without significant change (Ekdale & Bromely 1984). Sponges show little change across many mass extinction boundaries. Phos-

phatic inarticulate brachiopods are the disaster species of the basal Turonian (beginning of the C–T repopulation) in the WINA (Elder 1987*a*; Harries & Kauffman 1990). Agglutinated foraminifera show little change across the C–T and K–T boundaries in Europe (Koutsoukos & Hart 1988).

Reproductive

(i) Reproductive mechanisms. One of the key elements to survival is the ability to continue production of viable offspring during and immediately following mass extinction intervals. Listed below are some of the taxa to have developed different adaptations which can become especially beneficial in times of severe ecological and environmental stress (Jablonski & Lutz 1980).

(a) Taxa which are capable of producing an exceptionally high yield of larvae or offspring during periods of pronounced stress. The sheer number of offspring may help to insure that at least a fraction of them will survive. The Ostreidae are a group which utilize this reproductive strategy, and it is interesting to note that at a number of C–T (Elder 1987*a*; Harries & Kauffman, 1990; Harries, 1993) and K–T (Kauffman & Hansen in prep.) boundary sections, oysters are found in much greater abundances just prior to these extinction events and have high survivorship. Many ostreid lineage survived Mesozoic through Cenozoic mass extinction intervals.

(b) Young which are brooded, born at a more advanced juvenile stage, or develop from lecithotrophic larvae have adaptive advantages in that the young are more advanced when they settle than those requiring a long, exposed stage of metamorphosis. This is important because among ontogenetic stages the larval state is generally the least tolerant to stress in most groups, and mass extinction processes seem to affect the photic zone, where most planktotrophic larvae reside, more than deeper waters where brooding, lecithotrophism and short-lived nektonic larvae dominate. Hence those taxa which produce even slightly more developed offspring have a survival advantage. An interesting example of this strategy has been suggested for the pattern of ostracod extinction at the C–T boundary. The podocopids suffered drastic extinction during the mass extinction, whereas the platycopids were much less adversely affected. It has been hypothesized that the platycopids brooded their young for the first few instaars within their tests, releasing them into the environment when they were considerably less susceptible to the stressed conditions present during the mass extinction interval (Jarvis *et al.* 1988). Freshwater bivalves (Unionidae) and deeper marine Astartidae (Kauffman & Buddenhagen 1969) are known to brood their larvae, and they have very high rates of survivorship through Mesozoic–Cenozoic mass extinction intervals. Landman (1984) hypothesized that the reason for the extinction of the ammonites at the K–T boundary may be tied to their pelagic larvae, whereas the nautiloids, which survived this and many prior extinction events, employ an egg sack in which the young are protected during the early stages of development (Arnold 1987).

(c) Some taxa (especially deep benthic forms) reproduce only at long time intervals (tens of years), and the adult members are very long-lived and slow-growing, allowing them to pass through at least the most intense periods of the crisis in a semi-suspended physiological state. In this case, the individuals devote most of their resources and physiological energy to their own survival rather than to reproduction. But, once conditions ameliorate, the reproductive cycle is resumed. Deep-water molluscs (especially bivalves) employ this strategy (Turekian *et al.* 1975).

(d) Those taxa capable of sexual reproduction may have advantages over asexual reproduction in being able to accelerate the rate of potential evolution and adaptive change during stressed mass extinction intervals. This has been proposed to explain the greater survivorship of deeper water, large foraminifers (Braiser & Buxton 1988).

(e) Taxa which produce an egg resistant to desiccation and freezing, such as some of the ostracods (Whately 1990) may also be able to increase the number of offspring which survive.

(f) Another potential reproductive mechanism to promote survivorship is parthenogenesis. This requires that only one individual survive the extinction event, because each individual is furnished with the male and female reproductive organs and can reproduce alone. Certain groups of ostracods (the Cypridacea and the Darwinulacea) are capable of parthenogenesis (Whately, 1990).

(ii) Larval mechanisms. There are also variations in planktotrophic larval behaviour which may play a critical role in determining whether a species will be able to survive through a mass extinction crisis. Although Jablonski (1986c) saw no relationship between larval strategy and survivorship for Gulf Coast bivalves and gastropods, the parameters he used (solely planktotrophy v. non-planktotrophy) may not have been sensitive enough to discern the types of survival patterns listed below.

(a) Those larvae whose planktic habit involves a deep, rather than a shallow or surface position in the water column. The larvae's depth in the water column could help to buffer the effects of stress in cases where this was most intense at or near the air/water interface.

(b) Those larvae which are more eurytopic would also have a greater opportunity for survival.

(c) Larvae that have the ability to regulate their rate of metamorphosis could potentially increase their chances for survival. This implies that certain larvae may be able to speed up, delay or completely suspend their ontogenetic development until conditions are ameliorated or they reach a region that is less affected by stress. The bivalve *Mytilus edulis* employs this larval strategy and is able to drift for week in a physiologically suspended state under difficult environmental conditions before resuming metamorphosis when favourable planktic environs are reached (Kauffman 1975). Mytilidae, in general, have high survival rates at the species and lineage level during Mesozoic and Cenozcic mass extinction intervals.

Chance

By the luck of the draw, representatives of certain species are in the right place at the right moment for a few individuals to survive mass extinction intervals and to repopulate vacant ecospace rapidly following these crises (Fig. 10).

Summary

Given the various adaptive mechanisms (traits) for survival, combined with preliminary data on surviving taxa from a number of different mass extinction boundaries, the overall fabric of the survival and recovery intervals following a global biotic crisis is potentially much more complex than a scenario calling on explosive radiation from a few eurytopic and/or opportunistic groups. In a general sense, the combination of these numerous different survival mechanisms implies that repopulation can be extremely rapid and can involve genetically and ecologically more complex forms than previously hypothesized (Fig. 11).

The nature of the mass extinction event itself may have played a substantial role in determining the effectiveness of the various survival mechanisms. Data collected from the K–T boundary suggest that at least the main portion of the extinction events may have been essentially catastrophic (within a few Ka), whereas the data for the C–T and E–O boundaries suggest that the mass extinctions were step-wise or nearly so, and drawn out over 2–3 Ma. Thus, not only is the timing of the events different, but there is also considerable variation between the magnitude of the extinction and the overall survival and recovery patterns (Fig. 12). It should be stressed that these two models represent virtual end members of a spectrum, and that the majority of mass extinction events probably represent 'hybrid' extinctions.

Catastrophic extinctions are short-term, high intensity events that are much less selective and more pervasive than step-wise or gradual mass extinctions of similar magnitudes. Due to the shock effect of catastrophic forcing mechanisms, there are several predictions about the nature of survival that can be made. Of the mechanisms outlined above, the ones that may play major roles in survival through a short-term catastrophic mass extinction event are basically those that operate on ecological time scales. These include: chance, dormancy (which includes encystment as well as durable, protected germ cells), some opportunism, trophic strategies favouring taxa with predictable food resources (i.e. organic detritus), disaster species, and wide geographical dispersal. In addition, ecological generalism, certain larval strategies and chemosymbioses may also play important roles in determining the survivability of various taxa. Because these events are catastrophic and occur over a geologically (and perhaps even historically) short time-span, evolutionary mechanisms and abilities to withstand major population reduction should not be prominent. The resulting pattern of repopulation may be characterized by the increased span of survival and recovery intervals as well as their divisions (see Fig. 12).

However, step-wise or graded extinctions result in the breaching of a number of ecologically specific thresholds due to the lesser magnitude and greater selectivity of each extinction step in the progression. Step-wise extinctions contain partial normalization of environments and may even allow significant origination between extinction events, although few of these new taxa (crisis progenitors; see Kauffman & Harries this volume) range out of the extinction interval (Elder 1989). Extinction levels are not extensive enough to decimate the entire biota, and 'dead zones' following extinction steps are missing or very restricted in time and space (because there are more potential gene pools to provide new inhabitants of a devastated region).

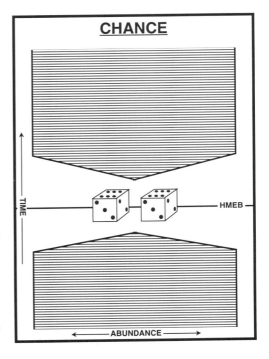

Fig. 10. Shows the relative temporal and abundance patterns for a species which survives owing to chance.

High survivorship and only partial ecosystem loss coupled with new origination damps the possibility for 'explosive' radiation following any major extinction steps because much ecospace remains filled after each step, some ecospace is progressively filled with new taxa between steps, and diversity may be rapidly built along old themes (see Fig. 12).

A large number of survival mechanisms on both ecological and evolutionary scales can potentially be used during longer step-wise to graded events than during catastrophic extinctions. Dominant survival mechanisms are opportunism, habitation of refugia and protected major habitats, the ability to migrate, the ability to withstand a major reduction in population size, rapid evolutionary rates, paedomorphism, ecological generalism, rapid, widespread dispersal, chemosymbioses, preadaption, and larval and reproductive mechanisms. It should also be noted that during gradual to step-wise extinction, origination rates may initially keep pace with extinction rates, but as more and more ecological thresholds are breached, the rate of origination is diminished and results in marked

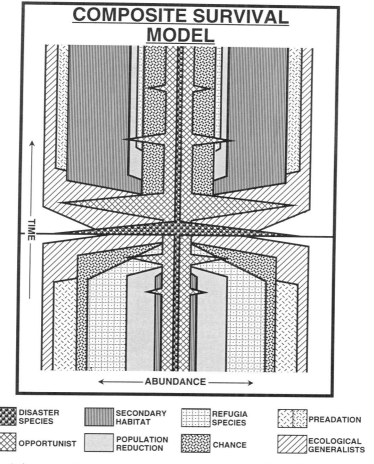

Fig. 11. Shows the relative temporal and abundance patterns and the manner in which the survival may proceed. Note that the initial colonizers are disaster species, followed by the opportunists and then the ecospace is filled by a number of species which represent not only a large number of complex strategies, but also the more stenotopic species.

decline in pre-extinction diversity. This implies that throughout the entire graded to step-wise mass extinction interval and ensuing survival intervals, new species are appearing but at a progressively lower rate until the recovery interval.

Another aspect that must be considered is the decoupling between the various climate zones and major habitats. The same extinction event may be manifested differently in the tropics than in the temperate or the polar climate zones. Because of much narrower adaptive ranges, tropical taxa may be highly susceptible to extinction, as clearly seen for reefs (Kauffman 1979, 1984; Fagerstrom 1987, Stanley 1988, Johnson & Kauffman 1990). Temperate taxa, however, may require considerably more stress in order for their extinction thresholds to be breached. At the K–T boundary, there is evidence that shallow-water and tropical planktic and benthic habitats experienced vastly increased rates of extinction relative to deep-water and temperate benthos (Kauffman 1979, 1984, 1988; Thomas 1990). At the C–T boundary the ammonites required far more time to become fully re-established than the epibenthic inoceramids (Elder, 1989; Harries & Kauffman, 1990). In addition, the terrestrial pollen, spore and plant megafossil record across the K–T boundary (at least for the WINA; Johnson & Hickey 1990; Fleming & Nichols 1990) show marked levels of extinction whereas the extinction record of terrestrial vertebrates (excluding the dinosaurs and elasmobranchs; Archibald &

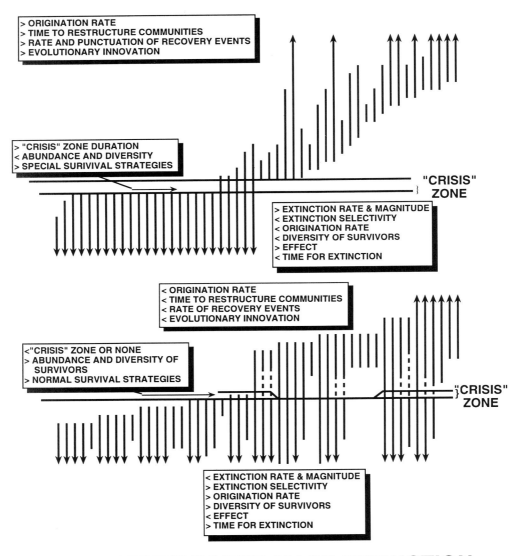

Fig. 12. Depicts the difference in survival patterns between two end members in extinction types, catastrophic and step-wise/graded.

Bryant 1988) is not nearly as dramatic.

Patterns in the fossil record

The diverse mechanisms (adaptive traits) for survival through intervals of mass extinction, proposed herein, represent possible explanations for the rapid rate and complex ecologic/genetic levels of recovery observed in the fossil record from early post-extinction intervals. The proof of a relationship between diverse survival mechanisms and evolutionary repopulation of the Earth lies in high-resolution (cm scale), palaeogeographically widespread and paleoenvironmentally diverse analyses of the fossil record during these short (500 Ka or less; Keller 1986; Harries & Kauffman 1990) repopulation intervals. Complete stratigraphic records of mass extinction or other intervals of this duration are thought to be quite rare (Dingus 1984), complicated by the fact that many mass extinctions occur at falling eustatic sea level, enhancing non-deposition of, restriction of, or erosion of epicontinental and continental marginal marine sections (Kauffman 1984). Nevertheless, the great interest in documenting the fabric of mass extinction has led to the search for and discovery of very complete boundary sections for most mass extinction intervals. From these, examples can be drawn for the rates and patterns of extinction, survival and recovery to test hypotheses proposed here.

One of the problems with the identification of survival mechanisms in the fossil record is that many of them will be difficult to differentiate, and, owing to sampling biases and rarefaction problems, small surviving populations will be difficult to find. Taxa which migrate to refugia and taxa whose population size is drastically reduced during the mass extinction event, will typically appear as Lazarus taxa in the fossil record. The ability to distinguish between these different mechanisms will hinge on the chance discovery and identification of the refugium or secondary habitat which a taxon exploits during the mass extinction, or rare individuals of *in situ* reduced and stranded populations. Ideally, if one could find one of these emigration/repopulation centres and could follow its development in terms of speciation and/or radiation the survival mechanisms would be very obvious. But given our ability to reconstruct and understand the biology and ecology of survivors this enables us to determine the survival mechanisms. Even without data on the nature of these different mechanisms within survival intervals, the existence of Lazarus taxa is well documented (Birkelund & Hakansson 1982; Flessa & Jablonski 1983; Surlyk & Johansen 1984; Jablonski 1986b; Harries & Kauffman 1990), and analyses of these species will provide valuable insight into survival mechanisms during mass extinction intervals. The implications of the survival mechanisms is that Lazarus taxa may represent groups which survived a given mass extinction event through a variety of different adaptations.

The recent emphasis on high-resolution stratigraphy, especially with respect to its applicability to the analyses of bio-events (Kauffman 1986, 1988a) indicates that in order for survival and recovery events to be accurately portrayed care must be taken to sample the events at centimetre scale intervals. In the few instances where this type of analysis has been performed (Birkelund & Hakansson 1982; Surlyk & Johansen 1984; Elder 1987a,b, 1989; Hansen et al. 1987, Hansen 1988; Hansen & Upshaw 1990; Harries & Kauffman 1990 and papers in this volume), a general pattern of survival and recovery can be seen. Following a major step of mass extinction there appears to be a period of approximately 50–100 Ka during which the diversity and abundance are extremely low. It is here that the dynamics of survival and early recovery need to be unravelled. Above this, taxa begin to reappear with increasing abundance and diversity, initially survivor species, then new species from surviving lineages, followed by new species from new lineages (see Harries and Kauffman 1990), and it is usual for them to do so in a punctuated fashion, i.e. several taxa reappear simultaneously, although this may be a function of the sampling technique (see Birkelund & Hakansson 1982; Surlyk & Johansen 1984; Harries & Kauffman 1990). This appears to indicate that repopulation occurs in a number of discrete events. Each step in this process is associated with the appearance of an increasingly specialized biota as well as an increase in overall diversity and ecological complexity. Several of the early colonizers (survival taxa) are short-lived after the extinction event (5–50 Ka), and this may be due to ecological competition which rapidly develops once environments are ameliorated to an extent where increasingly complex taxa return or evolve anew. This implies that extinction is also a regulating mechanism in the recovery intervals (e.g. those of the Early Turonian and Early Palaeocene; Kauffman 1984, 1986, 1988a; Elder 1987a,b, 1989; Hansen 1988a,b; Hansen & Upshaw 1990; Harries & Kauffman 1990). At this point, preliminary studies have produced a basic model of survival and recovery following mass extinction (Kauffman & Erwin 1995). However, in order for this model and the various survival mechanisms to be found, complete sections must be located representing a spectrum of different environmental conditions and ecological structure (i.e. physically- v. biologically-accommodated systems) for a number of mass extinction events. The variability between the records from these regions and time periods will allow a number of important questions to be answered. These include: is it possible to determine what regions served as the loci for immigration/ repopulation?; is it the same place(s) for all taxa or is there phyletic or some other form of taxonomic variability?; are these regions detectable given the spotty record which exists?; which taxa survive?; is there any similarity in terms of taxonomic affinities or ecological strategy between extinction events?; is there a correlation between the style of a mass extinction and the survival and recovery patterns?; how do the survival mechanisms vary depending on the cause(s) of a given mass extinction?; is this pattern the same globally, or are there differences based upon biogeographical dispersion?

Conclusion

Despite the large amount of data which have been collected in attempts to resolve and document the rates, patterns and causal factors for mass extinction and their manifestation in the geological record, the period of biotic survival and recovery following what appears to be a significant decimation of the global marine biota, has just begun to be studied. Obviously, certain taxa must survive these events in order for there to be only a short-term disruption in the overall Phanerozoic trend towards greater species diversity (Valentine 1970; Bambach 1977; Signor 1985). A 'clean slate' from which life itself had to re-evolve clearly was not a consequence of mass extinction given the speed at which high diversity and complex ecosystems were reestablished, normally within 250–500 Ka following one of these events, excluding reefal and other tropical communities.

Even from the preliminary data in hand, models for survival derived from them indicate that the traditional view of mass extinction survivors as limited to rare, generalized, ecologically eurytopic and opportunistic taxa cannot be logically or factually supported. Although it would be a relatively neat and uncumbersome solution if only one or two survival mechanisms could account for the continuation of life, evidence suggests that this is not the case. Based

on available data, the potential range of survival mechanisms involved in the post-extinction repopulation are numerous and certainly this is what one would expect given the complexity of biotic systems. In addition, the different nature of the Phanerozoic mass extinctions events and the manner in which the perturbations feed back into the environment suggest that the effectiveness of survival mechanisms will vary from event to event and from group to group depending upon their response to stress.

The mechanisms presented here should be seen more as a theoretical starting point from which to test the patterns of survival and repopulation seen in the geologic record. It is only as more detailed information, especially concerning the biology and ecology of survivors, becomes available from numerous mass extinctions and from a range of taxonomic groups that these mechanisms can be more fully tested.

We would like to thank the following people for their helpful discussions and careful criticisms rendered during various drafts of this paper: Claudia C. Johnson, Bradley B. Sageman, and several anonymous reviewers. Discussions with Paul Leary helped to give us insight into a number of survival mechanisms seen in the Cenomanian–Turonian microbiota. PJH would like to thank the Amoco Foundation for a scholarship which gave him the time initially to start working on the ideas contained within this paper and to the Fulbright Commission for the opportunity to finish it. He would also like to thank all the staff of the Institut and Staatssammlung für Paläontologie und historische Geologie for their hospitality during his stay in München, FRG. EGK's work was partially supported by NSF grant EAR 8411202.

References

ARCHIBALD, J. D. & BRYANT, L. J. 1988. Limitations on K–T mass extinction theories based on the vertebrate record. *LPI Contribution*, **673**, 4–5.

ARNOLD, J. M. 1987. Reproduction and embryology of Nautilus. *In:* SAUNDERS, W. B. & LANDMAN, N. H. (eds) *Nautilus: The Biology and Paleobiology of a Living Fossil.* Plenum, New York, 353–372.

ARTHUR, M. A., SCHLANGER, S. O. & JENKYNS, M. C. 1987. The Cenomanian–Turonian Oceanic Anoxic Event, II. Palaeoceanographic controls on organic-matter production and preservation. *In:* BROOKS, J. & FLEET, A. J. (eds) *Marine Petroleum Source Rocks,* Geological Society of London Special Publication, **26**, 401–420.

BAMBACH, R. K. 1977. Species richness in marine benthic habitats through the Phanerozoic. *Paleobiology*, **3**, 152–167.

BIRKELUND, T. & HAKANSSON, E. 1982. The terminal Cretaceous extinction in Boreal shelf seas – A multicausal event. *In:* SILVER, L. T. & SCHULTZ,

P. H. (eds) *Geological Implications of large Asteroid and Comets on the Earth.* Geological Society of America, Special Paper, **190**, 373–384.

BOUCOT, A. J. 1981. *Principles of Benthic Marine Paleoecology.* Academic, New York.

—— 1988. Silurian and pre-Upper Devonian bioevents. *The Third International Conference on Global Bioevents: Abrupt Changes in the Global Biota, Abstracts,* 10.

—— 1990. Phanerozoic extinctions: How similar are they to each other? *In:* KAUFFMAN, E. G. & WALLISER, O. H. (eds) *Extinction Events in Earth History,* Springer, Berlin, Lecture Notes in Earth History, **30**, 5–30.

BRAISER, M. D. & BUXTON, M. W. N. 1988. Varying response of larger benthic foraminifera in global biological events. *The Third International Conference on Global Bioevents: Abrupt Changes in the Global Biota, Abstracts,* 11.

BROUWERS, E. M. 1988. Late Maastrichtian–Danian faunas from northern Alaska: Reconstructions of environment and biogeography. *The Third International Conference on Global Bioevents: Abrupt Changes in the Global Biota, Abstracts,* 12.

CAVANAUGH, C. M. 1985. Symbioses of chemoautotrophic bacteria and marine invertebrates from hydrothermal vents and reducing sediments. *In:* JONES, M. L. (ed.) *Hydrothermal Vents of the Eastern Pacific: An Overview.* Bulletin of the Biological Society of Washington, **6**, 373–388.

CUVIER, G. 1812. *Recherches sur les ossemen fossiles, discourse preliminaire.*

DESBRUYERES, D. & LAUBIER, L. 1983. Primary consumers from hydrothermal vent animal communites. *In:* RONA, P. A., BOSTROEM, K., LAUBIER, L. & SMITH, J. K. L. (eds) *Hydrothermal Processes at Seafloor Spreading Centers.* NATO Conference Series in Marine Sciences, **12**, 711–734.

DINGUS, L. 1984. Effects of stratigraphic completeness on interpretations of extinction rates across the Cretaceous–Tertiary boundary. *Paleobiology,* **10**, 420–438.

EKDALE, A. A. & BROMELY, R. G. 1984. Sedimentology and ichnology of the Cretaceous–Tertiary boundary in Denmark: Implications for the causes of the terminal Cretaceous extinction. *Journal of Sedimentary Petrology,* **54**, 681–703.

ELDER, W. P. 1987a. *The Cenomanian–Turonian (Cretaceous) stage boundary extinctions in the Western Interior of the United States.* PhD Thesis, University of Colorado, Boulder.

—— 1987b. The paleoecology of the Cenomanian–Turonian (Cretaceous) stage boundary extinction at Black Mesa, Arizona. *Palaios,* **2**, 24–40.

—— 1989. Molluscan extinction patterns across the Cenomanian–Turonian stage boundary in the Western Interior of the United States. *Paleobiology,* **15**, 299–320.

FAGERSTROM, J. A. 1987. *The Evolution of Reef Communities.* Wiley, New York.

FELBECK, H., POWELL, M. A., HAND, S. C. & SOMERO, G. N. 1985. Metabolic adaptations of hydrothermal vent animals. *In:* JONES, M. L. (ed) *Hydrothermal Vents of the Eastern Pacific: An Overview.* Bulletin of the Biological Society of Washington, **6**, 261–272.

FLEMING, R. F. & NICHOLS, D. J. 1990. The fern-spore abundance anomaly at the Cretaceous–Tertiary boundary: a regional bioevent in North America. *In:* KAUFFMAN, E. G. & WALLISER, O. H. (eds) *Extinction Events in Earth History.* Springer, Berlin, Lecture Notes in Earth History, **30**, 347–349.

FLESSA, K. W. & JABLONSKI, D. 1983. Extinction is here to stay. *Paleobiology,* **9**, 315–321.

FRYXELL, G. A. 1983. *Survival Strategies of the Algae.* Cambridge University Press.

GRASSLE, J. F. 1983. Introduction to the biology of hydrothermal vents. *In:* RONA, P. A., BOSTROEM, K., LAUBIER, L. & SMITH, J. K. L. (eds) *Hydrothermal Processes at Seafloor Spreading Centers,* NATO Conference Series in Marine Sciences **12**, 665–675.

HANSEN, T. A. 1988. Early Tertiary radiation of marine molluscs and the long-term effects of the Cretaceous–Tertiary extinction. *Paleobiology,* **14**, 37–51.

—— & UPSHAW, B. 1990. Aftermath of the Cretaceous–Tertiary extinction: Rate and nature of the Early Paleocene molluscan rebound. *In:* KAUFFMAN, E. G. & WALLISER, O. H. (eds) *Extinction Events in Earth History.* Springer, Berlin, Lecture Notes in Earth History, **30**, 401–409.

——, FARRAND, R., MONTGOMERY, H. & BILLMAN, H. 1987. Sedimentology and extinction patterns across the Cretaceous–Tertiary boundary interval in East Texas. *Cretaceous Research,* **8**, 229–252.

HARRIES, P. J. 1993a. Dynamics of survival following the Cenomanian–Turonian (Upper Cretaceous) mass extinction event. *Cretaceous Research,* **14**, 563–583.

—— 1993b. *Patterns of repopulation following the Cenomanian–Turonian (Upper Cretaceous) mass extinction.* PhD Thesis, University of Colorado, Boulder.

—— & KAUFFMAN, E. G. 1990. Patterns of survival and recovery following the Cenomanian–Turonian (Late Cretaceous) mass extinction in the Western Interior Basin, United States. *In:* KAUFFMAN, E. G. & WALLISER, O. H. (eds) *Extinction Events in Earth History.* Springer, Berlin, Lecture Notes in Earth History, **30**, 277–298.

——, —— & HANSEN, T. A. 1987. *Biological patterns of survival and recovery following mass extinctions.* Geological Society of America, Abstracts with Programs, Rocky Mountain Section, **18E**.

——, —— & —— 1988. Models for survival and repopulation of the Earth following mass extinction. *The Third International Conference on Global Bioevents: Abrupt Changes in the Global Biota, Abstracts,* 20.

HARTMAN, W. D. & GOREAU, T. E. 1977. Jamaican coralline sponges; their morphology, ecology, and fossil representatives. *Journal Zoological Society of London Symposium,* **25**, 205–243.

HECKER, B. 1985. Fauna from a cold sulfur-seep in the Gulf of Mexico: Comparison with hydrothermal vent communities and evolutionary implication. *In:* JONES, M. L. (ed.) *Hydrothermal Vents of the Eastern Pacific: An Overview.* Bulletin of the Biological Society of Washington, **6**, 454–473.

HESSLER, R. R. & SMITHEY, W. M. JR. 1983. The distribution and community structure of megafauna at the Galapogos Rift hydrothermal vents. *In:* RONA, P. A., BOSTROEM, K., LAUBIER, L. & SMITH, J. K. L. (eds) *Hydrothermal Processes at Seafloor Spreading Centers*, NATO Conference Series in Marine Sciences, **12**, 735–770.

———, ——— & KELLER, C. H. 1985. Spatial and temporal variation of giant clams, tube worms and mussels, at deep-sea hydrothermal vents. *In:* JONES, M. L. (ed.) *Hydrothermal Vents of the Eastern Pacific: An Overview.* Bulletin of the Biological Society of Washington, **6**, 411–428.

HOWE, B. 1987. *Teepee buttes: A Petrological, paleontological, paleoenvironmental study of Cretaceous submarine spring deposits.* MSc Thesis, University of Colorado, Boulder.

——— & KAUFFMAN, E. G. 1986. Event communities of Teepee Buttes, Cretaceous submarine springs. *Fourth North American Paleontological Convention Abstract*, A-20.

HSÜ, K. J., McKENZIE, J. A. & HE, Q. X. 1982. Terminal Cretaceous environmental and evolutionary changes. *In:* SILVER, L. T. & SCHULTZ, P. H. (eds) *Geological Implications of large Asteroid and Comets on the Earth.* Geological Society of America, Special Paper, **190**, 317–328.

HUT, P., ALVAREZ, W., ELDER, W. P., HANSEN, T., KAUFFMAN, E. G., KELLER, G., SHOEMAKER, E. M. & WEISSMAN, P. 1987. Comet showers as a cause of mass extinctions. *Nature*, **329**, 118–126.

JABLONSKI, D. 1986a. Evolutionary consequences of mass extinctions. *In:* RAUP, D. M. & JABLONSKI, D. (eds) *Patterns and Processes in the History of Life*, 313–329.

——— 1986b. Causes and consequences of mass extinctions: A comparative approach. *In:* ELLIOT, D. K. (ed.) *Dynamics of Extinction*, 183–229.

——— 1986c. Background and mass extinctions: The alternation of macroevolutionary regimes. *Science*, **231**, 129–133.

——— & LUTZ, R. A. 1980. Molluscan larval shell morphology: Ecological and paleontological applications. *In:* RHOADS, D. C. & LUTZ, R. A. (eds) *Skeletal Growth of Aquatic Organisms: Biological Records of Environmental Change.* 323–377.

JACKSON, J. B. C. & WINSTON, J. E. 1982. Ecology of cryptic coral reef communities. I. Distribution and abundance of major groups of encrusting organisms. *Journal of Experimental Marine Biology*, **57**, 135–147.

———, GOREAU, T. F. & HARTMAN, W. D. 1971. Recent brachiopod–coralline sponge communities and their paleontological significance. *Science*, **143**, 623–625.

JARVIS, I., CARSON, G. A., COOPER, M. K. E., HART, M. B., LEARY, P. N., TOCHER, B. A., HORNE, D.

& ROSENFELD, A. 1988. Microfossil assemblages and the Cenomanian–Turonian (Late Cretaceous) Oceanic Anoxic Event. *Cretaceous Research*, **9**, 3–103.

JOHNSON, C. C. & KAUFFMAN, E. G. 1990. Originations, radiations and extinctions of Cretaceous rudisitid bivalve species in the Caribbean Province. *In:* KAUFFMAN, E. G. & WALLISER, O. H. (eds) *Extinction Events in Earth History.* Springer, Berlin, Lecture Notes in Earth History, **30**, 305–324.

JOHNSON, K. R. & HICKEY, L. J. 1990. Megafloral change across the Cretaceous–Tertiary boundary in the northern Great Plains and Rocky Mountains, U.S.A. *In:* SHARPTON, V. L. & WARD. P. W. (eds) *Global Catastrophes in Earth History: An Interdisciplinary Conference on Impacts, Volcanism, and Mass Mortality.* Geological Society of America, Special Paper, **247**, 433–444.

KAUFFMAN, E. G. 1975. Dispersal and biostratigraphic potential of Cretaceous benthonic Bivalvia in the Western Interior. *In:* CALDWELL, W. G. E. (ed.) *The Cretaceous System in the Western Interior of North America.* The Geological Association of Canada, Special Paper, **13**, 163–194.

——— 1977. Cretaceous extinction and collapse of marine trophic structure. *Journal of Paleontology*, **51** (supplement to no. 2, *North American Paleontological Convention II*, Abstracts of Papers) 16.

——— 1978. Evolutionary rates and patterns among Cretaceous Bivalvia. *Transactions of the Royal Society of London*, **B 284**, 277–304.

——— 1979. The ecology and biogeography of the Cretaceous–Tertiary extinction event. *In:* BIRKELUND, T. & CHRISTENSEN, W. K. (eds) *Cretaceous–Tertiary Boundary Events Symposium, II. Proceedings.* 29–37.

——— 1984. The fabric of Cretaceous marine extinctions. *In:* BERGGREN, W. A. & VAN COUVERING, J. A. (eds) *Catastrophes and Earth History*, 151–246.

——— 1986. High-resolution event stratigraphy: Regional and global bio-events. *In:* WALLISER, O. H. (ed.) *Global Bio-Events.* Springer, Heidelberg, Lecture Notes in Earth History, **8**, 279–335.

——— 1988a. The dynamics of marine stepwise mass extinction. *In:* LAMOLDA, M. A., KAUFFMAN, E. G. & WALLISER, O. H. (eds) *Paleontology and evolution: Extinction events; 2nd International Conference on Global Bioevents.* Revisita Espanola PaleontologiaI No. Extraordinario, 57–71.

——— 1988b. Concepts and methods of high-resolution event stratigraphy. *Annual Review of Earth and Planetary Science*, **16**, 605–654.

——— & BUDDENHAGEN, C. H. 1969. Protandric sexual dimorphism in Paleocene Astarte (Bivalvia) of Maryland. *In:* WESTERMANN, G. E. G. (ed.) *Sexual Dimorphism in Fossil Metazoa and Taxonomic Implications.* International Union of Geological Sciences, 76–93.

——— & ERWIN, D. H. Surviving mass extinctions. *Geotimes*, **14**, 14–17.

KELLER, G. 1986. Stepwise mass extinctions and impact events: Late Eocene and early Oligocene.

Marine Micropaleontology, **10**, 267–293.

—— 1988a. Extinction, survivorship and evolution of planktic foraminifera across the Cretaceous/Tertiary boundary at El Kef, Tunisia. *Marine Micropaleontology*, **13**, 239–263.

—— 1988b. Biotic turnover in benthic foraminifera across the Cretaceous/Tertiary boundary at El Kef, Tunisia. *Palaeogeography, Palaeoecology, Palaeoclimatology*, **66**, 153–171.

—— 1989. Extended period of extinction across the Cretaceous/Tertiary boundary in planktonic foraminifers of continental-shelf sections, implications of impact and volcanism theories. *Bulletin of the Geological Society of America*, **101**, 1408–1419.

—— & LINDINGER, M. 1989. Stable isotope, TOC, $CaCO_3$ record across the Cretaceous/Tertiary boundary at El Kef, Tunisia. *Paleogeography, Paleoecology, and Paleoclimatology*, **74**, 243–265.

KENNEDY, W. J. & COBBAN, W. A. 1990. *Rhamphidoceras saxatilis* n. gen. and sp., a micromorph ammonite from the Lower Turonian of Trans-Pecos Texas. *Journal of Paleontology*, **64**, 666–668.

KITCHELL, J. A., CLARK, D. L. & GOMBOS, A. M. JR. 1986. Biological selectivity of extinction: A link between background and mass extinction. *Palaios*, 504–511.

KOUTSOUKOS, E. A. M. & HART, M. B. 1988. Major foraminiferal changes and anoxic paleoenvironments in the mid-Cretaceous Northern South Atlantic. *The Third International Conference on Global Bioevents: Abrupt Changes in the Global Biota, Abstracts*, 24.

LANDMAN, N. H. 1984. Not to be or to be? *Natural History*, **93**, 4–41.

LEVINTON, J. S. 1970. The paleontological significance of opportunistic species. *Lethaia*, **3**, 69–78.

—— 1974. Trophic group and evolution of bivalve molluscs. *Palaeontology*, **17**, 579–585.

McLEAN, J. H. 1985. Preliminary report on the limpets at hydrothermal vents. *In:* JONES, M. L. (ed.) *Hydrothermal Vents of the Eastern Pacific: An Overview*. Bulletin of the Biological Society of Washington, **6**, 159–166.

MACLEOD, N. & KITCHELL, J. A. 1988. Inducement of heterochronic variation in a species of planktic foraminifera by a late Eocene impact event. *LPI Contribution*, **673**, 112.

NEWELL, N. D. 1967. *Revolutions in the history of life*. Geological Society of America, Special Paper, **89**, 63–91.

PALMER, A. R. 1984. The biomere problem: Evolution of an idea. *Journal of Paleontology*, **58**, 599–611.

PERCH-NIELSON, K. 1988. Uppermost Maastrichtian and lowermost Danian calcareous nannofossil assemblages. *The Third International Conference on Global Bioevents: Abrupt Changes in the Global Biota, Abstracts*, 29.

PERCIVAL, S. F. JR. & FISCHER, A. G. 1977. Changes in calcareous nannoplankton in the Cretaceous–Tertiary biotic crisis in Zumaya, Spain. *Evolutionary Theory*, **2**, 1–35.

PLAYFORD, P. E. 1980. Devonian 'Great Barrier Reef' of Canning Basin, Western Australia. *AAPG*

Bulletin, **64**, 814–840.

RAUP, D. M. 1984. Evolutionary radiations and extinctions. *In:* HOLLAND, H. D. & TRENDALL, A. F. (eds) *Report of the Dahlem Conference Workshop on Patterns of Change in Earth Evolution*, 5–14.

—— & SEPKOSKI, J. J. JR. 1986. Periodic extinctions of families and genera. *Science*, **231**, 833–836.

—— & JABLONSKI, D. 1993. Geography of end-Cretaceous marine bivalve extinctions. *Science*, **268**, 369–391.

ROY, J. M., MCMENAMIN, M. A. S. & ALDERMAN, S. E. 1990. Trophic differences, originations and extinctions during the Cenomanian and Maastrichtian stages of the Cretaceous. *In:* KAUFFMAN, E. G. & WALLISER, O. H. (eds) *Extinction Events in Earth History*. Springer, Berlin, Lecture Notes in Earth History, **30**, 299–303.

SEILACHER, A. 1990. Aberration in bivalve evolution related to photo- and chemosymbiosis. *Historical Biology*, **3**, 289–311.

SHEEHAN, P. M. & HANSEN, T. A. 1986. Detritus feeding as a buffer to extinction at the end of the Cretaceous. *Geology*, **14**, 868–870.

SIGNOR, P. W. III. 1985. Real and apparent trends in species richness through time. *In:* VALENTINE, J. W. (ed.) *Phanerozoic Diversity Patterns: Profiles in Macroevolution*. 129–150.

SOHL, N. F. 1987. Cretaceous gastropods: Contrasts between Tethys and the temperate provinces. *Journal of Paleontology*, **61**, 1085–1111.

STANLEY, G. D. JR. 1988. The history of Early Mesozoic reef communities: A three-step process. *Palaios*, **3**, 170–183.

SUESS, E., CARSON, B., RITGER, S. D., MOORED, J. C., JONES, M. L., KULM, L. D. & COCHRANE, G. R. 1985. Biological communities at vent sites along the subduction zone off Oregon. *In:* JONES, M. L. (ed.) *Hydrothermal Vents of the Eastern Pacific: An Overview*. Bulletin of the Biological Society of Washington, **6**, 475–484.

SURLYK, F. & JOHANSEN, M. B. 1984. End-Cretaceous brachiopod extinctions in the chalk of Denmark. *Science*, **223**, 1174–1177.

TAPPAN, H. 1979. Protistan evolution and extinction at the Cretaceous–Tertiary boundary. *In:* BIRKELUND, T. & CHRISTENSEN, W. K. (eds) *Cretaceous-Tertiary Boundary Events symposium, II. Proceedings*. 13–21.

THIERSTEIN, H. R. 1981. *Late Cretaceous nannoplankton and the change at the Cretaceous–Tertiary boundary*. Society of Economic Paleontologists and Mineralogists, Special Publication, **32**, 355–394.

THOMAS, E. 1990. Late Cretaceous–early Eocene mass extinctions in the deep sea. *In:* SHARPTON, V. L. & WARD, P. W. (eds) *Global Catastrophes in Earth History: An Interdisciplinary Conference on Impacts, Volcanism, and Mass Mortality*. Geological Society of America, Special Paper, **247**, 481–495.

TUREKIAN, K. K., COCHRAN, J. K., KHARKAR, D. P., CERRATO, R. M., VAISYNS, J. R., SANDERS, H. L., GRASSLE, J. F. & ALLEN, J. A. 1975. Slow growth rates of a deep-sea clam determined by ^{228}Ra chronology. *Proceedings of the National Academy of Science of the USA*, **72**, 2829–2832.

TURNCLIFFE, V., JUNIPER, S. K. & BURGH, M. E. D. 1985. The hydrothermal vent community on axial seamounds, Juan de Fuca Ridge. *In:* JONES, M. L. (ed.) *Hydrothermal Vents of the Eastern Pacific: An Overview*. Bulletin of the Biological Society of Washington, **6**, 453–464.

VALENTINE, J. W. 1970. How many marine invertebrate fossil species? A new approximation. *Journal of Paleontology*, **44**, 410–415.

VERMEIJ, G. J. 1986. Survival during biotic crises: The properties and evolutionary significance of refuges. *In:* ELLIOT, D. K. (ed.) *Dynamics of Extinction*, 231–246.

—— 1987. *Evolution and Escalation: An Ecological History of Life*, Princeton University Press.

WHATELY, R. 1990. The relationship between extrinsic and intrinsic events in the evolution of Mesozoic non-marine Ostracoda. *In:* KAUFFMAN, E. G. & WALLISER, O. H. (eds) *Extinction Events in Earth History*. Springer, Berlin, Lecture Notes in Earth History, **30**, 253–263.

WIEDMANN, J. 1973. Evolution or revolution of ammonoids at Mesozoic system boundaries. *Biological Reviews*, **48**, 159–194.

WILDE, P. & BERRY, W. G. N. 1986. The role of oceanographic factors in the generation of global bio-events. *In:* WALLISER, O. H. (ed.) *Global Bio-Events*. Springer, Heidelberg, Lecture Notes in Earth History, **8**, 75–91.

——, QUINBY-HUNT, M. S. & BERRY, W. B. N. 1990. Vertical advection from oxic or anoxic water from the main pycnocline as a cause of rapid extinction or rapid radiation. *In:* KAUFFMAN, E. G. & WALLISER, O. H. (eds) *Extinction Events in Earth History*. Springer, Berlin, Lecture Notes in Earth History, **30**, 85–98.

WORSELY, T. 1974. Cretaceous–Tertiary boundary events in the ocean. *In:* HAY, W. W. (ed.) *Studies in Paleo-Oceanography*, Society of Economic Paleontologists and Mineralogists, Special Publication, **20**, 94–125.

Recovery as a function of community structure

VALENTIN A. KRASSILOV

Palaeontological Institute, 123 Profsojusnaya, Moscow 117647, Russia

Abstract: According to the climax cut-off model, ecological succession can be truncated at an early stage by environmental factors thus causing elimination of climax species and their replacement by new dominants derived from successional species. Such profound restructurings are caused by prolonged stresses rather than by episodic impacts. Recovery of a dominant group is due to its representatives in the early stages of 'mixed' successions. In the 'graded' successions with the serial and climax species belonging to different evolutionary grades (e.g. dinosaurs and mammals in the terminal Cretaceous communities) such recoveries are less probable.

Recovery as an integral concept means a regaining of numbers, ranges, dominance and/or diversity after a temporal reduction of some or all of these parameters and applies to populations, species, higher taxa and communities. On the other hand, recovery as an analytical concept treats them differentially as separate kinds of recovery with their own causality.

Incidentally, population recovery is a part and parcel of the crash–founder–flush cycle (Carson 1975). The founder effect is in fact a recovery with change leading to punctuated speciation (Eldredge & Gould 1972), though recent genetic studies cast some doubt on the validity of the founder–flush speciation model (Galiama *et al.* 1993).

Recovery is likewise a normal issue of perpetually ongoing successional processes in biotic communities. Not only natural disasters such as hurricane or fire but also the fall of a large tree can set out ecological succession starting with population growth in pioneer species – their recovery after a reduction in the period of relative ecological stability – and leading, through a sequence of several species, to the recovery of dominant climax species.

Yet the question is to what extent these short-term events can be used as a model of the geological scale processes deduceable from the fossil record, such as near extinction and reappearance of species and higher taxa.

Climax cut-off model

The evolutionary significance of ecological succession is underscored in the climax cut-off model (Krassilov 1992*a, b*) which maintains that under environmental stresses a succession can be halted at an early stage, never reaching the potential climax. In effect, recovery of the climax species is postponed and, in the long sustained disclimax situations, would never occur. Major ecological crises truncate successions over a wide range of terrestrial and marine habitats, thus leading the climax species to extinction (this would explain mass extinctions of the fossil record which are simultaneous disappearances of apparently unrelated dominant species and higher taxa in a number of different environments). With stabilization a new climax phase is formed by derivatives of polymorphic successional species which experience speciation bursts in the process.

For instance, rudist bivalves evolved as the climax stage dominants of Cretaceous reefal communities starting with bryozoans (Kauffman 1974). The end-Cretaceous extinction of rudists and the simultaneous rise of bryozoans may evidence a climax cut-off event.

The climax cut-off model predicts that even mild long-duration impacts would produce larger and less recoverable changes than occasional short-term catastrophes. Moreover, their effects would be qualitatively different: episodic stresses endanger rare species while prolonged stresses induce replacement of dominant species. Incidentally, glaciations in Europe caused periodic reduction of beech forests followed by their rapid range recoveries while some exotic elements, such as *Cathaya*, presently a far-eastern conifer, were shed in the process. In contrast, the non-glaciated areas, such as southern Primorye in the Russian Far East, have preserved a number of Arcto-Tertiary relics, yet the beech forests, a dominant plant formation in the Miocene, perished with the steady lowering of winter temperatures (Krassilov & Alexeyenko 1977).

An additional feature of the model is that,

From Hart, M. B. (ed.), 1996, *Biotic Recovery from Mass Extinction Events*,
Geological Society Special Publication No. 102, pp. 61–63

insofar as phylogenetically ancient plants and animals may survive as pioneers of biotic communities dominated by phylogenetically younger groups, cut-off of the latter would bring them to the fore. Their rapid increase would then appear as recovery. For example. *Selaginella*, phylogenetically the most ancient extant genus appearing in the Carboniferous as a common herbaceous plant, was rare through the Mesosoic and Tertiary, but then emerged as a dominant in the stunted Pleistocene tundra–steppe communities.

There was mass extinction in the dominant Mesozoic gymnosperm groups, cycadophytes, ginkgophytes and czekanowskialeans in the mid-Cretaceous. Some of them survived, however, as accessory components of riparian and delta plain marsh communities. Characteristically, they became more prominent again in the terminal Cretaceous just before the great floristic turnover at the K–T boundary. *Czekanowskia* was found, after a long period of non-occurrence, in the uppermost dinosaur bed of the Amur Basin. In the K–T boundary tuff assemblage of the Augustovka River Section in western Sakhalin there are two cycadophyte genera, *Pterophyllum* and *Cycadites*, giving it a definitely Mesozoic aspect (Krassilov 1979) while *Encephalartites* is common in the Maastrichtian dinosaur beds of Chukotsk Region in the northeast (Krassilov *et al.* 1990). Such recoveries of the long-surviving relics usually presage a major ecological crisis.

Recoveries in mixed and graded communities

As for the dominant taxa, their recoveries seem to depend on their ecological roles as the climax or successional components or both. Schematically, there are two kinds of successions: the 'mixed' and the 'graded', differing in the distribution of major taxa among the stages.

In the mixed successions, a dominant taxon participates in all or most stages. For instance, in broadleaf forests dicotyledonous angiosperms occur as both successional and climax trees, e.g. hazel and lime. In the heavily polluted environments lime recovery is hampered but dicotyledons as a taxon are not endangered.

It can be different in the plant communities where early successional species are predominantly monocots, as in the forest–steppe ecotonal zone, or even ferns, as in *Nothofagus* rain forests. Such successions are graded in the sense that successional and climax species belong in different evolutionary grades. Here the climax cut-off would endanger the dominant higher taxon as a whole.

The fossil examples are all interpretations, but they seem plausible at least in the cases of the mid-Cretaceous cycadophyte and the end-Cretaceous lowland redwood extinctions. The Mesozoic cycadophyte-dominated communities have their closest extant analogue in the *Encephalartos* shrub savanna where the upper storey is formed by a phylogenetically primitive cycad grade while advanced monocots prevail in the lower storey. The Early Cretaceous angiosperms were small shrubs or herbs of cycadophyte shrublands which turned into angiosperm shrublands following the cut-off of the dominant cycadophyte species.

The late Cretaceous lowland evergreen redwoods dominated by *Sequoia reichenbaechii*, *Cupressinocladus cretaceous* and related conifer species were likewise graded, with angiosperms, notably small-leaved *Trochodendroides*, as successional shrubs or low trees. After the K–T climax cut-off they disappeared as a major lowland plant formation surviving in the scattered upslope refugia alone. At the same time *Trochodendroides* emerged as a dominant tree of the newly formed flood plain deciduous hardwood forests. The Palaeocene *Trochodendroides* species show exceptionally high polymorphism of their leaf and fructification characters far exceeding the range of variation in the earlier Cretaceous and the later Eocene species. They were used as an example of cyclic speciation (Krassilov 1989, 1992*a*, *b*) in which marcropolymorphism results from a decrease of selection pressure and release of latent genetic potentials in disclimax ecosystems, the macropolymorphic populations serving then as a source of subsequent bursts of adaptive radiation.

Dinosaur recoveries after the Jurassic–Cretaceous and mid-Cretaceous extinctions and the K–T non-recovery extinction could also be related to the mixed v. graded community structures. Each of these events correlates with a certain vegetational change, i.e. (1) reduction of fern marshes at the Jurassic–Cretaceous boundary followed by their partial recovery; (2) replacement of the macrophyllous cycadophyte shrub communities by the microphyllous early angiosperm shrubs resulting in a considerable foliage mass reduction; and (3) spread of the low capacity deciduous forests at the expense of the more resilient plant communities (Krassilov 1981).

However, in the recovery cases, dinosaur communities might have been of mixed type with, for example, smaller protoceratopsians and related groups as putative pioneer forms, while in the terminal Cretaceous non-recovery case, graded types prevailed, with dinosaurs as the climax dominants and mammals as the pioneer-successional forms (locally also as the climax codominants which perished with dinosaurs).

Giants recovery

Admittedly a much more detailed palaeoecological analysis is needed to overcome a plain schematism of the above reasoning. A further problem related to the dinosaur story is the recovery of gigantism among phytophagous vertebrates. It was repeatedly suggested that dinosaurs were poisoned with angiosperm metabolite chemicals. In that they survived in the angiosperm-dominated surroundings for at least 30 Ma, poisoning is not feasible as a cause of extinction.

There is, however, another little-explored aspect to herbivore intoxication. It was suggested (Bryant *et al.* 1991) that satiation marks a threshold beyond which further eating is dangerous because of excessive accumulation of unpalatable matter. Satiation is, thus, a meeting point of the plant–animal coevolution.

Since larger animals need more food, they have to be tolerant of a larger dose of intoxication. Their potential size would increase with tolerance to habitual plant secondary metabolites and decrease with new ones. It follows that any vegetational change in the succession of diplodocian–indricotherian–elephantid life forms relates to the mutual physiological adjustments of plants and plant-eaters.

This work was partly supported by Project 6.5.4. of the Russian State Programme 'Global change of natural environments and climate'. I thank Dr Jane Francis for her comments on the manuscript.

References

BRYANT, J. P., PROVENZA, F. D., PASTOR, J., REICHARDT, P. B. & CLAUSEN, T. P. 1991. Interaction between woody plants and browsing mammals mediated by secondary metabolites. *Ann. Rev. Ecol. Syst.*, **22**, 431–446.

CARSON, H. L. 1975. The genetics of speciation at the diploid level. *American Naturalist*, **109**, 83–92.

ELDREDGE, N. & GOULD, S. I. 1972. Punctuated equilibria: an alternative to phyletic gradualism. *In:* SCHOPF, T. M. (ed.) *Models in Paleobiology*. San Francisco, 82–115.

GALIAMA, A., MOGA, A. & AYALA, F. J. 1993. Founder–flush speciation in *Drosophila pseudoobscura* – a large-scale experiment. *Evolution*, **47**, 432–444.

KAUFFMAN, E. G. 1974. Cretaceous assemblages, communities and associations, western interior United States and Caribbean Islands. *Sedimenta*, **4**, 1–27.

KRASSILOV, V. A. 1979. *Cretaceous flora of Sachalin*. Moscow [in Russian].

—— 1981. Changes of Mesozoic vegetation and the extinction of dinosaurs. *Palaeogeography, Palaeoclimatology, Palaeoecology*, **34**, 207–224.

—— 1989. Vavilov's species concept and the evolution of variation. *Evolutionary Theory*, **9**, 37–44.

—— 1992a. Ecosystem theory of evolution. *Rivista Biologia*, **85**, 243–245.

—— 1992b. [*Nature protection: principles, problems and priorities*.] Institute for Natural Conservation, Moscow [in Russian].

—— & ALEXEYENKO, T. M. 1977. succession of plant assemblages in the Palaeogene and Neogene of southern Primorye. *In:* KRASSILOV, V. A. (ed.) *Palaeobotany in the Far East*. Far-Eastern Scientific Centre, Vladivostok, 7–17.

—— GOLOVENEVA, L. B. & NESSOV, L. A. 1990. Cycadophyte from the Late Cretaceous dinosaur locality in northern Koryakia. *In:* KRASSILOV, V. A. (ed.) [*Continental Cretaceous of USSR*.] Academy of Sciences, Far-East Branch, Vladivostok, 213–215 [in Russian].

Insect origination and extinction in the Phanerozoic

E. A. JARZEMBOWSKI[1] & A. J. ROSS[2]

[1] *Postgraduate Research Institute for Sedimentology, University of Reading, PO Box 227, Whiteknights, Reading RG6 2AB and Maidstone Museum and Art Gallery, St Faith's Street, Maidstone ME14 1LH, UK*

[2] *Booth Museum of Natural History, 194 Dyke Road, Brighton BN1 5AA and Department of Palaeontology, Natural History Museum, Cromwell Road, London SW7 5BD, UK*

Abstract: Insects (Superclass Hexapoda) are the most successful group of living terrestrial arthropods and the richness of their fossil record is only just beginning to be realized with the recent publication of two extensive databases. Hexapods first appeared in the Early Devonian, post-dating the Cambrian 'explosion' of marine arthropods by some 140 Ma. The earliest Hexapoda belong to primitively wingless taxa; however, these 'Apterygota' comprise less than 1% of all hexapod species. The appearance of winged insects (Pterygota) in the mid-Carboniferous was accompanied by a major adaptive radiation of non-holometabolous hexapods at ordinal level. This was supplemented in the Permian by the radiation of insects with complete metamorphosis (Holometabola). The number of insect orders present in this period was similar to that at the present day. The family data suggest four major periods of origination in the Phanerozoic, with peaks in the Permo-Carboniferous, Early Jurassic, Early Cretaceous and Eocene. Unlike the Tertiary, the Palaeozoic and Cretaceous peaks are accompanied by considerable turnover of families; they are followed by reduced palaeodiversity in the Early Triassic and Late Cretaceous and Palaeocene. The former decline may be linked with the general extinction at the Permian/Triassic boundary and the latter with the rise of the angiospermous flowering plants. Insect generic data for the Phanerozoic reflect the pattern shown by families but not orders. In general, insect diversity may be explained by an overall trend towards low extinction and steady origination at a sub-ordinal level since the Palaeozoic.

The success of insects (Superclass Hexapoda), as expressed by species diversity, is such that some recent reports have suggested that insects may be immune to mass extinction (Labandeira & Sepkoski 1993). Our work, based on Ross & Jarzembowski (1993) and Appendix 1 herein, suggests that this is not so. In this paper we briefly review pattern and process in our developing understanding of the Phanerozoic insect record.

Diversity patterns

Hexapods (Early Devonian–Recent) are represented by 1167 families in the fossil record (see Appendix 1). The general family trend in the Phanerozoic has been one of cumulative origination coupled with constant but low extinction (Fig. 2b). However, the trend at ordinal level differs in that origination emerges as slightly negative, reflecting the major adaptive radiation of hexapods in the late Palaeozoic (Fig. 1b). Nevertheless, the extinction trend at ordinal level has also been fairly constant and low. More detailed examination of family data reveals diversity peaks in the Late Permian (P2), Early Cretaceous (K1) and Mio-Pliocene (Fig. 2a). The peaks show progressively less turnover and origination greatly exceeds extinction by the Tertiary. The troughs in the Early Triassic (Tr1) and Palaeocene (Pal) could reflect extinctions at the Permo-Triassic and Cretaceous/Tertiary boundaries, but the narrow trough in the Middle Jurassic (J2) is probably an artefact of under-recording due to stratigraphic problems at the Lower Jurassic (J1)/Middle Jurassic (J2) boundaries in Asia. The same graph identifies origination peaks in the Late Carboniferous (C2)/P2, J1, K1 and Eocene (Eoc).

Ordinal data show that numerically, insects were as diverse in P2 as they are today (Fig. 1a); a shallow trough in the Triassic (bottoming out in the Middle Triassic (Tr2)) is followed by gradual recovery. A generic graph for the Phanerozoic (Fig. 3) resembles that based on families with three progressively larger peaks

From Hart, M. B. (ed.), 1996, *Biotic Recovery from Mass Extinction Events*, Geological Society Special Publication No. 102, pp. 65–78

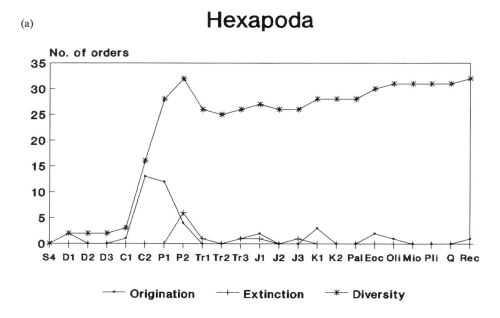

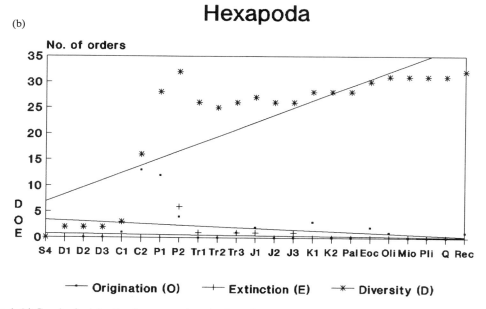

Fig. 1. (a) Graph of origination (appearance), extinction (disappearance) and diversity (total number) of hexapod (insect) orders in the Phanerozoic. Abbreviations in this and subsequent figures are: S(ilurian); D(evonian); C(arboniferous); P(ermian); Tr(iassic); J(urassic); K(Cretaceous); Pal(aeocene); Eoc(ene); Oli(gocene); Mio(cene); Pli(ocene); Q(uaternary); Rec(ent); 1, 2, 3 = Early, Middle, Late. (b) Trends in the origination, extinction and diversity of hexapod orders in the Phanerozoic (based on Ross & Jarzembowski (1993) and Appendix 1).

(a)
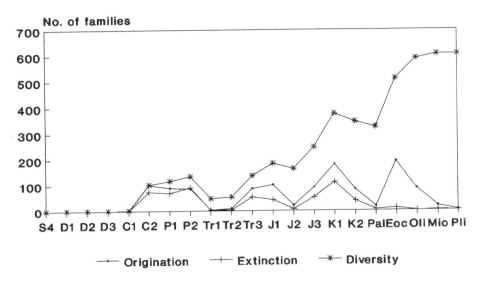

(b)
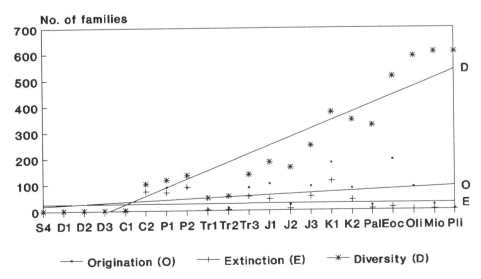

Fig. 2. (a) Origination, extinction and diversity of hexapod families in the Phanerozoic. (b) Trends in origination, extinction and diversity of hexapod families in the Phanerozoic (based on Ross & Jarzembowski (1993) and Appendix 1).

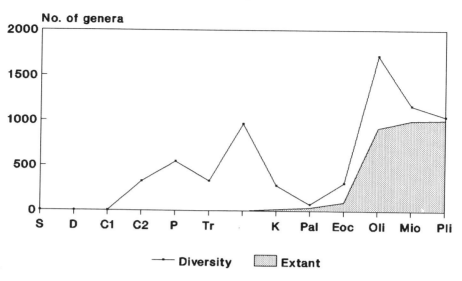

Fig. 3. Diversity of hexapod genera in the Phanerozoic. Based on Carpenter (1992) and excluding doubtfully assigned Coleoptera species. (White area under graph represents extinct genera.)

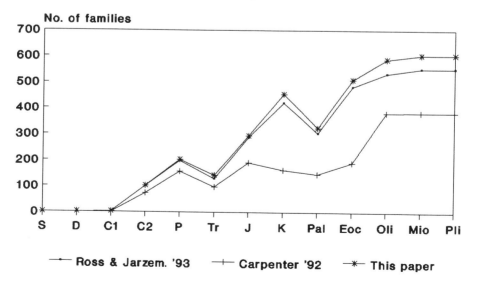

Fig. 4. Comparison of hexapod family diversity based on *Fossil Record 2*, the *Treatise on Invertebrate Paleontology*, and this paper.

Fig. 5. Earliest 'apterygote': *Rhyniella praecursor* Hirst & Maulik, 1926 (Hexapoda: Collembola), Lower Devonian, UK. Reconstruction from Jarzembowski (1989a). Specimen length 1.5 mm.

Fig. 6. Archeopteran: *Stenodictya* spp. (Hexapoda: Palaeodictyoptera), Upper Carboniferous, France. Reconstruction from Kukalová (1970). Wingspan c. 170 mm.

Fig. 7. 'Polyneopteran': *Elisama* sp. (Hexapoda: Blattodea). Lower Purbeck Beds, Durlston Bay, Dorset. Sedgwick Museum registration no. X24676a. Length 14 mm.

of Baltic amber from Early Oligocene to late Eocene has the effect of smoothing the Palaeogene curve.

There were four main events in the early (Palaeozoic) evolution of insects (Jarzembowski 1989a). These were the rise of: (1) primitively wingless insects (Fig. 5); (2) primitively winged insects not capable of wing folding (Fig. 6); (3) winged insects capable of wing folding (Fig. 7); (4) insects with complete metamorphosis (Fig. 8, right). These events correspond to the appearance of the following major clades: Pterygota and Palaeoptera (2), Neoptera (3) and Oligoneoptera (=Endopterygota, =Holometabola) (4), and an apterygote grade (1).

The 'apterygotes' (non-pterygote hexapods) comprise several monophyletic lineages including the earliest hexapods (e.g. Collembola; Fig. 5) as well as the sister group of the Pterygota; however, they comprise less than 1% of all Recent species and only some 21 families in the fossil record (Fig. 9). The highly successful pterygotes appeared in the late Early Carboniferous (Namurian A) at the beginning of the late

and two troughs. However, the most recently published generic database is only up to 1983 (Carpenter 1992). Some deficiencies may be indicated by comparing the family curves based on Carpenter and Ross & Jarzembowski (1993) (Fig. 4). This shows the effect of a considerable increase in our knowledge of Cretaceous insect life during the past decade. Also, the re-dating

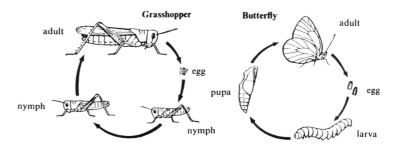

Fig. 8. Oligoneopteran (right) and 'polyneopteran' (left) life cycles compared. The Oligoneopteran pupal stage enables complete metamorphosis so that adult and larva are effectively different organisms, as well as providing a convenient resting stage during adverse environmental conditions. From Chinery (1993). (Nymph = larva).

Palaeozoic radiation of the hexapods. The two constituent clades of this subclass, Neoptera and Palaeoptera (winged insects capable and not capable of wing folding respectively) also arose in the Namurian (Ross & Jarzembowski 1993). The Phanerozoic record of palaeopterous families shows a prominent Palaeozoic peak due to the evolution of the Archeoptera ('palaeodictyopteroids'), a distinct group of Carboniferous–Permian essentially plant-sucking insects (Figs 6, 10). The Archeoptera became extinct by the Early Triassic (Tr1) and the subsequent history of the Palaeoptera was that of the orders Odonata (dragonflies) and Ephemeroptera (mayflies) alone.

The large cohort Neoptera is traditionally divided into three superorders in phylogenetically ascending order: 'Polyneoptera', Paraneoptera and Oligoneoptera. The first of these is now generally considered to be unnatural and is likely to be replaced by two or three monophyletic groups based on Orthoptera (grasshoppers, crickets and locusts), Plecoptera (stoneflies) and even cockroaches (Blattodea). However, it may be noted that the family graph for these insects shows an interesting resemblance to that for Palaeoptera (cf. Figs 10, 11), although the Palaeozoic peak of 'Polyneoptera' is facilitated by 'Protorthoptera' and other extinct orders, e.g. the beetle-like Protelytroptera.

Just as the record of 'Polyneoptera' may be compared with Palaeoptera, so can Paraneop-

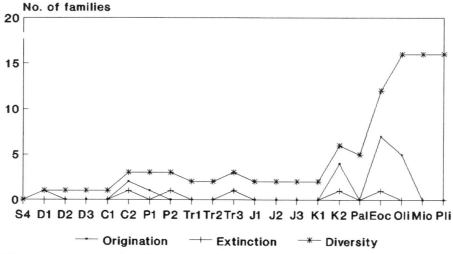

Fig. 9. Phanerozoic origination, extinction and diversity of 'apterygotes' (based on Ross & Jarzembowski (1993) and Appendix 1).

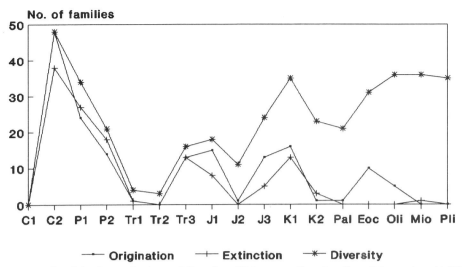

Fig. 10. Phanerozoic origination, extinction and diversity of Palaeoptera (based on Ross & Jarzembowski (1993) and Appendix 1).

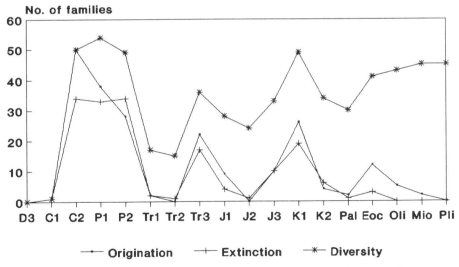

Fig. 11. Phanerozoic origination, extinction and diversity of 'Polyneoptera' (based on Ross & Jarzembowski (1993) and Appendix 1).

tera (hemipteroid orders) be compared with Oligoneoptera (Figs 12, 13). Unlike the former pair, the latter radiated after the Carboniferous with a low Palaeozoic peak in the Late Permian (P2). Paraneoptera shows Late Cretaceous–Palaeocene decline, especially in Hemiptera (true bugs) (Jarzembowski & Ross 1993, figs 9 (graphics), 10 (caption); 1994, fig. 4), which is discussed below. In the Oligoneoptera, two major orders, Hymenoptera (wasps, ants and bees) and Lepidoptera (moths and butterflies, Jarzembowski & Ross 1993, figs 10 (graphics),

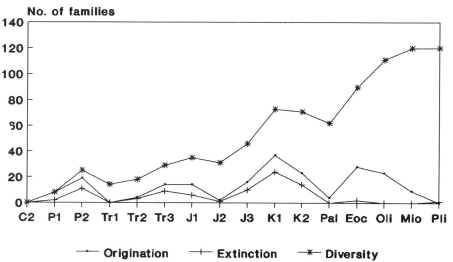

Fig. 12. Phanerozoic origination, extinction and diversity of Paraneoptera (based on Ross & Jarzembowski (1993) and Appendix 1)

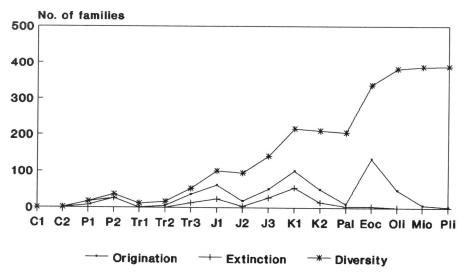

Fig. 13. Phanerozoic origination, extinction and diversity of Oligoneoptera (based on Ross & Jarzembowski (1993) and Appendix 1).

11 (caption); 1994, fig. 5) originated after the Middle Triassic (Tr2).

From the above discussion, it would appear that the main steps in hexapod evolution were accomplished prior to the Permo-Triassic extinction providing the basis for the great diversity of present-day insects. However, in addition to general increases in diversity on land and in freshwater, several important developments came later, such as the rise of social insects (ants, termites) in the Early Cretaceous (K1) and various types of parasitic insects in the

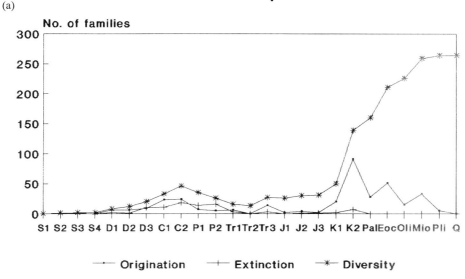

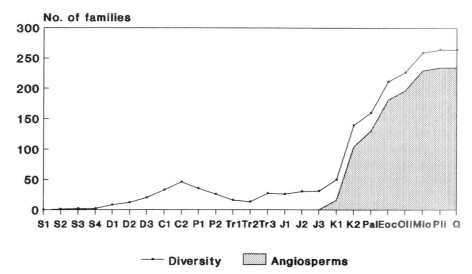

Fig. 14. (a) Origination, extinction and diversity of vascular plants in the Phanerozoic; (b) family diversity of angiosperms and other vascular plants in the Phanerozoic. (Based on Cleal 1993 *a, b*; Collinson *et al.* 1993)

Mesozoic and Tertiary. Important though they are biologically, these developments have been of greatest consequence at lower taxonomic level, contributing only five small orders (Isoptera, Phthiraptera, Zoraptera, Siphonaptera and Strepsiptera) with 16 families to the fossil record.

Possible processes

It has been claimed that vertebrate predation has been a major influence on insect evolution (Downes 1987), and that geological events have affected important developments, e.g. rising sea-level in the Early Triassic led to habitat loss and

Hexapoda

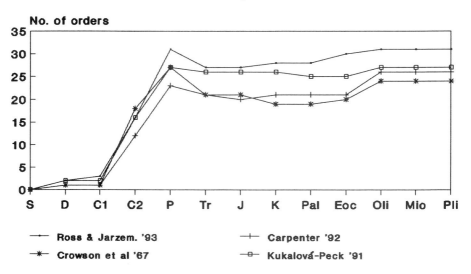

Fig. 15. Ordinal diversity of hexapods in the Phanerozoic after various authors (see text).

corresponding insect extinction (Erwin 1994). However, most workers consider that the evolution of plants has been the single most important influence on insect diversity (Gullan & Cranston 1994). The absence of insect phytophagy in the form of plant damage until the Carboniferous suggests that coevolution may have played a part in the adaptive radiation of insects at this time (Jarzembowski & Ross 1993). The decline of the Archeoptera and rise of Hemiptera in the Permian may be viewed as a change in plant-feeding strategy associated with floral change, favouring the new Mesophytic Flora (DiMichele & Hook 1992). Most importantly, the adaptive radiation of the angiosperms in the Cretaceous is often considered as the major cause of Recent insect diversity. This assumption has been questioned by Labandeira & Sepkoski (1993) who concluded that the reverse is true i.e. insects may have forced plant evolution in the Mesozoic. Their argument is based on the modern aspect of Cretaceous families, and wide range of mouthpart types in hexapods – and thus trophic possibilities – by the Middle Jurassic. Some such 'preadaptation' is not surprising considering the earlier diversification of orders and associated phylogenetic groundplans. However, our data suggest that the angiospermous radiation may have had an initial deleterious effect on insect families. This included a temporary decline in Hemiptera (plant bugs) between K1 and Eoc, and delayed radiation of the Lepidoptera until the Tertiary (Jarzembowski & Ross 1993, 1994; Jarzembowski 1995). Nevertheless, we agree that the end-Palaeozoic extinction was significant at a higher taxonomic level than that at the end of the Mesozoic when there was no loss of insect orders. Indeed, infra-Cretaceous extinction of families was greater than end-Cretaceous (Jarzembowski 1989b). The family diversity of vascular plants, like hexapods, shows a notable trough in Tr2 accompanied by low turnover, with recovery in Tr3 (Fig. 14a). However, hexapods differ by showing more pronounced expansion and contraction in the Permian and Tr1 respectively (Fig. 2a). The Cretaceous/Tertiary extinction of hexapod families appears spread over the Late Cretaceous (K2) and Palaeocene (Pal), followed by recovery from Eoc onwards accompanied by low extinction, unlike in the Palaeozoic and Mesozoic radiations (Fig. 2a). In contrast, the adaptive radiation of plants which commenced in K1 only levelled off in the late Cenozoic (Fig. 14b). Mitter et al. (1988) concluded that there is a significant association between diversity in living insects and the adoption of phytophagy. This effect was evidently delayed in the Cretaceous angiosperm-dominated radiation. Indeed, the high turnover and extinction of insects at this time (Fig. 2a) may reflect disruption of ecosystems. Insects, like many other life forms, are facing major extinction in the 21st century through anthropogenic causes. However, the

INSECTS IN THE PHANEROZOIC

accompanying loss of terrestrial habitats makes it unlikely that there will be recovery on the scale of the Tertiary.

Labandeira & Sepkoski (1993) considered that family data provide a good extinction signal, possibly better than species diversity, and that paraphyly is not a problem. Unfortunately, there is no published database of hexapod species to test the former, but the latter would appear to negate the general cladistic emphasis on monophyletic groups. Our database is essentially a compilation from global literature and, though selective, is ultimately based on a mixture of systematic practices. We have therefore compared ordinal diversity with three other comprehensive works (Fig. 15). The compilation by Crowson *et al.* (1967) is from *The Fossil Record* and the classification is essentially traditional, although phenetic methods, especially numerical, were beginning to influence taxonomy at the time. Carpenter (1992) employed a traditional and essentially conservative classification but, in contrast, Kukalová-Peck (1991) is a cladistic one, albeit essentially qualitative and morphological. Inspection shows that there is a good degree of correspondence between the curves, all picking up the Palaeozoic radiation, the ensuing shallow trough, and recovery by the Late Cenozoic. Our data and Kukalová-Peck's only differ in the reversed asymmetry of the trough. Thus taxonomic 'bias' need not be a major problem, at least at ordinal level. Interestingly, and by contrast, more detailed data on plant species suggest a Permo-Triassic decline in diversity, but significant and sustained increase since the Early Cretaceous, supporting the family data (Knoll 1986).

Conclusion

Much research remains to be done on improving the fossil record of hexapods, ranging from description of many new species to advanced interpretation of existing data, including testing 'ghost' ranges, sample size and taphonomic factors. Better stratigraphic resolution of the various non-marine, insect-bearing sediments would help bring new information to bear on extinctions and other geologically short-lived events. Such information probably already exists in geological sources, but is simply unused by systematists and taxonomists. In the meantime, we hope that the above framework will stimulate further research and contribute to hypothesis-testing in a superclass which does reflect the vicissitudes of Phanerozoic history.

We are indebted to Dr J. Kukalová-Peck (Carleton University) for comments on insect phylogeny, Mr P. Austen (Seaford) for compiling plant data. Mr P. Crabb (NHM) for Fig. 7 and Michael Chinery/Collins for Fig. 8 by Denys Ovenden. This is PRIS contribution no. 435 for EAJ.

References

CARPENTER, F. M. 1992. Superclass Hexapoda. *Treatise on Invertebrate Paleontology*, (R. Arthropoda 4) **3 & 4.**

CHINERY, M. 1993. *Insects of Britain and northern Europe.* 3rd edition, HarperCollins, London.

CLEAL, C. J. 1993a. Pteridophyta. *In:* BENTON, M J. (ed.) *The Fossil Record 2.* Chapman & Hall, London, 779–794.

—— 1993b. Gymnospermophyta. *In:* BENTON, M. J. (ed.) *The Fossil Record 2.* Chapman & Hall, London, 795–808.

COLLINSON, M. E., BOULTER, M. C. & HOLMES, P. L. 1993. Magnoliophyta ('Angiospermae'). *In:* BENTON, M. J. (ed.) *The Fossil Record 2.* Chapman & Hall, London, 809–841.

CROWSON, R. A., SMART, J. & WOOTTON, R. J, 1967. Class Insecta. *In:* HARLAND, W. B., HOLLAND, C. H., HOUSE, M. R., HUGHES, N. F., REYNOLDS, A. B., RUDWICK, M. J. S., SATTERTHWAITE, G. E., TARLO, L. B. H. & WILLEY, E. C. (eds) *The Fossil Record.* Geological Society, London, 508–528.

DIMICHELE, W. A. & HOOK, R. W. 1992. Paleozoic terrestrial ecosystems. *In:* BEHRENSMEYER, A. K., DAMUTH, J. D., DIMICHELE, W. A., POTTS, R., SUES, H.-D. & WING, S. L. (eds) *Terrestrial ecosystems through time.* University of Chicago Press, Chicago, 206–326.

DOWNES, W. L. JR. 1987. The impact of vertebrate predators on early arthropod evolution. *Proceedings of the Entomological Society of Washington,* **89** (3), 389–406.

ERWIN, D. H. 1994. The Permo-Triassic extinction. *Nature,* **367,** 231–236.

GULLAN, P. J. & CRANSTON, P. S. 1994. *The insects: an outline of entomology.* Chapman & Hall, London.

HARLAND, W. B., COX, A. V., LLEWELLYN, P. G., PICKTON, C. A. G., SMITH, A. G. & WALTERS, R. 1982. *A geologic time scale.* Cambridge University Press.

JARZEMBOWSKI, E. A. 1987. The occurrence and diversity of Coal Measure insects. *Journal of the Geological Society, London,* **144,** 507–511.

—— 1989a. A century plus of fossil insects. *Proceedings of the Geologists' Association,* **100,** 433–449.

—— 1989b. Cretaceous insect extinction. *Mesozoic Research,* **2** (1), 25–28.

—— 1995. Early Cretaceous insect faunas and palaeoenvironment. *Cretaceous Research,* **16** (6).

—— & ROSS, A. 1993. Time flies: the geological record of insects. *Geology Today,* **9,** 218–223.

—— & —— 1994. Progressive palaeoentomology. *Antenna,* **18,** 123–126.

KNOLL, A. H. 1986. Patterns of change in plant

communities through geological time. *In:* DIA-MOND, J. M. & CASE, T. J. (eds) *Community ecology.* Harper & Row, New York, 126–141.

KUKALOVÁ, J. 1970. Revisional study of the order Palaeodictyoptera in the Upper Carboniferous shales of Commentry, France. Part III. *Psyche, Cambridge,* **77**, 1–44.

KUKALOVÁ-PECK, J. 1991. Fossil history and the evolution of hexapod structures. In: CSIRO. *The insects of Australia* (2nd edn). Melbourne University Press, Carlton, **1**, 141–179.

LABANDEIRA, C. C. & SEPKOSKI, J. J. JR. 1993. Insect diversity in the fossil record. *Science,* **261**, 310–315.

MITTER, C., FARRELL, B. & WIEGMANN, B. 1988. The phylogenetic study of adaptive zones: has phytophagy promoted insect diversification? *American Naturalist,* **132**, 107–128.

POINAR, G. O. JR. 1992. *Life in amber.* Stanford University Press, Stanford, California.

ROSS, A. J. & JARZEMBOWSKI, E. A. 1993. Arthropoda (Hexapoda; Insecta). *In:* BENTON, M. J. (ed.) *The Fossil Record 2.* Chapman & Hall, London, 363–426.

Appendix 1

This is an update of Ross & Jarzembowski (1993) using information from papers published between January 1992 and December 1993, and from *Paleontological Journal* **27** (1A) published January 1994. Also included is information from pre-1992 papers that were not seen in time for the publication of *The Fossil Record 2.* The age abbreviations follow Harland *et al.* (1982) except T (for Tertiary) and the European notation for the Carboniferous is used. Details of first and last occurrences and references are available on request from AJR.

Some insect-bearing deposits have been re-dated and the family ranges that are affected by this are given below. Thus Poinar (1992) gives a probable Hauterivian age for Lebanese amber (an Aptian age was used in *The Fossil Record 2*) and a Middle Eocene age for Burmese amber (an Oligocene age was used in *The Fossil Record 2*). The Insect Bed (Bembridge Marls) of the Isle of Wight (England) is more likely to be Late Eocene (Priabonian) than Early Oligocene (Rupelian) as given in *The Fossil Record 2.*

Families with an asterisk were not included in *The Fossil Record 2* or were included under other families. Families with an r have a range change or are given a more accurate age. Families in *The Fossil Record 2* that are now considered to belong to other families are given in parentheses. Abbreviations: O, order; F, family.

O. Diplura
F. Japygidae T(Cht)–Rec *

F. Procampodeidae T(Oli)–Rec *

O. Archaeognatha
F. Meinertellidae T(Oli)–Rec *

O. Zygentoma
F. Nicoletiidae T(Oli)–Rec *

O. Diaphanopterodea
F. Paruraliidae P(Kun) *

O. Ephemeroptera
F. Leptophlebiidae K?(Apt)–Rec r
F. Palaeoanthidae K(San) *

O. Megasecoptera
F. Anchineuridae C(Ste B) r
F. Aykhalidae C2 *
F. Hanidae P(Art) r

O. Odonata
F. Aeschnidiidae J3–K(Cen) r
F. Amphipterygidae J3–Rec r
F. Archithemistidae J1 r
F. Calopterygidae T(Oli)–Rec r
F. Campterophlebiidae (Karatawiidae) Tr(Rht)–J3
F. Congquingiidae K1 *
F. Cordulegastridae T(Rup)–Rec *
F. Eosagrionidae J(Toa) *
F. Epiophlebiidae J(Toa)–Rec *
F. Euphaeidae T(Prb)–Rec r
F. Heterophlebiidae J1 r
F. Isophlebiidae J(Bth)–K2 r
F. Liassogomphidae J(Toa) r
F. Mesophlebiidae Tr3 r
F. Myopophlebiidae J1 *
F. Permagrionidae P2 r
F. Petaluridae J(Tth)–Rec r
F. Platycnemidae T(Prb)–Rec r
F. Progonophlebiidae J1 *
F. Selenothemistidae J(Toa) *
F. Triassolestidae Tr3 r
F. Triassoneuridae Tr *
F. Turanothemistidae J r
F. Zacallitidae T(Prb)–Rec r

O. Palaeodictyoptera
F. Eubleptidae C(WesD) r
F. Hypermegethidae C2 *
F. Mongolodictyidae P2 *

Super O. Palaeoptera: O. Uncertain
F. Miracopteridae P(Art–Kun) *

O. Blattodea
F. Blattulidae J1–K(Vlg) r
F. Cainoblattinidae T(Eoc) *
F. Polyphagidae (Corydiidae) K(Ber)–Rec *

O. Dermaptera
F. Labiduridae T(Eoc2)–Rec r

O. Embioptera
F. Anisembiidae T(Oli)–Rec *
F. Burmitembiidae T(Eoc2) r
F. Teratembiidae T(Oli)–Rec *

O. Grylloblattodea P1–Rec r
F. Bajanzhargalanidae J3 *
F. Liomopteridae (transferred from O. Protorthoptera) P1–P2 r
F. Megakhosaridae (transferred from O. Protorthoptera) P2–Tr3 r
F. Mesojabloniidae Tr3 *
F. Perloblattidae Tr3 *
F. Phenopteridae (transferred from O. Protorthoptera) P1–P(Kaz) r
F. Tillyardembiidae (transferred from O. Protorthoptera) P(Kun) r
F. Tologopteridae P2 *

O. Isoptera
F. Mastotermitidae K(Cen)–Rec r
F. Termopsidae K(Hau)–Rec *

O. Mantodea
F. Amorphoscelidae K1–Rec *
F. Baissomantidae K1 *
F. Cretomantidae K1–K(San) *
F. Mantidae T(Prb)–Rec r

O. Orthoptera
F. Baissogryllidae J3–K1 r

O. Phasmatodea P2–Rec r
F. Aerophasmatidae J(Sin)-K2 r
F. Permophasmatidae P2 *

O. Plecoptera
F. Capniidae J?(Toa)–Rec r

O. 'Protorthoptera'
F. Ampelipteridae C(Nam A) *
F. Geraridae (Omaliidae) C(Nam B–Ste B) r
F. Paoliidae C2 r
F. Stygnidae C(Nam B) r

O. Hemiptera
F. Bernaeidae K(Hau) r
F. Cixiidae K(Hau)–Rec r
F. Coreidae Tr3–Rec r
F. Corixidae Tr3–Rec r
F. Cretamyzidae K(Cmp) *
F. Dactylopiidae T(Cht)–Rec *
F. Dipsocoridae K(Hau)–Rec r
F. Discolomidae T(Cht)–Rec *
F. Enicocephalidae K(Hau)–Rec r
F. Gerridae T(Tha)–Rec r
F. Hydrometridae T(Tha)–Rec r
F. Ipsviciidae Tr(Ans)–K1 r
F. Karajassidae J2–K1 r
F. Mesozoicaphididae K(Cmp) *
F. Palaeontinidae Tr3–K1 r
F. Schizopteridae T(Oli)–Rec *
F. Stenoviciidae P2–K(Hau) r
F. Termitaphididae T(Cht)–Rec *
F. Thaumastellidae K(Hau)–Rec r
F. Thaumastocoridae T(Oli)–Rec *

O. Psocoptera

The Mesopsocidae and Polypsocidae do not have a fossil record.
F. Amphientomidae K(San)–Rec r
F. Cladiopsocidae T(Oli)–Rec *
F. Dolabellopsocidae T(Oli)–Rec *
F. Elipsocidae K(San)–Rec r
F. Lachesillidae K(San)–Rec r
F. Lepidopsocidae T(Oli)–Rec r
F. Lophioneuridae P1–K(San) r
F. Myopsocidae T(Cht)–Rec r
F. Pachytroctidae T(Eoc2)–Rec r
F. Peripsocidae T(Cht)–Rec r
F. Psocidae K(San)–Rec r
F. Psoquillidae T(Oli)–Rec *
F. Psyllipsocidae K(San)–Rec r
F. Ptiloneuridae T(Oli)–Rec *
F. Sphaeropsocidae K(San)–Rec r
F. Troctopsocidae T(Oli)–Rec *
F. Trogiidae K(San)–Rec r
F. Ulyanidae K1 *

O. Thysanoptera
F. Aeolothripidae K(San)–Rec r
F. Jezzinothripidae K(Hau) *
F. Phlaeothripidae K(San)–Rec r
F. Stenurothripidae K(Hau)–Rec *
F. Thripidae K(San)–Rec r

O. Coleoptera
F. Biphyliidae T(Oli)–Rec *
F. Brachypsectridae T(Oli)–Rec *
F. Buprestidae J(Bth)–Rec r
F. Byturidae K(Ber)–Rec r
F. Chrysomelidae Tr?(Ans)–Rec r
F. Cistelidae T(Cht)–Rec *
F. Colymbotethidae Tr3 *
F. Coptoclavidae Tr3?–K1 r
F. Cossonidae T(Cht)–Rec *
F. Dermestidae K?(Cmp)–Rec r
F. Endomychidae K1–Rec r
F. Euglenidae T(Oli)–Rec *
F. Inopeplidae T(Oli)–Rec *
F. Languriidae T(Oli)–Rec *
F. Leiodidae K1–Rec r
F. Limulodidae T(Oli)–Rec *
F. Mycteridae T(Oli)–Rec *
F. Obrieniidae Tr(Crn)–J3 *
F. Pedilidae T(Oli)–Rec *
F. Permosynidae (Ademosynidae) Tr(Ans)–K1 *
F. Salpingidae K(Hau)–Rec r

O. Diptera
F. Alinkidae Tr3 *
F. Anthomyiidae T(Eoc2)–Rec r
F. Architipulidae (Diplopolyneuridae) Tr3–Rec *
F. Ceratopogonidae K(Ber?)–Rec r
F. Chloropidae K(Hau)–Rec r
F. Corethrellidae T(Oli)–Rec *
F. Culicidae K(Cmp)–Rec r
F. Cylindrotomidae T(Tha)–Rec *
F. Glossinidae T(Rup)–Rec r
F. Hyperpolyneuridae Tr3 r
F. Periscelidae T(Oli)–Rec r
F. Procramptonomyiidae Tr3–J3 r

F. Protopleciidae (Dyspolyneuridae) J1–K1
F. Sciaridae K(Hau)–Rec r
F. Tipulidae s.s. K(Cen?)–Rec r

O. Glosselytrodea P1–J3 r
F. Permoberothidae P1 r

O. Hymenoptera
F. Chalcididae K(Hau)–Rec r
F. Colletidae T(Oli)–Rec *
F. Diapriidae K2–Rec r
F. Evaniidae T(Eoc2)–Rec r
F. Formicidae K(Hau?)–Rec r
F. Mymaridae K(Hau)–Rec r
F. Myrmiciidae J2–J(Tth) r
F. Scolebythidae K(Hau)–Rec *
F. Scoliidae K(Apt)–Rec r
F. Sphecidae K(Ber)–Rec r

O. Lepidoptera
F. Acrolophidae T(Oli)–Rec *
F. Arctiidae T(Prb)–Rec *
F. Argyresthiidae T(Prb)–Rec *
F. Blastobasidae T(Oli)–Rec *
F. Copromorphidae T(Prb)–Rec r
F. Cosmopterygidae T(Oli)–Rec *
F. Cossidae T(Prb?)–Rec r
F. Eriocraniidae T(Eoc 2)–Rec *
F. Ethmiidae T(Cht)–Rec *
F. Gelechiidae T(Prb)–Rec *
F. Geometridae T(Prb)–Rec r
F. Heliozelidae T(Prb)–Rec *
F. Incurvariidae K(Hau)–Rec *
F. Mnesarchaeidae K(San)–Rec *
F. Nymphalidae T(Prb)–Rec r
F. Scythrididae T(Prb)–Rec *
F. Sphingidae T(Prb)–Rec r
F. Symmocidae T(Prb)–Rec *
F. Thyrididae T(Prb)–Rec *
F. Walshiidae T(Cht)–Rec *
F. Yponomeutidae T(Prb)–Rec *

O. Mecoptera
F. Bittacidae J2–Rec r
F. Boreidae J3–Rec *

O. Megaloptera P(Kun)–Rec r
F. Parasialidae (transferred from O. Miomoptera) P(Kun–Kaz) r

O. Neuroptera
F. Babinskaiidae K1 r
F. Berothidae K(Hau)–Rec r
F. Neurorthidae T(Prb)–Rec r

O. Raphidioptera
F. Huaxiaraphidiidae K1 *
F. Jilinoraphidiidae K1 *

O. Siphonaptera
F. Rhopalopsyllidae T(Oli)–Rec *

O. Trichoptera P(Ass)–Rec r
F. Microptysmatidae P(Art–Tat) r

F. Polycentropodidae K1–Rec r
F. Protomeropidae (transferred from O. Mecoptera) P(Ass–Tat) r

Super O. Oligoneoptera: O. Uncertain
F. Dictyodipteridae (transferred from O. Diptera) J1
F. Strashilidae J3 *

Note added in proof. Since this paper was submitted, two relevant articles have appeared:

1. An additional database:

LABANDEIRA, C. C. 1995. A compendium of fossil insect families. *Milwaukee Public Museum Contributions in Biology & Geology*, **88** (December 31, 1994).

This supplements Ross & Jarzembowski (1993) by providing detailed ranges and primary references for those families for which we only gave a minimum amount of information (because they were not discussed in the post-1983 literature). Labandeira claims a 98% stratigraphic resolution to stage for families and criticizes us for not giving stage-level ranges for every family. Unfortunately, such data are not available for every first and last record; indeed, for some Chinese localities there is uncertainty as to which epoch they belong to and we consider that such a high level of stratigraphic certainty is not possible at the present time. Labandeira gives 1272 families in the fossil record compared with our 1167 families. However, we have adopted a more conservative classification and synonymized some families that certain entomologists would keep separate. Also, we have omitted a few families e.g. those recorded in 1994. However, there are 23 families listed in this paper not given by Labandeira.

2. High atmospheric oxygen levels have been linked with the important late Palaeozoic insect radiations, with gigantism in Protodonata (giant dragonflies) as supporting evidence:

GRAHAM, J. B., DUDLEY, R., AGUILAR, N. M. & GANS, C. 1995. Implications of the late Palaeozoic oxygen pulse for physiology and evolution. *Nature*, **375**, 117–120.

Unfortunately, Protodonata also includes small taxa of 'normal' size in the Permo-Carboniferous. Also, giant dragonflies lack the more advanced odonatan specializations of nodus, arculus and pterostigma implying less aerial mobility and need for bursts of high metabolic rate. Moreover, Permian insect faunas are *distinguished* from Carboniferous ones by the adaptive radiation of holo-metabolous insects and respiratory (spiracular) function in insects is stimulated by CO_2 (and not O_2). The undoubtedly selective gigantism of terrestrial arthropods needs to be considered too in the context of predation (or lack of it), development, climatic and other palaeoenvironmental factors, and cannot be viewed as a simple function of hyperoxia.

Reef ecosytem recovery after the Early Cambrian extinction

ANDREY Yu. ZHURAVLEV

Palaeontological Institute, Russian Academy of Sciences, Profsouyznaya 123, Moscow 117647, Russia

Abstract: Revised and new stratigraphic data indicate a complex extinction event in the late Early Cambrian which consisted of two, temporally separate, but related phases. The earliest phase (mid-Botomian, Sinsk Event) may be related to widespread anoxia due to eutrophication and phytoplankton blooms. The later event (early Toyonian, Hawke Bay Event), is connected to a world wide regression. This double extinction event severely injured the reefal biota which has undergone a further rejuvenation during the remainder of the Cambrian. As a result of the lowering of grazing pressure and unhealthy metazoan–calcimicrobial interactions, the remaining metazoan reef-builders were eliminated by the end of the Early Cambrian. Consequent reduction of space heterogenity led to the decline of calcified microbes (*Renalcis*, etc.) in both diversity and abundance which gave way to the thrombolite–stromatolite community. The recovery of the reefal biota occurred during the very end of the Late Cambrian–Early Ordovician and may be attributed to Elvis-taxa, with the exception of the spicular demosponges. This recovering biota intruded into the thrombolite–stromatolite community and created a space for the more successful encrusting reef dwellers of Middle–Late Ordovician time.

The reduction of the whole Cambrian period to about 35–45 Ma even including the Nemakit–Daldynian earlier stage makes the distinct Cambrian fauna (*sensu* Sepkoski 1981) and flora (*sensu* Chuvashov & Riding 1984) a very ephemeral consortium. The Cambrian fauna evolved and almost completely disappeared during the Early Cambrian epoch as a result of the severe Sinsk (mid-Botomian) and Hawke Bay (early Toyonian) extinction events (Zhuravlev & Wood 1994). These were just about 20 Ma after their insertion. The Cambrian flora survived beyond this but was eliminated at the beginning of the Late Cambrian, about 10 Ma later.

The record of the Cambrian reefs gives us some indications of why it has happened so quickly. Cambrian reefs provide information qualitatively and quantitatively comparable with that provided by the best Lagerstätten, mainly because of the significant role of ealry cementation in their diagenesis. This process immures *in situ* some of the organisms which would usually only be preserved in 'soft' Lagerstätten (e.g. coeloscleritophorans).

Although it is a commonly-held belief that the Cambrian reefs which were an archaeocyathan–algal consortium vanished by the end of the Early Cambrian (e.g. Fagerstrom 1987; Stanley 1992; Copper 1994), this is not the case. They persisted after the archaeocyathan extinction and the biggest Cambrian framework reefs are those of the Middle Cambrian (McIlreath 1977; Savitskiy 1979; Astashkin *et al.* 1984) although the metazoan role in their construction was negligible.

The evolution of the reef ecosystem through the Cambrian–Early Ordovician has had a very complicated pattern. When calcified microbes gradually intruded into the 'Precambrian' thrombolitic build-ups, the first framework reefs were formed by archaeocyaths in the very beginning of the Tommotian Stage. The appearance of the *Epiphyton*-group of calcified microbes made the Cambrian reefs an unsuitable place for metazoan proliferation and only a few encrusting animals could survive the competition. The disappearance of those few exceptions had occurred by the middle Botomian and Early Toyonian extinction events, gave carte-blanche to calcified microbes which suppressed the rest of the metazoan consortium at the very beginning of the Middle Cambrian. However, the lowering of grazing pressure due to the same extinction events and the complete absence of reef-building metazoans providing additional space for calcimicrobial settlement, led to a rejuvenation of the reef ecosystem and the loss of diversity within the calcified microbes. This steady loss of diversity of the calcified microbes during the Middle–Late Cambrian was in turn expressed as a reduction of habitats until the almost complete elimination of the framework reefs at the beginning of the Late Cambrian and

From Hart, M. B. (ed.), 1996, *Biotic Recovery from Mass Extinction Events,*
Geological Society Special Publication No. 102, pp. 79–96

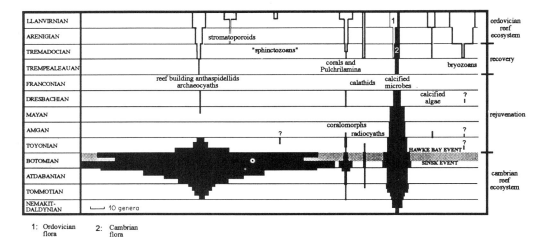

Fig. 1. Generic diversity of the reef-building organisms through the Cambrian–early Middle Ordovician. 1 & 2 respectively Cambrian and Ordovician floras of Chuvashov & Riding (1984).

in the stromatolitic–thrombolitic reef resurgence. The latter proliferated until the Middle Ordovician when new groups of calcified microbes and true calcified algae became widespread. However, there was a major intrusion of metazoans into the stromatolitic–thrombolitic consortium during the Early Ordovician. The success of the new Ordovician metazoan reef-builders in the algal–cyanobacterial framework reefs lies in their modular (high integrated) nature. However, the influence on the Late Cambrian–Early Ordovician events of certain abiotic factors such as sea-level and ocean-chemistry changes cannot be completely excluded, although the biotic factors, by themselves, played the major role.

The Siberian and Laurentian time scales are chosen for Lower–Middle and Upper Cambrian, respectively (Fig. 1) as the bulk of Cambrian reefs are ascribed to their subdivisions.

Microbial carbonate build-ups are subdivided in this account (following Aitken 1967; Kennard & James 1986; Riding 1991a) into dendrolites, thrombolites and stromatolites. Stromatolites are laminated microbial build-ups, thrombolites are clotted ones and dendrolites are dendritic in structure. Stromatolites and thrombolites are built by a non-calcified community of bacteria (and algae?) and dendrolites are formed by calcified microbes such as *Epiphyton*, *Proaulopora*, *Tharthinia*. *Razumovskia*, etc. Any of them can contain diverse reef-dwellers, even stromatolites (Zhuravleva 1960; Runnegar *et al.* 1979; Wood *et al.* 1993). Thrombolites are not burrowed stromatolites (*sensu* Walter & Heys 1985) because the later ones being burrowed still retain their laminated fabric.

Reefal biota crisis

Sinsk Event

The late–Early Cambrian extinction has long been recognized as the first Phanerozoic event (Newell 1972; Burrett & Richardson 1978; Signor 1992). The poorly constrained stratigraphic correlation, enhanced by taxonomic problems, has hindered a detailed study of this event. A new biostratigraphic framework now makes such a study possible (Zhuravlev in press). The critical interval from the latest Atdabanian until the Middle Toyonian is especially well calibrated by assemblages of different fossils (e.g. trilobites, archaeocyaths, small shelly fossils, etc.) on a zonal level, which allows a reasonable correlation of sections in the principal Cambrian basins.

Quantitative analysis of both global generic diversity and species diversity for the studied regions shows striking changes of character through the Early Cambrian. The major extinction events are noted in the mid-Botomian and Early Toyonian. Each of these events eliminated more than 50% of the genera extant at the time of the event. Almost all groups of organisms display prominent losses of diversity and enhanced extinction rates during the mid-Botomian, Sinsk Event (Zhuravlev & Wood 1994). This event is marked by the distribution, during the Early–Middle Botomian, of varved 'black shale' facies within the Sinsk Formation, its correlative strata on the Siberian Platform, contemporaneous formations in South Australia, on the Yangtze Platform and in some sections in Iran and parts of the Altay Sayan Foldbelt, Transbaikalia, Russian Far East, Kazakhstan and Mongolia.

The type facies of the Sinsk Formation include bituminous limestone, chert, argillaceous, siliceous and calcareous (often sapropelic) black shales, all with a high content of primary pyrite framboids. The common textural feature is a thin well-expressed lamination on a submillimetre scale, lack of bioturbation and a black colour arising from the high organic carbon content (Bakhturov et al. 1988; Clarke 1990; Steiner et al. 1993). Calcite laminae are enriched by monospecific acritarchs of the Heliosphaeridium-group. They are separated by brown clay layers. The brown colour of the clay laminae is probably due to finely disseminated organic matter. The present study has indicated that the bulk of the organic matter in the Sinsk Formation and its correlatives is composed of acritarchs, of marine phytoplankton origin (Zhegallo et al. 1994). The biota has features of the anaerobic (Siberian Platform) or dysaerobic (Yangtze Platform, South Australia) fauna of Byers (1977) (Erdtmann et al. 1990; Chen & Lindström 1991; Zhuravlev & Wood 1995).

Varved microfacies, similar to those observed in the Sinsk Formation, have been reported from a number of present-day eutrophied marine and lacustrine basins (Hollander et al. 1992; Anderson et al. 1994; Brodie & Kemp 1994). Such a rhythmic lamination of organic-rich limestone facies usually results from a regular alternation of spring/summer phytoplankton bloom which facilitated the calcareous precipitation by CO_2 extraction (calcite, phytoplankton-containing laminae) and winter falls (clay laminae). Their repetitive accumulation is normally under anoxic bottom conditions. The anoxia, by itself, is due to a biological process of oxygen consumption (Southam et al. 1982; Pedersen & Calvert 1990). The enrichment of certain trace elements (Cu, Ni, U, V, Mo) which is observed in strata of the Siberian and Yangtze Platforms (Bakhturov et al. 1988; Coveney & Nansheng 1991; Zhuravlev & Wood 1995) is indicative of a biotic 'high productivity' water column anoxia under eutrophic conditions rather than abiotic bottom anoxia (Calvert & Pedersen 1993).

Hypertophy which results from continuous eutrophication, is a common phenomenon of Recent marine and lacustrine basins characterized by a persistent phytoplankton bloom which causes a significant biota depauperization (Brag-

inskiy et al. 1968; Aubert 1988; Cruzado 1988; Stachowitsch & Avčin 1988). Apart from the direct elimination of biota by anoxia, the phytoplankton bloom affects are less likely to asphyxiate groups by cutting the trophic web. Phytoplankton eliminates bacterial plankton during bloom periods due to the production of extra-cellular metabolites and a direct competition for the growth limiting substrate (Guillard & Hellebust 1971; Karl et al. 1991). There was a severe elimination of archaeocyathan genera (Fig. 1), whose diet was presumably (like modern sponges) mostly bacterial (Wood et al. 1993).

Hawke Bay Event

The Hawke Bay Event is related to the Hawke Bay Regression of Palmer & James (1979) and might be a direct consequence of the Sinsk Event. The increased carbon burial due to overproductivity and anoxia could weaken greenhouse conditions by oxygen release to the atmosphere and the subsequent polar ice cap formation could cause a regression which is widely seen in basal Toyonian strata. The features of this regression are worldwide and recognized by Skolithos-facies, bird-eye dolostones, etc., in the circum-Iapetus region (Bergström & Ahlberg 1981), on the Baltic Platform (Brangulis et al. 1986), in Spain (Liñan & Gámez-Vintaned 1993), Morocco (Geyer 1989), western Laurentia (Fritz et al. 1991; Mansy et al. 1993), Australia (Gravestock & Hibburt 1991) and on the Siberian Platform (Bakhturov et al. 1988). The continued $\delta^{13}C$ negative shift (Brasier et al. 1994) and positive $\delta^{13}S$ shift (Claypool et al. 1980) both could be due to the oxidation of organic matter on the continental shelves during the regressive episode. The severe elimination of the reefal biota (archaeocyaths, coralomorphs, cribricyaths, calcified microbes) during the Hawke Bay Event (Fig. 1) and the almost complete indifference of other groups to it (Zhuravlev & Wood 1995) confirms that this extinction event could be due to a regression and a resultant restriction of shelf areas.

The time of the formation of carbonate–siltstone couplets in varved strata varies from 1 to 8–16 years depending on the character of basin (Hollander et al. 1992; Brodie & Kemp 1994). Assuming that the bioturbants producing pellets on the surface sediment reduce the couplet thickness and number, and taking into account a mean couplet thickness (c. 0.001 m) and the total thickness of the Sinsk Formation in its distal less-disrupted by storm events facies (40 m), the duration of the Sinsk Event can be estimated as 0.05–0.3 Ma or even less. These data are rather comparable to the duration of the Cretaceous–Tertiary (0.03 Ma) rather than the Permian–Triassic (5–10 Ma) extinction events (Holser & Magaritz 1992). The whole episode (Sinsk + Hawke Bay Events) which eliminated the Cambrian Fauna *sensu lato* could last less than 5 Ma.

Further rejuvenation

Organism interactions in the Cambrian reefs

Among the main causes of the complete elimination of the Cambrian reef ecosystem, are particular organism interactions. Two groups of reef-builders can be recognized in the Cambrian. Calcified autotrophs (*Renalcis*, *Epiphyton*, etc.) were the most important and metazoans commonly played a surbordinate role (Zhuravleva 1972; Zadorozhnaya 1974; Zamarreño 1977; James & Debrenne 1980; Selg 1986; Debrenne *et al.* 1989*a*,*b*; Rees *et al.* 1989). Although the nature of these autotrophs is still under discussion, their cyanobacterial, rather than algal, affinity is preferable (Riding 1991*b*). Rare, possibly calcified, algae are known from the Amgan Stage of the Siberian Platform (*Amgaella*) but their affinity with red or green algae is equivocal. The metazoan, reef-building component is basically attributed to archaeocyaths, but other reef-builders such as radiocyaths and coralomorphs could be prominent (Pratt 1990; Kennard 1991; Kruse 1991; Lafuste *et al.* 1991; Wood *et al.* 1993; Zhuravlev *et al.* 1993).

Archaeocyaths are a group of calcified sponges, probably most closely related to the Class Demospongiae (Debrenne & Zhuravlev 1994). Radiocyaths are compared to calcified algae (Nitecki & Debrenne 1979) but some features of their growth pattern and ecological responses indicate their possible affinity to filter-feeders (Zhuravlev 1986; Wood *et al.* 1993).

Coralomorphs are a heterogeneous group of problematic animals of probable cnidarian grade of organization (e.g. *Khasaktia*, *Hydroconus*, *Flindersipora*) (Fig. 2b–d). Bivalve, mollusc-like, stenothecoids were facultative reef-builders and formed the latest (early Amgan) 'metazoan' reefs after the archaeocyathan–radiocyathan–coralomorph demise. Spicular sponges (Calcarea, rarely Hexactinellida) and problematic calcareous microfossil cribricyaths (Fig. 2a) contributed a little to reef-building and were mainly cryptobionts. Chancelloriids, orthothecimorph hyoliths, calciate brachiopods and, subsequently, echinoderms were common sessile reef-dwellers and were often immured *in situ* during primary, marine, cement formation.

There appears to have been an inverse relationship between calcimicrobes and metazoans in the Early Cambrian reef communities. Reefs were either dominantly microbial or, more rarely, dominated by metazoans (Zamarreño 1977; Gandin & Debrenne 1984; Selg 1986; Stepanova 1986; Debrenne *et al.* 1989*a*, *b*; Rees *et al.* 1989; Debrenne & Gravestock 1990; James & Gravestock 1990; Wood *et al.* 1993; Moreno-Eiris 1994). These reef-building metazoan–calcimicrobial interactions probably caused the complete elimination of the metazoan component from the reef biota. The author's observations indicate that these interactions were not very healthy for the metazoans (Riding & Zhuravlev 1994).

Archaeocyaths and other sessile metazoans, preferentially attached to non-microbial substrates (Fig. 2a, c–e), and often formed long chains of individuals one on top of the other (Fig. 2b). There are several reasons for such a pattern of metazoan/calcimicrobial distribution.

Fig. 2. Organisms' interactions in the Early Cambrian reefs. (**a**) Thalamid archaeocyaths (*Clathricoscinus infirmus* Zhuravleva) encrust skeletal fragments (arrowed) and another thalamid archaeocyath *Polythalamia perforata* (Vologdin) (bottom left in a cryptic cavity), ×3, Palaeontological Institute of the Russian Academy of Sciences (PIN) 4451/69, Sukhie Solontsy Valley, Kuznetskiy Alatau, Russia, Botomian Stage. (**b**) Chain of organisms in a cryptic cavity formed by coralomorph *Hydroconus mirabilis* Korde (top), two individuals of archaeocyath *Molybdocyathus* sp. and thalamid archaeocyath *P. perforata* (bottom), ×7, PIN, Western Sayan, Russia, Botomian Stage. (**c**) Two archaeocyaths *Neoloculicyathus sibiricus* (Sundukov) encrust coralomorph *Khasaktia vesicularis* Sayutina in a cryptic cavity, ×7, PIN, Oymuraan village, middle Lena River, Siberian Platform, Russia, Atdabanian Stage. (**d**) Archaeocyath *Dictyosycon gravis* Zhuravleva encrusts calcified microbes *Gordonophyton* and coralomorph *K. vesicularis* in a cryptic cavity, ×3, PIN, same locality and age as for above. (**e**) Chaetetid archaeocyath *Dictyofavus araneosus* (Gravestock) encrusts a brachiopod shell, ×10, PIN, Wilkawillina Gorge, Flinders Ranges, South Australia, Atdabanian Stage. (**f**) Several individuals of archaeocyath *Archaeolynthus polaris* (Vologdin) are sealed by calcified microbes *Renalcis jacuticus* Korde in a cryptic cavity, ×6, PIN, Byd'yangaya Creek, middle Lena River, Siberian Platform, Russia, Tommotian Stage. (**g**) Archaeocyath *Ajacicyathina* gen. and sp. indet. is encrusted by calcified microbes *Gordonophyton durum* (Korde) Korde, ×30, PIN, same locality and age as for (a).

REEF ECOSYSTEM RECOVERY IN THE M. CAMBRIAN

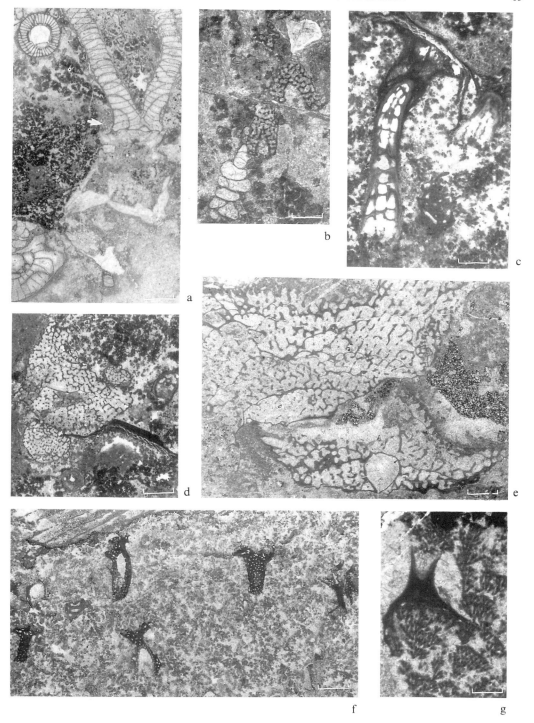

(i) Production of extracellular toxic slimes and nitrogen fixation by bacteria including cyanobacteria (Gerdes & Krumbein 1984; Carmichael 1994) are significant in metazoan dwarfing. Archaeocyaths and other animals growing in the pure *Epiphyton*-group (*Epiphyton, Gordonophyton, Tubomorphophyton*) bioherms are often small and produce a large amount of exotheca (James & Debrenne 1980; Debrenne & Zhuravlev 1992) (Fig. 2g). Such structures could be a response to chemical irritation by bacteria (Reitner 1993). The attribution of metazoan decline to oxygen-depletion (Monty 1973) or to poisoning (Edhorn 1979) of the milieu by microbes, however, is speculative.

(ii) Calcified microbes readily encrust sessile metazoans and commonly sealed them (Figs 2f, 3a, d). Only modular encrusting archaeocyaths and coralomorphs could out-compete calcified microbes because of higher growth rates and opportunistic spreading rather than polarized growth form (Copper 1974; Wood *et al.* 1992b) (Figs 2c–d, 3b–c).

(iii) Escape from the algal substratum by sponges and corals is a common phenomenon, observed temporally in Recent seas, and is due to out-competing by algae for larval substrates (Shlesinger & Loya 1985; Ilan & Loya 1988; Kielman 1993). In the absence of grazers such a pattern would be permanent rather than temporal (Dart 1972; Borowitzka 1981; Sammarco 1987). However, there are no indications of prominent grazing pressure, at least for the Early–Middle Cambrian.

Possible Cambrian algal croppers have been noted by Edhorn (1977) from the Tommotian–Atdabanian Bonavista Group of eastern Newfoundland and by Read & Pfeil (1983) from the Toyonian Upper Shady Dolomite of Virginia. The latter are erosional pits of an unknown origin. It is very probable that the 'algal croppers' of Edhorn (1977) are orthothecimorph hyoliths 'Ladatheca' of Landing (1993) which are described from the same stratigraphic interval and the same area. *In situ* immured sessile orthothecimorph hyoliths are common in Early Cambrian reefs (Riding & Zhuravlev 1995).

Principal grazer groups, such as teleost fishes, chitons, limpets, insects, crustaceans and sea urchins capable of consuming calcified algae are a relatively recent (Late Mesozoic–Cenozoic) addition to the marine biota (Gerdes & Krumbein 1984; Lewis 1986; Shunula & Ndibalema 1986; Stenek 1986). Even heavily armed parrot fish prefer to consume turf algae on substrates infested with endolithic algae whereas crustose coralline algae are avoided (Bruggemann 1995). The significance of the meiofauna (nematodes, turbellarians, gastrotrichs, gnathostomulids, rotatorians, foraminiferans, ciliates, etc.) is negligible. The nematodes, for instance, though living in great abundance within the stromatolites are not able to disturb the lamination because of their small size in relation to the sediment particles (Schwarz *et al.* 1975). Comparatively few small species possess the necessary adaptation for consuming filamentous cyanobacteria (Pratt 1982; Farmer 1992) which were the major constructors of Proterozoic and Cambrian stromatolites (Riding 1991a). Thus, meiofaunal grazers could not be responsible for stromatolite decline (cf. Walter & Heys 1985; Awramik 1991). Cambrian echinoderms and crustaceans were mainly suspension-feeders (Smith 1990; Guensburg & Sprinkle 1993; Butterfield 1994). Cambrian molluscs were small and together with other possible Cambrian grazers (some trilobites, tommotiids and vagrant coeloscleritophorans) were to suffer seriously during the Sinsk Event (Zhuravlev & Wood 1995) which is reflected also in the drop of trace fossil diversity (Crimes & Droser 1992). In addition, malacostracans, polychaetes and non-limpet gastropods are not throught to have had a major effect on benthic algae due to an inability of their appendages to exert a sufficiently srong downward force. Thus, even slight grazing pressure was reduced by the beginning of Toyonian as a result of the extinction events. The lowering of grazing pressure allowed calcified microbes to

---

Fig. 3. Organisms' interactions in the Early and Middle Cambrian reefs. (**a**) Archaeocyath *Spirocyathella kyzlartauense* Vologdin is terminated by calcified microbes *Razumovskia uralica* Vologdin (arrowed), ×10, PIN, South Urals, Russia, Botomian Stage, Early Cambrian. (**b**) Archaeocyath *Okulitchicyathus* sp. encrusts coralomorph *Hydroconus tenuis* (Vologdin) and calcified microbes *Gordonophyton nodosum* Drozdova, ×30, PIN, Nuur Mogoy, Khasagt Khairkhan Ridge, Mongolia, Atdabanian Stage, Early Cambrian. (**c**) Archaeocyath *Archaeocyathina* gen. and sp. indet. encrusts calcified microbes *G. durum,* ×30, Batenevskiy Ridge, Kuznetskiy Alatau, Russia, Botomian Stage, Early Cambrian. (**d**) Calcified microbes *G. durum* encrust archaeocyath *Nalivkinicyathus cyroflexus* (Boyarinov and Osadchaya) (arrowed), ×15, PIN, Sukhie Solontsy Valley, Kuznetskiy Alatau, Russia, Atdabanian Stage, Early Cambrian. (**e**) Fragment of a reef built by anthaspidellid sponge *Rankenella* ex gr. *mors* (Gatehouse), ×3, Iranian Geological Survey, Elburz Mountains, Iran, Mayan Stage, Middle Cambrian.

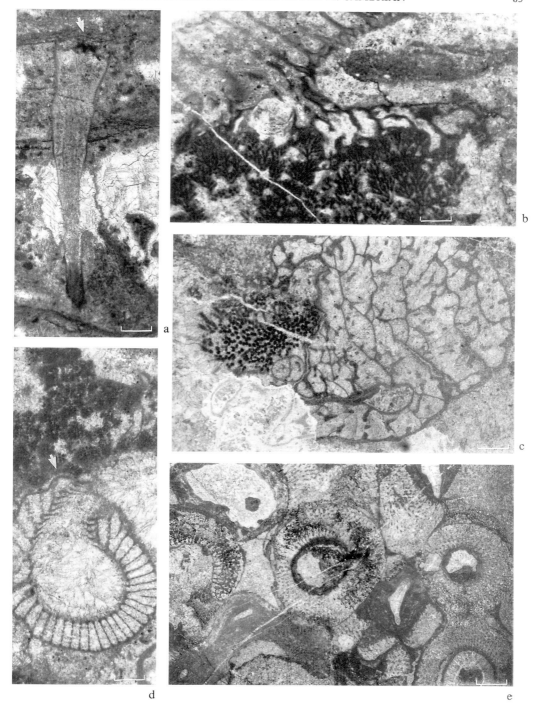

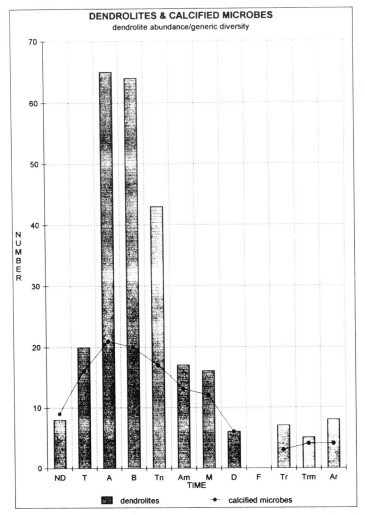

Fig. 4. Generic diversity of calcified microbes plotted against dendrolite abundance which is approximated as a number of dendrolite containing formations per stage interval. Lower Cambrian stages: ND, Nemakit-Daldynian; T, Tommotian; A, Atdabanian; B, Botomian; Tn, Toyonian. Middle Cambrian stages: Am, Amgan; M, Mayan. Upper Cambrian stages: D, Dresbachian; F, Franconian; Tr Trempealeauan. Lower Ordovician series: Trm, Tremadocian; Ar, Arenigian.

force out the rest of the sessile metazoans having lost their encrusting representatives, in accordance with the cropping principle (Stanley 1973).

Middle Cambrian reefs

In spite of the common belief that the Cambrian reef-building phase ended with the disappearance of the archaeocyaths (Stanley 1992), it was the Middle Cambrian when Cambrian reefs as lithological phenomena achieved their acme. The biggest Cambrian reefs are known from the Amgan Stage of Laurentia (Orr Fm, Lohmann 1976; Cathedral Fm, McIlreath 1977), Siberian Platform (Amga, Nel'gaka and Udachny Formations, Korde 1961; Golosheykin et al. 1978; Stepanova 1979; Luchinina & Stepanova 1983; Astashkin et al. 1984).

The continuous decline of calcimicrobial diversity that started in the Early Botomian (Fig. 4) could be, however, a direct consequence of the elimination of both potential tiny grazers and reef-building metazoans during the Sinsk and Hawke Bay events. This is because it is the

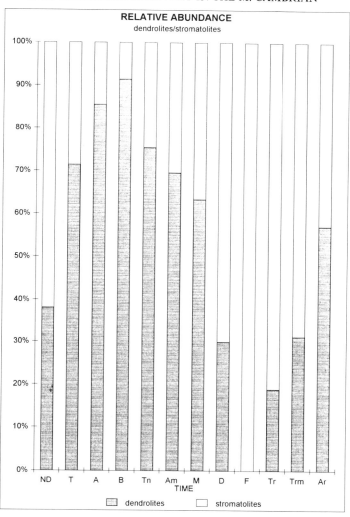

Fig. 5. Relative abundance of dendrolite and stromatolite containing formations through the Cambrian–Lower Ordovician. Stage indexes as in Fig. 4.

lowering of grazing pressure which produces more homogeneous rather than heterogeneous assemblages, especially in the case of microbial communities (Stanley 1973; Gebelein 1974; Schidlowski et al. 1992). Besides, any species-rich ecosystem needs a disturbance to maintain the diversity (Phillips et al. 1994). Metazoan reef-builders are also responsible for creating a spatial heterogenity which supports a diverse calcimicrobial community (Valentine 1973). It should be noted that even Early Cambrian microburrowers inhabiting reefs preferred the more sheltered, stable, niches provided by a metazoan framework (Wood et al. 1993; Kruse et al. 1995). For example, obligate and semi-obligate coelobiontic calcimicrobes (*Chabakovia, Sajania, Wetheredella* and branching forms of *Renalcis*) are only known from metazoan reefs and disappeared together with them. The ability of large sessile metazoans (radiocyaths, some archaeocyaths and coralomorphs) to persist in conditions of higher sedimentation rates and siliciclastic input (Gandin & Debrenne 1984; Narbonne & Arbuckle 1989; Wood et al. 1993; Riding & Zhuravlev 1995) has also modified conditions making them suitable for calcimicrobial settlement. The latest reef-building metazoans were Early Amgan stenothecoids and very sporadic Middle and early Late Cambrian archaeocyaths, coralomorphs and anthaspidellid demosponges (Fritz & Howell 1955; Debrenne *et al.* 1984; Wood *et al.* 1992a; Hamdi *et al.* 1995). Some sessile Cambrian reef dwellers, such as chancelloriids, persisted into the early Middle Cambrian (Dresbachian) and disappeared together with the last abundant dendrolites. Commonly this group of coeloscleritophorans is ascribed to sponge megasclers (King & Chafetz 1983).

In turn, reduction in spatial heterogenity led to the shrinking of morphological diversity of calcified microbes and, as a result, to the loss of habitats which is directly indicated by the congruence of calcimicrobial generic diversity curve and plot of dendrolite abundance (Fig. 4). The latter is compiled from reports of formations containing dendrolites per stage. No dendrolites are known from the Franconian Stage and only a few dendrolites have been described from the Trempealeauan Stage (Wilberns Fm, Ahr 1971; Cow Head Gr, James 1981; James & Stevens 1986). However, those of Trempealeauan age rather represented a recovery interval of the reefal ecosystem.

Late Cambrian reefs

As a sequence of the dendrolite decline, stromatolite (Fig. 5) and thrombolite (Kennard & James 1986, fig. 8) resurgence is observed. Huge Late Cambrian stromatolite build-ups occupied sites extending from the supratidal zone to infra- and subtidal ones and ranged from low-energy lagoons to high-energy tidal channels on Laurentia (Goldring 1938; Stubblefield 1960; Oder & Bumgarner 1961; Aitken 1967, 1989; Ahr 1971; Lochman-Balk 1971; Palmer 1971a, b; Chafetz 1973; Reinhardt & Hardie 1976; Cook & Taylor 1977; Runnegar *et al.* 1979; Markello & Read 1981), South America (Keller *et al.* 1989), Siberian Platform (Astashkin *et al.* 1984, 1991), Australia (Shergold *et al.* 1985) and this pattern does not fit closely with the predicted progressive stromatolite decline (Awramik 1971; Monty 1973) but rather confirms the interpretation of stromatolites as post-mass extinction disaster forms (Schubert & Bottjer 1992).

However, the availability of space was probably the only reason for the new spread of stromatolites. The increase in the depth and extent of bioturbation (Bottjer & Ausich 1986; Crimes & Droser 1992), and increase of tiering in infaunal suspension-feeding communities (Ausich & Bottjer 1982), together all indicate a new wave of grazers just at the same time. The first probable chitons also appeared during the Late Cambrian (Franconian–Trempealeauan) (Bergenhayan 1960; Stinchcomb Darrough

1995). Some of these animals have been attributed to stromatolite grazers by ecological observations (Runnegar *et al.* 1979; Stinchcomb & Darrough 1995).

In any case, any attempt to explain stromatolite decline/proliferation by the development of favourable/unfavourable conditions for their grazers fails because even the most unfavourable conditions, such as restricted lagoons, sabkhas and thermal springs do not exclude grazing pressure (Javor & Castenholz 1984; Wickstrom & Castenhol 1985). Moreover, such conditions first eliminate predators that consume grazers; and species adapted to such conditions can dramatically inhibit the development of stromatolites (Gerdes & Krumbein 1984). In normal marine conditions there are many more opportunities to escape grazing. Various ecophysiological adaptations, such as O_2 and H_2S co-existence in alternating laminae of oxygenic and anoxygenic photosynthetic bacteria and cyanobacteria, production of extracellular slimes and nitrogen fixation, are more significant (Gerdes & Krumbein 1984).

Rebound

The stromatolite–thrombolite consortium continued to dominate during the Early and early Middle Ordovician in Laurentia (Pitcher 1964; Young 1970; Toomey 1981; Pratt & James 1982; Cecile 1989; Pratt 1989a, b) and on the Siberian Platform (Myagkova 1973; Luchinina 1988), the only regions from which build-ups of this age are known. Earliest reef-building metazoans of the Ordovician ecosystem appeared in stromatolitic build-ups during the latest Middle and Late Cambrian (Fig. 3e) (Cloud & Barnes 1948; Wilson 1950; Hamdi *et al.* 1995) and Early Ordovician (Toomey & Nitecki 1979; Cecile 1989). Probably these biotopes were more easily colonized by metazoans than the remaining dendrolites. In this case, stromatolites were not only disaster forms but pioneer forms at the same time. The intrusion of metazoans again created spatial heterogenity convenient for calcified microbes, especially cavities (Pratt & James 1982, 1989a). A new surge of calcimicrobe diversity started (Fig. 1) which is known as the Ordovician Flora of Chuvashov & Riding (1984).

Pioneer reef-building skeletal colonizers of the latest Late Cambrian–Early Ordovician (Tremadocian–Arenigian) stromatolites–thrombolites include calcified microbes, demosponges, tabulate corals, stromatoporoid-like problematics *Pulchrilamina* and bryozoans (Rigby 1966,

REEF ECOSYSTEM RECOVERY IN THE M. CAMBRIAN

1971; Riding & Toomey 1972; Myagkova 1973; Sokolov & Tesakov 1975; Toomey & Nitecki 1979; Toomey 1981; Webby 1984a; Cecile 1989; Pratt 1989a, b; Pratt & James 1989a, b). Among these, demosponges (mainly Anthaspidellidae), calathids and *Pulchrilamina* were commonly the principal framework builders (Finks & Toomey 1969; Myagkova 1973; Toomey & LeMone 1977).

Ecologically, morphologically and probably taxonomically this new reefal biota resembled that of the Early Cambrian (Wood *et al.* 1993, table 1). Radiocyaths and calathids, which could be related filter-feeders (Zhuravlev 1986; Church 1991; Rozanov & Zhuravlev 1992; Wood *et al.* 1993), represented large solitary and branching bafflers in the Early Cambrian and Early Ordovician reefal communities, respectively. Archaeocyaths which might be close to demosponges (Debrenne & Zhuravlev 1994) are comparable with anthaspidellid demosponges, as both were small solitary and low modular bafflers. The Early Cambrian coralomorphs *Flindersipora* and *Khasaktia* were repeated in encrusting Early Ordovician tabulates and *Pulchrilamina* (Webby 1986; Lafuste *et al.* 1991). The Early Cambrian and Early Ordovician calcified microbes were so similar that they are described under the same generic names (*Girvanella*, *Gordonophyton*, *Renalcis*, etc.). However, with exception of *Girvanella*, all of them, as well as the radiocyath–calathid group and coralomorphs, had a discrete stratigraphic record with a hiatus in several stages (Fig. 1). The relative ease of biomineralization in microbes and lower metazoans suggests that the rebound of the reefal community may be attributed to Elvis-taxa *sensu* Erwin & Droser (1993) rather than to Lazarus-taxa. Sphinctozoans known from the Toyonian (Pickett & Jell 1983) and the Middle Ordovician (Rigby & Potter 1986), can be added to this list but their taxonomic unity, as well as the affinity of the Cambrian genera to sponges, are doubtful (Reitner & Engeser 1985; Wood 1990). The only relatively new group were the anthaspidellids, in spite of the fossil record of demosponges from the late Atdabanian at least (Gruber & Reitner 1991; Rozanov & Zhuravlev 1992). They did not come onto shallower carbonate platform areas from deeper slope environments until after the demise of archaeocyaths (Webby 1984b), but colonized the same biotope after calcimicrobial demise.

Calcified algae (e.g. *Cyclocrinites*, *Dimorphosiphon*, *Vermiporella*, *Rhabdoporella*) and new groups of calcified microbes (e.g. *Hedstroemia*, *Rothpletzella*), diverse corals and chaetetid and stromatoporoid sponges were subsequently added during the Middle Ordovician and diversified by the end of Ordovician (Sokolov & Tesakov 1975; Webby 1984a, b, 1994; Pratt 1989c; Desrochers & James 1989; James & Cuffey 1989; James & Klappa 1989). Stromatolites and thrombolites were gone again by about the Middle Ordovician but probably not because of grazing pressure (Garrett 1970) but rather because of simply displacement through competition for space by calcified cyanobacteria–algae–animals, as has been noted for thrombolites by Kennard & James (1986). It was the mainly modular nature of most metazoans that appeared during the Middle Ordovician that allowed them actively to coexist with calcified microbes and sufficiently reorganize the reefal ecosystem.

D. Erwin and M. Hart are thanked for the kind invitation to present this paper at the IGGP Project 335 Meeting in Plymouth; R. Riding, R. Wood and the anonymous referee for helpful discussion; and P. Taylor for data on Ordovician bryozoans. This research was supported by a Royal Society Postdoctoral Fellowship during the author's visit to the Department of Earth Sciences, University of Wales, Cardiff. This paper is a contribution to IGCP Projects 335 on Biotic Recovery from Mass Extinctions and 366 on Ecological Aspects of the Cambrian Radiation.

References

AHR, W. M. 1971. Paleoenvironment, algal structures and fossil algae in the Upper Cambrian of Central Texas. *Journal of Sedimentary Petrology*, **41**, 205–216.

AITKEN, J. D. 1967. Classification and environmental significance of cryptalgal limestones and dolomites with illustrations from the Cambrian and Ordovician of southwestern Alberta. *Journal of Sedimentary Petrology*, **37**, 1163–1178.

——— 1989. Cambrian reefs and mounds. *In:* GELDSETZER, H. H. J., JAMES, N. P. & TEBBUTT, E. (eds) *Reefs, Canada and Adjacent Area.* Canadian Society of Petroleum Geologists Memoir, **13**, 135–138.

ANDERSON, R. F., LYONS, T. W. & COWIE, G. L. 1994. Sedimentary record of a shoaling of the oxic/anoxic interface in the Black Sea, *Marine Geology*, **116**, 373–384.

ASTASHKIN, V. A., PEGEL, T. V., REPINA, L. N., ROZANOV, A. YU., SHABANOV, YU. YA., ZHURAVLEV, A. YU., SUKHOV, S. S. & SUNDUKOV, V. M. 1991. *The Cambrian System on the Siberian Platform, Correlation Chart and Explanatory Notes.* International Union of Geological Sciences Publication, **27**.

———, VARLAMOV, A. I., GUBINA, N. K., EKHANIN, A. E., PERELADOV, V. S., ROMENKO, V. I.,

SUKHOV, S. S., UMPEROVICH, N. V., FEDOROV, A. B., FEDYANIN, A. P., SHISKIN, B. B. & KHOBNYA, E. I. 1984. [*Geology and Prospects of Oil-gas-bearing of the Cambrian Reef Systems of the Siberian Platform.*] Nedra, Moscow [in Russian].

AUBERT, M. 1988. Théorie générale de l'eutrophication. *Unesco Reports in Marine Sciences*, **49**, 91–94.

AUSICH, W. I. & BOTTJER, D. J. 1982. Tiering in suspension-feeding communities on soft substrata throughout the Phanerozoic. *Science*, **216**, 173–174.

AWRAMIK, S. M, 1971. Precambrian columnar stromatolite diversity: Reflection of metazoan appearance. *Science*, **174**, 825–827.

—— 1991. Archaean and Proterozoic stromatolites. *In:* RIDING, R. (ed.) *Calcareous Algae and Stromatolites*. Springer-Verlag, Berlin, 289–304.

BAKHTUROV, S. F., EVTUSHENKO, V. M. & PERELADOV, V. S. 1988. [*Kuonamka Bituminous Formation.*] Trudy Instituta Geologii i Geofiziki Sibirskogo Otdeleniya Akademii Nauk SSSR, 671 [in Russian].

BENGTSON, S., CONWAY MORRIS, S., COOPER, B. J., JELL, P. A. & RUNNEGAR, B. N. 1990. *Early Cambrian Fossils from South Australia*. Association of Australasian Palaeontologists, Memoir, **9**.

BERGENHAYAN, J. B. M. 1960. Cambrian and Ordovician loricates from North America. *Journal of Paleontology*, **34**, 168–178.

BERGSTRÖM, J. & AHLBERG, P. 1981. Uppermost Lower Cambrian biostratigraphy in Scania, Sweden. *Geologiska Föreningens i Stockholm Förhandlingar*, **103**, 193–214.

BOROWITZKA, M. A. 1981. Algae and grazing in coral reef ecosystems. *Endeavour, New Series*, **5**, 99–106.

BOTTJER, D. J. & AUSICH, W. I. 1986. Phanerozoic development of tiering in soft substrata suspension-feeding communities. *Paleobiology*, **12**, 400–420.

BRAGINSKIY, L. P., BEREZA, V. D., VELICHKO, I. M., GRIN, V. G., GUSYNSKAYA, S. L., DENISOVA, A. I., LITVINOVA, M. A. & SYSUEVA-ANTIPCHUK, A. F. 1968. [Water blooming spots, surged masses casting ashore of blue-green algae and biological processes occurring in them.] *In:* TOPACHEVSKIY, A.V. (ed.) [Water Blooming.] Naukova Dumka Kiev, 92–149 [in Russian].

BRANGULIS, A., MURNIEKS, A., NATLE, A. & FRIDRIHSONE, A. 1986. [Middle Baltic facies profile of the Vendian and Cambrian.] *In:* PIRRUS, E. A. (ed.) [*Facies and Stratigraphy of the Vendian and Cambrian of the western East-European Platform.*] Akademiya nauk Estonskoy SSR, Tallin, 24–33, [in Russian].

BRASIER, M. D., ROZANOV, A.Yu., ZHURAVLEV, A.Yu., CORFIELD, R. M. & DERRY, L. A. 1994. A carbon isotope reference scale for the Lower Cambrian series in Siberia: Report of IGCP Project 303. *Geological Magazine*, **131**, 767–783.

BRODIE, I. & KEMP, A. E. S. 1994. Variations in biogenic and detrital fluxes and formation of laminae in late Quarternary sediments from the Peruvian coastal upwelling zone. *Marine Geology*, **116**, 385–398.

BRUGGEMANN, J. S. 1995. *Parrotfish Grazing on Coral Reefs: A Trophic Novelty*. Rijksuniversitet Groningen, Utrecht.

BURRETT, C. F. & RICHARDSON, R. G. 1978. Cambrian trilobite diversity related to cratonic flooding. *Nature*, **272**, 717–719.

BUTTERFIELD, N. J. 1994. Burgess Shale-type fossils from a Lower Cambrian shallow-shelf sequence in northwestern Canada. *Nature*, **369**, 477–479.

BYERS, C. W. 1977. Biofacies patterns in euxinic basins: A general model. *In:* COOK, H. E. & ENAS, P. (eds) *Deep-Water Carbonate Environments*. Society of Economic Paleontologists and Mineralogists, Special Publication, **25**, 5–17.

CALVERT, S. E. & PEDERSEN, T. F. 1993 Geochemistry of Recent oxic and anoxic marine sediments: Implications for the geological record. *Marine Geology*, **113**, 67–88.

CARMICHAEL, W. W. 1994. The toxins of cyanobacteria. *Scientific American*, January, 78–86.

CECILE, M. P. 1989. Ordovician reefs and organic build-ups. *In:* GELDSETZER, H. H. J., JAMES, N. P. & TEBBUTT, E. (eds) *Reefs, Canada and Adjacent Area*. Canadian Society of Petroleum Geologists, Memoir, **13**, 171–176.

CHAFETZ, H. S. 1973. Morphological evolution of Cambrian algal mounds in response to a change in depositional environment. *Journal of Sedimentary Petrology*, **43**, 435–446.

CHEN JUN-YUAN & LINDSTRÖM, M. 1991. A Lower Cambrian soft-bodied fauna from Chegjiang, Yunnan, China. *Geologiska Foreningens i Stockholm Förhandlingar*, **113**, 79–81.

CHURCH, S. B. 1991. A new lower Ordovician species of Calathium and skeletal structure of western Utah calathids. *Journal of Paleontology*, **65**, 602–610.

CHUVASHOV, B. I. & RIDING, R. Principal floras of Palaeozoic marine calcareous algae. *Palaeontology*, **27**, 487–500.

CLARKE, J. D. A. 1990. Slope facies deposition and diagenesis of the Early Cambrian Parara Limestone, Wilkawillina Gorge, South Australia. *In:* JAGO, J. B. & MOORE, P. S. (eds) *The Evolution of a Late Precambrian–Early Palaeozoic Rift Complex: The Adelaide Geosyncline*. Geological Society of Australia, Special Publication, **16**, 230–246.

CLAYPOOL, G. E., HOLSER, W. T., KAPLAN, I. R., SAKAI, H. & ZAK, I. 1980. The age curves of sulfur and oxygen isotopes in marine sulfate and their mutual interpretation. *Chemical Geology*, **28**, 199–260.

CLOUD, P. E. JR. & BARNES, V. E. 1948. The Ellenburger Group of central Texas. *Texas University Bureau of Economic Geology, Geological Circular*, **4621**, 1–473.

COOK, H. E. & TAYLOR, M. E. 1977. Comparison of continental slope and shelf environments in the Upper Cambrian and lowest Ordovician of Nevada. *In:* COOK, H. E. & ENAS, P. (eds) *Deep-Water Carbonate Environments*. Society of Economic Paleontologists and Mineralogists, Special Publication, **25**, 51–81.

COPPER, P. 1974. Structure and development of early Palaeozoic reefs. *Proceedings of the Second International Coral Reef Symposium*, **1**, 365–386, Great Barrier Reef Committee, Brisbane.

—— 1994. Ancient reef ecosystem expansion and collapse. *Coral Reefs*, **13**, 3–11.

COVENEY, R. M. & NANSHENG, C. 1991. Ni-Mo-PGE-Au-rich ores in Chinese black shales and speculations on possible analogues in the United States. *Mineralium Deposita*, **26**, 83–88.

CRIMES, T. P. & DROSER, M. L. 1992. Trace fossils and bioturbation: The other fossil record. *Annual Review of Ecology and Systematics*, **23**, 339–360.

CRUZADO, A. 1988. Eutrophication in the pelagic environment and its assessment. *Unesco Reports in Marine Sciences*, **49**, 57–66.

DART, J. K. G, 1972. Echinoids, algal lawn and coral recolonization. *Nature*, **239**, 50–51.

DEBRENNE, F. & GRAVESTOCK, D. I. 1990. Archaeocyatha from the Sellick Hill Formation and Fork Tree Limestone of Fleurie Peninsula, South Australia. *In:* JAGO, J. B. & MOORE, P. S. (eds) *The Evolution of a Late Precambrian–Early Palaeozoic Rift Complex: The Adelaide Geosyncline*. Geological Society of Australia, Special Publication, **16**, 290–309.

—— & ZHURAVLEV, A. 1992. *Irregular Archaeocyaths*. Cahiers de Paléontologie, CNRS Editions, Paris.

—— & —— 1994. Archaeocyathan affinity: How deep can we go into the systematic affiliation of an extinct group? *In:* VAN SOEST, R. W. M., VAN KEMPEN, T. M. G. & BRAEKMAN, J. C. *Sponges in Time and Space: Biology, Chemistry, Paleontology*. Balkema, Rotterdam, 3–12.

—— GANDIN, A. & GANGLOFF, R. A. 1990. Analyse sedimentologique et paléontologie de calcaires organogènes du Cambrian inférieur de Battle Mountain (Nevada, U.S.A.). *Annales de Paléontologie*, **76**, 73–119.

——, —— & PILLOLA, G. L. 1989a. Boistratigraphy and depositional setting of Punta Manna Member type section (Nebida Formation, Lower Cambrian, S.W. Sardinia, Italy). *Rivista Italiana della Paleontologia i Stratigrafia*, **94**, 1–22.

——, —— & ROWLAND, S. M. 1989b. Lower Cambrian bioconstructions in northwestern Mexico (Sonora). Depositional setting, paleoecology and systematics of archaeocyaths. *Géobios*, **22**, 137–195.

——, ROZANOV, A. Yu. & WEBERS, G. F. 1984. Upper Cambrian Archaeocyatha from Antarctica. *Geological Magazine*, **121**, 291–299.

DESROCHERS, A. & JAMES, N. P. 1989. Middle Ordovician (Chazyan) bioherms and biostromes of the Mingan Islands, Quebec. *In:* GELDSETZER, H. H. J., JAMES, N. P. & TEBBUTT, E. (eds) *Reefs, Canada and Adjacent Area. Canadian Society of Petroleum Geologists*, Memoir, **13**, 183–191.

EDHORN, A-S. 1977. Early Cambrian algae croppers. *Canadian Journal of Earth Sciences*, **14**, 1014–1020.

—— 1979. Girvanella in the "Button Algae" Horizon of the Forteau Formation (Lower Cambrian), western Newfoundland. *Bulletin des Centres de Recherches Exploration–Production Elf-Aquitaine*, **3**, 557–567.

ERDTMANN, B. D., HUTTEL, P. & CHEN JUNYUAN. 1990. Depositional environment and taphonomy of the Lower Cambrian soft-bodied fauna at Chengjiang, Yunnan Province, China. *Third International Symposium on the Cambrian System, 1–9 August 1990, Novosibirsk, USSR, Abstracts*, 91.

ERWIN, D. H. & DROSER, M. L. 1993. Elvis taxa. *Palaios*, **8**, 623–624.

FAGERSTROM, J. A. 1987. *The Evolution of Reef Communities*. Wiley, New York.

FARMER, J. D. 1992. Grazing and bioturbation in modern microbial mats. *In:* SCHOPF, W. J. & KLEIN, C. (eds) *The Proterozoic Biosphere: A Multidisciplinary Study*. Cambridge University Press, 295–297.

FINKS, R. M. & TOOMEY, D. F. 1969. Trip F: The paleoecology of Chazyan (lower Middle Ordovician) "reefs" or "mounds". *New York State Geological Association, Guidebook to Field Excursion, 41st Annual Meeting*. State University College of Arts and Science, Plattsburgh, New York, 93–120.

FRITZ, M. A. & HOWELL, B. F. 1955. An Upper Cambrian coral from Montana. *Journal of Paleontology*, **29**, 181–183.

FRITZ, W. H., CECILE, M. P., NORFORD, B. S., MORROW, D. & GELDSETZER, H. H. J. 1991 Cambrian to Middle Devonian assemblages. *In:* GABRIELSE, H. & YORATH, C. J. (eds) *Geology of the Cordilleran Orogeny in Canada*. Geological Survey of Canada, Geology of Canada, **4**, 151–218.

GANDIN, A. & DEBRENNE, F. 1984. Lower Cambrian bioconstructions in Southwestern Sardinia (Italy). *Géobios, Memoire speciale*, **8**, 231–240.

GARRETT, P. 1970. Phanerozoic stromatolites: Noncompetitive ecologic restriction by grazing and burrowing animals. *Science*, **169**, 171–173.

GEBELEIN, C. D. 1974. Biologic control of stromatolite microstructure: Implications for Precambrian time stratigraphy. *American Journal of Sciences*, **274**, 575–598.

GERDES, G. & KRUMBEIN, W. E. 1984. Animal communities in Recent potencial stromatolites of hypersaline origin. *In:* COHEN, Y., CASTENHOLZ, R. W. & HALVORSON, D. (eds) *Microbial Mats: Stromatolites*. Alan R. Liss Inc., New York, 59–83.

GEYER, G. 1989. Late Precambrian to early Middle Cambrian lithostratigraphy of southern Morocco. *Beringeria*, **1**, 115–143.

GOLDRING, W. 1938. Algal barrier reefs in the lower Ozarkian of New York with a chapter on the importance of coralline algae as reef builders through the ages. *New York State Museum Bulletin*, **315**, 5–75.

GOLOSHEYKIN, A. B., ASTASHKIN, V. A. & EGOROVA, L. I. 1978. [Structure and age of the Tangha

Formation of the Cambrian Amgan Stage stratotype area.] *Trudy Sibirskogo Nauchno-Issledovatel'skogo Instituta Geologii, Geofiziki i Mineral'nogo Syr'ya*, **260**, 37–41 [in Russian].

GRAVESTOCK, D. I. & HIBBURT, J. E. 1991. Sequence stratigraphy of the eastern Officer and Arrowie Basins: A framework for Cambrian oil search. *The APEA Journal*, 177–190.

GRUBER, G. & REITNER, J. 1991. Isolierte Mikro- und Megaskleren ton Poriferen aus dem Untercampan von Hover (Norddeutschland) und Bemerkungen zur Phylogenie der Geodiidae (Demospongiae). *Berliner Geowissenschaft (A)*, **134**, 107–117.

GUENSBURG, T. E. & SPRINKLE, J. 1992. Rise of echinoderms in the Paleozoic evolutionary fauna: Significance of palaeoenvironmental controls. *Geology*, **20**, 407–410.

GUILLARD, R. R. L. & HELLEBUST, J. A. 1971. Growth and the production of extracellular substances by two strains of *Phaeocystis pouchetii*. *Journal of Phycology*, **7**, 330–338.

HAMDI, B., ROZANOV, A. YU. & ZHURAVLEV, A. YU. 1995. Latest Middle Cambrian metazoan reef from northern Iran. *Geological Magazine*, **132**, 367–373.

HOLLANDER, D. J., MCKENZIE, J. A. & LO TEN HAVEN, H. 1992. A 200 year sedimentary record of progressive eutrophication in Lake Greifen (Switzeland): Implications for the origin of organic-carbon-rich sediments. *Geology*, **20**, 825–828.

HOLSER, W. T. & MAGARITZ, M. 1992. Cretaceous/Tertiary and Permian/Triassic boundary events compared. *Geochimica et Cosmochimica Acta*, **56**, 3297–3309.

ILAN, M. & LOYA, Y. 1988. Reproduction and settlement of the coral reef sponge *Niphates* sp. (Red Sea). *Proceedings of the 6th International Coral Reef Symposium, Australia, 1988*, **2**, 745–749.

JAMES, N. P. 1981. Megablocks of calcified algae in the Cow Head Breccia, western Newfoundland: Vestiges of a Cambro-Ordovician platform margin. *Geological Society of America Bulletin*, Pt 1, **92**, 799–811.

—— & CUFFEY, R. J. 1989. Middle Ordovician coral reefs, western Newfoundland. *In:* GELDSETZER, H. H. J., JAMES, N. P. & TEBBUTT, E. (eds) *Reefs, Canada and Adjacent Area*. Canadian Society of Petroleum Geologists, Memoir, **13**, 192–195.

—— & DEBRENNE, F. 1980. Lower Cambrian bioherms: Pioneer reefs of the Phanerozoic. *Acta Palaeontologica Polonica*, **25**, 655–668.

—— & GRAVESTOCK, D. I. 1990. Lower Cambrian shelf and shelf-margin build-ups, Flinders Ranges, South Australia. *Sedimentology*, **37**, 455–480.

—— & KLAPPA, C. F. 1989. Lithistid sponge bioherms, early Middle Ordovician, western Newfoundland. *In:* GELDSETZER, H. H., JAMES, N. P. & TEBBUTT, E. (eds) *Reefs, Canada and Adjacent Area*. Canadian Society of Petroleum Geologists, Memoir, **13**, 196–200.

—— & STEVENS, R. K. 1986. Stratigraphy and correlation of the Cambro-Ordovician Cow Head Group, western Newfoundland. *Geological Society of Canada Bulletin*, **366**, 1–143.

JAVOR, B. J. & CASTENHOLZ, R. W. 1984. Invertebrate grazers of microbial mats, Laguna Guerrero Negro, Mexico. *In:* COHEN, Y., CASTENHOLTZ, R. W. & HALVORSON, D. (eds) *Microbial Mats: Stromatolites*. Liss, New York, 59–83.

KARL, D. M., HOLM-HANSEN, O., TAYLOR, G. T., TIEN, G. & BIRD, D. F. 1991. Microbial mass and productivity in the western Bransfield Strait, Antarctica during the 1986–87 austral summer. *Deep-Sea Research*, **38**, 1029–1055.

KELLER, M., BUGGISCH, W. & BERCOWSKI, F. 1989. Facies and sedimentology of Upper Cambrian shallowing-upward cycles in the La Flecha Formation (Argentine Precordillera). *Zentralblatt für Geologie und Paläontologie*, Teil I, 5/6, 999–1011.

KENNARD, J. M. 1991. Lower Cambrian archaeocyath build-ups, Todd River Dolomite, northeast Amadeus Basin, central Australia: Sedimentology and diagenesis. *In:* KORSCH, R. J. & KENNARD, J. M. (eds) *Geological and Geophysical Studies in the Amadeus Basin, Central Australia*. Bulletin of the Bureau of Mineral Resources, **236**, 195–225.

—— & JAMES, N. P. 1986. Thrombolites and stromatolites: Two distinct types of microbial structures. *Palaios*, **1**, 492–503.

KIELMAN, M. 1993. Reef communities along the NE coast of Colombia and Curaçao related to abiotic circumstances. *4th International Porifera Congress, Book of Abstracts*. Institute of Taxonomic Zoology, University of Amsterdam, Amsterdam.

KING, D. T. & CHAFETZ, H. S. 1983. Tidal-flat to shallow-shelf deposits in the Cap Mountain Limestone Member of the Riley Formation, Upper Cambrian of central Texas. *Journal of Sedimentary Geology*, **53**, 261–273.

KORDE, K. B. 1961. [Cambrian algae of the south-east of the Siberian Platform.] *Trudy Paleontologicheskogo Instituta Akademii Nauk SSSR*, **89**, 1–148 [in Russian].

KRUSE, P. D. 1991. Cyanobacterial–archaeocyathan–radiocyathan bioherms in the Wirrealpa Limestone of South Australia. *Canadian Journal of Earth Sciences*, **28**, 601–605.

——, ZHURAVLEV, A. YU. & JAMES, N. P. 1995. Primordial metazoan-calcimicrobial reefs: Tommotian (Early Cambrian) of the Siberian Platform. *Palaios*, **10**, 291–321.

LAFUSTE, J. G., DEBRENNE, F., GANDIN, A. & GRAVESTOCK, D. I. 1991. The oldest tabulate coral and the associated Archaeocyatha, Lower Cambrian, Flinders Ranges, South Australia. *Géobios*, **24**, 697–718.

LANDING, E. 1993. In situ earliest Cambrian tube worms and the oldest metazoan-constructed biostrome (Placentian Series, southeastern Newfoundland). *Journal of Paleontology*, **67**, 333–342.

LEWIS, S. M. 1986. The role of herbivorous fishes in the organization of a Caribbean reef community. *Ecological Monographs*, **56**, 183–200.

LIÑAN, E. & GÁMEZ-VINTANED, A. 1993. Lower

Cambrian palaeogeography of the Iberian Peninsula and its relations with some neighbouring European areas. *Bulletin de la Société géologique de France*, **164**, 831–842.

LOCHMAN-BALK, C. 1971. The Cambrian of the craton of the United States. *In:* HOLLAND, C. H. (ed.) *Cambrian of the New World.* Wiley, London, 79–167.

LOHMANN, K. C. 1976. Lower Dresbachian (Upper Cambrian) platform to deep shelf transition in eastern Nevada and western Utah: An evaluation through lithologic cycle correlation. *Brigham Young University Studies*, **23**, 111–132.

LUCHININA, V. A. 1988. [Calcareous algae in stromatolite build-ups of the lower Palaeozoic of the Siberian Platform.] *In:* DUBATOLOV, V. N. & MOSKALENKO, T. A. (eds) [*Calcareous Algae and Stromatolites (Sistematics, Biostratigraphy, Facies Analysis).*] Nauka, Novosibirsk, 139–145 [in Russian].

—— & STEPANOVA, M. V. 1983. [Algae of the boundary strata of the Lower and Middle Cambrian in Siberia.] *Trudy Instituta geologii i geofiziki Sibirskogo otdeleniya Akademii nauk SSSR*, **548**, 118–120 [in Russian].

MCILREATH, I. 1977. Accumulation of a Middle Cambrian deep water limestone debris apron adjacent to a vertical submarine carbonate escarpment, southern Rocky Mountains, Canada. *In:* COOK, H. E. & ENAS, P. (eds) *Deep-Water Carbonate Environments.* Society of Economic Paleontologists and Mineralogists, Special Publication, **25**, 113–124.

MANSY, J. L., DEBRENNE, F. & ZHURAVLEV, A. YU. 1993. Calcaires à archéocyathes du Cambrien inferieur du Nord de la Colombie brItannique (Canada). Implications paléogéographiques et précisions sur l'extension du continent Américano-Koryakien. *Géobios*, **26**, 643–683.

MARKELLO, J. R. & READ, J. F. 1981. Carbonate ramp-to-deeper shale transition of an Upper Cambrian intrashelf basin, Nolichucky Formation, Southwest Virginia Appalachians. *Sedimentology*, **28**, 573–597.

MORENO-EIRIS, E. 1994. Lower Cambrian reef mounds of Sierra Morena (SW Spain). *Courier Forschungsinstitut Senkenberg*, **172**, 185–192.

MONTY, C. L. V. 1973. Precambrian background and Phanerozoic history of stromatolitic communities, an overview. *Annales de la Société géologique de Belgique*, **96**, 585–624.

MYAGKOVA, E. I. 1973. [On the ecology of Early Ordovician soanitids.] *In:* BETEKHTINA, O. A. & ZHURAVLEVA, I. T. (eds) [*Environment and Life in the Geological Past (Late Precambrian and Palaeozoic of Siberia).*] Trudy Instituta Geologii i Geofiziki Sibirskogo Otdeleniya Akademii Nauk SSSR, **169**, 65–69 [in Russian].

NARBONNE, G. M. & ARBUCKLE, S. M. 1989. Lower Cambrian algal-archaeocyathan reef mounds from the Wernecke Mountains, Yukon Territory, *In:* GELDSETZER, H. H., JAMES, N. P. & TEBBUTT, E. (eds) *Reefs, Canada and Adjacent Area.* Canadian Society of Petroleum Geologists, Mem-

oir, **13**, 156–160.

NEWELL, N. D. 1972. The evolution of reefs. *Scientific American*, **226**, 54–65.

NITECKI, M. H. & DEBRENNE, F. 1979. The nature of radiocyathids and their relationship to receptaculitids and archaeocyathids. *Géobios*, **12**, 5–27.

ODER, C. R. L. & BUMGARNER, J. G. 1961. Stromatolitic bioherms in the Maynardville (Upper Cambrian) Limestone, Tennessee. *Geological Society of America Bulletin*, **72**, 1021–1028.

PALMER, A. R. 1971a. The Cambrian of the Great Basin and adjacent areas, western United States. *In:* HOLLAND, C. H. (ed.) *Cambrian of the New World.* John Wiley & Sons, London, 1–78.

—— 1971b. The Cambrian of the Appalachian and eastern New England regions, eastern United States. *In:* HOLLAND, C. H. (ed.) *Cambrian of the New World.* John Wiley & Sons, London, 169–217.

—— & JAMES, N. P. 1979. The Hawke Bay event: A circum-Iapetus regression near the Lower Middle Cambrian boundary. *The Caledonids in the U.S.A., I.G.C.P.* 1979, 15–18, Blacksburg, Virginia.

PEDERSEN, T. F. & CALVERT, S. E, 1990. Anoxia vs. productivity: What control the formation of organic-carbon-rich sediments and sedimentary rocks? *AAPG Bulletin*, **74**, 454–466.

PHILLIPS, O. L., HALL, P., GENTRY, A. H., SAWYER, S. A. & VÁSQUEZ, R. 1994. Dynamics and species richness of tropical rain forest. *Proceedings National Academy of USA*, **91**, 2805–2809.

PICKETT, J. C. & JELL, P. A. 1983. Middle Cambrian Sphinctozoa (Porifera) from New South Wales. *Memoir of the Association of Australasian Palaeontologists*, **1**, 85–92.

PITCHER, M. 1964. Evolution of Chazyan (Ordovician) reefs of eastern United States and Canada. *Canadian Petroleum Geology Bulletin*, **12**, 632–691.

PRATT, B. R. 1982. Stromatolite decline – a reconsideration. *Geology*, **10**, 512–515.

—— 1989a. Continental margin reef tract of Early Ordovician age, Broken Skull Formation, Mackenzie Mountains, northwestern Canada. *In:* GELDSETZER, H. H., JAMES, N. P. & TEBBUTT, E. (eds) *Reefs, Canada and Adjacent Area.* Canadian Society of Petroleum Geologists Memoir, **13**, 208–212.

—— 1989b. Early Ordovician cryptalgal-sponge reefs, survey Peak Formation, Rocky Mountains, Alberta. *In:* GELDSETZER, H. H., JAMES, N. P. & TEBBUTT, E. (eds) *Reefs, Canada and Adjacent Area.* Canadian Society of Petroleum Geologists, Memoir, **13**, 213–217.

—— 1989c. Small early Middle Ordovician patch reefs, Laval Formation (Chazy Group), Caughnawaga, Montreal area, Quebec. *In:* GELDSETZER, H. H., JAMES, N. P. & TEBBUTT, E. (eds) *Reefs, Canada and Adjacent Area.* Canadian Society of Petroleum Geologists Memoir, **13**, 218–223.

—— 1990. Lower Cambrian reefs of the Mural Formation, southern Canadian Rocky Mountains, *13th International Sedimentological Con-*

gress, *Nottingham, U.K., August 1990, Abstracts of Papers*, 436.

—— & JAMES, N. P. 1982. Cryptalgal–metazoan bioherms of early Ordovician age in the St. George Group, western Newfoundland. *Sedimentology*, **29**, 543–569.

—— & —— 1989a. Coral–*Renalcis*–thrombolite reef complex of early Ordovician age, St. George Group, western Newfoundland. *In:* GELDSETZER, H. H., JAMES, N. P. & TEBBUTT, E. (eds) *Reefs, Canada and Adjacent Area*. Canadian Society of Petroleum Geologists, Memoir, **13**, 224–230.

—— & —— 1989b. Early Ordovician thrombolite reefs St. George Group, western Newfoundland. *In:* GELDSETZER, H. H., JAMES, N. P. & TEBBUTT, E. (eds) *Reefs, Canada and Adjacent Area*. Canadian Society of Petroleum Geologists, Memoir, **13**, 231–240.

READ, J. F. & PFEIL, R. W. 1983. Fabrics of allochthonous reefal blocks, shady Dolomite (Lower to Middle Cambrian), Virginia Appalachians. *Journal of Sedimentary Petrology*, **53**, 761–778.

REES, M., PRATT, B. R. & ROWELL, A. J. 1989. Early Cambrian reefs, reef complexes, and associated lithofacies of the Shackleton Limestone, transantarctic Mountains. *Sedimentology*, **36**, 341–361.

REINHARDT, J. & HARDIE, L. A. 1976. Selected examples of carbonate sedimentation, Lower Paleozoic of Maryland. *Maryland Geological Survey Guidebook*, **5**, 1–53.

REITNER, J. 1993. Modern cryptic microbialite/metazoan facies from Lizard Island (Great Barrier Reef, Australia). Formation and concept. *Facies*, **29**, 3–40.

—— & ENGESER, T. 1985. Revision der Demospongier mit einem thalamiden, aragonitishen Basalskelett und trabekularer Interstruktur ('Sphinctozoa' pars). *Berliner Geowissenschaft (A)*, **60**, 151–193.

RIDING, R. 1991a. Classification of microbial carbonates. *In:* RIDING, R. (ed.) *Calcareous Algae and Stromatolites*. Springer, New York, 21–51.

—— 1991b. Cambrian calcareous cyanobacteria and algae. *In:* RIDING, R. (ed.) *Calcareous Algae and Stromatolites*. Springer, New York, 305–334.

—— & TOOMEY, D. F. 1972. The sedimentological role of *Epiphyton* and *Renalcis* in Lower Ordovician mounds, southern Oklahoma. *Journal of Paleontology*, **46**, 509–519.

—— & ZHURAVLEV, A. YU. 1994. Cambrian reef builders: Calcimicrobes and archaeocyaths. *Terra abstracts, Abstract supplement No. 3 to Terra nova*, **6**, 7.

—— & —— 1995. Structure and diversity of oldest sponge-microbe reefs: Lower Cambrian, Aldan River, Siberia. *Geology*, **23**, 649–652.

RIGBY, J. K. 1966. Evolution of Lower and Middle Ordovician sponge reefs in western Utah. *Special Papers of the Geological Society of America*, **87**, 1–137.

—— 1971. Sponges and reef and related facies through time. *Proceedings of the North American Paleontological Convention (Chicago)*, Part J,

1374–1388.

—— & POTTER, A. W. 1986. Ordovician sphinctozoan sponges from the Klamath Mountains, northern California. *Memoires of the Paleontological Society*, **20**, 1–47.

ROZANOV, A. YU. & ZHURAVLEV, A. YU. 1992. The Lower Cambrian fossil record of the Soviet Union. *In:* LIPPS, J. H. & SIGNOR, P. W. *Origin and Early Evolution of the Metazoa*. Plenum, New York, 205–282.

RUNNEGAR, B., POJETA, J. JR, TAYLOR, M. E. & COLLINS, D. 1979. New species of the Cambrian and Ordovician chitons *Matthevia* and *Chelodes* from Wisconsin and Queensland: Evidence for the early history of polyplacophoran mollusks. *Journal of Paleontology*, **53**, 1374–1394.

SAMMARCO, P. W. 1987. A comparison of some ecological processes of the Caribbean and the Great Barrier Reef. *Unesco Reports in Marine Sciences*, **46**, 127–166.

SAVITSKIY, V. E. 1979. *[Geology of the Cambrian Reef Systems in Western Yakutia.] Trudy Sibirskogo Nauchno-Issledovatel'skogo Instituta Geologii, Geofiziki i Mineral'nogo Syr'ya*, 270 [in Russian].

SCHIDLOWSKI, M., GORZAWSKI, H. & DOR, I. 1992. Experimental hypersaline ponds as model environments for stromatolite formation 2. Isotopic biogeochemistry. *In:* SCHIDLOWSKI, M. (ed.) *Early Organic Evolution: Implications for Mineral and Energy Resources*. Springer, Berlin; Heidelberg.

SCHUBERT, J. K. & BOTTJER, D. J. 1992. Early Triassic stromatolites as post-mass extinction disaster forms. *Geology*, **20**, 883–886.

SCHWARZ, H.-U., EINSELE, G. & HERM, D. 1975. Quartz-sandy, grazing-countoured stromatolites from coastal embayment of Mauritania, West Africa. *Sedimentology*, **22**, 539–561.

SELG, M. 1986. Algen als Faziesindikatoren: Bioherme und Biostrome in Unter-Cambrium van SW-Sardinien. *Geologische Rundschau*, **75**, 693–702.

SEPKOSKI, J. J. JR. 1981. A factor analytic description of the Phanerozoic marine fossil record. *Paleobiology*, **7**, 36–53.

SHERGOLD, J., JAGO, J., COOPER, R. & LAURIE, J. 1985. *The Cambrian System in Australia, Antarctica and New Zealand. Correlation Charts and Explanatory Notes*. International Union of Geological Sciences Publication, 19.

SHLESINGER, Y. & LOYA, Y. 1985. Coral community reproductive patterns: Red Sea versus the Great Barrier Reef. *Science*, **228**, 1333–1335.

SHUNULA, J. P. & NDIBALEMA, V. 1986. Grazing preferences of *Diadema setosum* and *Heliocidaris erythrogramma* (echinoderms) on an assortment of marine algae. *Aquatic Botany*, **25**, 91–95.

SIGNOR, P. W. 1992. Taxonomic diversity and faunal turnover in the Early Cambrian: Did the most severe mass extinction of the Phanerozoic occur in the Botomian Stage? *North American Paleontological Convention, 5th, Abstracts with Programs*, 272.

SMITH, A. B. 1990. Evolutionary diversification of echinoderms during the early Palaeozoic. *In:* TAYLOR, P. D. & LARWOOD, G. P. (eds). *Major*

Evolutionary Radiations. Systematic Association, 42, Clarendon, Oxford, 265–286.

SOKOLOV, B. S. & TESAKOV, YU. I. 1975. *[Ordovician Stratigraphy of the Siberian Platform.]* Trudy Instituta Geologii i Geofiziki Sibirskogo Otdeleniya Akademii Nauk SSSR, **200**.

SOUTHAM, J. R., PETERSON, W. H. & BRASS, G. W. 1982. Dynamics of anoxia. *Palaeogeography, Palaeoclimatology, Palaeoecology*, **40**, 183–198.

STACHOWITSCH, M. & AVČIN, A. 1988. Eutrophication-induced modifications of benthic communities. *Unesco Reports in Marine Sciences*, **49**, 67–80.

STANLEY, G. D. JR. 1992. Tropical reef ecosystems and their evolution. *In:* NIERENBERG, W. A. (ed.) *Encyclopedia of Earth System Sciences.* Academic, San Diego, **4**, 375–388.

STANLEY, S. M, 1973. An ecological theory for the sudden origin of multicellular life in the Late Precambrian. *Proceedings of the National Academy of Sciences of the USA*, **70** (5), 1486–1489.

STEINER, M., MEHL, D., REITNER, J. & ERDTMANN, B. D. 1993. Oldest entirely preserved sponges and other fossils from the Lowermost Cambrian and a new facies reconstruction of the Yangtze platform (China). *Berliner Geowissenschaft (E)*, **9**, 293–329.

STENEK, R. S. 1986. The ecology of coralline algal crust: convergent pattern and adaptive strategies. *Annual Review of Ecology and Systematics*, **17**, 273–303.

STEPANOVA, M. V, 1979. [Middle Cambrian algae in organogeneous build-ups of the Sinsk-Botoma Facies Region on the Siberian Platform.] *In:* SAVITSKIY, V. E. (ed.) *[Geology of the Cambrian Reef System in Western Yakutia.]* Trudy Sibirskogo Nauchno-Issledovatel'skogo Instituta Geologii, Geofiziki i Mineral'nogo Syr'ya, **270**, 105–109 [in Russian].

—— 1986. [Dependence of the taxonomic composition of algal communities on facies environments on the example of the Lower Cambrian stratotype section of the Siberian Platform.] *In:* KRASNOV, V. D. (ed.) *[Palaeoecological and Lithological-Facies Analyses for the Grounds of Scrutiny of the Stratigraphic Charts.]* Sibirskiy Nauchno-Issledovatel'skiy Institut Geologii, Geofiziki i Mineral'nogo Syr'ya, Novosibirsk, 22–30, [in Russian].

STINCHCOMB, B. L. & DARROUGH, G. 1995. Some molluscan problematica from the Upper Cambrian–Lower Ordovician of the Ozark Uplift. *Journal of Paleontology*, **69**, 52–65.

STUBBLEFIELD, C. J. 1960. Sessile marine organisms and their significance in pre-Mesozoic strata. *Quarterly Journal of the Geological Society of London*, **116**, 219–238.

TOOMEY, D. F. 1981. Organic-build-up constructional capability in Lower Ordovician and late Paleozoic mounds. *In:* GRAY, J., BOUCOT, A. J. & BERRY, W. B. N. (eds) *Communities of the Past.* Hutchinson Ross Publishing Company, Stroudsburg, Pennsylvania, 35–68.

—— & LEMONE, D. 1977. Some Ordovician and Silurian algae from selected areas of the southwestern United States. *In:* FLUGEL, E. (ed.) *Fossil Algae.* Springer, Berlin; Heidelberg, 351–359.

—— & NITECKI, M. H. 1979. Organic build-ups in the Lower Ordovician (Canadian) of Texas and Oklahoma: *Fieldiana*, new ser., **2**, 1–181.

VALENTINE, J. W. 1973. *Evolutionary Paleoecology of the Marine Biosphere.* Prentice-Hall, New Jersey.

WALTER, M. R. & HEYS, G. R. 1985. Links between the rise of the Metazoa and the decline of stromatolites. *Precambrian Research*, **29**, 149–174.

WEBBY, B. D, 1984a. Ordovician reefs and climate: a review. *In:* BRUTON, D. L. (ed.) *Aspects of the Ordovician System.* Paleontological Contribution from the University of Oslo, **8**, 87–98.

—— 1984b. Early Phanerozoic distribution patterns of some major groups of sessile organisms. *Proceedings of the 27th International Geological Congress*, **2**, 193–208, VNU Science Press, Utrecht.

—— 1986. Early stromatoporoids. *In:* HOFFMAN, A. & NITECKI, M. H. (eds) *Problematic Fossil Taxa.* Oxford University Press, New York; Clarendon, Oxford, 148–166.

—— 1994. Evolutionary trends in Ordovician stromatoporoids. *Courier Forschungsinstitut Senkenberg*, **172**, 373–380.

WICKSTROM, C. E. & CASTENHOLZ, R. W. 1985. Dynamics of cyanobacteria–ostracod interactions in an Oregon hot spring. *Ecology*, **66**, 1024–1041.

WILSON, J. L. 1950. An Upper Cambrian pleospongiid (?). *Journal of Paleontology*, **24**, 591–593.

WOOD, R. 1990. Reef-building sponges. *American Scientist*, **78**, 224–235.

—— EVANS, K. R. & ZHURAVLEV, A. YU. 1992a. A new post-early Cambrian archaeocyath from Antarctica. *Geological Magazine*, **129**, 491–495.

—— ZHURAVLEV, A. YU. & DEBRENNE, F. 1992b. Functional biology and ecology of Archaeocyatha. *Palaios*, **7**, 131–156.

——, —— & CHIMED TSEREN, A. 1993. The ecology of Lower Cambrian buildups from Zuune Arts, Mongolia: implications for early metazoan reef evolution. *Sedimentology*, **40**, 829–858.

YOUNG, L. M. 1970. Early Ordovician sedimentary history of Marathon Geosyncline, Trans-Pecos, Texas. *AAPG Bulletin*, **54**, 2303–2316.

ZADOROZHNAYA, N. M, 1974. [Early Cambrian organogenous build-ups of the eastern part of the Altay Sayan Foldbelt]. *Trudy Instituta Geologii i Geofiziki Sibirskogo Otdeleniya Akademii Nauk SSSR*, **84**, 158–186 [in Russian].

ZAMARREÑO, I. 1977. Early Cambrian algal carbonates in southern Spain. *In:* FLUGEL, E. (ed.) *Fossil Algae.* Springer, Berlin; Heidelberg, 360–365.

ZHEGALLO, E. A., ZAMIRAYLOVA, A. G. & ZANIN, YU. N. 1994. [Microorganisms in the composition of the Kuonamka Formation rocks of the Lower–Middle Cambrian on the Siberian Platform (Molodo River).] *Litologiya i poleznye iskopaemye*, **5**, 123–127 [in Russian].

ZHURAVLEV, A. YU. 1986. Radiocyathids. *In:* HOFFMAN, A. & NITECKI, M. H. (eds) *Problematic Fossil Taxa.* Oxford University Press, New York; Clarendon, Oxford, 35–44.

—— 1995. Preliminary suggestions on the global

Early Cambrian zonation. *Beringeria*, Special Issue, **2**, 147–160.

—— & WOOD, R. A. 1994. Early Cambrian extinctions. *Terra abstracts, Abstract supplement No. 3 to Terra nova*, **6**, 9.

—— & —— 1995. Anoxia and eutrophication as the case of the mid-Early Cambrian (Botomian) extinction event. *Geology*, in press.

——, DEBRENNE, F. & LAFUSTE, J. 1993. Early Cambrian microstructural diversification of cnidaria. *Courier Forschungsinstitut Senkenberg*, **164**, 365–372.

ZHURAVLEVA, I. T. 1960. [Archaeocyaths of the Siberian Platform.] USSR Academy of Sciences Publishing House, Moscow [in Russian].

—— 1972 [Early Cambrian facies assemblages of archaeocyaths (Lena River, middle courses).] *In:* ZHURAVLEVA, I. T. (ed.) [*Problems of the Lower Cambrian Biostratigraphy and Palaeontology of Siberia.*] Nauka, 31–109 [in Russian].

Ostracode speciation following Middle Ordovician extinction events, north central United States

F. M. SWAIN

Department of Geology and Geophysics, University of Minnesota, Minneapolis, MN 55455, USA

Abstract: Studies by several workers show that major extinctions of trilobites, echinoderms, brachiopods and gastropods (but not conodonts) occurred at the horizon of a middle Ordovician K-bentonite (Deicke metabentonite) in the north central United States. Extinctions of Ostracoda are also represented at the Deicke horizon. The metabentonite, a Rocklandian stage ash fall the age of which is approximately 454 Ma, middle Caradocian, originated in the eastern United States.

About 160 species of Ostracoda have been recorded in the middle and upper Ordovician of the north central USA. Ostracode speciation was low prior to the Deicke extinction event, consisting of a few eoleperditiids, leperditellids, drepanellids, aparchitids and bairdiids of the Glenwood and Platteville formations (Blackriveran–early Rocklandian). Following the extinction event, during the next approximately 1.5 Ma, 85 ostracode species appeared in the uppermost Platteville Limestone and in the overlying Decorah Shale (late Rocklandian–early Shermanian). The known Ostracoda species durations averaged about 2 Ma, although some are only half that length of time.

Additional episodes of ash falls, following the Deicke and associated episodes, occurred in the next approximately 7.5 Ma in the region. The preceding Decorah ostracode fauna disappeared from the area and about 50 recorded new or immigrant taxa appeared following the last volcanic event in the late Edenian and Maysvillian stages (Dubuque and Maquoketa formations). Species duration during this time interval apparently averaged less than in the Decorah, about 1 Ma.

The volcanic episodes appear to have provided nutrient-rich habitats that resulted in evolutionary spurts in the Ostracoda, following at least two of the episodes. Duration of the species was perhaps related to the total nutrient budget provided by the ash falls as well as to salinity regimes, palaeoclimates and other factors.

Middle and upper Ordovician marine rocks in Minnesota, Iowa and Illinois consist of about 200 m of clastic and carbonate rocks ranging in age from about 458 Ma to about 438 Ma. They represent late Champlanian through Cincinnatian Series of the United States and middle Caradoc through Ashgill series of western Europe. Invertebrate phyla are abundantly represented in the strata but many species become abruptly extinct at certain levels in the succession. Altered volcanic ash beds (metabentonite) occur at several of the major extinction horizons and appear to be in part responsible for the extinctions (Sloan 1987).

Other, more gradual patterns of extinctions appear to be related to environmental changes such as shoaling, salinity changes, varying rates of terrestrial sedimentation and food supplies.

The middle and upper Ordovician rocks of the north central United States have yielded approximately 160 species of marine Ostracoda (Swain 1987). In this paper the stratigraphic distribution of the Ostracoda will be discussed in relationship to the volcanic ash horizons and the extinction patterns of the other invertebrate groups.

Metabentonite horizons

Fifteen metabentonite layers, typically 2 or 5 cm thick, originating from volcanic eruptions in the present eastern part of the North American Plate were deposited during the middle and late Ordovician in the north central United States (Fig. 1; Sloan 1987). Various aspects of the volcanic ash beds, their correlation, dating by fission track and argon 40/argon 39 methods, and rare element fingerprinting are discussed elsewhere (Catalani 1987; Sloan 1987).

The lower two (unnamed) ash layers lie in the lower Platteville/upper Glenwood formations and probably formed between 454 and 455 Ma. The Deicke is followed in the lower part of the Decorah Shale by the Millbrig K-bentonite at slightly less than 454 Ma. The Millbrig is succeeded by the Elkport and Dickeyville K-bentonite in the middle Decorah Shale at about 453.25–453.5 Ma (Sloan 1987). These six ash

From Hart, M. B. (ed.), 1996, *Biotic Recovery from Mass Extinction Events*, Geological Society Special Publication No. 102, pp. 97–104

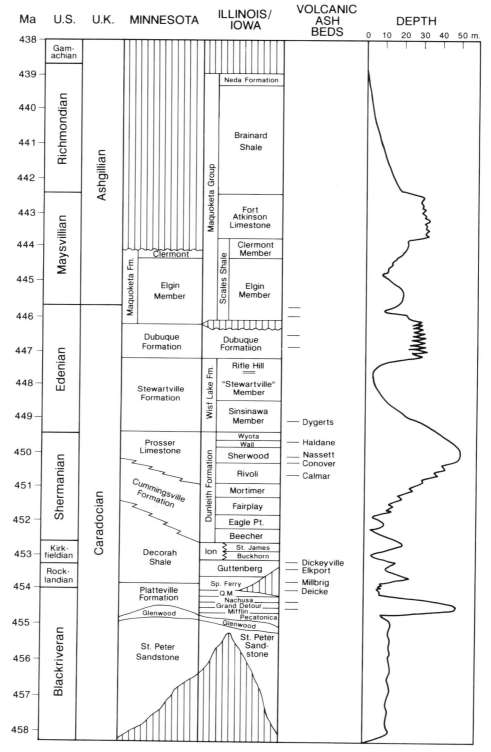

Fig. 1. Summary correlation chart of Middle and Upper Ordovician time-stratigraphic units, nomenclature of Minnesota, Iowa and Illinois and estimated depth of deposition in Minnesota. Modified from Sloan (1987).

OSTRACODE SPECIATION, USA

Table 1. *Stratigraphic distribution of Ordovician Ostracoda in Upper Mississippi Valley. Short vertical lines at top of page indicate horizons of metabentonite layers: a, Deicke; b, Millbrig; c, Dygerts*

Species	Glenwood	Platteville	Decorah	Cummingsville	Prosser	Stewartville	Dubuque	Maquoketa
Leperditella mundula (Ulrich)	—							
Krausella variata Kraft	——							
Aparchites fimbriatus (Ulrich)	—————————————————————————?
Drepanella summitensis Johnson et al.	—							
Tvarenella? sp.	—							
Conchoprimites? glenwoodensis Johnson et al.	—							
Bullatella granilabiata (Ulrich)	————							
Aparchites posterotumida Swain & Cornell	————							
Bythocypris? robusta Ulrich		—						
Drepanella bigeneris Ulrich		—						
Eoleperditia fabulites (Conrad)		—						
Leperditella canalis Ulrich		—						
Schmidtella crassimarginata Ulrich		—						
Eurychilina reticulata Ulrich		————						
Krausella inaequalis Ulrich		————————————————————						
Macronotella scofieldi Ulrich		———						
Eurychilina subradiata Ulrich			—					
Krausella arcuata Ulrich			—					
Rigidella cannonensis Swain & Cornell			—					
Pedomphalella subovata Swain & Cornell			————					
Eurychilina incurva Ulrich			————					
Eurychilina ventrosa Ulrich			————————					
Bromidella depressa Kay			—					
Bromidella rhomboides Kay			—					
Krausella curtispina Kay			————					
Bythocypris? curta Ulrich			————					
Bairdiocypris granti (Ulrich)			————					
Punctaparchites rugosus (Jones)			————					
Aparchites ellipticus Ulrich			———					
Aparchites macrus (Ulrich)			————					
Aparchites paratumidus Swain & Cornell			————					
Cytherella? rugosa Jones		?——————————————?					
Byrsolopsina ovata (Kay)			————————				
Paraschmidtella uphami (Ulrich)			————					
Saccelatia arcuamuralis Kay			————					
Saccelatia arrecta (Ulrich)			————					
Elliptocyprites paracylindrica Swain & Cornell			————					
Parabolbina antecedans Kay			————					
Parabolbina acarnifera Kay			————					
Parabolbina staufferi Kay			————					
Maratia sp.			————					
Bollia subaeuata Ulrich			————————				
Bellornatia tricollis Kay			———					
Ceratopsis humilinoda Kay			————					
Dicranella bicornis Ulrich			————					
Dicranella spinosa Ulrich			————					
Thomasatia falcicosta Kay			————					
Hesperidella initialis (Ulrich)			————					
Winchellatia lansingensis Kay			————					
Winchellatia longispina Kay			————					
Winchellatia minnesotensis Kay			————					
Bassleratia typa Kay			————					

Species	Glenwood	Platteville	Decorah	Cummingsville	Prosser	Stewartville	Dubuque	Maquoketa
	‖	a ‖ ‖	b ‖‖‖		‖‖ ‖	c ‖	‖	‖
Cryptophyllus sulcatus Levinson			---------					
Euprimitia linepunctata (Kay)			---------					
Euprimitia? celata (Ulrich)			---------					
Euprimitia sanctipauli (Ulrich)			-------					
Hallatia duplicata (Ulrich)			---------					
Hallatia magnipunctata Kay			---------					
Laccoprimitia fillmorensis (Ulrich)			---------					
Leperditella macra Ulrich			---------					
Leperditella millepunctata (Ulrich)			---------					
Leperditella persimilis Ulrich			---------					
Primitiella constricta Ulrich			---------					
Primitiella plattevillensis Kay			---------					
Primitiella sp.			---------					
Schmidtella affinis Ulrich			---------					
Eoschmidtella umbonata (Ulrich)			---------					
Primitiopsis? bella Kay			--------					
Byrsolopsina centipunctata (Kay)			------					
Byrsolopsina planilateralis			-----					
Parenthatia punctata (Ulrich)			---					
Aechmina ionensis Kay			-----					
Tetradella ellipsilira Kay			------					
Byrsolopsina normella Swain & Hansen			---					
Saccelatia angularis (Ulrich)			---					
Saccelatia cletifera Kay			-----					
Maratia mara Kay			----					
Nodambichilina symmetrica (Ulrich)			----	--------				
Primitia tumidula Ulrich			----	--------------------------------				
Cryptophyllus oboloides (Ulrich & Bassler)			----	------------				
Eurychilina partifimbriata Kay			----					
Scofieldia bilateralis (Ulrich)			----					
Parenthatia camerata Kay			----					
Ceratopsis quadrifida (Jones)			----					
Dicranella marginata Ulrich			----					
Dicranella simplex (Ulrich)			----					
Dicranella typa Ulrich			----					
Raymondatia goniglypta Kay			----					
Rigidella sp.			----					
Euprimitia labiosa (Ulrich)			----	----------				
Pseudoprimitiella unicornis (Ulrich)			----	------------------				
Hallatia convexa Kay			----					
Schmidtella brevis Ulrich			----					
Pseudulrichia simplex (Ulrich)			---					
Macronotella arcta (Ulrich)			---					
Opikatia emaciata Kay			--					
Opikatia rotunda Kay			-----					
Maratia micula (Ulrich)			---					
Hallatia particylindrica Kay			--					
Eurychilina minutifoveata Kay			--					
Ceratopsis chambersi (Miller)				? --------	--------	--------	--------	?
Macrocyproides trentonensis (Ulrich)				? --------	?			
"Bollia" unguloidea Ulrich				? --------	?			
Schmidtella subaequalis Ulrich				? --------	?			
Tetradella lunatifera Ulrich				? --------	?			
Ctenobolbina obliqua Ulrich				? --------	?			

OSTRACODE SPECIATION, USA

Species	Glenwood	Platteville	Decorah	Cummingsville	Prosser	Stewartville	Dubuque	Maquoketa
	‖	a ‖	b ‖‖	‖‖	‖‖	c ‖	‖	‖
"Jonesella" obscura Ulrich				?---------		?		
Primitia mucila Ulrich							--------	
Primitia mammata Ulrich		?------?						
Punctaparchites splendens Keenan							--------	
Leperditella sp.							-- ------	
Ulrichia nodosa (Ulrich)							--	
Ulrichia saccula Burr & Swain							-- ------	
"Bollia" regularis (Emmons)								---
Ningulella paucisulcata (Burr & Swain)								---
Schmidtella lacunosa Burr & Swain								---
Zygobolboides grafensis Spivey								------
Americoncha marginata (Ulrich)								--------
Milleratia cincinnatiensis (Miller)								---
Aparchites barbatus Keenan								---
Paraschmidtella irregularis Keenan								---
Eukloedenella richmondensis Spivey								---
Beyrichia irregularis Spivey								---
Zygobolboides calvini Spivey								---
Zygobolboides iowensis Spivey								---
Zygobolboides thomasi Spivey								-----
Bollia ruthae Spivey								---
Ctenobolbina emaciata (Ulrich)								---
Ctenobolbina maquoketensis Spivey								---
Leperditella fryei Spivey								---
Primitia gibbera Ulrich								---
Primitiella bellvuensis Spivey								---
Primitiella carli Spivey								---
Primitiella milleri Spivey								-----
Ellesmeria scobeyi Spivey								---
Macrocyproides clermontensis Spivey								---
Quasibollia ridicula Keenan								--
Aechmina cuspidata Jones & Holl								--
Aechmina maquoketensis Keenan								--
Aechmina taurea Keenan								---
Ceratopsis trilobis Keenan								---
Tetradella carinata Keenan								---
Tetradella septinoda Keenan								---
Ctenobolbina bispinosa Ulrich								---
Kiesowia binoda Keenan								---
Kiesowia insolens Keenan								---
Laccoprimitia elegantula (Keenan)								---
Eridoconcha gibbera (Ulrich)								---
Eridoconcha? punctata Keenan								---
Euprimitia floris Keenan								---
Euprimitia minuta Keenan								---
Leperditella? dorsicornis Ulrich								---
Leperditella sacceliformis Keenan								---
Schmidtella latimarginata Keenan								---

falls took place during the late Blackriveran, Rocklandian and early Kirkfieldian Provincial stages.

A quiescent period of a little more than 2 Ma was followed by a second set of ash falls recorded in the upper Dunleith (Prosser) and lower Wise Lake (Stewartville) formations of Illinois, Iowa and Minnesota. These lie in the late Shermanian and early Edenian Provincial stages. The resulting metabentonites are named, in ascending order, Calmar, Conover, Nassett, Haldane and Dygerts, and formed approximately 450.5–449 Ma.

After another quiescent interval of about 2 Ma, a third set of (unnamed) ash falls is recorded in the late Edenian Dubuque Formation and in the Elgin Member of the Maquoketa Formation, approximately 447–445.7 Ma.

Extinction horizons of invertebrate species

Studies by Sloan (1987), Rice (1987), De Mott (1987) and Kolata *et al.* (1987) have shown that seven of 18 species of brachiopods became extinct at the horizon of the Deicke K-bentonite. Nine of 10 species of trilobites became extinct at that horizon. As to gastropods, 48 of 60 species and all of the echinoderms became extinct. The conodonts, however, show extinction of only two natural or residual form species at the Deicke horizon (Webers 1966), possibly as a result of their nektonic character. The Deicke horizon lies within *Phragmodus undatus* conodont zone of Sweet (1987). Other ash horizons associated with the Deicke may also involve invertebrate extinctions but the Deicke is the most important (Sloan 1987). Catalani reviewed (1987) the stratigraphic distribution of the cephalopods across the Platteville–Decorah boundary which approximates to the position of the cluster of Deicke and associated metabentonites. He found that of 67 cephalopod species in the upper Platteville, only four occur in younger Ordo-vician rocks of the area. There is no clear explanation as to why the cephalopods, in contrast to the conodonts, did not have a better survival rate.

Seventeen ostracode species have been recorded from the Platteville Limestone (Ulrich 1894) and uppermost Glenwood Formation (Johnson *et al.* 1991). Six of these become extinct at or just above the Deicke horizon. Five other species become extinct at or just above the lower Platteville/upper Glenwood ash horizons. The other six species range upward into the Decorah Shale.

The next higher cluster of metabentonites in the late Shermanian Prosser/Dunleith and early Edenian Stewartville formations was marked by the extinction of at least six ostracode species and possibly six more (Table 1) out of the 19 species recorded from the Prosser Formation. Sloan (1987) noted a marked extinction of invertebrate species in the lower Stewartville Sinsinawa Member and ascribed this event to marine shoaling, possibly resulting from a global eustatic sea-level drop. The cluster of ash falls should also be taken into consideration as a contributing cause of the extinction event.

The third cluster of (unnamed) ash falls in the late Edenian Dubuque and lower Maquoketa formations is marked by the extinction of four out of six ostracode species recorded from that interval. The Dubuque–Maquoketa boundary is an unconformity in Illinois but is comfortable or paraconformable in Minnesota and northern Iowa, where the ostracodes occur.

Noteworthy in the ostracode extinctions of 32 species in or near the four ash-fall clusters is the relatively small number of total species involved (38 species): eight in the Glenwood, seven in the Platteville, 16 in the Prosser, and seven in the Dubuque, in comparison with the 165 ostracode species recorded from the Middle and Upper Ordovician in the north central United States. Environmental conditions were not favourable in the pre-ash fall carbonates for the development of rich ostracode populations. These conditions may have included elevated salinities (as proposed for the earlier Glenwood Formation (Fraser 1976)), declining food supplies and unfavourable substrates, or a combination of factors.

Environmental conditions affecting the ostracodes and other invertebrates, including the brachiopods, did not apply to the trilobites (De Mott 1987), which were more numerous in the Platteville Limestone than in the overlying Decorah Shale (53 species v. 16 species), or to the cephalopods discussed above.

Ostracode speciation following extinction events

The uppermost shaly part of the Platteville Limestone and the Decorah Shale are abundantly fossiliferous with brachiopods (Rice & Hedblom 1987), bryozoans (Karklins 1987), ostracodes (Ulrich 1894; Kay 1934, 1940; Swain *et al.* 1961; Swain & Cornell 1987), crinoids (Brower 1987), cystoids and other echinoderms (Kolata *et al.* 1987), gastropods (Sloan & Webers 1987), conodonts (Webers 1966; Sweet 1987) and some bivalves (Pojeta 1987). The diversity of these groups is generally greater than in the underlying Platteville Limestone. Trilobites and cephalopods on the other hand do not exhibit as great diversity in the Decorah Shale as in the Platteville Limestone (Catalani 1987; De Mott 1987). A vagrant mode of life of the latter two groups does not completely explain the differences in diversity, because the cystoids, conodonts, gastropods and bivalves are also vagrant.

In nearly all phyla except the conodonts (Sweet 1987), repopulation following the Deicke and associated ash falls was gradual rather than abrupt (Sloan 1987).

In the case of the Ostracoda, during and following the deposition of the Deicke and associated ash falls, 85 species appeared in the uppermost Platteville Limestone and the Decorah Shale (Table 1). These did not all appear at the same time but developed gradually over 1.5–2 Ma. Twelve species appeared following the Deicke K-bentonite in the uppermost shaly Platteville (lower Spechts Ferry Member). Nine species appeared in the lowermost Decorah Shale. Thirty-two species appeared a little higher in the Decorah Shale (upper Spechts Ferry Member). The remaining species developed in groups throughout the rest of the Decorah. Distribution of the Decorah ostracodes permits recognition of three ostracode assemblage zones in the formation (Swain *et al.* 1961; Swain & Cornell 1987). The duration of each of these zones appears to have been less than 1 Ma. The span of individual ostracodes species in the Decorah Shale is estimated to have been about 2 Ma, but some were apparently less.

The episodes of ash fall, in addition to terrestrial clastic influx, provided a renewed source of nutrients to the Decorah area of deposition, following the Platteville carbonates, that enhanced the development of immigrant species as well as of endemic forms in Rocklandian and early Shermanian times.

Very few of the Decorah Shale ostracode species continued through the overlying Cummingsville formation and into the Prosser Limestone (Table 1); most became extinct before the advent of late Shermanian–early Edenian ash falls of the Dygerts set of metabentonites. Thus these ash falls do not, to an important degree, appear to have been responsible for the extinction of the Decorah ostracodes. Instead, depleted nutrients in the shelf environment and perhaps other unknown environmental factors, as well as evolutionary changes in the individual species are suggested as having been responsible for the disappearance of the Decorah Shale fauna.

The ostracodes of the Prosser Limestone are known only from the early work of Ulrich (1894) and Ulrich & Bassler (1908) who described seven species. This fauna may have developed in response to a renewal of nutrients provided by the Dygerts set of ash falls, but the Prosser assemblage needs to be studied further. The Prosser carbonates do not seem to have been favourable for the development of any fauna except for sparse ostracode populations.

The next evolutionary spurt in ostracode populations in the region was in the late Edenian–early Maysvillian Maquoketa formation (Table 1). The preceding Dubuque formation contains a restricted ostracode fauna (Burr & Swain 1965). The basal Maquoketa Argo–Kay bed ('Depauperate Zone') is a thin dark grey phosphatic euxinic shale with pyritized and phosphatized ostracodes and molluscs. The Depauperate Zone fauna developed during the ash fall episodes of the unnamed third group in the late Edenian. Following the ash falls 43 species of ostracodes appeared in the Edenian and Maysvillian provincial stages.

As in the case of the Decorah Shale ostracode fauna, the appearance of the species was not simultaneous in the Maquoketa Formation. It was to some degree stepwise and may also have involved punctuated evolution (Table 1). Many of the middle upper Maquoketa Formation forms have short ranges, perhaps less than 1 Ma.

Conclusions

Each of the three clusters of middle and upper Ordovician metabentonites in the Upper Mississippi Valley region shows evidence of having influenced or caused the extinction of a small group of ostracode species. Other invertebrate classes were also affected to a greater or lesser degree than the ostracodes.

The Rocklandian and late Edenian extinction events were followed in the case of the Ostracoda by gradual repopulation that resulted in the development of much greater numbers of species than were present in the region prior to each of the two events. The intervening Shermanian–early Edenian event was characterized by lithofacies only marginally conducive to support of ostracode populations. The marked increase in ostracode populations in the Rocklandian–Kirkfieldian and late Edenian–early Maysvillian resulted from favourable shaly facies as well as from an increase of nutrients provided by the ash falls. Decline in the nutrient supplies as well as in the return of mainly carbonate lithofacies, depleted nutrients in the shelf habitats and evolutionary changes of the species themselves, contributed to the reduction in numbers of ostracodes in the middle Shermanian and in the Richmondian.

As a general statement regarding these middle and upper Ordovician ostracode faunas, the effects of associated ash falls were less the cause of extinction of large numbers of species than the provider of nutrient-rich habitats for the development of new populations.

Robert E. Sloan provided useful information on the stratigraphic aspects of the paper. Michelle Elston typed the manuscript.

References

BROWER, J. C., 1987. The Middle Ordovician crinoid fauna of the Twin Cities area. *In:* SLOAN, R. E. (ed.) *Middle and Late Ordovician Lithostratigraphy and Biostratigraphy of the Upper Mississippi Valley.* Minnesota Geological Survey, Report of Investigations, 35, 177–178.

BURR, J. H. JR. & SWAIN, F. M. 1965. *Ostracoda of the Dubuque and Maquoketa Formations of Minnesota and northern Iowa.* Minnesota Geological Survey, Special Publication Series, SP-3.

CATALANI, J. A. 1987. Biostratigraphy of the Middle and late Ordovician cephalopods of the Upper Mississippi Valley area. *In:* SLOAN, R. E. (ed.) *Middle and Late Ordovician Lithostratigraphy and Biostratigraphy of the Upper Mississippi Valley.* Minnesota Geological Survey, Report of Investigations, 35, 187–189.

DE MOTT, L. L. 1987. Platteville and Decorah trilobites from Illinois and Wisconsin. *In:* SLOAN, R. E. (ed.) *Middle and Late Ordovician Lithostratigraphy and Biostratigraphy of the Upper Mississippi Valley.* Minnesota Geological Survey, Report of Investigations, 35, 63–98.

FRASER, G. S. 1976. Sedimentology of a quartz–arenite–carbonate transition (Middle Ordovician) in the Upper Mississippi Valley. *Geological Society of America Bulletin*, 86, 833–845.

JOHNSON, J. D., BENOLKIN, L. V. & SWAIN, F. M. 1991. Ostracoda from the Glenwood Shale (Ordovician–Middle Caradocian) of Minnesota. Revista Española de Micropaleontológia, 28, 141–152.

KARKLINS, O. L. 1987. Bryozoa from Rocklandian (Middle Ordovician) rocks of the Upper Mississippi Valley. *In:* SLOAN, R. E. (ed.) *Middle and Late Ordovician Lithostratigraphy and Biostratigraphy of the Upper Mississippi Valley.* Minnesota Geological Survey, Report of Investigations, 35, 173–176.

KAY, G. M. 1934. Mohawkian Ostracoda: species common to Trenton faunules from the Hull and Decorah Formations. *Journal of Paleontology*, 8, 328–343.

—— 1940. Ordovician Mohawkian Ostracoda: lower Trenton Decorah fauna. *Journal of Paleontology*, 14, 234–268.

KOLATA, D. R., BROWER, J. C. & FREST, J. J. 1987. Upper Mississippi Valley Champlanian and Cincinnatian echinoderms. *In:* SLOAN, R. E. (ed.)

Middle and Late Ordovician Lithostratigraphy and Biostratigraphy of the Upper Mississippi Valley. Minnesota Geological Survey, Report of Investigations, 35, 179–181.

RICE, W. F. 1987. The systematics and biostratigraphy of the Brachiopoda of the Decorah Shale at St. Paul, Minnesota. *In:* SLOAN, R. E. (ed.) *Middle and Late Ordovician Lithostratigraphy and Biostratigraphy of the Upper Mississippi Valley.* Minnesota Geological Survey, Report of Investigations, 35, 136–166.

—— & HEDBLOM, E. P. 1987. Brachiopods and trilobites of the Sardeson beds in the Twin Cities. *In:* SLOAN, R. E. (ed.) *Middle and Late Ordovician Lithostratigraphy and Biostratigraphy of the Upper Mississippi Valley.* Minnesota Geological Survey, Report of Investigations, 35, 131–135.

SLOAN, R. E. 1987. *Middle and Late Ordovician Lithostratigraphy and Biostratigraphy of the Upper Mississippi Valley.* Minnesota Geological Survey, Report of Investigations, 35.

—— & WEBERS, G. F. 1987. Stratigraphic ranges of Middle and Late Ordovician gastropoda and monoplacophora of Minnesota. *In:* SLOAN, R. E. (ed.) *Middle and Late Ordovician Lithostratigraphy and Biostratigraphy of the Upper Mississippi Valley.* Minnesota Geological Survey, Report of Investigations, 35, 183–186.

SWAIN, F. M. 1987. Middle and Upper Ordovician Ostracoda of Minnesota and Iowa. *In:* SLOAN, R. E. (ed.) *Middle and Late Ordovician Lithostratigraphy and Biostratigraphy of the Upper Mississippi Valley.* Minnesota Geological Survey, Report of Investigations, 35, 99–101.

—— & CORNELL, J. R. 1987. Ostracoda of the Superfamilies Drepanellacea, Hollinacea, Leperditellacae, and Healdacea from the Decorah Shale of Minnesota. *In:* SLOAN, R. E. (ed.) *Middle and Late Ordovician Lithostratigraphy and Biostratigraphy of the Upper Mississippi Valley.* Minnesota Geological Survey, Report of Investigations, 35, 102–130.

——, —— & HANSEN, D. L. 1961. Ostracoda of the Families Aparchitidae, Aechminidae, Leperditellidae, Drepanellidae, Eurychilinidae, and Punctaparchitidae from the Decorah Shale of Minnesota. *Journal of Paleontology*, 35, 345–372.

SWEET, W. C. 1987. Distribution and significance of conodonts in Middle and Upper Ordovician strata of the Upper Mississippi Valley Region. *In:* SLOAN, R. E. (ed.) *Middle and Late Ordovician Lithostratigraphy and Biostratigraphy of the Upper Mississippi Valley.* Minnesota Geological Survey, Report of Investigations, 35, 167–172.

ULRICH, E. O. 1894. *The Lower Silurian Ostracoda of Minnesota.* Minnesota Geology and Natural History Survey, Final Report, 3, 629–693.

—— & BASSLER, R. S. 1908. New American Paleozoic Ostracoda: Preliminary revision of the Beyrichiidae, with descriptions of new genera. *US National Museum Proceedings*, 35, 277–340.

WEBERS, G. F. 1966. *The Middle and Upper Ordovician conodont faunas of Minnesota.* Minnesota Geological Survey, Special Publication Series, SP-4.

Biotic recovery after mass extinction: the role of climate and ocean-state in the post-glacial (Late Ordovician–Early Silurian) recovery of the conodonts

H. A. ARMSTRONG

Department of Geological Sciences, University of Durham, South Road, Durham DH1 3LE, UK

Abstract: The pattern of recovery in conodonts following the Late Ordovician mass extinction does not conform to the classical adaptive radiation model of transgression, shelf area expansion and cladogenesis. An alternative hypothesis is proposed in which progenitor 'survivor' species evolved in the bathyal ecozone during Ashgill global cooling. Their appearance in low latitude shallow water environments occurred as the result of two phases of migration from the bathyal ecozone. The first, during the *extraordinarius* Biozone sea-level fall, included species attributed to the *Ozarkodina* and *Oulodus*? assemblages. This migration is considered to have been a response to the upward movement of the permanent thermocline, as high latitude climate and ocean conditions developed at low latitudes. The second phase of emergence occurred during the *persculptus* Biozone transgression when species of the *Dapsilodus–Distomodus* assemblage, appeared ahead of advancing anoxia.

Mean rates of per taxon origination and extinction, measured from an Upper Ordovician to Lower Silurian (low latitude) graphical composite reference section, were unequal at 0.09 and 0.19 taxa per standard time unit. The mean value for net rate of change in diversity was −0.66. Species originations and extinctions were non-random, a feature consistent with the dramatic changes in environment at this time.

The Late Ordovician mass extinction has been recognized as one of the five statistically significant mass extinction events (Raup & Sepkoski 1982). As currently understood, it was a composite event made up of separate waves of extinction in the late Caradoc, early Ashgill and late Ashgill (Brenchley 1984, 1988). The earlier episodes of extinction were probably related to the closure of the Iapetus Ocean, which allowed interchange and competition between benthic faunas (McKerrow & Cocks 1976). The last episode, restricted to a short period represented by the Hirnantian stage of the Late Ordovician, has been correlated with a major phase of Gondwana glaciation (Berry & Boucot 1973; Sheehan 1973; Brenchley & Newell 1984; see also *Modern Geology* 1994, **20** for a thematic set of papers on the Late Ordovician mass extinction). Various hypotheses have attributed extinction to reduction in shelf area during glacio-eustatic sea-level fall (Berry & Boucot 1973; Jaanusson 1879), dramatic changes in temperature (Berry & Boucot 1973; Sheehan 1973, 1979; Boucot 1975) and complex changes in ocean states (Armstrong 1994; Brenchley *et al.* 1994).

A conceptual model to explain biotic recovery after mass extinction has been proposed as a starting point for IGCP Project 335 (Fig. 1). In this model the survivors of the mass extinction event include long-term survivors, Lazarus taxa (shallow water, shelf species surviving in refugia) and new progenitor species. The latter result from cladogenesis during mass extinction and diversification during adaptive recolonization of previously vacated niches. The Upper Ordovician to Lower Silurian conodont record provides an ideal test for this adaptive radiation model, as glacial regression during the Late Ordovician had removed large areas of shelf habitat.

Conodonts were a group of marine, Late Cambrian–latest Triassic vertebrates, bearing a mineralized feeding apparatus of conodont elements (Briggs *et al.* 1983; Aldridge *et al.* 1986, 1993a; Sansom *et al.* 1992). Whole animal morphology and species distribution studies have suggested that the majority of conodonts with complex, platform-bearing apparatuses had a predominantly nektobenthonic mode of life. Conodonts were widely distributed, abundant and have an excellent fossil record. The taxonomy of the group is now sufficiently advanced to consider multi-element species as biological entities. These characteristics make the conodonts one of the premier groups of biostratigraphically useful fossils in the Palaeozoic, and

From Hart, M. B. (ed.), 1996, *Biotic Recovery from Mass Extinction Events,*
Geological Society Special Publication No. 102, pp. 105–117

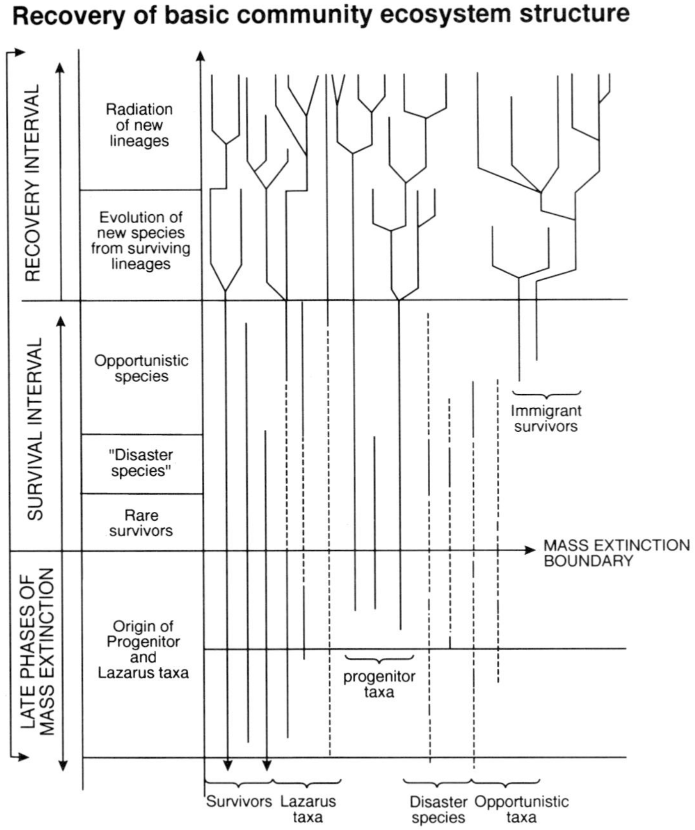

Recovery of basic community ecosystem structure

Fig. 1. Conceptual model developed for IGCP Project 335, *Biotic recovery after mass extinction*. This paper considers the role of glaciation and changing ocean states in the origin and radiation of Late Ordovician, progenitor conodont species.

an ideal group in which to test models of evolutionary palaeobiogeography.

In a study of Late Ordovician mass extinction, Armstrong (1994) suggested the post-glacial conodont fauna contained few if any Lazarus taxa, pelagic species with a long pre-glacial evolutionary history and new species. He assumed an evolutionary locus in shelf settings and postulated that at low latitudes, environmental instability during the glaciation was the main stimulus to cladogenesis. The period of cladogenesis must have been short-lived as lower Llandovery carbonate platform sequences typically contain a low diversity fauna of species which originated during the glaciation (e.g. Armstrong 1990, figs 6, 7, 9, 10). This pattern of low diversity was punctuated by species diversity peaks in the lower Aeronian and upper Telychian, attributed to changes in ocean-state (Jeppsson 1990; Aldridge *et al.* 1993b). The pattern of post-glacial recovery of the conodonts

therefore comprised three phases: Late Ordovician cladogenesis, the rapid migration of new species into shelf areas and a period of post-migration evolutionary stasis.

Comparison of the Upper Ordovician–Lower Silurian sea-level curve with the appearances of new species shows that migration was initiated during sea-level fall and continued as sea-level rose. The adaptive radiation model is thus rejected as a primary explanation for conodont recovery. In this paper an alternative hypothesis is proposed in which nektobenthonic, progenitor species had their evolutionary origins in the bathyal ecozone. Migrations during the developing regression were driven by the upward migration of the permanent thermocline during the onset of the glacial maximum. The rapid return to ocean anoxia and sea-level highstand, in response to global warming, drove even deeper water biofacies into shelf areas. The absence of Lazarus taxa in Lower Silurian

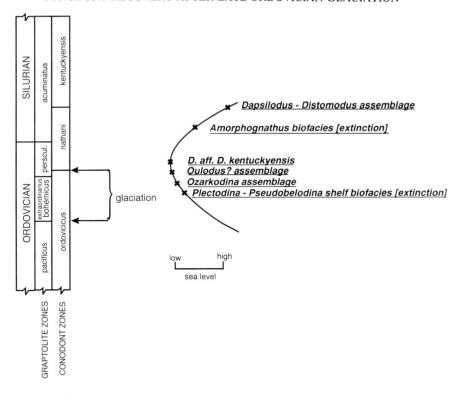

Fig. 2. Superimposed assemblage appearances and global sea-level curve (from Brenchley *et al.* 1994). *Plectodina–Pseudobelodina* and *Amorphognathus* biofacies as defined by Sweet & Bergström 1989). New assemblages are in bold, normal type marks the final extinction of Late Ordovician biofacies. The vertical time scale is in standard time units taken from Fig. 6.

sequences suggests that the majority of pre-glacial, shallow water, shelf species became extinct during the glaciation.

Timing

A review of Ordovician–Silurian boundary biostratigraphy shows that in many parts of the world sections are incomplete or poorly known (Barnes & Bergström 1988). The best documented, most complete sections occur in areas marginal to the Laurentian craton. Graphical correlation of sections containing conodonts and graptolites (Armstrong 1994) showed that the Ordovician–Silurian boundary, as defined at the base of the *acuminatus* Biozone, lies within the *O? nathani* conodont biozone. Graptolite and conodont extinctions occurred within the Late Ordovician. New conodont species first start to appear in the *extraordinarius* Biozone, before the influx of new graptolite taxa in the *persculptus* Biozone (Armstrong 1994). Major glaciation was apparently confined to the Hirnantian, which lasted 0.5–1 Ma and was associated with a 45–65 m fall in sea-level (Brenchley *et al.* 1994). The glaciation came to an abrupt end during the early *persculptus* Biozone (Armstrong 1994; Brenchley *et al.* 1994).

Ocean-state model

Global oceanography responds to variations in climate (Wilde & Berry 1984, 1986). Jeppsson (1990) linked climate, ocean-state and biotic changes in an 'oceanic model'. He defined two end member conditions in the model, an S-state (e.g. Fig. 4, stage 1) during which atmospheric carbon dioxide levels are high and greenhouse conditions prevail. During the S-state the oceans are salinity stratified and poorly ventilated (Bralower & Thierstein 1984). The opposite condition, the P-state (e.g. Fig. 4, stage 3), occurs when high latitude temperature falls to 5°C and dense, cold, oxygenated waters descend into the deep oceans (Schopf 1980). During the

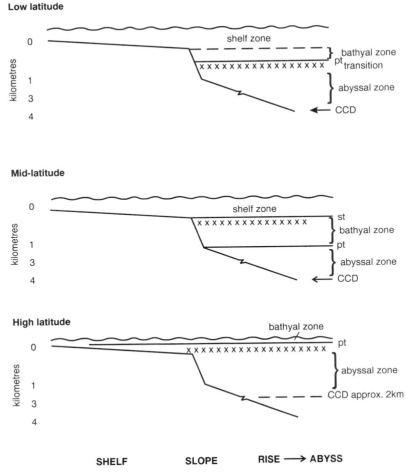

Fig. 3. Oceanographic parameters in the modern North Atlantic based upon data in Gage & Tyler (1992). The position of the CCD is taken from Leggett (1985). st, seasonal thermocline; pt, permanent thermocline; py, pycnocline; ha, halocline.

P-state the oceans are thermally stratified and the abyssal ecozone is oxic. The Late Ordovician was a time for dramatic environmental change during which the Earth passed from a prolonged S-state, through a short-lived P-state (the *extraordinarius* Biozone) back into an S-state (Armstrong 1994).

The ocean-state model has been used to explain evolutionary patterns in Silurian shelf conodonts (Jeppsson 1990; Aldridge *et al.* 1993*b*) and conodont extinction during the Late Ordovician (Armstrong 1994). No attempt has yet been made to assess its evolutionary effects in the deep-sea. The model has been refined here using oceanographic parameters and faunal distribution patterns seen in the modern North Atlantic Ocean.

The fauna of the North Atlantic is clearly separated into highly distinct shelf, bathyal and abyssal ecological depth ecozones. The fundamental boundaries between these ecozones lie at the shelf break ('mud-line' of Murray 1988) and permanent thermocline. At mid-latitudes the former coincides with the seasonal thermocline (Fig. 3) and at high latitudes the permanent thermocline reaches the surface. The bathyal ecozone occupies a distinct physio-chemical and hydrodynamic area of the continental slope and at the present day coincides with the oxygen minimum zone (Gage & Tyler 1992).

The causes of faunal zonation on the continental shelves are relatively well known and are related to physical and trophic changes along hydrodynamic gradients. These led to a mosaic

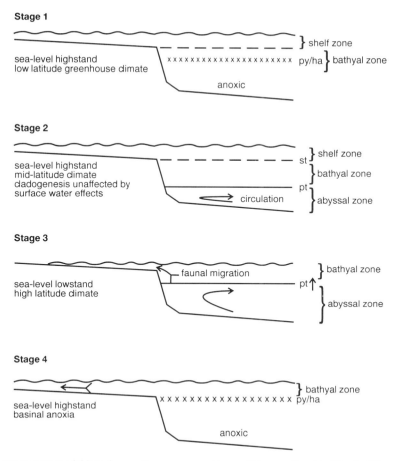

Fig. 4. Refined ocean-state model based upon the oceanographic conditions outlined in Fig. 2. Abbreviations as for Fig. 3.

of biofacies which can be mapped across the continental shelf. In comparison, the controls on faunal zonation in the deep-sea are only poorly understood. Controls postulated in the literature include physical effects (temperature, topography, substrate, etc.), effects of deep boundary currents, changes in trophic strategy and biological interactions (Wilson & Hessler 1987).

The micro- and meiofauna on the bathyal ecozone exhibit greater depth differentiation than the bathypelagic macrofauna, which shows a pattern of gradual change down the continental slope. Closely related members of the bathypelagic fauna tend to have overlapping but slightly different ecological ranges, and sibling species are common (Wilson & Hessler 1987). Limited data suggest that the genetic heterozygosity in the deep-sea macrofauna is low (Wilson & Hessler 1987). The deep-sea is an area of low population densities but also of high diversity, locally approaching values found in tropical rainforests (Gage & Tyler 1992). At the present day there is little species differentiation between shelf ecozone and bathyal faunas at high latitudes.

Figure 4 illustrates how oceanographic conditions might change at low latitudes as the Earth passed through a glaciation. The predicted biotic changes are summarized in Fig. 5.

Stage 1 (S-state). During a long stable period of 'greenhouse' climate, low latitude, oceanographic conditions would be significantly different from those found at the present day. Higher average global temperatures would lead to salinity stratification in the oceans with a well developed pycnocline (density barrier) and coincident halocline (salinity barrier). The pycno-

Ocean-state model	Shelf	Slope-Rise-Abyss
Stage 1 "S-state" low latitude climate, high pycnocline/halocline.	low diversity, warm, stenotopic specialists in low productivity, narrow niches selection of K-strategists in stable, favourable environments low rate of evolution Lazarus effects in response to Milankovitch cycles	low rates of evolution in constricted bathyal ecozone selection of opportunists in stable, unfavourable environments no abyssal ecozone
Stage 2 "transitional" S to P mid-latitude climate, seasonal thermocline and deep permanent thermocline.	extinctions as shelf ecozone contracts and mid- to low latitudes cool migration into refugia Lazarus effects relating to Milankovitch cycles	expansion of the bathyal ecozone high rates of evolution due to decrease in population densities, increase in local demes, restricted gene flow, cladogenesis with many sibling species increased bathymetric zonation selection of opportunists in stable, unfavourable environments
Stage 3 "P-state" high latitude climate and high permanent thermocline.	final extinctions in warm, stenotopic taxa appearance of eurytopic bathyal species, with pre-adapted opportunistic life strategies. (ideal for unstable glacial environments) high plankton and benthic productivity associated with upwelling and continental run-off	constriction of the bathyal ecozone migration upwards of eurytopic species (zoned during stage 2) development of an abyssal fauna decline in distinction between shelf and bathyal ecozone faunas
Stage 4 "return to S-state"	migration of perched bathyal ecozone species low inherited genetic heterozygosity and eurytopic nature restrict adaptive radiation and produce a low diversity shelf fauna Stage 1 develops as surface effects cause a differentiation of the shelf and bathyal ecozones.	upward movement of bathyal species following transgression return of anoxia, constriction of the bathyal zone. ? extinctions in bathyal zone extinctions in abyssal zone species unable to migrate into bathyal zone.

Fig. 5. Summary of biotic changes associated with changing ocean-state. Predicted changes in shelf species are taken from Jeppsson (1990).

cline/halocline would lie closer to the surface at low latitudes and increase in depth in mid-latitudes. In the absence of a deep, cold circulation the deep-sea would become anoxic (e.g. Railsback *et al.* 1990 for the Late Ordovician ocean) and abyssal species unable to migrate into the bathyal ecozone would become extinct.

Stage 2 ('transitional state' from S to P). This spans the period of pre-glacial, global cooling when a mid-latitude climate and ocean-state

developed at low latitudes. In this transitional period a seasonal thermocline would develop, isolating the shelf and the bathyal ecozones and the pycnocline/halocline is replaced by a permanent thermocline. The bathyal ecozone would expand down the slope.

During the expansion of the bathyal ecozone, already low population densities would decrease further, facilitating isolation, restricted gene flow, formation of demes and allopatric speciation (see Wilson & Hessler 1987 for a discussion on the mechanisms of speciation in the deep-

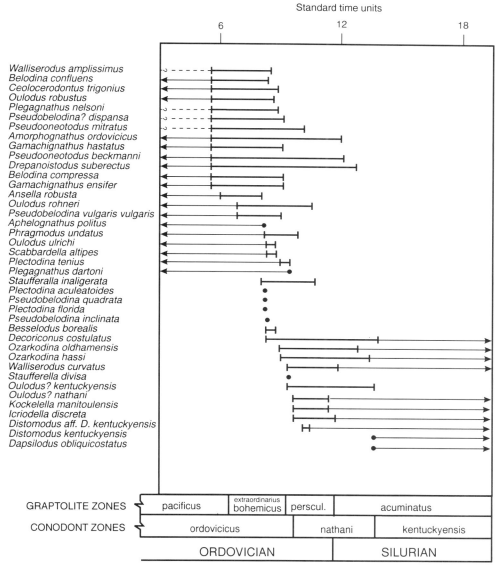

Fig. 6. CSRS range chart. Species ranges of nektobenthonic conodonts taken from Armstrong (1994). Range bars and ● indicate species occurring in the CSRS. Arrows signify species known to range below and above the stratigraphic range of the CSRS. The 11 included sections are all taken from Laurentia which lay close to the equator during the Late Ordovician

sea). Cladogenesis in, and increased zonation of, the bathyal ecozone would occur during Stage 2.

Stage 3 (P-state). During glacially induced regression, shelf species would migrate towards the shelf break and in the absence of a permanent thermocline the shelf ecozone would extend down the slope. However, during the change to high latitude conditions the permanent thermocline would move up through the water column, constricting the previously broad bathyal ecozone and acting as a barrier to the down slope migration of stenothermal shelf species.

With the establishment of high latitude, thermal stratification patterns, eurytopic species

Palaeo-environment	Lower Ashgill biofacies	Upper Ashgill assemblages	Modern ocean
continental shelf	Aphelognathus-Oulodus Pseudobelodina Plectodina	Plectodina-Pseudobelodina	shelf
continental slope	Phragmodus Amorphognathus	Ozarkodina Oulodus?	bathyal
continental rise - abyssal plain	Dapsilodus-Periodon	Distomodus-Dapsilodus	abyssal

Fig. 7. Correlation of *velicuspis* Chronozone (late Caradoc–early Ashgill) depth biofacies (from Sweet & Bergström 1989), upper Ashgill depth assemblages and the modern ocean depth ecozones. The latter is based upon the assumption that the seasonal and permanent thermoclines define the bathyal ecozone and mark the major changes in the deep-sea fauna. In the conodont record it appears that these changes are at the generic level.

previously restricted to beneath the seasonal thermocline, would be able to invade shelf habitats and compete with increasingly stressed shelf taxa. A combination of global cooling, environmental instability and the absence of lower latitude (i.e. warmer) refugia would cause mass extinction in stenothermal shelf species during Stage 3. As the permanent thermocline rose, species from progressively deeper areas of the bathyal ecozone would be able to migrate onto the shelf and the distinction between shelf and bathyal ecozone faunas would decline.

Stage 4 (return to S-state). The return of deep-sea anoxia with global warming would isolate eurytopic bathyal species which would become 'perched' in shallow water settings. During sea-level rise, the perched bathyal fauna would expand onto the shelf and deeper bathyal species migrate into deep shelf areas. During Stage 4 any shallow water species which might have survived in refugia would appear in the fossil record.

Conodont species recovery during the Late Ordovician and Early Silurian as a test of the ocean-state model

Biotic responses to environmental change are easily documented for individual sections or sedimentary basins. However, local facies effects have a strong control on species distributions and may lead to complex Lazarus effects. If models are to be developed to explain evolutionary responses to global phenomena, it is crucial that local effects are minimized. This can be achieved by employing numerical biostratigraphic methods to compile global composite species ranges (Armstrong 1994).

Dataset

The graphical method (Shaw 1964) has been used to produce a composite standard reference section (CSRS) from geographically disparate sections in different facies from Canada and North Greenland (Armstrong 1994). These areas occupied a significant part of Laurentia which, for much of the Lower Palaeozoic, lay in tropical and sub-tropical latitudes (Trench & Torvik 1992). For details of the graphical correlation method refer to the reviews of Sweet (1984), Edwards (1984) and Armstrong (1994). The ranges of 40 putatively nektobenthic conodont species have been extracted from the upper Ashgill and lower Llandovery CSRS (Armstrong 1994) and are shown in Fig. 6.

Conodont species appearances

The CSRS range chart shows new taxa appearing as a series of steps through the Late Ordovician. This pattern can be subjectively divided into four generically distinct assemblages defined as follows.
Plectodina–Pseudobelodina assemblage. Plectodina aculeatoides Sweet, *P. florida* Sweet, *Pseudo-*

	Shelf Ecozone	Bathyal Ecozone
Stage 2 - global cooling (early regression)	**extinction in shallowest water species** [*Aphelognathus* biofacies]	cladogenesis **Specialist (archaic) ancestors** (eg. *Amorphognathus, Gamachignathus* etc.) **Opportunist ("progenitor") descendants** [*Ozarkodina, Oulodus?, Dapsilodus-Distomodus* assemblages]
Stage 3 (end regression)	**further decline in deeper shelf species** [*Plectodina-Pseudobelodina, Phragmodus* biofacies] upwards migration of progenitor species [*Ozarkodina, Oulodus?* assemblage]	**archaic species remain in slope refugia, may expand into abyssal ecozone.** [*Amorphognathus* biofacies]
Stage 4-global warming (transgression)	**appearance of deeper bathyal ecozone taxa** [*Dapsilodus-Distomodus* assemblage]	**extinction amongst archaic species** [*Amorphognathus* biofacies]

Fig. 8. Evolutionary scenario during the Late Ordovician showing the differentiation of bathyal ecozone species into specialist (archaic, ancestors) and opportunist (progenitor, descendants). Archaic taxa become trapped in the bathyal and ?abyssal ecozones, suffering extinction with the return of greenhouse climate and oceanic anoxia. Progenitor species expand into shelf settings.

belodina quadrata Sweet, *P. inclinata* (Branson & Mehl), *Besselodus borealis* Nowlan & McCracken, *Staufferella inaligerata* McCracken & Barnes and *Decoriconus costulatus* (Branson & Mehl). The majority of species within the *Plectodina–Pseudobelodina* assemblage became extinct during the *extraordinarius* Biozone and are thought to have been stenothermal, shallow water specialists (see Sweet & Bergström 1989).

Ozarkodina assemblages. Base at the first appearance of *Ozarkodina*, includes *Ozarkodina oldhamensis* (Rexroad) and *Ozarkodina hassi* (Pollock, Rexroad & Nicoll).

Oulodus? assemblages. Base at the first appearance of *Oulodus?* Includes two subgroups of taxa, the earlier comprises *Oulodus? kentuckyensis* (Branson & Branson), *Staufferella divisa* Sweet and *Walliserodus curvatus* (Branson & Branson). The later subgroup contains *Oulodus? nathani* McCracken & Barnes, *Kockelella manitoulinensis* (Pollock, Rexroad & Nicoll) and *Icriodella discreta* Pollock, Rexroad & Nicoll.

Dapsilodus–Distomodus assemblage. The definition of the base of this assemblage is difficult as there is a significant stratigraphic gap between the first appearance of *Distomodus* aff. *D. kentuckyensis* of McCracken & Barnes in the lowermost *persculptus* Biozone and *Distomodus kentuckyensis* Branson & Branson and *Dapsilodus obliquicostatus* (Branston & Mehl) in the lower *acuminatus* Biozone. *Distomodus* spp. first appear early in the transgression and for convenience of discussion are grouped into one assemblage.

In the absence of precursor lineages in shallow water sequences, the stepwise pattern of appearances in the ocean-state model must reflect phylogenetic emergence of deep water species. Little is known about the ecological tolerances of Late Ordovician conodont species. However, depth related conodont biofacies have been documented from the upper Caradoc–lower Ashgill (*velicuspis* chronozone) of Laurentia (Sweet & Bergström 1989; Fig 7). At this time shelf ecozones were dominated by members of the *Aphelognathus* and *Pseudobelodina–Plectodina* biofacies. The *Phragmodus* biofacies lay close to the shelf break. The *Amorphognathus* biofacies occupied the slope and the *Dapsilodus–Periodon* biofacies, commonly associated with cherts beneath the CCD, was found in lower slope-rise and abyssal settings.

The upper Ashgill *Plectodina–Pseudobelodina*

stu interval	D	S	E	rs	re	rd	rdD
14	10	0	0	0	0	0	0
13	9	2	1	0.22	0.11	0.11	0.99
12	12	0	3	0	0.25	-0.25	-3
11	12	0	0	0	0	0	0
10	14	1	3	0.07	0.21	-0.41	-1.96
9	15	9	10	0.6	0.66	-0.06	-0.9
8	22	7	14	0.32	0.64	-0.32	-7
7	22	0	0	0	0	0	0
6	22	0	0	0	0	0	0
5	22	0	0	0	0	0	0
Mean				0.09	0.19	-0.66	

Fig. 9. Species diversity data from the CSRS range chart. D, original taxonomic diversity; S, additional new taxa in the time interval; E, number of taxa becoming extinct during the time interval; rs, per taxon rate of originations; re, per taxon rate of extinctions; rd, net rate of change in diversity; rdD, rate of change in the total diversity in the clade.

assemblage is considered to be equivalent to the lower Ashgill *Aphelognathus* and *Pseudobelodina–Plectodina* biofacies. The *Ozarkodina, Oulodus*? and *Dapsilodus–Distomodus* assemblages would, from comparison with the lower Ashgill, to be equivalent to the outer shelf to abyssal biofacies (Fig. 7). This suggestion is supported by the occurrence of early species of *Ozarkodina* in Caradoc deep water limestones. For example *Ozarkodina*? *pseudofissilis* (Lindström), the ancestor of *O. oldhamensis* occurs in the Crug Limestone of Wales (Orchard 1980) and the Ireleth Limestone of the southern Lake District (Armstrong unpublished data).

Phylogenetic emergence of eurytopic, progenitor species from the bathyal ecozone is predicted in the ocean-state model during regression (Figs 3 & 8, Stage 3) and transgression (Figs 3 & 8, Stage 4). During regression, outer shelf and upper bathyal ecozone species would appear as the permanent thermocline moved up through the water column. Plotting assemblage and species appearances onto the global sea-level curve (from Brenchley *et al.* 1994; Fig 2) shows that the shallower *Ozarkodina* and *Oulodus*? assemblages appeared sequentially during the later part of the regression and that the deeper *Dapsilodus–Distomodus* assemblage appeared during the early part of the deglacial transgression.

In the ocean-state model ancestral stocks are stenotopic, having evolved in a narrow, greenhouse bathyal ecozone. The stable, relatively hospitable environment would favour the selection of species with a competitive (K-) life strategy, with competition for resources between adults. Expansion of the bathyal zone during Stage 2 leads to a stable, relatively inhospitable environment. Under these circumstances selection pressures on descendants would favour tolerant (r-) life strategies. The high degree of specialization in the archaic ancestors means that in order to survive extinction during Stage 2 and/or Stage 3 they must migrate into refugia (Fig. 8).

Per taxon rates of extinction and origination in Late Ordovician and Early Silurian nektobenthic conodont species

In the ocean-state model standing diversity comprises two components, extinctions in shelf species and originations in the bathyal ecozone. If the bathyal ecozone fauna is impoverished relative to the shelf then a decline in taxonomic diversity would occur as bathyal species came to dominate during Stages 3 and 4. With reduced cladogenesis in Stages 3 and 4, post-glacial standing diversity would approach that of the glacial bathyal ecozone. The ocean-state model also predicts that the greatest rate of change in diversity will occur, during the later parts of Stage 3 (8–10 stu in the composite range chart), as shelf species are replaced by a low density bathyal fauna and Stage 4 (12–13 stu) as extinctions occur in archaic species and further migrations of progenitor taxa occur.

Measurements of species diversity are notoriously difficult to make and interpret. The problems of an incomplete fossil record and difficulties in correlating disparate sections can lead to over-splitting of lineages and over-estimates of species diversity. A species range in a CSRS approaches the true phylogenetic range of that species and diversity measurements made from a CSRS range chart should more closely reflect biological standing diversities. In most analyses of diversity, higher taxonomic categories are used as a proxy of relative species numbers. However, this becomes unnecessary where reliable estimates of species numbers can be made from a good fossil record.

Per taxon rates of species origination and extinction in the taxonomic diversity calculations (Fig. 9) exclude, possibly, pelagic taxa which have a less well developed taxonomy and may have had different environmental requirements. For full details of the methodology refer to Sheldon & Skelton (1993).

Through any given time interval the average rate of origination of new taxa within a clade is

$$S/\Delta t \qquad (1)$$

where S is additional new taxa, and Δt the time interval. However, this rate is also dependent upon the original taxonomic diversity, D, within the clade. In a post-extinction setting D will vary significantly between clades and a closer measure of diverse changes and taxonomic overturn is given by the per taxon rate of origination (rs) and extinction (re):

$$rs = (S/\Delta t)/D \qquad (2)$$

$$re = (E/\Delta t)/D \qquad (3)$$

where E is the number of taxa becoming extinct during the time interval. The net rate of change in diversity is

$$rd = rs-re \qquad (4)$$

and the rate of change in total diversity in a clade, per interval of time is

$$rdD. \qquad (5)$$

If species origination and extinction are considered to be random processes, then over an extended period the means of rs and re should be approximately equal. If the mean values of rs and re are equal then the mean value of rd would be 0.

Per taxon rates of origination and extinction are measured here against the standard time unit (stu) scale of the CSRS and are not values in real time (Fig. 9). The advantages of the stu timescale are that it is applicable for periods of time less than a million years (usually beyond the resolution of radiometric dating), can be used where reliable radiometric dates are not available and, unlike the relative timescale, stu intervals are equal in length.

Over the time period of the CSRS (approxi-
mately 1 Ma) rates of per taxon origination and extinction, in nektobenthonic conodonts, were unequal at 0.09 and 0.19 taxa per stu. The mean value for net rate of change in diversity is –0.65. This suggests that conodont evolution through this time period was not random and deterministic explanations for that must be sought. The post-glacial standing diversity is approximately half that of the pre-glacial situation. This supports the conclusion that either the bathyal ecozone had a low diversity fauna or that only part of the fauna is being sampled.

Traditionally, the view has been that the deep sea was impoverished in species owing to harsh conditions for life and an environment where, because of the presumed constancy of the environment, evolution has proceeded much more slowly than in shallow water (Carter 1961). However, the discovery of high species richness in the modern deep sea show it is a site of prolific speciation (Hessler & Wilson 1983; Gage & Tyler 1992). The low diversity bathyal fauna of the Late Ordovician could in part reflect the short time available for cladogenesis during Stage 2.

Conclusions

The graphically generated CSRS provides an ideal dataset on which to test palaeogeographical models. The current Upper Ordovician to Lower Silurian graphical composite, though restricted to low latitude sections, highlights a number of major paradoxes in conodont evolution. New 'progenitor' conodont species first appeared on the shelf during glacial regression and, conflicting with the predictions of the adaptive radiation model, remained at a relatively low diversity for much of the Llandovery. Biotic recovery in the conodonts predated that of the graptolites, suggesting different ecological and evolutionary controls on the zooplankton and nektobenthos at this time.

It is proposed that the evolution of the conodonts through the Late Ordovician glaciation is best explained in terms of changes in climate and ocean-state. Global cooling and the imposition of high latitude climatic and oceanic conditions in low latitudes restricted the number of potential refugia for stenothermal, shelf species which became extinct. The appearance of new progenitor species during glaciation regression is consistent with an evolutionary ancestry in the bathyal ecozone and a rising permanent thermocline. The rapid return of a greenhouse climate and oceanic anoxia stranded eurytopic bathyal species of the *Ozarkodina Oulodus*? and *Dapsilodus–Distomodus* assemblages in shelf settings. Eustatic sea-level rise and expansion of new shelf niches enabled these faunas to expand. The apparent absence of speciation during this radiation may reflect the eurytopic habit and low genetic heterozygosity in bathyal species.

Conodonts came close to extinction during the Late Ordovician glaciation. Their ultimate survival was a result of the presence of a reservoir of eurytopic taxa within the bathyal ecozone. The cryptic ancestry of many Silurian conodont lineages may be explained by evolution within the bathyal ecozone; the search for stem phylogenies should be directed towards deep water biofacies.

Dr R. J. Aldridge and Dr C. T. Scrutton are kindly thanked for discussing ocean-states and improving an early edition of this manuscript. Karen Atkinson drew the diagrams.

References

ALDRIDGE, R. J., BRIGGS, D. E. G., CLARKSON, E. N. K. & SMITH, M. P. 1986. The affinities of conodonts – new evidence from the Carboniferous of Edinburgh, Scotland. *Lethaia*, **19**, 279–291.
——, ——, SMITH, M. P., CLARKSON, E. N. K. & CLARK, N. D. L. 1993a. The anatomy of conodonts. *Philosophical Transactions of the Royal Society of London B*, **340**, 405–421.
——, JEPPSSON, L. & DORNING, K. 1993b. Early Silurian episodes and events. *Journal of the Geological Society, London*, **150**, 501–513.
ARMSTRONG, H. A. 1990. Conodonts from the Upper Ordovician–Lower Silurian carbonate platform of North Greenland. *Gronlands geologiske Undersogelse, Bulletin*, **159**.
—— 1994. High-resolution biostratigraphy (conodonts and graptolites) of the Upper Ordovician and Lower Silurian – evaluation of the late Ordovician mass extinction. *Modern Geology*, **20**, 1–28.
BARNES, C. R. & BERGSTRÖM, S. M. 1988. Conodont biostratigraphy of the uppermost Ordovician and lowermost Silurian. *British Museum of Natural History (Geology), Bulletin*, **43**, 325–343.
BERRY, W. B. N. & BOUCOT, A. J. 1973. Glacioeustatic control of Late Ordovician–Early Silurian platform sedimentation and faunal changes. *Geological Society of America, Bulletin*, **84**, 275–284.
BOUCOT, A. J. 1975. *Evolution and Extinction Rate Controls*. Elsevier, Amsterdam.
BRALOWER, T. J. & THIERSTEIN, H. R. 1984. Low productivity and slow deep-water circulation in mid-Cretaceous oceans. *Geology*, **12**, 614–618.
BRENCHLEY, P. J. 1984. Late Ordovician extinctions and their relationship to the Gondwana glaciation. *In:* BRENCHLEY, P. (ed.) *Fossils and Climate*. Wiley, London, 291–315.

—— 1988. Environmental changes close to the Ordovician–Silurian boundary. *British Museum of Natural History (Geology), Bulletin*, **43**, 377–385.
—— & NEWALL, G. 1984. Late Ordovician environmental changes and their effect on faunas. *In:* BRUTON, D. L. (ed.) *Aspects of the Ordovician System*. Palaeontological Contributions from the University of Oslo. **295**, 65–79.
——, MARSHALL, J. D., CARDEN, G. A. F., ROBERTSON, D. B. R., LONG, D. G. F., MEIDLA, T., HINTS, L. & ANDERSON, T. F. 1994. Bathymetric and isotopic evidence for a short-lived Late Ordovician glaciation in a greenhouse climate. *Geology*, **22**, 295–298.
BRIGGS, D. E. G., CLARKSON, E. N. K. & ALDRIDGE, R. J. 1983. The conodont animal. *Lethaia*, **16**, 1–14.
CARTER, G. S. 1961. Evolution in the deep sea. *In:* SEARS, M. (ed.) *Oceanography*. American Association for the Advancement of Science, Washington, DC, 229–238.
EDWARDS, L. E. 1984. Insights on why graphical correlation (Shaw's method) works. *Journal of Geology*, **92**, 583–597.
GAGE, J. D. & TYLER, P. A. 1992. *Deep-sea biology: A natural history of organisms at the deep-sea floor.* Cambridge University Press.
HESSLER, R. R. & WILSON, G. D. F. 1983. The origin and biogeography of the malacostracan crustaceans in the deep sea. *In:* SIMS, R. W., PRICE, J. H. & WHALLEY, P. E. S. (eds) *Evolution, Time and Space: the Emergence of the Biosphere*. Academic, London, 227–254.
JAANUSSON, V. 1979. Ordovician. *In:* ROBINSON, R. A. & TEICHERT, C. (eds) *Treatise on Invertebrate Palaeontology. A. Introduction, Fossilification (Taphonomy), Biogeography and Biostratigraphy*. Geological Society of America and University of Kansas Press, Lawrence, A136–A166.
JEPPSSON, L. 1990. An oceanic model for lithological and faunal changes tested on the Silurian record. *Journal of the Geological Society, London*, **147**, 663–675.
LEGGETT, J. K. 1985. Deep-sea pelagic sediments and palaeoceanography: a review of recent progress. *In:* BRENCHLEY, P. J. & WILLIAMS, B. P. J. (eds) *Sedimentology: Recent Developments and Applied Aspects*. Geological Society, London, Special Publication, **18**, 95–121.
McKERROW, W. S. & COCKS, L. R. M. 1976. Progressive faunal migration across the Iapetus Ocean. *Nature*, **263**, 304–306.
MURRAY, J. W. 1988. Neogene bottom water-masses and benthic Foraminifera in the NE Atlantic Ocean. *Journal of the Geological Society, London*, **145**, 125–132.
ORCHARD, M. J. 1980. Upper Ordovician conodonts from England and Wales. *Geologica Palaeontologica*, **14**, 9–44.
RAILSBACK, L. B., ACKERLEY, S. C., ANDERSON, T. F. & CISNE, J. L. 1990. Paleontological and isotope evidence for warm saline deep water in Ordovician oceans. *Nature*, **343**, 156–159.
RAUP, D. M. & SEPKOSKI, J. J. 1982. Mass extinctions

in the marine fossil record. *Science*, **215**, 1501–1503.

SANSOM, I. J., SMITH, M. P., ARMSTRONG, H. A. & SMITH, M. M. 1992. Presence of the earliest vertebrate hard tissues in conodonts. *Science*, **256**, 1308–1311.

SCHOPF, T. M. J. 1980. *Palaeoceanography*. Harvard University Press.

SHAW, A. B. 1964. *Time in Stratigraphy*. McGraw-Hill, New York.

SHEEHAN, P. M. 1973. The relation of the Late Ordovician glaciation to the Ordovician–Silurian changeover in North American brachiopod faunas. *Lethaia*, **6**, 147–154.

—— 1979. Swedish late Ordovician marine benthic assemblages and their bearing on brachiopod zoogeography. *In:* GRAY, J. & BOUCOT, A. J. (eds) *Historical Biogeography, Plate Tectonics and the changing environment*. Oregon State University Press, 61–73.

SHELDON, P. & SKELTON, P. 1993. Phylogenetic patterns. *In:* SKELTON, P. (ed.) *Evolution – a biological and palaeontological approach*. Addison-Wesley in association with the Open University, Wokingham, 743–841.

SWEET, W. C. 1984. Graphic correlation of upper Middle and Upper Ordovician rocks, North American Midcontinent Province, U.S.A. *In:* BRUTON, D. L. (ed.) *Aspects of the Ordovician System*. Palaeontological Contributions from the University of Oslo, **295**, 23–35.

—— & BERGSTRÖM, S. M. 1989. *Conodont provinces and biofacies of the Late Ordovician*. Geological Society of America, Special Paper, **196**, 69–87.

TRENCH, A. & TORSVIK, T. H. 1992. The closure of the Iapetus Ocean and Tornquist Sea: new palaeomagnetic constraints. *Journal of the Geological Society, London*, **149**, 867–870.

WILDE, P. & BERRY, W. B. N. 1984. Destabilization of oceanic density structure and its significance to marine 'Extinction' events. *Palaeogeography, Palaeoclimatology, Palaeoecology*, **48**, 143–162.

—— & —— 1986. The role of oceanographic factors in the generation of global bio-events. *In:* WALLISER, O. H. (ed.) *Global bio-events*. Springer, Berlin, 75–91.

WILSON, G. D. F. & HESSLER, R. R. 1987. Speciation in the deep-sea. *Annual Review of Ecology and Systematics*, **18**, 185–207.

Recovery of post–Late Ordovician extinction graptolites: a western North American perspective

WILLIAM B. N. BERRY

Department of Geology & Geophysics, University of California, Berkeley, CA 94720, USA

Abstract: Late Ordovician to Early Silurian graptolite-bearing sequences have been recognized in three primary western North American areas: southeastern Alaska, northern Canadian Cordillera and the Great Basin (Nevada–Idaho). A spectrum of marine environments is represented in strata in these sequences, ranging from those of volcanic islands and deep oceans to those of shelves and shelf margins. During the time of glacial maximum, the time of the *extraordinarius* zone, sea-level lowered and oxic conditions prevailed on the ocean floor. The prominent Late Ordovician near-extinction among graptolites denotes the *pacificus–extraordinarius* zone boundary. Normalograptids appear to be the only taxa in the *extraordinarius* zone. They are the prominent taxa in the superjacent *persculptus* and *acuminatus* zones. Normalograptids are joined by a few diplograptids in the *persculptus* zone, and by a few more diplograptids as well as *Parakidograptus* and *Cystograptus* in the *acuminatus* zone. The first monograptids appear in the *atavus* zone in western North America. The *atavus* zone fauna is primarily diplograptids. Significant sea-level rise and transgression across the shelf following after deglaciation did not occur in western North America until the *acinaces* zone. At that time, new monograptids, including *Lagarograptus*, *Pristiograptus* and *Pribylograptus*, appeared as did many new diplograptid taxa. Stratigraphic occurrence data indicate that the major post near-extinction reradiation took place in western North America as shelf seas expanded and shelf sea habitats became increasingly available. That pattern of new taxa appearing during transgression continued into the *gregarius* zone.

In their discussions of the Late Ordovician graptolite near-extinction, both Melchin & Mitchell (1991) and Koren (1991) indicated a need to know precise patterns of graptoloid occurrences in the extinction interval on local and regional scales. Melchin & Mitchell (1991, p. 147) drew attention to different stratigraphic range patterns among Late Ordovician graptolites observed in different areas of the world. Recent field work in central Nevada by S. C. Finney and his students and by Finney and the author together has revealed new graptolite occurrences relevant to the pattern of near-extinction and reradiation after that biotic crisis in that area. The Late Ordovician biotic crisis observed in studies in the Selwyn Basin, northwestern Canada has been summarized by Wang *et al.* (1993). Melchin reported on graptolite occurrences before and after the near-extinction in the area covered by that study (see Melchin in Wang *et al.* 1993). This new information, some of it unpublished, and certain latest Ordovician–Early Silurian graptolite collections obtained from precisely measured stratigraphic sections in southeastern Alaska (Churkin & Carter 1970), southcentral Nevada (Berry in Mullens 1980), central Idaho (Dover *et al.* 1980), and the

Northern Canadian Cordillera (Lenz 1982) comprise the data set for this review of graptolite recovery after the Late Ordovician near-extinction from the western North American perspective. This review presents data from recent collecting and aims at a regional synthesis of the biotic recovery among graptolites seen in the context of tectonic controls on graptolite-bearing strata. Cocks & Scotese (1991) discussed global Silurian geography, indicating that the area under discussion in this review lay close to a subduction zone. That proximity to an area in which one plate was being subducted under another may be reflected in occurrences of Early Silurian graptolites in western North America.

The physical setting

The rock sequences bearing western North American Late Ordovician Early Silurian graptolites developed on or close to the margin of the Laurentian plate (Cocks & Scotese 1991; Fig. 1). That plate margin lay within a tropical, westerly-flowing open ocean current system (Wilde *et al.* 1991). Wilde *et al.* (1991) suggested that an ocean current (the North Subpolar Current of Wilde *et al.* 1991, figs 3 & 4) flowed south-

From Hart, M. B. (ed.), 1996, *Biotic Recovery from Mass Extinction Events*, Geological Society Special Publication No. 102, pp. 119–126

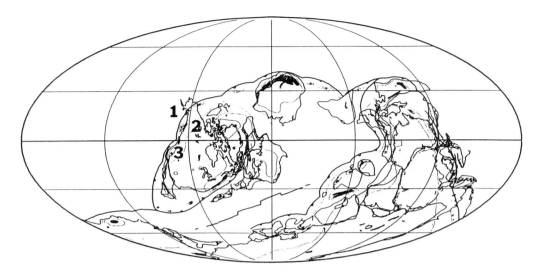

Fig. 1. Llandovery (Early Silurian) palaeogeography. Map provided by C. R. Scotese. Dark stipple indicates probable highlands. Light stipple indicates land areas of low relief. Light areas on plates adjacent to lands were sites of shelf marine environments. Lines with teeth indicate likely subduction zones. Note proximity of Laurentian plate (North America) margin to subduction zone. Latest Ordovician into Early Silurian localities discussed are 1, southeastern Alaska; 2, Northern Canadian Cordillera; 3, the Great Basin (Nevada and Idaho).

westerly and westerly across the western side of the Laurentian plate. Such a current would have created upwelling conditions along the plate, and continental shelf margin during the Late Ordovician and Early Silurian. If upwelling did take place along the western margin of the Laurentian plate, then rock sequences that formed there under the upwelling waters should be relatively highly fossiliferous. The very species-rich Llandovery graptolite faunas described by Lenz (1982) in stratigraphic sections in the Northern Canadian Cordillera and the coeval rich graptolite faunas in the Canadian Arctic described by Melchin (1989) are consistent with accumulation of the organic-rich rocks bearing them under oceanic upwelling conditions.

Sea level changes during the Silurian were reviewed and described by Johnson et al. (1991). That review documents a significant global drop in sea-level in the Late Ordovician. That sea-level drop may be correlated with southern hemisphere glaciation on Gondwanaland. Johnson et al. (1991) indicated that sea-level rose relatively rapidly in the Early Llandovery, during the *Parakidograptus acuminatus* into *Coronograptus cyphus* zone interval. The Late Ordo-vician sea-level fall would have resulted in the presence of those waters in which most graptolites lived over the outermost part of the Laurentian plate shelf and its slope. Sea-level rise in the Early Silurian would have expanded graptolite habitats as waters spread across shelf areas that had been exposed or had been sites of shallow marine environments.

Western North America Late Ordovician–Early Silurian graptolites lived in environments that changed significantly as a consequence of relative fall then rise of sea-level and as a consequence of the proximity of the Laurentian plate margin to a subduction zone. Preservation of the rock sequences bearing these graptolites also reflects changes resulting from sea-level fall and rise and tectonism.

Late Ordovician–Early Silurian graptolites have been found in two primary areas in the Great Basin region of the western United States. One of the two is in a black, organic-rich shale and argillite succession in central Idaho that formed close to the plate margin. The second includes two types of occurrences in rock suites in central Nevada. One of the two is in fault-bounded thrust slices. These rocks appear to have been deposited in oceanic settings at some distance from the Laurentian plate. The second type is in rocks that formed in platform environments on the plate. Coeval graptolites found in southeast Alaska seem to have lived in waters close to volcanic islands that were situated in the open ocean at a significant distance from the shelf portion of the plate. Northwest Canadian Late Ordovician–Early Silurian graptolites occur in dark shales and thinly-bedded limestones that accumulated in plate platform and slope environments.

The Great Basin

Late Ordovician and Early Silurian graptolites occur in the Phi Kappa Formation in central Idaho (Berry in Dover *et al.* 1980). Ross & Berry (1963) described Late Ordovician graptolites found in dark argillites and Berry (in Dover *et al.* 1980) identified *Atavograptus atavus* and possible *Coronograptus cyphus* zone faunas from stratigraphically higher shales.

Glyptograptus (or possibly *Normalograptus*) *persculptus* and possible *Atavograptus atavus* zone graptolites found in dark, organic-rich limestones in the Copenhagen Canyon succession in central Nevada were described by Berry (1986). Mullens (1980) and Murphy (1989) discussed Late Ordovician–Silurian strata in Copenhagen Canyon in the Monitor Range, central Nevada. The latest Ordovician–Early Silurian graptolite-bearing limestones there seem to have developed in relatively deep-water platform environments. Murphy (1989, fig. 128) pointed out that sandstones deposited in shallow marine environments underlie the *G. persculptus* zone strata. Murphy (1989) suggested that these shallow water sandstones were deposited during lowstand of sea-level related to Late Ordovician glaciation. The latest Ordovician–Early Silurian graptolites in the Monitor Range and other central Nevada localities (see Mullens 1980) occur in thinly-bedded dark grey limestones interbedded with cherts and dark bioclastic limestones that are debris flow deposits.

In the course of mapping a large area in central Nevada, the Cortez Quadrangle, Gilluly & Masursky (1965) recognized two graptolite-bearing rock suites bounded by thrusts. One of them, the Fourmile Canyon Formation, includes cherts, argillites and siltstones that bear a fauna of Early Silurian normalograptids with a possible *Glyptograptus* (Berry and Ross & Berry in Gilluly & Masursky 1965, table 2). The second, the Elder Sandstone, is predominantly feldspathic sandstones and siltstones that bear monograptids suggestive of the *C. cyphus* zone (Berry and Ross & Berry in Gilluly & Masursky 1965, table 2).

The most impressive sequence of Ordovician graptolite faunas in the Great Basin has been recorded by Finney (Finney & Perry 1991; Finney & Ethington 1992). These faunas have come from a wide spectrum of rock types that collectively are termed the Vinini Formation in central Nevada (see Finney & Perry 1991; Finney & Ethington 1992). Finney's studies have resulted in recognition of a relatively complete succession of Ordovician graptolites in a stratigraphic section in the Roberts Mountains. Collecting in that section by Finney and Finney & Berry together since the cited publications has resulted in recognition of *Paraorthograptus pacificus* zone faunas followed, in superjacent strata, by faunas of the *Normalograptus extraordinarius* and *Glyptograptus persculptus* zones. The zonal faunas are characterized by presence of the name-bearers. *P. pacificus* zone taxa occur in dark shales and argillites whereas the normalograptids that characterize the superjacent zones are found in calcareous shales and thin bedded limestones. Argillites and limestones in the *N. extraordinarius* zone bear evidence of bioturbation as well as sole markings suggestive of benthic organism activity. Vinini Formation strata occur in fault-bounded slices that were piled on top of shelf carbonates during mid-Palaeozoic tectonism. The Vinini Formation rock suite appears to have accumulated in oceanic environments at some, perhaps not too great, distance from the Laurentian plate. As Finney has pointed out (see Finney & Perry 1991; Finney & Ethington 1992), debris flows derived from shallow shelf environments occur in part of the Vinini Formation interbedded with autochthonous shales. At least some of the Vinini Formation rocks appear to have accumulated in oceanic environments on the margins of the Laurentian plate.

Northern Canadian Cordillera, northwest Canada

Latest Ordovician and Early Silurian (Llandovery) graptolites and the strata bearing them have been discussed by Lenz (1979, 1982). Lenz (1982, p. 1) stated that 'graptolite-bearing strata encompassed at least in part within the Road River Formation are widespread throughout the northern Canadian Cordillera'. The lithologic aspects of the Road River Formation, Lenz (1979, p. 141) pointed out, 'vary greatly, ranging from almost entirely chert with a few shale interbeds to almost totally dark shale, with minor chert. In still other areas, the unit consists of interbedded shales and thin band of evenly bedded limestones'. Environments of deposition of the strata included within the Road River Formation appear to range from those of the Laurentian plate continental slope to those on the continental shelf. In some areas, the Road River Formation strata pass laterally into carbonates that accumulated in shallow marine environments (Lenz 1979, p. 141). Lenz (1982, p. 3) noted that although the *G. persculptus* zone has not been recognized with certainty in the northern Canadian Cordillera, no obvious stratigraphic discontinuity has been seen between strata bearing *P. acuminatus* zone faunas and those bearing *P. pacificus* zone faunas. As described by Lenz (1979, 1982) this interval between Late Ordovician and Early Silurian graptolite-bearing strata appears to be relatively thin and indicative of a condensed stratigraphic section. The condensed strata succession suggests low stand of sea level at the time of accumulation. Recently, Melchin (in Wang *et al.* 1993, p. 1879) cited the occurrence of *Normalograptus* cf. *N. persculptus*, *N. normalis*, *Glyptograptus*? *laciniosus* and *Glyptograptus* n. sp.? in strata 1 m above those bearing *Dicellograptus* sp. aff. *D. minor* in Road River Formation shales exposed in the Selwyn Basin, northwest Canada. The *N. extraordinarius* zone position is represented by about 0.5 m of strata in this stratigraphic section. This condensation of the stratigraphic thickness in this interval is consistent with a lowered sea-level at the time of the *N. extraordinarius* zone. Wang *et al.* (1993) indicated that the depositional environment at the time was relatively shallower than it had been previously and they suggested that Cerium concentrations in this stratigraphic interval indicated 'a short period of basin ventilation in the otherwise anoxic Selwyn Basin'. They (Wang *et al.* 1993) discussed other geochemical features of the same stratigraphic interval, including changes in the concentration of C^{13}.

Lenz (1982) cited the stratigraphic occurrences of all Llandovery graptolites found in the northern Canadian Cordillera and described many of the taxa. He (Lenz 1979, 1982) reviewed the zonal distributions of the Llandovery graptolites in the area. His data indicate that this succession of Llandovery graptolites is one of, if not the, richest in terms of number of species in the world. The rocks bearing the richest of the Early Llandovery faunas were deposited in a suite of shelf environments that were spread widely across a margin of the Laurentian plate.

Southeastern Alaska

Churkin & Carter (1970) reviewed latest Ordovician–Early Silurian graptolites and their stratigraphic positions in southeastern Alaska. The faunas occur in the Descon Formation. Churkin & Carter (1970) indicated that the Descon Formation is composed of coarse greywackes, conglomerates, some black chert and siliceous shale and basaltic volcanic rocks. The volcanics include pillow basalts and flow breccias that are, at least locally, interbedded with graptolite-bearing shales. The graptolite faunas indicate that 'repeated volcanism occurred from different centres during Early Ordovician through Early Silurian time' (Churkin & Carter 1970, p. 3). Clearly, the depositional environments in the area of accumulation of these graptolite-bearing strata were oceanic and in the environs of long-term, although sporadic, volcanism. The prominence of graptolite taxa closely similar to those found in coeval strata in the Great Basin and the northern Canadian Cordillera suggests that the volcanism took place in oceanic settings relatively close to the Laurentian plate margin. Churkin & Carter (1970, fig. 4) indicated that volcanic flow rocks intervene in strata bearing Late Ordovician, *P. pacificus* zone, taxa and those bearing *P. acuminatus* zone taxa.

Latest Ordovician into Early Silurian graptolite-bearing strata in the northwestern part of North America accumulated in a spectrum of plate margin (shelf seas and slope) environments as well as in a variety of open ocean basinal settings. The range in depositional environments may be broader than in other areas of comparable size and age.

The evidence from the Great Basin and northern Canadian Cordillera indicates that sea-level fell and the sea floor was ventilated coincident with the most prominent extinction of Late Ordovician graptolites. That extinction marks the top of the *P. pacificus* zone. The Selwyn Basin and Copenhagen Canyon localities provide data indicating that the time of maximum sea-level lowering and basinal ventilation was that of the *N. extraordinarius* zone. In Copenhagen Canyon, strata subjacent to those identified by Murphy (1989, fig. 128) as having accumulated at the time of glaciation, bear *P. pacificus* zone taxa, including *P. pacificus*, in abundance. This collection was made by the author subsequent to the collection of *N.* (or *G.*) *persculptus* zone taxa from strata superjacent to those deposited during glaciation (Berry 1986). Vinini Formation *N. extraordinarius* zone strata show evidence of marine benthic activity whereas subjacent strata, those bearing *P. pacificus* zone taxa, do not. Wang *et al.* (1993) discussed the consequence of ocean ventilation at the same time interval.

Volcanic activity took place in southeastern Alaska in the latest Ordovician–Earliest Silurian (Churkin & Carter 1970). Structural evidence in the Phi Kappa Formation latest Ordovician–earliest Silurian strata suggests possible tectonism in the depositional environment at that time.

The physical evidence from latest Ordovician–earliest Silurian rocks in the northwestern part of North America reveals not only changes in sea-level and ocean chemistry but also tectonism potentially linked to plate motions. That plate motion-related tectonism may be suggestive of proximity to a subduction zone as indicated in Cocks & Scotese (1991; Fig. 1). The changes in sea-level, ocean ventilation and ocean chemistry altered marine habitats in the latest Ordovician. The western North American graptolite occurrence data indicate that these physical environmental changes were linked to major graptolite extinctions in the Late Ordovician. Graptolite recovery from that extinction was linked to rising sea-level and re-establishment of many habitats over the continental shelf and shelf margin during the early part of the Silurian. That recovery may be seen in review of the prominent developments among graptolites collected from stratigraphic sections in northwestern North America, especially those from the Road River Formation.

Post-*P. pacificus* zone extinction graptolite recovery

Post-*P. pacificus* zone graptolite occurrences (Fig. 2) are seen primarily in stratigraphic sections in southeastern Alaska (Churkin & Carter 1970) and the northern Canadian Cordillera (Lenz 1982). *N. extraordinarius* zone graptolites have been found only in the Vinini Formation, and *persculptus* zone faunas have been recovered from three sites. *Persculptus* zone taxa are known in the Selwyn Basin, Canadian Cordillera (Melchin in Wang *et al.* 1993), and Copenhagen Canyon sequence and the Vinini Formation, both in Nevada.

N. extraordinarius zone

N. extraordinarius zone taxa include only normalograptids. Although the zone is characterized by the presence of the name-bearer, *N. miserabilis* is more common.

N. persculptus zone

The presence of the name-bearer permits recognition of the zone. The zonal fauna includes almost entirely normalograptids. The relationships of *Glyptograptus laciniosus* Churkin & Carter (1970) are uncertain as are small specimens of *Glyptograptus*? sp. found in the zone. Possibly, glyptograptid species join normalograptids in the zonal fauna.

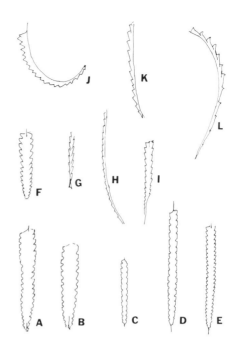

Fig. 2. Diagrammatic sketches of selected latest Ordovician–Early Silurian graptolites illustrating general features of colony form among post–Late Ordovician extinction graptolites. A, B, *Normalograptus extraordinarius*; C, *Normalograptus miserabilis*; D, *Normalograptus normalis*; E, *Glyptograptus laciniosus*; F, *Glyptograptus ? persculptus* group form; G, *Parakidograptus acuminatus*; H, *Atavograptus atavus*; I, *Dimorphograptus confertus swanstoni*; J, *Coronograptus gregarius*; K, *Pristiograptus fragilis*; L, *Lagarograptus acinaces*. All except L are ×1.6; L is ×3.3.

Parakidograptus acuminatus zone

The *P. acuminatus* zone is denoted by the appearance of the name-bearer in an association rich in normalograptids, glyptograptids and diplograptids. Lenz (1982) cited more than 20 taxa from only a few localities in the northern Canadian Cordillera. *Cystograptus vesiculosus* and *Orthograptus eberleini* appear for the first time in the zone. The zonal fauna is represented in southeast Alaska by *P. acuminatus*, *Normalograptus triflis* and other normalograptids and a possible *Orthograptus* (Churkin & Carter 1979). The species and numerical richness of the zonal fauna are seen in the platform localities in the northern Canadian Cordillera. Although both *Parakidograptus* and *Cystograptus* are new phyletic elements of the zone fauna, the fauna is numerically predominantly species in the

124 W. B. N. BERRY

Table 1. *Prominent developments among western North American latest Ordovician–Early Silurian graptolites*

Zone	Prominent features
Gregarius	Appearance of coronograptids and monograptids with lobate thecae
Acinaces	Influx of diverse monograptids – *Lagarograptus, Pristiograptus, Pribylograptus*. *Metaclimacograptus* appears
Atavus	Appearance of first monograptids, *Atavograptus*, and of Dimorphograptids. Primarily diplograptid fauna
Acuminatus	Primarily diplograptid fauna. Normalograptids prominent.
Persculptus	Normalograptids, some glyptograptids
Extraordinarius	Normalograptids
Extinctions	
Pacificus	*Paraorthograptus pacificus, Dicellograptus ornatus* group. Orthograptids of the *O. truncatus* group, climacograptids? of the hastatus and supernus groups, *Arachniograptus*.

genera *Normalograptus, Glyptograptus* and *Diplograptus*.

Atavograptus atavus *zone*

Rich *Atavograptus atavus* zone faunas have been recovered from only a few localities in the Road River Formation in the northern Canadian Cordillera. The number of species and the prominence of diplograptids in the fauna remains about the same as in the *P. acuminatus* zone. The new feature is, of course, the appearance of the first monograptids in the western North American successions. Three species of *Atavograptus* mark the introduction of the monograptids in the succession where they are joined by the first species of *Dimorphograptus* to appear. *Atavograptus* occurs not only in the platform succession but also in the rocks deposited in open ocean settings. Species of *Normalograptus, Glyptograptus* and *Diplograptus* remain the numerically abundant taxa.

Lagarograptus acinaces *zone*

The zonal fauna is characterized by a marked increase in new taxa of monograptids. In addition to *Lagarograptus*, species of *Pristiograptus* and *Pribylograptus* appear in both platform and oceanic depositional settings. *Metaclimacograptus* makes its initial appearance in the western North American successions as do new species of *Glyptograptus* and *Dimorphograptus*. Although collected from only a few western North American localities, the fauna of the zone is characterized by at least forty-five taxa of which about half are new to the succession. The prominent features of the zone fauna are the new

monograptid genera and speciations among glyptograptids. Normalograptid prominence wanes in the zone. The zone fauna has been found in more platform environments than the faunas of the prior post-extinction zones.

Coronograptus gregarius *zone*

The *C. gregarius* fauna is recognized by the sudden appearance of a new monograptid rhabdosome form, that of the curved coronograptids and of the proximally curved *M. revolutus*. Proximal thecae in the latter species are hooked or lobate. Species of diverse monograptids become numerically more prominent in the *C. gregarius* zone and the diplograptids are significantly less prominent. Both normalograptids and glyptograptids occur but in less abundance than in prior post-extinction zones. By this, the last zone within the Early Llandovery, the monograptid diversification is in full development and diplograptids have waned. The number of taxa in the zone fauna seen in western North America remains about the same as in the *L. acinaces* zone. The collections bearing the greatest number of species are those in shales formed in platform environments.

Major features of post-*P. pacificus* zone recovery

Collecting from a spectrum of ocean basin and shelf environments has resulted in certain general conclusions (Table 1). One of them involves a difference between occurrences in strata deposited in ocean basins and those deposited on the Laurentian plate continental shelf margin. Even though the number of specimens of a given taxon may be the same in collections from basin and platform strata, the number of taxa is far richer in the platform rock suites than in ocean basin strata. The collections in which the number of different taxa present is great and the number of specimens of each taxon present is high came from strata that formed close to the platform margin. The number of taxa present in ocean basin strata may be enhanced by the presence of gravity flow rocks in the basinal suites. Gravity flow rocks may bear numbers of different taxa that have been swept along by the flow from shelf environments to be deposited in the basin.

The marked radiations among the post-extinction faunas are seen in strata that developed at the time of post-glaciation sea-level rise and resultant spread across the shelf margin. Only a few taxa were present before sea-levels appear to have begun to rise significantly. Those taxa are mostly normalograptids.

The occurrences of atavograptids on the Avalonian plate two zones prior to occurrences of atavograptids in western North America suggest that a period of about 3–4 Ma may have been necessary for the atavograptids to be transported around the Laurentian plate to western North America.

Comparison of the rich Llandovery graptolite faunas described in the Cape Phillips Formation in the Canadian Arctic by Melchin (1989) with those in western North America reveals marked similarities among them. The Arctic faunas include a number of taxa described from Siberia that have not been found in western North America. Potentially, the Canadian Arctic part of the Laurentian plate lay more directly in the path of ocean currents that flowed westerly past the Siberian plate towards the Laurentian.

Post-extinction recovery was marked by developments among normalograptids initially. Diverse glyptograptids and possibly certain species of *Diplograptus* appeared. Diplograptid faunas developed significantly. They were joined first by *Parakidograptus* and *Cystograptus*, and secondly by the first monograptids, atavograptids. Not until the *L. acinaces* zone did monograptid diversification appear. At the same time, however, the diplograptid fauna remained prominent. With the appearance of the curved and proximally coiled monograptids in the *C. gregarius* zone where they joined monograptids continuing from the *A. atavus* and *L. acinaces* zones, the monograptid diversification was well developed. At that time, the diplograptids that had been prominent in the post-extinction faunas began to disappear. The first steps in post-extinction recovery were among normalo-

graptids and other diplograptids. Not until monograptids entered the western North American Llandovery environments and became diversified in them did they become more prominent than the diplograptids.

Normalograptids demonstrably tolerated a wide range of environmental conditions for they have been found in rock suites that formed under a range of shelf and ocean basin settings. They even occur in rocks in the Vinini Formation in which benthic organisms left traces of their activities, suggesting that the rhabdosomes found were not food for some benthic organisms. Graptolites are seen to become more diverse when those depositional environments returned that led to anoxic bottom conditions and the preservation in dark, organic rich mud rocks.

The patterns in post-extinction recovery appear to reflect changes in oceanic habitats following deglaciation and establishment of a well-developed oxygen-poor water mass that moved with rising sea-level across the outer part of the Laurentian plate shelf. Environmental changes at the time of maximum or near-maximum glaciation seem to have so altered habitats in which graptolites lived that they nearly became extinct. Wilde *et al.* (1990) suggested certain mechanisms that could result in extinctions among organisms that result from vertical advection of toxic waters from the main ocean pycnocline. Upward advection of toxic waters could have been a significant influence on Late Ordovician graptolite extinctions. Possibly only those taxa living highest in the oceanic water column survived for they lived above that part of the ocean into which toxic waters were advected. Wang *et al.* (1993) also drew attention to ocean circulation and chemical changes in oceanic habitats as factors influencing Late Ordovician graptolite extinctions.

Conclusion

Western North America post-Late Ordovician graptolite near-extinction reflects ocean ventilation and chemical changes in graptolite habitats at the time of glaciation. Tectonism at the Laurentian plate margin influenced accumulation and preservation of the rock suites that bear latest Ordovician–Early Silurian graptolites. The presence of several taxa that seem to be endemic to the area (many of these are diplograptids described initially by Churkin & Carter 1970) suggests that the area was relatively isolate from major ocean circulation systems until about the *Atavus* or *Acinaces* zones. Changes in ocean circulation following from lowered sea-level and plate margin tectonism both may have contributed to regional endemism. The western North American regional pattern of graptolite recovery after the post–Late Ordovician near-extinction differs in certain features (primarily presence or absence of certain species in some zones) from that seen in other regions. Inter-regional syntheses are needed to understand the post-near extinction pattern of recovery more fully.

References

BERRY, W. B. N. 1986. Stratigraphic significance of *Glyptograptus persculptus* group graptolites in central Nevada, U.S.A. *In:* HUGHES, C. P. & RICKARDS, R. B. (eds) *Palaeoecology and Biostratigraphy of Graptolites.* Geological Society, London, Special Publication, **20**, 135–143

CHURKIN, M. JR & CARTER, C. 1970. *Early Silurian Graptolites from southeastern Alaska and their correlation with graptolitic sequences in North America and the Arctic.* United States Geological Survey, Professional Paper, **653**.

COCKS, L. R. M. & SCOTESE, C. R. 1991. The Global Biogeography of the Silurian Period. *In:* BASSETT, M. G., LANE, P. D. & EDWARDS, D. (eds) *The Murchison Symposium. Proceedings of an International Conference on the Silurian System.* Special Papers in Palaeontology, **44**, 109–122.

DOVER, J. H., BERRY, W. B. N. & ROSS, R. J. JR 1980. *Ordovician and Silurian Phi Kappa and Trail Creek Formations, Pioneer Mountains, Central Idaho – Stratigraphic and structural revisions, and new data on graptolite faunas.* United States Geological Survey, Professional Paper, **1090.**

FINNEY, S. C. & ETHINGTON, R. L. 1992. Whiterockian graptolites and conodonts from the Vinini Formation, Nevada. *In:* WEBBY, B. D. & LAURIE, J. R. (eds) *Global Perspectives on Ordovician Geology.* Balkema, Rotterdam, 153–162.

—— & PERRY, B. D. 1991. Depositional setting and paleogeography of Ordovician Vinini Formation, central Nevada. *In:* COOPER, J. D. & STEVENS, C. H. (eds) *Paleozoic Paleogeography of the Western United States – II.* Pacific Section Society of Economic Paleontologists and Mineralogists, **67**, 747–766.

GILLULY, J. & MASURSKY, H. 1965. *Geology of the Cortez Quadrangle, Nevada.* United States Geological Survey, Bulletin, **1175**.

JOHNSON, M. E., KALJO, D. & RONG, J.-Y. 1991. Silurian Eustasy. *In:* BASSETT, M. G., LANE, P. D. & EDWARDS, D. (eds) *The Murchison Symposium. Proceedings of an International Conference on the*

Silurian System. Special Papers in Palaeontology, **44**, 145–163.

KOREN, T. N. 1991. Evolutionary crisis of the Ashgill graptolites. *In:* BARNES, C. R. & WILLIAMS, S. H. (eds) *Advances in Ordovician Geology.* Geological Survey of Canada, paper **90–9**, 157–164.

LENZ, A. C. 1979. Llandoverian graptolite zonation in the Northern Canadian Cordillera. *Acta Palaeontologica Polonica.* **24**, 137–153.

—— 1982. *Llandoverian graptolites of the Northern Canadian Cordillera:* Petalograptus, Cephalograptus, Rhaphidograptus, Dimorphograptus, *Retiolitidae,* and *Monograptidae.* Life Sciences Contributions, Royal Ontario Museum, **130**.

MELCHIN, M. J. 1989. Llandovery graptolite biostratigraphy and paleobiogeography, Cape Phillips Formation, Canadian Arctic Islands. *Canadian Journal of Earth Sciences*, **26**, 1726–1746.

—— & MITCHELL, C. E. 1991. Late Ordovician extinction in the Graptoloidea. *In:* BARNES, C. R. & WILLIAMS, S. H. (eds) *Advances in Ordovician Geology.* Geological Survey of Canada, paper **90-9**, 143–156.

MULLENS, T. E. 1980. *Stratigraphy, petrology, and some fossil data of the Roberts Mountains Formation, North-Central Nevada.* United States Geological Survey, Professional Paper, **1063**.

MURPHY, M. A. 1989. Central Nevada. *In:* HOLLAND, C. H. & BASSETT, M. G. (eds) *A Global Standard for the Silurian System.* National Musem of Wales, Geological Series, **9**, 171–177.

ROSS, R. J. JR & BERRY, W. B. N. 1963. *Ordovician graptolites of the Basin Ranges in California, Nevada, Utah and Idaho.* United States Geological Survey, Bulletin, **1134**.

WANG, K., CHATTERTON, B. D. E., ATTREP, M. JR & ORTH, C. J. 1993. Late Ordovician mass extinction in the Selwyn Basin, northwestern Canada: geochemical, sedimentological, and paleontological evidence. *Canadian Journal of Earth Sciences*, **30**, 1870–1880.

WILDE, P., BERRY, W. B. N. & QUINBY-HUNT, M. S. 1991. Silurian oceanic and atmospheric circulation and chemistry. *In:* BASSETT, M. G., LANE, P. D. & EDWARDS, D. (eds) *The Murchison Symposium. Proceedings of an International Conference on the Silurian System.* Special Papers in Palaeontology, **44**, 123–143.

——, QUINBY-HUNT, M. S. & BERRY, W. B. N. 1990. Vertical advection from oxic or anoxic water from the main pycnocline as a cause of rapid extinction or rapid radiations. *In:* KAUFFMAN, E. G. & WALLISER, O. H. (eds) *Extinction Events in Earth History.* Springer, Berlin, Lecture Notes in Earth Sciences, 85–98.

Diachronous recovery patterns in Early Silurian corals, graptolites and acritarchs

DIMITRI KALJO

Institute of Geology, Estonian Academy of Sciences, 7 Estonia Ave, EE0100 Tallinn, Estonia

Abstract: The extinctions that occurred in the latest Ordovician and earliest and latest Wenlock, were the most impressive in the biotic history of this period. They were probably caused by glaciations. Recovery processes, more or less, follow the classical scenario: extinction–low-diversity survival interval–recovery through radiation events. Rather often there occurs diachrony of the phases of this scenario in the different groups discussed. Graptolite recovery was the most rapid with a diversity maximum in the mid-Llandovery. Their main extinction and low-diversity interval were in the latest Ordovician. The evolution of corals and acritarchs was slower – after a few lower-scale origination events a diversity burst was reached in the late Llandovery. An analogous but lower-scale pattern was noted at the very beginning and in the late Wenlock. The difference was caused by evolutionary and ecological reasons. A good correlation between diversity changes and terrestrial environmental events (glaciations, sea-level movements, stable isotope records) is noted.

Biotic recovery is understood in this contribution as restoration of taxonomic diversity of biota (or part of it) after a crisis, to a higher, but not necessarily to a pre-crisis level. Biotic progress, the acquiring or forming of new characteristics (innovations), the adopting of new habitats etc., are favourable factors for recovery, but do not belong to the last concept.

Different general diversity data of Silurian biota have been published repeatedly, e.g. by Sepkoski 1986. In more detail the topic was discussed by Kaljo *et al.* (1995) in the final monograph of the IGCP Global Bio-event project. In summary, the diversity of Silurian biota was strongly influenced by extinctions that occurred in the latest Ordovician, earliest and latest Wenlock and late Ludlow.

All these and a number of less significant Silurian bio-events are generally believed to be caused by terrestrial, environmental (climate including glaciations, oceanic conditions, sea-level and facies changes, nutrient supply) and biotic reasons (Kaljo *et al.* 1995). However, a remote or indirect influence of cosmic agents, like Milankovitch cycles, cannot be neglected (Jeppsson 1990).

The same set of environmental factors should also be considered when analysing different biotic recovery scenarios. In this paper the diversity dynamics of corals (sessile macrobenthos), graptolites (macroplankton) and acritarchs (microphytoplankton) will be discussed, compared with each other and with some parameters of the early Silurian environment.

Coral scenario

Corals experienced a severe crisis at the end of the Ordovician with 62 out of 90 genera, i.e. nearly 70%, becoming extinct. At the species level only a few (in Estonia only 5% of species; Nestor *et al.* 1991) passed into the early Silurian. The ensuing renaissance of the coral fauna proceeded in successive phases, with some accleration of origination in the second half of the Rhuddanian and especially in the Telychian. The latter can be considered some kind of a burst of corals. This is well illustrated by more detailed Baltic tabulate data (Klaamann 1986): in the early Rhuddanian (lower G_{1-2}, stratigraphic indices see Fig. 1) only one new genus (*Macleodia*) appears; a step higher (upper G_{1-2}), two more (*Halysites, Ramusculipora*); in the late Rhuddanian and Aeronian (G_3), four (*Favosites, Parastriatopora*, etc.); and in the Telychian (H), eight new genera (*Thecia, Subalveolites, Placocoenites, Angopora*, etc.) come in. In summary (Fig. 1), the most energetic diversity rise occurred in the Llandovery (84 new genera appeared, i.e. 75% of the total fauna). Later, origination became slower, but due to only a moderate extinction rate in the Wenlock, at that period the Silurian coral fauna reached its maximum diversity (Scrutton 1988, 1989; Kaljo & Märss 1991).

The end-Ordovician mass extinction occurred in Estonia (Nestor *et al.* 1991) in two main steps (within the limits of the precision available). The first one was at the end of the pre-Hirnantian

From Hart, M. B. (ed.), 1996, *Biotic Recovery from Mass Extinction Events*, Geological Society Special Publication No. 102, pp. 127–133

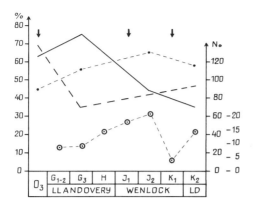

Fig. 1. Coral diversity changes. Solid line – origination rate; percentage of new genera per unit. Dashed line – extinction rate; percentage of genera in a unit which are absent in a succeeding one. Dashed line with black dots – the number of coral genera per unit. Circle with a point – the number of tabulate genera per unit from Estonia only. Scale for Estonian tabulates is the first from the right. Units: O_3, late Ordovician; S_2, Ludlow. Stages: RHUD, Rhuddanian; AER, Aeronian; TEL, Telychian; SHEIN, Sheinwoodian; HOMER, Homerian. Estonian stages: G_{1-2}, Juuru; G_3, Raikküla; H, Adavere; J_1, Jaani; J_2, Jaagarahu; K_1, Rootsiküla; K_2, Paadla. Black arrows, levels of glaciations suggested by isotope studies.

Pirgu Stage and the second at the very end of the Ordovician (among tabulate corals the extinction rate was 60 and 67 per cent correspondingly). During the last time interval, the first corals of 'Silurian type' (e.g. dissepimentate *Paliphyllum* and *Strombodes* among Rugosa) appear. This appearance can be considered a pre-recovery or an innovation event, which created morphological preconditions for the following recovery. During the Rhuddanian these new forms remain relatively rare and therefore the genera of 'Ordovician type' play an important role in the coral assemblage of the early and partly also of the mid-Llandovery.

Among this early post-crisis fauna there are many corals of small size (*Densiphyllum*, some *Paleofavosites* etc.), which show the 'Lilliput' phenomenon, as discussed by Urbanek (1993) in graptolite evolution. In our case these dwarfs are especially striking as they contrast with some of the latest Ordovician gigantic rugose and tabulate corals, which were rather common: the well-known *Grewinkia buceros* is about 30 cm high while a corallum of *Mesofavosites dualis* from the Porkuni quarry (Estonia) is more than 1 m in diameter.

The early Rhuddanian coral survival period with lilliputs was an interval of only relatively low diversity, not comparable, for example, with the severe diversity drop in graptolites in the latest Ordovician.

Coral origination intensity was highest in the Telychian and began to fall in the Wenlock (Fig. 1; Kaljo & Märss 1991). Scrutton (1988) demonstrated that this was true owing to the predominance of Rugosa; the tabulate coral origination maximum occurred before the Late Ordovician mass extinction and never recovered to the pre-crisis level.

Origination–extinction patterns in corals during the transition from the Ordovician to the Silurian, seems to be correlated with the changing environment (Hirnantian glaciation with accompanying sea-level drop and succeeding climate amelioration, transgression, etc.). Later in the early Silurian succession, environmental and coral changes are not so clearly connected. Of course, insufficiently detailed global scale data on coral distributions hinder making far-reaching conclusions. Therefore, most observations of this kind are easier to explain by local ecological conditions (e.g. in Fig. 1 a diversity low of Estonian tabulates in the late Wenlock is strongly influenced by local factors), though there may also be a global component of the process.

Graptolite scenario

Graptolites have a detailed biostratigraphy and their rapid evolution might be much better correlated with different environmental changes.

Morphological innovations and diversity dynamics of graptolites were recently summarized by Koren (in Kaljo *et al.* 1995). Therefore, without going into details and using only a graptolite diversity curve compiled by her (Fig. 2A), we can note some interesting concurrent changes in biota and environment.

1. Graptolite diversity increase was most pronounced in the Rhuddanian after a short post-crisis low-idversity interval (*extraordinarius* and *persculptus* zones, latest Ordovician, see Table 1). This process has roots in a fundamental innovation, the 'uniserial event' (origination of a monograptid colony; Rickards 1988) in *persculptus* time and in a series of radiation events in the earliest Silurian (Berry *et al.* 1990; Koren in Kaljo *et al.* 1995). Strikingly the graptolite diversity increase is correlated with post-glaciation warming of the climate (Spiroden Secundo Episode of Aldridge *et al.* 1993, which may nevertheless have been relatively cool) and rapid sea-level rise and corresponding changes in the state of oceanic waters.

2. The maximum diversity of Silurian graptolites was reached at the beginning of the Aeronian (*gregarius* Zone). Later, especially in the early Telychian (*turriculatus–crispus* and *griestoniensis* zones), the extinction rate was much higher than origination (Fig. 2B) and therefore the diversity curve (Fig. 2A) was falling drastically. A low stand was reached in the early Wenlock (*riccartonensis* Zone), and for the second time in the *nassa* Zone after the *lundgreni* extinction event.

Brenchley *et al.* (1994) gave bathymetric and carbon and oxygen isotopic evidence, showing that the end-Ordovician glaciation, confined to the first half of the Hirnantian, was a short episode (0.5–1 Ma) in a long greenhouse period. The Hirnantian is usually correlated with the *extraordinarius* and *persculptus* graptolite zones, characterized by a low-diversity assemblage. The main graptolite extinction occurred in the preceding *pacificus* Zone (Koren 1987), which should coincide with the very beginning of the glaciation marked by sea-level drop in the earliest Hirnantian. At least in this case we can see clear environmental reasons for graptolite diversity change. Berry *et al.* (1990), referring to data from South China, suggested a diachronous glaciation event starting in the *pacificus* time in high latitudes and reaching a low latitude area later in the *extraordinarius* time.

Brenchley *et al.* (1994) suggest that during the glacial event the pre-Hirnantian warm saline deep waters changed to those with a strong circulation of cold well-oxygenated bottom waters and upwelling bringing a rich influx of nutrients to the surface waters. High bioproductivity and carbon sedimentation caused the high $\delta^{13}C$ values identified in a number of sections in Baltoscandia, North America and South China (Brenchley *et al.* 1994).

The correlation of these data with the diversity curve of graptolites shows that they prefer to live in warm waters, which are not well-oxygenated and stratified. Berry *et al.* (1990) added that optimal conditions for graptolites were bacteria-rich waters.

Early Silurian glaciations have been under discussion for some time. Recently Grahn & Caputo (1992) gave evidence for four of them: the O/S boundary interval, the *gregarius* Zone, early Telychian (possibly starting in the late Aeronian) and the most wide-spread tillites marking a glaciation from the latest Llandovery to earliest Wenlock (Fig. 2A). Johnson & McKerrow (1991) also advocated a late Wenlock glaciation, which caused a considerable lowering of global sea-level.

In order to find independent data, geochem-

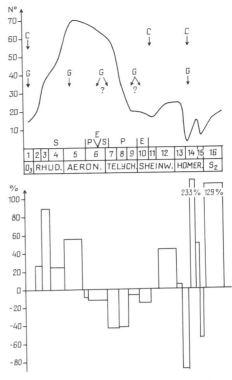

Fig. 2. Diversity dynamics of Silurian graptolites. (**A**) The curve shows the number of species per zone; (**B**) the histogram shows the percentage of summary changes of diversity from one unit to the next. In zones 5, 13, 14, 15 changes are shown separately from the lower and upper parts of the zone. The curve is taken from Koren in Kaljo *et al.* 1995. Standard graptolite zones are shown by numbers: 1, *extraordinarius–persculptus*; 2, *acuminatus*; 3, *vesiculosus*; 4, *cyphus*; 5, *gregarius*; 6, *convolutus–sedgwickii*; 7, *turriculatus–crispus*; 8, *griestoniensis*; 9, *crenulata*; 10, *centrifugus–murchisoni*; 11, *riccartonensis*; 12, *rigidus–ellesae*; 13, *lundgreni*; 14, *nassa–deubeli*; 15, *ludensis*; 16, *nilssoni*. Other units see Fig. 1 Letters with arrows: C, levels of isotope data ($\delta^{13}C$); G, levels of glaciations (see text). Letters: oceanic state episodes according to Aldridge *et al.* 1993: S, secundo and P, primo episodes; E, events.

istry and stable isotopes of Baltic Silurian rocks were also studied at the Institute of Geology, Estonian Academy of Sciences (Kaljo *et al.* 1994). The results of the study will be published in full elsewhere, but here it would be appropriate to note the main conclusion. We have no data on the isotopes in the Llandovery so far, but in the Wenlock two very distinctive $\delta^{13}C$ peaks occur in the Ohesaare borehole. One is in the early Wenlock (*riccartonensis* Zone and the beginning of the *rigidus–ellesae* Zone) and the other in the late Wenlock (*nassa* Zone) reaching +4.2 and 4.6‰ levels respectively. Both peaks were observed also in a graptolite mudstone section in Latvia (Priekule). Jux & Steuber (1992) recorded high $\delta^{13}C$ level values (+4.97 and 4.69‰) in the Högklint and Tofta beds (corresponds roughly to the *riccartonensis–rigidus* peak), but not in the upper Wenlock. Corfield *et al.* (1992) in turn established a $\delta^{13}C$ peak and depletion in the upper Wenlock of Britain (*nassa* Zone).

As mentioned above, the *riccartonensis* and *nassa* zones are post-crisis low-diversity intervals of graptolite evolution, which precede the radiation events and diversity rise. The comparison of those with the described end-Ordovician–earliest Silurian extinction–survival–recovery scenario shows great similarity in many aspects. This suggests that all three were directed by the same factors – climate, oceanic state and bioproductivity. Depending on the evolutionary (phenotypic) situation, the resulting diversity curve may differ to some extent.

Acritarch scenario

Acritarch diversity was recently analysed by Le Herisse (in Kaljo *et al.* 1995). According to his data (used below), near the Ordovician–Silurian boundary there was no drastic acritarch extinction, but many genera and species disappeared during a longer period in the Late Ordovician. The same can be seen from data provided by Uutela & Tynni (1991). Because of this the Ordovician and Silurian acritarch floras differ considerably and the few 'Silurian'-type genera that appeared in the Hirnantian do not change the above conclusion.

In the early Llandovery (*acuminatus* Zone) a low-diversity interval occurred, where several genera passing from the Ordovician are represented by small-sized species like 'lilliputs'.

The first radiation event among acritarchs occurred in the middle Rhuddanian (*vesiculosus* Zone), where six very distinctive genera appeared. Acritarch diversity then increases step by step through the mid- and late Llandovery to a maximum level in the Telychian and at the very beginning of the Wenlock (*centrifugus–murchisoni* Zone), especially on the species level (Fig. 3).

This was followed by a major extinction event and a diversity low in the middle Sheinwoodian (*riccartonensis* Zone and the beginning of the *rigidus–ellesae* Zone). Analogous diversity changes, but on the lower level, occurred also in the late Sheinwoodian and Homerian with a

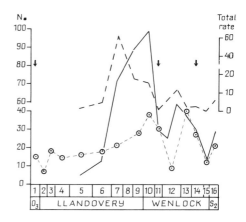

Fig. 3. Diversity of early Silurian acritarchs. Circle with a point – the number of genera per unit; solid line – the number of species per unit; dashed line – total rate of appearing species (number of appearing taxa per 1 Ma). Both species level curves are based on data from the Silurian of Gotland. Data for the figure are taken from Le Herisse in Kaljo *et al.* 1995. Black arrows – levels of glaciations suggested by isotope studies. Standard graptolite zones are shown by numbers, for explanation see Fig. 2.

maximum in the *ellesae* (species) and *lundgreni* (genera) zones and with a minimum in the *ludensis* Zone.

The acritarch diversity dynamics has much in common with that of corals in the Llandovery, i.e. the early Rhuddanian low diversity interval, the first radiation in the middle Rhuddanian and diversity burst in the Telychian. Wenlock curves are different, but bearing in mind the falling appearance rates in both groups, the general process seems to be the same.

The most important difference between acritarchs and graptolites is the very early diversity peak of the latter in the early Aeronian. Wenlock curves of both groups are more or less parallel.

Summary and conclusions

The above-discussed patterns of diversity dynamics of three ecologically different groups of biota show some striking similarities and differences. In Table 1 these are correlated with environmental parameters. Using these data we may deduce the following.

Mass extinction of corals (E_1) and graptolites (E_{max}) occurred at the end of a warm greenhouse period or at the very beginning of a glacial event (end of the *pacificus* Zone). At that time (and until the end of the Ordovician) acritarchs

Table 1. *Correlation of biotic events and environmental parameters*

Series	Stages Standard	Stages Estonian	Graptolite zones	Corals	Graptolites	Acritarchs	$\delta^{13}C$ (‰)	Climate	Sea-level	Ocean state
Llandovery	Rhuddanian	Raikküla	cyphus	O_3	O_4					
		Juuru	vesiculosus	O_2	O_3	O_1		Greenhouse		As pre-Hirnantian, perhaps cooler
			acuminatus	LD	O_2	LD	0. . .–1	Warming	Rapid rise	
Ashgill	Hirnantian	?	persculptus		LD O_1	"small" extinctions	+4. . . +7	Glacial event	Low stand	Bottom waters cold, well oxygenated, strong upwelling
		Porkuni	extraordinarius	E_{sum} giants $E_{tab2}O_1$	LD			Cooling		
		?	pacificus		E_{mass}		? 0. . .–2	? Greenhouse	? High	Deep waters, warm, weak circulation
	Rawtheyan	Pirgu		E_{tab1}						

Notes: E_{sum}, summary coral extinction; $E_{tab1, 2}$, tabulate extinction events (see text), E_{mass}, mass extinction; o_1. . . o_2 or O_1. . . O_4, originations, correspondingly a few or many new taxa; LD, low diversity interval; climate, sea-level and ocean state according to Brenchley *et al.* (1994).

experienced normal background or stepwise extinction. The extinction of corals accelerated in the Hirnantian icehouse period concurrently with the cooling of the ocean, although the water was rich in oxygen and nutrients. The shallow shelf sea, which was a habitat for giant corals, might still have remained warm owing to tropical sunlight in the equatorial areas (e.g. Baltica). During this period graptolites survived at a low diversity level. All three groups, however, displayed different morphological innovation and origination of a few new taxa. At the very beginning of the Silurian, graptolites demonstrated a very rapid diversity rise (Fig. 2), whereas corals and acritarchs experienced a brief low-diversity period and much slower origination of new taxa.

This correlation allows us to conclude the following.

1. As is commonly known, corals preferred to live in well-aerated waters rich in food. Their diversity drop therefore seems to have been caused by the lowering of temperature at the outset of glaciation (E_1 extinction; Table 1) and, perhaps, by the reduction or even loss of suitable habitats (E_2) due to a rapid sea-level rise at the very beginning of the Silurian. The giants recorded in the Porkuni Stage in Estonia and elsewhere were evidently inhabiting warmer niches.

2. Graptolites were expected to show low diversity in the glacial period, because as shown by Berry *et al.* (1990), they usually occur in oxygen-poor waters.

3. Most biomass is usually produced by planktic, especially microplanktic organisms. Owing to their relatively low diversity in the latest Ordovician, the share of the groups discussed in the summary bioproduction seems to have been insufficient for ensuring high values of δ^{13}C (Brenchley *et al.* 1994; Table 1). The real reason for this still remains obscure, especially because there is no direct dependence between diversity and bioproduction. A good example is the *Monograptus riccartonensis* low-diversity assemblage producing a considerable amount of biomass due to the mass occurrence of the index-species.

The end-Ordovician–Early Silurian sequence of bioevents was discussed in more detail in order to get an idea of the pattern followed by different groups through the extinction–survival–recovery scenario. An analogous diachronous pattern is observed at the Llandovery–Wenlock junction (the extinction of graptolites began just before or at the very beginning of the glaciation; Fig. 2; Melchin 1994; a strong diversity drop of acritarchs followed a step later,

etc., see above) and in a less pronounced form also in the late Wenlock.

In general, proceeding from the above data and discussion, we can draw the following conclusions:

1. After the Late Ordovician, early and late Wenlock, probably glaciation- (climate-) triggered extinctions, the recovery processes of corals, graptolites and acritarchs took place more or less according to the classical scenario: extinction–survival–recovery, but are different in detail as explained above.

2. Diachrony of phases of this scenario in different groups discussed occurs frequently: main extinctions might be coinciding (sometimes partly) or successive or show a stepwise pattern: survival or low-diversity interval and recovery of these groups began correspondingly at slightly different time levels. The latter might be rapid in a short interval (graptolites) or occur as a slow prolonged rise of diversity (corals), sometimes with a burst in a certain time (acritarchs).

3. The differences in diversity dynamics seem to depend on evolutionary and ecological characteristics of the group of organisms involved.

This could account for the rapid response of graptolites (early diversity peak) to the changing environment with the profound innovations as pointed out above. Surprising is the parallelism in coral and acritarch dynamics.

4. For understanding mutual relations and influence in an ecosystem (organisms + environment), exact dating of different events, isotope excursions, etc., seems to be crucial.

The author thanks P. J. Brenchley for his helpful comments and suggestions. His colleagues T. Kiipli, T. Martma and A. Noor are thanked for their help. The study was partly supported by International Science Foundation (grant No. LC 4000) and Estonian Science Fund (grant No. 314).

References

ALDRIDGE, R. J., JEPPSSON, L. & DORNING, K. J. 1993. Early Silurian oceanic episodes and events. *Journal of the Geological Society, London*, **150**, 501–513.

BERRY, W. B. N., WILDE, P. & QUINBY-HUNT, M. S. 1990. Late Ordovician graptolite mass mortality and subsequent Early Silurian re-radiation. *In:* KAUFFMAN, E. G. & WALLISER, O. H. (eds) *Extinction Events in Earth History.* Springer, Berlin, Lecture Notes in Earth Sciences, **30**, 115–123.

BRENCHLEY, P. J., MARSHALL, J. D., CARDEN, G. A. F., ROBERTSON, D. B. R., LONG, D. G. F., MEIDLA, T., HINTS, L. & ANDERSON, T. F. 1994.

Bathymetric and isotopic evidence for a short-lived Late Ordovician glaciation in a greenhouse period. *Geology*, **22**, 295–298.

CORFIELD, R. M., SIVETER, D. J., CARTLIDGE, J. E. & McKERROW, W. S. 1992. Carbon isotope excursion near the Wenlock–Ludlow (Silurian) Boundary in the Anglo-Welsh Area. *Geology*, **20**, 371–374.

GRAHN, Y. & CAPUTO, M. V. 1992. Early Silurian glaciation in Brazil. *Palaeogeography, Palaeoclimtology, Palaeoecology*, **99**, 9–15.

JEPPSSON, L. 1990. An oceanic model for lithological and faunal changes tested on the Silurian record. *Journal of the Geological Society, London*, **147**, 663–674.

JOHNSON, M. E. & McKERROW, W. S. 1991. Sea level and faunal changes during the latest Llandovery and earliest Ludlow (Silurian). *Historical Biology*, **5**, 153–169.

JUX, U. & STEUBER, T. 1992. C_{carb-} und C_{org-} Isotopenverhältnisse in der silurischen Schichtenfolge Gotlands als Hinweise auf Meeresspiegelschwankungen und Krustenbewegungen. *Neues Jahrbuch für Geologie und Paläontologie, Monatshefte*, **7**, 385–413.

KALJO, D. & MÄRSS, T. 1991. Pattern of some Silurian bioevents. *Historical Biology*, **5**, 145–152.

——, BOUCOT, A. J., CORFIELD, R. M., KOREN, T. N., KRIZ, J., LE HERISSE, A., MÄNNIK, P., MÄRSS, T., NESTOR, V., SHAVER, R. H., SIVETER, D. J. & VIIRA, V. 1995. Silurian bio-events. *In:* WALLISER, O. H. (ed.) *Global events and event-stratigraphy in the Phanerozoic*. Springer, Berlin, 173–226.

——, KIIPLI, T. & MARTMA, T. 1994. Geochemical and isotope (^{13}C) event markers through the Wenlock–Pridoli sequence in Ohesaare (Estonia). *In:* JOACHIMSKI, M. (ed.) *Geochemical event markers in the Phanerozoic. Abstracts and Guidebook*. Erlangen Geologische Abhandlungen, **122**, 36.

KLAAMANN, E. R. 1986. Soobshchestva i biozonalnost tabulatomorfnykh korallov Pribaltiki. The tabulate communities and biozones of the East Baltic Silurian. *In:* KALJO, D. & KLAAMANN, E. (eds)

Theory and practice of ecostratigraphy. Valgus Publishing, Tallinn, 80–98 [in Russian].

KOREN, T. 1987. Graptolite dynamics in Silurian and Devonian time. *Geological Society of Denmark Bulletin*, **35**, 149–159.

MELCHIN, M. J. 1994. Graptolite extinction at the Llandovery–Wenlock boundary. *Lethaia*, **27**, 285–290.

NESTOR, H. E., KLAAMANN, E. R., MEIDLA, T. R., MÄNNIK, P. E., MÄNNIL, R. P., NESTOR, V. V., NÕLVAK, J. R., RUBEL, M. P., SARV, L. J. & HINTS, L. M. 1991. Dinamika fauny v Baltijskom basseine na granitse ordovika i silura [Faunal dynamics in the Baltic basin at the Ordovician–Silurian boundary]. *In:* KALJO, D., MODZALEVSKAYA, T. & BOGDANOVA, T. (eds) *[Major biological events in Earth history.]* All-Union Paleontological Society Transactions of the XXXII session, Tallinn, 79–86 [in Russian].

RICKARDS, R. B. 1988. Anachronistic, heraldic and echoic evolution: new patterns revealed by extinct planktonic hemichordates. *In:* LARWOOD, G. P. (ed.) *Extinction and survival in the fossil record*. Systematics Association, Special Volume, **34**, Clarendon, Oxford, 211–230.

SCRUTTON, C. T. 1988. Patterns of extinction and survival in Palaeozoic corals. *In:* LARWOOD, G. P. (ed.) *Extinction and survival in the fossil record*. Systematics Association, Special Volume, **34**, Clarendon, Oxford, 65–88.

—— 1989. Corals and stromatoporoids. *In:* HOLLAND, G. H. & BASSETT, M. G. (eds) *A global standard for the Silurian System*. National Museum of Wales, Cardiff, Geological Series, **9**, 228–230.

SEPKOSKI, J. J. 1986. Global bioevents and the question of periodicity. *In:* WALLISER, O. H. (ed.) *Global bio-events*. Springer, Berlin, Lecture Notes in Earth Sciences, **8**, 47–61.

URBANEK, A. 1993. Biotic crises in the history of Upper Silurian graptoloids: a palaeobiological model. *Historical Biology*, **7**, 29–50.

UUTELA, A. & TYNNI, R. 1991. Ordovician acritarchs from the Rapla borehole, Estonia. *Geological Survey of Finland Bulletin*, **353**, 1–135.

Searching for extinction/recovery gradients: the Frasnian–Famennian interval, Mokrá Section, Moravia, central Europe

PETR ČEJCHAN & JINDŘICH HLADIL

Geological Institute, Czech Academy of Sciences, Rozvojová 135, CZ-16502 Praha 6-Suchdol, Czech Republic

Abstract: A series of ancient seafloors colonized by diverse organisms has been documented from the Upper Devonian rocks of the Western Mokrá Quarry. Situated in the southern tectonic closure of the Moravian Karst, the Frasnian–Famennian shallow carbonate ramps exhibit both Rhenish and Ukrainian affinities. Reconstruction of palaeo-sea floor horizons results in a series of 28 quadrats sufficient for further evaluation. Eighty-five taxa involved were scrutinized for abundance, occupied area, skeletal mass production and biomass production. The aim of the study was to determine whether the observed sequence of quadrats can be distinguished from a random one, and to discover any possible unidimensional gradient as a latent control. Monte Carlo simulations and a graph theoretical approach were utilized. Although the raw data seemed chaotic, the simulations demonstrated the observed sequence is not random. A significant influence of a hidden control is thus suggested. Fifteen characteristics of quadrats (e.g. diversity, number of taxa, vertical stratification of community, number of patches) were utilized for final interpretation. The gradient reconstructed by TSP algorithm reveals a significant crisis within the uppermost part of the *Amphipora*-bearing limestone.

The Mokrá section represents one of the final Devonian carbonate ramps with an abundant reef-dwelling fauna. Its shallow-water record of the Frasnian–Famennian (F–F) Kellwasser events differs from that of other, usually open-marine sequences. In the Mokrá section, the typical dark Kellwasser shale intervals were represented by stratigraphic omission surfaces. Light-coloured limestones with scattered accumulations of *Amphipora* stems prevail. The onset of deposition of nodular limestone is delayed. The fairly exceptional position of this section relative to many other global F–F sections makes it difficult to locate accurately the Kellwasser events here. Accessible colonization surfaces yield a good data base for quantitative evaluation of bio-events as recorded in the Mokrá section. However, the composition of assemblages of benthic organisms vary extremely among individual palaeo-sea floors. As no clear pattern emerged from visual inspection, quantitative methods were used to elucidate the F–F history of this section. A new tool based on graph theory was developed and successfully tested for its ability to reveal the underlying pattern of extinction/recovery processes.

Geology, facies and stratigraphy

The Mokrá quarries are situated about 10 km ESE of Brno. This site occurs within the southern tectonic closure of the Moravian Karst and contains Frasnian/Famennian limestone sequences related to shallow carbonate ramps of both Rhenish and Ukrainian affinity. Palaeogeographically, the Mokrá area was situated on the northern margin of a belt of Devonian back-arc or related basins. This belt of extensional basins originated during Late Emsian to Eifelian times and can be traced from SW England, across Europe towards Russia, cutting both the accreted Avalonian segments and southern margins of Laurussia. After maximum extension during the late Middle Devonian–early Late Devonian, closure began during the Late Frasnian. The sequence of strata studied herein represents one of the latest carbonate ramps worldwide with a moderately abundant reef fauna. During the F–F interval, light coloured limestones with rhythmic occurrences of *Amphipora* stems prevailed. These facies differ from normal pelagic facies and the typical dark Kellwasser shale intervals of the F–F boundary, and this interval is normally represented by stratigraphic omission surfaces in parts of the Mokrá Quarries.

The Mokrá quarries have attracted considerable interest among palaeontologists because they possibly record survival of Frasnian communities into the Early Famennian. This selective survivorship has been discussed in many papers (Friáková *et al.* 1985; Dvořák *et al.* 1987;

From Hart, M. B. (ed.), 1996, *Biotic Recovery from Mass Extinction Events,*
Geological Society Special Publication No. 102, pp. 135–161

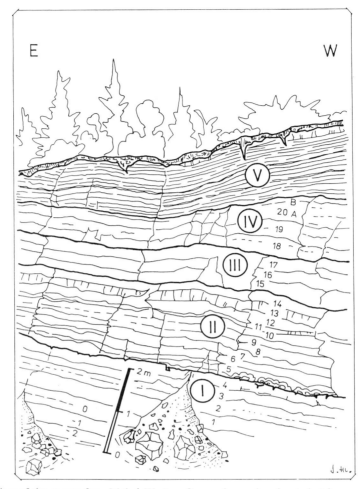

Fig. 1. General view of the quarry face: Mokrá Western Quarry (upper bench, at electric-power transformer). Drawn from photograph. The section is conserved for further investigation thanks to the exploitation company 'Cementárny a vápenky Mokrá'.

Hladil et al. 1989, 1991). Whereas many data support the idea of selective survivorship, opposing evidence has also been presented.

Different types of F–F facies crop out in the Mokrá quarries as a result of intensive Variscan tectonic thrusting that modified the original positions of the sedimentary basins (Hladil et al. 1991). Three types of the F–F boundary sequences have been identified:

1. A sequence related to a trough-shaped tectonic sag consists of well-bedded dark limestones, primarily calciturbidite beds with fewer background basinal clay facies and geostrophically reworked mass flow deposits. However, the relative clay and siliceous silt components increase upsection towards the Famennian.

2. Stratigraphic omission surfaces spanning the F–F boundary are characteristic over elevated palaeoridges. Thick-bedded limestones with amphiporans and solenoporaceans were deeply eroded down to Upper Frasnian levels. During sea-level highstands, numerous neptunian dykes originated on the slopes of these ridges and were filled with both subaeral and marine sediments and cements. The Lower Famennian is biostratigraphically determined only in cave infillings. Overlying stratigraphic cover is light-coloured and largely detrital. Conodonts of the *Palmatolepis marginifera* Zone are associated with '*Megalodon*', gastropod shells, and very rare corals.

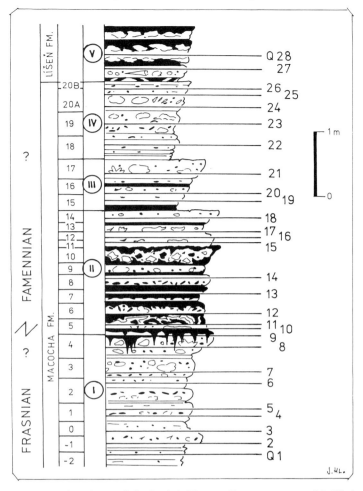

Fig. 2. Simplified stratigraphic column of the Mokrá Western Quarry sequence. Modified after Hladil *et al.* (1989). Names of formations, symbols of beds and stratigraphic intervals (left) are correlated with the levels where quadrats were reconstructed (right).

3. The third type of F–F sequence is the subject of the present study (Figs 1 and 2). This section is located near the entrance to the upper bench of the Western Mokrá Quarry, behind the electric power transformer. The section contains a prominent hardground subsequently overlain by facies indicating fluctuating dysoxia, superimposed on 'Milankovitch-type' cyclicity in the upper part of the *Amphipora*-bearing limestone sequence. Clay admixture is locally present. Carbonate microfacies analysis indicates a sheltered ramp environment. The uppermost beds contain *Amphipora*, other stromatoporoids and coral skeletons consisting of almost pure calcite. Sedimentary cover conformably overlying these beds consists of a few centimetres of fine-detrital material overlain by nodular micritic limestones.

The change of sedimentary environments between the coral-bearing limestone and beds overlying this facies is distinct and corresponds to final depletion of coral assemblages. Conodont assemblages of *Palmatolepis crepida* Zone were found both below and above this boundary.

Stratigraphic discussion

Despite extensive investigation, the stratigraphy of the Western Mokrá Quarry, a Type 3 sequence, remains puzzling. Three interpretations (A–C below) have been suggested.

A. Sedimentation of the coral-rich limestones continues from the Late Frasnian into the lower

P. ČEJCHAN & J. HLADIL

Table 1. *Data utilized for calculations*

Quadrat	Taxon	Abundance	Area occupied	Skeletal mass prod.	Biomass prod.
1	Cyanobacteria sp. 1	18	209	411	2680
1	Gastropoda sp. 1	7	1	0.2	1.9
1	Issinella sp. 2	1720	52	54	126.5
1	Multiseptida corallina	450	5	0.09	0.3
1	Podocopa sp. 1	690	10	0.15	0.5
1	Polychaeta sp. 2	75	4	3.2	24.8
1	Pseudopemmatites sp.	4	2	1.5	5.8
1	Sphaerocodium sp.	4	4	4.5	7.6
1	Tikhinella fringa	2450	13	0.4	1.5
2	Amphipora hanimedi	870	51	169	563
2	Amphipora moravica	450	23	105	329
2	Issinella sp. 3	3300	95	12.4	26.2
2	Kamaena spp.	10600	40	3.4	5.3
2	Solenoporaceae sp.	16	2	3.2	7.5
3	Issinella sp. 2	2900	83	54	139.5
3	Kamaena spp.	17900	112	4	6.4
3	Multiseptida corallina	1890	9	0.4	1.5
3	Nanicella porrecta	400	8	0.8	0.6
3	Rhynchonellidae sp. 2	48	4	1.9	21.6
4	Amphipora hanimedi	420	16	80	264.2
4	Amphipora moravica	6020	298	1391	4430
4	Monactin sp. 2	69	35	10.9	57.7
4	Podocopa sp. 3	2730	39	0.64	1.5
4	Scoliopora vassinoensis	1	1	0.64	0.4
5	Amphipora moravica	1220	68	283	890.1
5	Atrypidae ? sp. 1	770	55	50.3	315.6
5	Disphyllum sp.	1	2	0.64	4.2
5	Issinella sp. 3	710	79	16.7	33.5
5	Rhynchonellidae sp. 1	65	10	7.1	105.2
5	Scoliopora denticulata	3	2	1.93	1.2
5	Spiriferidae sp. 1	63	9	2.4	25.2
6	Amphipora hanimedi	3550	127	858	2911
6	Amphipora moravica	240	12	52	162
6	Cynobacteria sp. 2	5	31	38.5	1550
6	Holothuroidea sp.	7	14	0.6	246
6	Kamaena spp.	4400	20	1.3	2
7	Actinostroma sp.	7	12	201.3	517
7	Alaiophyllum jana	1	3	0.4	1.4
7	Amphipora moravica	430	25	109.6	348
7	Multiseptida corallina	3910	23	0.9	3.1
7	Solenoporaceae sp.	270	14	30.4	71
7	Stachyodes lagowiensis	53	23	37.7	55
7	Stromatopora sp. 2	7	8	52.9	118
7	Syringostroma tenuilaminatum	10	7	88.2	199
7	Syringostroma vesiculosum	26	16	171.2	416
7	Taleastroma sp.	13	8	53.7	133
7	Tikhinella fringa	3060	34	0.6	2.2
8	Actinostroma sp.	4	8	79.6	242
8	Amphipora hanimedi	750	29	175.5	590
8	Amphipora moravica	23100	231	4387	13940
8	Issinella sp. 1	410	15	15.2	39.1
8	Nanicella porrecta	1350	18	2.7	2.3
8	Scoliopora kaisini	1	10	5.8	3.5
8	Stachyodes lagowiensis	28	35	35.5	48.1
8	Vicinesphaera sp.	3200	40	0.06	0.4
9	Amphipora moravica	1160	68	257.9	813
9	Bacteria sp. 2	7	89	64.2	2970
9	Nanicella porrecta	2010	10	3.9	3.3
9	Scoliopora vassinoensis	16	9	8.3	5.2

EXTINCTION/RECOVERY GRADIENTS, MORAVIA

9	Spiriferidae sp. 2	790	13	87.7	430
10	Amphipora moravica	95	5	20.3	64.8
10	Amphipora tschussovensis	60	4	11.1	39.5
10	Aulostegites sp.	7	16	8.6	14.2
10	Kamaena spp.	4000	17	1.3	1.9
10	Porifera sp. 1	16	6	0.9	3.9
10	Spiriferidae sp. 1	3500	187	187	1103
11	Amphipora moravica	610	42	156	498
11	Labechia cumularis	28	23	445	2940
11	Spiriferidae sp. 3	630	35	78.5	491
11	Syringopora volkensis	4	292	1744	3896
11	Tienodictyon sp.	5	61	238	2999
12	Amphipora moravica	1300	75	300	958
12	Kamaena spp.	13700	76	3.2	10.9
12	Polychaeta sp. 1	17	83	2.8	73.1
13	Amphipora moravica	2400	81	514	1632
13	Ichnia sp. (Bivalvia)	14	11	2.4	16.9
13	Porifera sp. 2	20	17	1.3	8.7
13	Stachyodes sp.	8	9	4.1	4.4
14	Amphipora moravica	640	22	128	4.6
14	Amphipora tschussovensis	1550	85	305	1050
14	Jansaella sp.	1510	36	14.3	40.2
14	Monactin sp. 2	45	21	4.9	25.6
14	Scoliopora rachitiforma	18	27	38.5	24.8
14	Stachyodes sp.	6	7	3.9	3.8
15	Kamaena spp.	16600	73	4.9	17.5
15	Multiseptida corallina	8800	30	1.7	6
15	Palaeoplysina sp.	8	8	2.8	21.8
15	Vicinesphaera sp.	8400	54	1.1	3.9
16	Amphipora moravica	3080	88	642	2040
16	Multiseptida corallina	1950	7	0.3	1.1
16	Vicinesphaera sp.	14300	55	2.1	7.9
17	Actinostroma sp.	12	13	235	715
17	Amphipora moravica	6000	29	1391	4420
17	Amphipora tschussovensis	1350	52	260	919
17	Girvanella sp.	240	6	4.5	39.5
17	Kamaena spp.	1620	7	1.9	7
17	Labechia cumularis	19	26	278	1820
17	Praewagenoconcha sp.	28	6	2.1	13.2
17	Vicinesphaera sp.	6000	33	0.1	0.4
18	Alaiophyllum jana	5	7	3.2	10.2
18	Amphipora sp.	920	25	235	735
18	Monactin sp. 1	25	8	3.9	14.6
18	Rugosa sp.	2	3	0.9	2.5
18	Stromatopora sp. 1	1	1	0.6	2.1
18	Stromatoporella sp.	4	6	102	188
19	Atelodictyon sp.	1	1	1.9	4.4
19	Gastropoda sp. 2	17	5	23.5	37
19	Kamaena spp.	13500	170	4	6.6
19	Labechia cumularis	1	2	1.9	12.7
19	Polychaeta sp. 4	1	1	0.1	2
20	Alaiophyllum jana	4	5	4.7	17.2
20	Amphipora moravica	480	30	103	331
20	Amphipora tschussovensis	260	15	52.4	180
20	Atrypidae sp. 3	48	9	33.4	139
20	Calcisphaera sp.	840	56	162	455
20	Cribrosphaeroides sp.	650	28	1.3	6.1
20	Disphyllum veronica	4	12	9.6	43.7
20	Habrostroma incrustans	1	2	21.4	32.8
20	Monactin sp. 1	27	12	7.9	30
20	Natalophyllum perspicuum	3	7	103	62.7
20	Pentagonostipes ? sp.	6	13	12.4	29
20	Scoliopora rachitiforma	7	6	118	77.5
20	Scoliopora vassinoensis	4	17	156	94.2

P. ČEJCHAN & J. HLADIL

20	Stromatopora sp. 1	3	8	10.9	34.6
20	Tikhinellidae sp. 1	900	23	0.2	0.6
20	Tikhinellidae sp. 2	330	18	0.1	0.2
21	Alaiophyllum jana	2	1	1.1	3.9
21	Disphyllum veronica	6	4	1.9	8.3
21	Habrostroma incrustans	5	15	135	207
21	Multiseptida corallina	6300	30	1.1	3.5
21	Natalophyllum perspicuum	3	8	137	83.2
21	Parathurammina div. ssp.	5760	48	0.4	1.5
21	Scoliopora rachitiforma	3	3	52.9	33.7
21	Scoliopora vassinoensis	4	3	64.2	39
21	Solenoporaceae sp.	860	66	161	368
21	Stachyodes sp.	14	11	439	431
21	Tabulophyllum maria	14	14	52.4	323
21	Tournayellidae sp.	670	17	0.02	0.1
22	Actinostroma sp.	7	9	150	455
22	Amphipora moravica	940	47	235	825
22	Amphipora tschussovensis	1560	65	321	1208
22	Archaesphaera sp.	17700	8	0.2	0.6
22	Atrypidae ? sp. 2	65	11	28.9	108
22	Cribrosphaeroides sp.	3300	37	3.9	14.9
22	Labechia cumularis	10	24	154	1037
22	Multiseptida corallina	14500	73	3	9.1
22	Natalophyllum perspicuum	1	9	76.6	44.4
22	Scoliopora vassinoensis	8	6	62.1	37.7
22	Solenoporaceae sp.	85	34	161	375
22	Tetractin sp. 2	7	2	6.4	14.6
22	Tournayellidae sp.	16400	39	0.2	0.7
23	Alaiophyllum jana	4	4	1.7	6.1
23	Amphipora moravica	260	16	68.5	223
23	Amphipora tschussovensis	2400	153	470	1760
23	Archaesphaera sp.	18200	93	2.1	7.2
23	Cribrosphaeroides sp.	1450	49	1.7	6.6
23	Palaeoplysina sp.	8	11	4.5	33.4
23	Scoliopora denticulata	5	10	90	52.9
23	Vicinesphaera sp.	2800	15	0.4	1.4
24	Amphipora moravica	1270	79	235	759
24	Amphipora tschussovensis	2770	132	461	1720
24	Calcisphaera sp.	700	27	103	288
24	Multiseptida corallina	9500	50	2.8	8.4
24	Scoliopora rachitiforma	10	10	4.7	2.9
24	Solenoporaceae sp.	560	37	94.2	222
25	Actinostroma sp.	4	5	111.3	427
25	Amphipora tschussovensis	490	58	113.4	485
25	Atelodictyon sp.	4	30	44.3	103
25	Habrostroma incrustans	4	17	152	230
25	Labechia sp.	2	8	19.9	139
25	Multiseptida corallina	8755	43	1.9	7.2
25	Natalophyllum perspicuum	2	3	91	50.4
25	Parathurammina sp. 1	55100	184	2.6	9.6
25	Scoliopora rachitiforma	7	3	13.5	8.2
25	Tabulophyllum maria	1	5	7.7	50.2
26	Amphipora tschussovensis	630	42	115	275
26	Archaesphaera sp.	7160	70	0.1	0.2
26	Bacteria sp. 1	35	58	4.3	453
26	Calcisphaera sp.	480	96	8.1	22.8
26	Cyclocyclicus sp.	2	2	4.5	15.6
26	Stromatopora sp. 1	1	4	81	258
26	Tabulophyllum maria	1	3	6	39
26	Tetractin sp. 1	28	4	7	26.2
27	Bacteria sp. 1	16	20	4.1	430
27	Cyclocyclicus sp.	18	18	15	40.1
27	Irregularina sp.	15040	188	2.4	19.5
27	Kamaena spp.	19840	62	7.5	10.8

27	Palaeoplysina sp.	5	2	4.3	34
27	Podocopa sp. 2	3200	20	0.4	5.9
27	Rauserina sp.	68000	43	1.5	5.5
27	Tetractin sp. 1	50	35	51.4	190
28	Archaesphaera sp.	62100	345	10.3	56
28	Entomozoa sp. 1	20280	13	2.8	14.6
28	Podocopa sp. 2	5200	29	1.1	3.4
28	Polychaeta sp. 3	3	2	1.3	9.5
28	Rauserina sp.	25000	74	1.7	5
28	Tetractin sp. 1	30	24	24	82

Data are rough estimates based on slabs, etched surfaces and thin sectioned samples. Variables and their units are explained in 'Data' section.

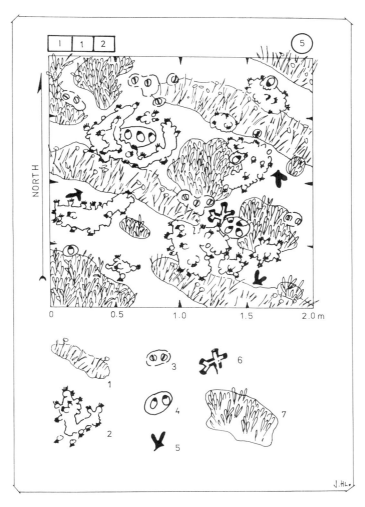

Fig. 3. Illustration of reconstructed sea floor: Quadrat 5 (interval I, bed 1, level 2). 1, *Issinella* with thick-walled tube segments of thalli (Algae); 2, Atrypid with 'spiny' shells (Brachiopoda); 3, Rhynchonellid with large shells (Brachiopoda); 4, Spiriferid with smooth shells (Brachiopoda); 5, *Scoliopora denticulata denticulata* (Tabulata); 6, *Disphyllum* 'Older Form' (Rugosa); 7, *Amphipora moravica* (Stromatoporoidea).

Table 2. *Inspected overall variables, related to the quadrats 1–28*

(i)

quadrat	total number of patches	number of isolated patches	ratio all patches/ isolated pateches	log 10 of 'patch' ratio
1	59	16	3.69	0.57
2	46	24	1.92	0.28
3	30	13	2.31	0.36
4	22	1	22	1.34
5	32	7	4.57	0.66
6	27	9	3	0.48
7	104	48	2.17	0.34
8	35	4	8.75	0.94
9	54	26	2.08	0.32
10	24	15	1.6	0.2
11	46	3	15.33	1.19
12	54	24	2.25	0.35
13	31	28	1.11	0.04
14	36	16	2.25	0.35
15	32	18	1.78	0.25
16	12	10	1.2	0.08
17	52	19	2.74	0.44
18	28	22	1.27	0.11
19	15	9	1.67	0.22
20	52	13	4	0.6
21	49	8	6.5	0.81
22	39	2	19.5	1.19
23	47	2	23.5	1.37
24	33	2	16.5	1.22
25	33	1	33	1.52
26	40	3	13.33	1.23
27	44	4	11	1.04
28	19	1	19	1.28

(ii)

quadrat	number of taxa	benthos stratification, max. number	Shannon index of diversity	total biomass
1	9	2	1.5	831
2	5	2	1.88	306
3	5	3	1.5	58
4	5	3	1.15	1559
5	7	3	2.05	434
6	5	2	1.67	1455
7	11	5	3.21	653
8	8	5	2.02	4892
9	5	3	1.74	1161
10	6	1	1.15	364
11	5	4	1.62	3372
12	3	2	1.58	337
13	4	2	1.37	546
14	6	3	2.23	511
15	4	2	1.7	15
16	3	2	1.19	673
17	8	3	2.63	2527
18	6	4	2.04	325
19	5	2	0.37	24
20	16	6	3.62	582
21	12	5	2.9	637
22	8	4	2.19	683
23	13	7	3.22	1333
24	6	6	2.23	976
25	10	4	2.1	517
26	8	5	2.21	329
27	8	6	2.28	206
28	6	4	1.39	53

EXTINCTION/RECOVERY GRADIENTS, MORAVIA

(iii)

quadrat	biomass of colonies	% of biomass of colonies from total biomass	biomass of retro-colonies	% of biomass of retro-colonies from bm of colonies
1	778	93	0	0
2	294	96	292	99
3	0	0	0	0
4	1559	100	1541	99
5	295	68	295	100
6	1393	96	996	72
7	651	100	115	18
8	4877	100	4773	98
9	1030	89	268	26
10	41	11	40	98
11	3230	96	164	5
12	314	93	314	100
13	541	99	537	99
14	495	97	472	95
15	6	42	0	0
16	671	100	671	100
17	2521	100	1748	69
18	325	100	247	76
19	5	22	0	0
20	372	64	188	51
21	635	100	98	15
22	678	99	633	93
23	1290	97	647	50
24	875	90	794	91
25	512	99	164	32
26	322	98	109	34
27	178	86	0	0
28	27	50	0	0

(iv)

quadrat	% of biomass of bacteria, etc. from total biomass	skeleton mass production	mass % of skeleton production from total biomass production	% of quadrat area occupied by bottom colonizers
1	93	119	14	75
2	0	73	24	53
3	0	15	26	54
4	1	371	24	97
5	0	91	21	56
6	27	238	16	51
7	0	187	29	43
8	0	1175	24	97
9	65	106	9	47
10	0	57	16	59
11	0	666	20	100
12	0	77	23	59
13	1	130	24	29
14	2	124	24	50
15	0	3	20	41
16	0	161	24	38
17	0	543	22	43
18	2	87	27	13
19	0	8	33	45
20	2	199	34	65
21	0	261	41	55
22	0	160	23	88
23	0	301	23	91
24	0	225	23	84
25	0	140	27	89
26	37	57	17	70
27	83	22	11	97
28	50	10	19	100

(i) Data on patches (on the full square of 4 m^2); (ii) shows number of taxa, benthos stratification, Shannon index diversity (based on patches, with unit of dm^2) and total amount of produced biomass (g/m^2/year). (iii) and (iv) describe amount of produced biomass and ratios among specific groups of organisms.

Table 3. *Correlation coefficients between rank (average, modal, best) and involved variables related to quadrats*

Variables	rank			Category
	average	modal	best	
Shannon (patches, dm^2)	0.54	0.64	0.58	g
number of taxa	0.55	0.62	0.55	g
max floors	0.56	0.53	0.45	g
% isolated patches	−0.44	−0.37	−0.31	g
% biomass production of colonies from total	0.24	0.35	0.37	b
total number of patches	0.13	0.25	0.21	b
% skeleton production	0.28	0.33	0.2	b
number of isolated patches	−0.22	−0.12	−0.1	b
% area occupied	0.21	0.13	0.09	b
total biomass production	−0.21	−0.13	−0.09	b
biomass production colonies	−0.18	−0.11	−0.07	b
% biomass production retro-colonies from colonies	−0.07	0.02	0.06	n
% biomass production of bacteria etc.	0.02	−0.05	0.05	n
production of skeleton	−0.12	−0.05	−0.04	n
biomass production of retro-colonies	−0.04	−0.05	-0.01	n

Rough categories of correlation: g, good; b, bad; n, none.

Famennian (Friáková *et al.* 1985; Dvořák *et al.* 1987; Hladil *et al.* 1989, 1991). Indications supporting this interpretation are as follows:

(a) A weathered hardground at the boundary between beds 4 and 5 is found 4 m below the onset of nodular limestone deposition. Relative sea-level fall recorded by this stratigraphic omission surface may correspond to a significant global event.

(b) Facies overlying this hardground record fluctuating dysoxic conditions.

(c) *Nanicella* (foraminifer) and *Scoliopora kaisini* (tabulate coral) became extinct at the boundary between beds 4 and 5.

(d) Coral-stromatoporoid assemblages changed at this level. Survivors were restricted to only a few, often thinly skeletonized species.

(e) Pelagic conodont assemblages of the *Pa. crepida* Zone have been found in the highest coral-bearing interval at several levels, i.e. in beds 16 and 20A by Olga Friáková, and in bed 20A in several parts of the quarry by Zuzana Krejčí-Kesslerová and Jiří Kalvoda. The assemblages consist predominantly of juvenile specimens. Some undetermined conodont sections were observed to be embedded in primary skeletal debris of bed 20A. This may support the idea that these assemblages were at least partially autochthonous. No older conodont fauna was found.

(f) A conformable contact between beds 20A and 20B coincides with the extinction of corals and is remarkably concordant. This flat contact is traceable for 100 m in the Western Mokrá

Quarry, and was also found in a block that cropped out in the northwestern corner of the Middle Mokrá Quarry, 400 m to the east. No erosional irregularities are visible on this contact; bed 20B is regionally very continuous. This does not support placing the F–F boundary omission surface at the bed 20A/20B contact.

(g) The Janovice-7 and Křtiny HV-105 boreholes also indicate similar survival patterns. The first borehole shows gradual change from *Amphipora*-bearing banks to nodular limestone. The earliest conodont fauna of this borehole, of *Pa. crepida* age, appears in the upper part of the *Amphipora*-bearing limestone. The second borehole was drilled in the gently inclined reef slope of the Moravian Karst where conodont data were continuous from Early Frasnian to Late Famennian time. The light grey coral-bearing limestone occurs up to the *Pa. crepida* Zone.

The observations a–d suggest correlation of the bed 4/5 boundary with the upper Kellwasser event, while e–g indicate its placement somewhere below the bed 20A or 16, respectively.

B. The occurrence of corals within the Lower Famennian is restricted to reworked Frasnian pebbles embedded in younger rocks (O. H. Walliser, pers. comm. 1986). There is a single supporting indication only: the coral–stromatoporoid assemblage corresponds to lagoon or sheltered ramp environments, whereas the conodont assemblage represents a pelagic environment. This would suggest that the Late Frasnian carbonate platform coral fauna was resedimen-

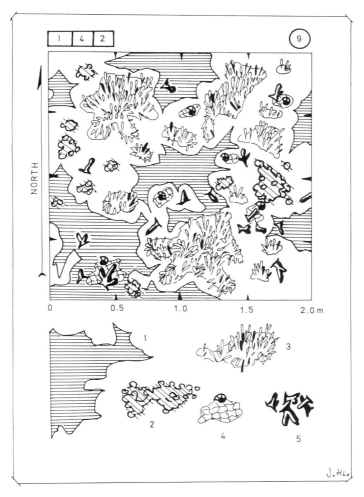

Fig. 4. Quadrat 9 (interval I, bed 4, level 2). 1, Ferroan coating assigned to bacteria (Bacteria); 2, small spiriferids with smooth shells (Brachiopoda); 3, *Amphipora moravica* (Stromatoporoidea); 4, *Nanicella porrecta* (Foraminifera); 5, *Scoliopora denticulata vassinoensis* (Tabulata).

ted into Lower Famennian beds. However, there is no sedimentological evidence for this type of reworking.

C. The *Amphipora*- or coral-bearing limestones belong to the eroded top of the Frasnian sequence (G. Racki, M. Joachimski, pers. comm. 1993). Indications supporting this interpretation are as follows:

(a) The top of the bed 20A is slightly eroded, some cemented coenostea of stromatoporoids are truncated.

(b) Scattered fissures exist in bed 20A and are filled with clay minerals; rare and undetermined conodont elements are present. This could mean that some younger conodonts were incorporated.

(c) Large portions of bed 20A lack conodonts. This observation corroborates somewhat the possible incorporation of conodont elements from above.

(d) A tectonic thrust block cropping out 0.5 km to the southeast has a pronounced stratigraphic gap at the F–F boundary, representing a Type 2 sequence boundary (see above). There, only cave infills and neptunian dykes contain strata of Early Famennian age.

(e) There is similarity of the Western Mokrá Quarry Type 3 section to the Galezice–Ostrówka section of the Polish Holy Cross Mountains. In

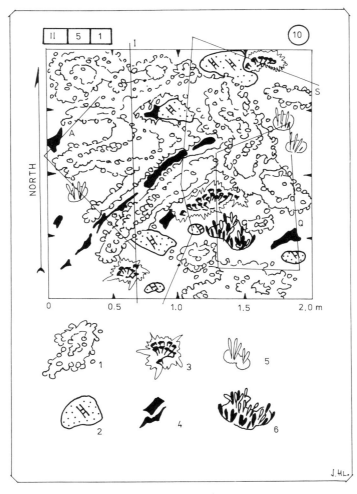

Fig. 5. Quadrat 10 (interval II, bed 5, level 1). 1, Spiriferids with relatively smooth shells (Brachiopoda) [some similar species were probably undistinguished and they were accounted to this population]; 2, *Kamaena* spp. (Algae); 3, *Aulostegites* (Tabulata); 4, thin, plaster-shaped coatings of sponges (Calcispongia); 5, *Amphipora moravica* (Stromatoporoidea); 6, *Amphipora tschussovensis* (Stromatoporoidea).

that section, the conodont-sterile *Amphipora*-bearing limestone is weathered and its surface is truncated. Strata overlying this discontinuity consist of very condensed *Pa. marginifera* cephalopod limestones (Racki *et al.* 1993). However, the last *Amphipora*-bearing beds there contain *A. moravica, A. tschussovensis* and *Multiseptida corallina* (J. Hladil, unpublished data). Thus, an Early Famennian age cannot be entirely excluded for this assemblage.

(f) Increasing levels of heavy carbon isotope values in carbonates below the top of the *Amphipora*-bearing limestone were documented by Michael Joachimski (pers. comm. 1994). The shape of the $\delta^{13}C$ curve resembles that of the standard $\delta^{13}C$ curve just below the Kellwasser crisis.

(g) It is commonly believed that the Lower Famennian lacks any amphiporan or coral survivors.

Yet, a number of puzzling questions concerning the F–F boundary have developed as the number of studies of the Mokrá sections has increased. Interpretation B conflicts with the

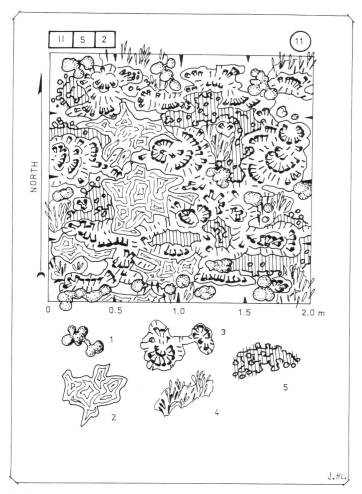

Fig. 6. Quadrat 11 (interval II, bed 5, level 2). 1, *Labechia cumularis* (Stromatoporoidea); 2, *Tienodictyon* (Stromatoporoidea); 3, *Syringopora volkensis* (Tabulata); 4, *Amphipora moravica* (Stromatoporoidea); large smooth shells of spiriferids (Brachiopoda).

microfacial evidence, while each of the remaining interpretations, A and C, should be seriously considered.

Data

A series of sea floor colonization surfaces were documented between 1983 and 1994. Some buried palaeo-sea floors are preserved locally; other parts of these surfaces were either disturbed by storm resedimentation or were subsequently cannibalized, even after their full cementation. Assembled (as opposed to real) $4 m^2$ quadrats of the palaeo-sea floor were reconstructed from fragments. Although the basic biological data were obtained from direct observation of these palaeo-sea floor fragments, the data were also supplemented by numerous thin-sections and dissolved samples, which enhanced each of the reconstructions. The assignment to contemporaneous colonization was carried out as carefully as possible, using skeletal growth layers and sedimentary lamina. Twenty-eight reconstructed quadrats were assembled (Figs 3–11) and scrutinized for several variables for each of the 85 taxa involved. The assessed variables were: abundance, measured as the number of individuals for solitary organisms or as the number of colonies for clonal organisms; occupied area (dm^2); estimated skeletal mass

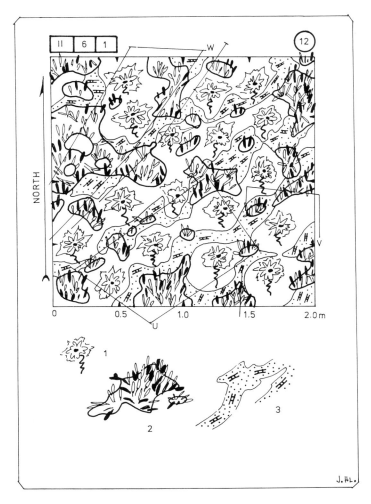

Fig. 7. Quadrat 12 (interval II, bed 6, level 1). 1, Thick sedentary worm, with vertical spiral tube and a pit in sediment (Polychaeta); 2, *Amphipora moravica* (Stromatoporoidea); 3, *Kamaena* spp. (Algae).

produced per year (g/m²/year); and estimated biomass produced per year (g/m²/year). Estimated values are listed in Table 1.

Quadrats were described by the following characteristics: number of documented taxa that occurred in each quadrat; total number of patches in each quadrat; number of isolated patches not in contact with surrounding patches; percentage of isolated patches among all patches (the ratio of their numbers); vertical stratification of benthic communities (number of bioconstructional floors); percentage of occupied surface relative to the total area of the reconstructed quadrat; Shannon–Wiener index of diversity (calculated from patches where units were dm² of the occupied area); total biomass production (= biomass production of all documented benthic organisms (see Appendix 1 for terminology), (g/m²/year); biomass production of colonies (g/m²/year); percentage of biomass production of colonies relative to the total biomass production; percentage of biomass production of non-stromatoporoid sponges and bacteria relative to total biomass production; biomass production of retrocolonies (g/m²/year); percentage of biomass production of retrocolonies relative to the biomass production of all colonies; total mass of skeleton production (= skeleton production of all documented benthic organisms) (g/m²/year); mass percentage

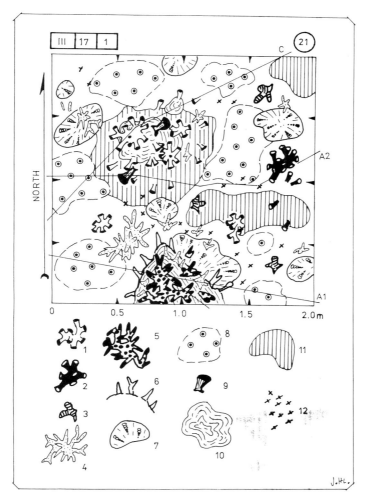

Fig. 8. Quadrat 21 (interval III, bed 17, level 1). 1, *Tabulophyllum maria* (Rugosa); 2, *Disphyllum veronica* (Rugosa); 3, *Scoliopora denticulata rachitiforma* (Tabulata); 4, *Stachyodesl* 'Late Form' (Stromatoporoidea); 5, *Natalophyllum perspicuum* (Tabulata); 6, *Scoliopora denticulata vassinoensis* (Tabulata); 7, *Multiseptida corallira* (Foraminifera); 8, Solenoporaceae (Algae); 9, *Alaiophyllum jana* (Rugosa); 10, *Habrostroma incrustans* (Stromatoporoidea); 11, *Parathurammina* (Foraminifera); 12, Tournayellidae (Foraminifera).

of skeleton production relative to total biomass production. Overall characteristics are listed in Table 2.

Aims and questions

On first examination, the data seem to be chaotic and show no pronounced pattern (Figs 3–11). We hypothesized that a simple, unidimensional gradient approximating the recovery process, something like a successional series, may exist. For the purpose of this study, we asked the following three questions:

1. Can the observed sequence of quadrats be considered non-random?
2. Is a meaningful gradient involved in the data?
3. Is this gradient relevant to processes of recovery from mass extinction?

Methods and tools

The primary objective of our analysis was to determine whether or not the observed sequence

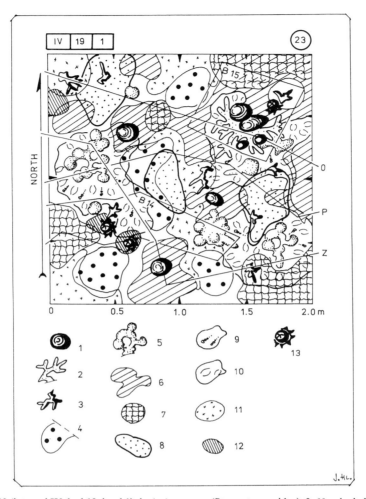

Fig. 9. Quadrat 23 (interval IV, bed 19, level 1). 1, *Actinostroma* (Stromatoporoidea); 2, *Natalophyllum perspicuum* (Tabulata); 3, *Scoliopora denticulata vassinoensis* (Tabulata); 4, Solenoporaceae (Algae); 5, *Labechia cumularis* (Stromatoporoidea); 6, *Amphipora tschussovensis* (Stromatoporoidea); 7, *Amphipora moravica* (Stromatoporoidea); 8, *Cribrosphaeroides* (Foraminifera); 9, *Multiseptida corallina* (Foraminifera); 10, ?Atrypidae with large ribs (Brachiopoda); 11, Tournayellidae (Foraminifera); 12, *Archaesphaera* (Foraminifera); 13, Solitary massive sponge colonies, Tetractinidae (Silicispongia).

of 28 quadrats was random. We used a classical probability theory approach to this problem, testing the null hypothesis that the observed sequence is a random one. We have chosen the sum of squared Euclidean distances (calculated for each of four variables separately) between consecutive quadrats as the criterion to test this hypothesis. We assumed that a sequence of quadrats which would reflect gradual changes in faunal composition along a gradient should be 'shorter' (in terms of the sum of inter-quadrat compositional distance = dissimilarity) than the majority of random sequences of the same quadrats. Consequently, we used a one-tailed probability test. We implemented this test as a Monte Carlo generation of a large number of random sequences, constructing their length frequency distribution. From this distribution a risk of rejecting the null hypothesis, when valid, was derived. The null hypothesis of randomness should be rejected when this risk is sufficiently low.

A second objective was to extract some meaningful gradient from the data, i.e. to get a

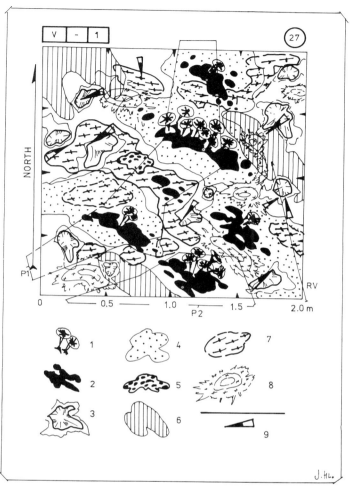

Fig. 10. Quadrat 27 (interval V, bed not labelled, level 1). 1, *Cyclocyclicus* (Crinoidea); 2, *Tetractinidae* (Silicispongia); 3, bacteria poisoning of surface ('Bacteria Killer'); 4, *Irregularina* (Foraminifera); 5, *Palaeoplysina* (?Algae); 6, *Rauserina* (Foraminifera); 7, *Kamaena* spp. (Algae); 8, thin-shelled, bean-shaped Podocopa (Ostracoda); 9, Conical cephalopod shells 'not accounted, only for illustration'.

sequence of quadrats that reflects some simple, directional underlying process. We assumed that a 'short enough' sequence could represent the gradient. Simplified relations among the quadrats were approximated by squared Euclidean distances. A graph theory has been used. Briefly, a graph is a set of items called vertices that are connected by edges. The edge can have a number associated with it, called a length. A path is a sequence of vertices which are connected by edges. When each vertex of a graph is involved in a path just once, it is called a Hamiltonian Path.

In this case, we have a complete graph (i.e. every vertex is connected with each other by an existing edge) whose vertices are quadrats, and the lengths of edges are inter-quadrat distances (measured as the squared Euclidean distances). A complete sequence of quadrats can thus be considered to be a Hamiltonian Path. The goal is then to find the shortest Hamiltonian Path. This task is well known in graph theory as the 'Travelling Salesman Problem' (TSP). Unfortunately the search for the shortest Hamiltonian Path is not an easy task. The TSP belongs to a class of so-called NP – complete tasks, that very

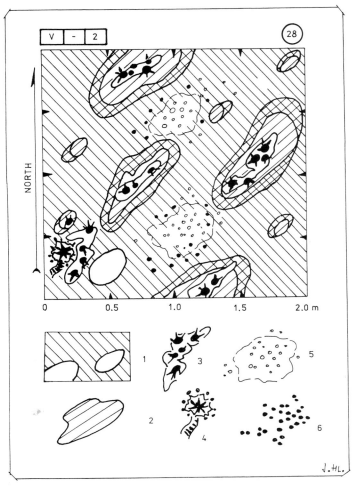

Fig. 11. Quadrat 28 (interval V, bed not labelled, level 2). 1, *Archaesphaera* (Foraminifera); 2, *Rauserina* (Foraminifera); 3, Tetractinidae (Silicispongia); 4, buried sedentary worm (Polychaeta); 5, thin-shelled, bean-shaped Podocopa (Ostracoda); 6, Entomozoa with shells ornamented by long and thin lamellae (Ostracoda).

rapidly become constrained by computing time with increasing number of vertices. The number of all possible (undirected) different sequences of n vertices is $n!/2$; this number increases very rapidly with increasing n; in this case, the number of all possible sequences is approximately 1.5×10^{29}. It is evident that such a great number of sequences cannot be inspected in real time. Under these circumstances, heuristics are usually used instead of exact algorithms. To obtain a sufficiently short sequence, we have developed a heuristic algorithm that manipulates blocks of the sequence to shorten its length. The algorithm works iteratively until no improvement is achieved. There is, of course, no guarantee that the resulting sequence is the shortest one; the algorithm can be trapped into some local minimum that it cannot overcome. However, this shortcoming is common to all heuristics for the TSP. Our algorithm has worked well on artificial testing examples. With real data it resulted in better sequences than a so-called greedy algorithm. However, getting the shortest path alone does not necessarily provide reason to claim it as a reflection of a discovered gradient without knowledge of other short enough paths. The shortest path must exist in any case, even in completely random data. The

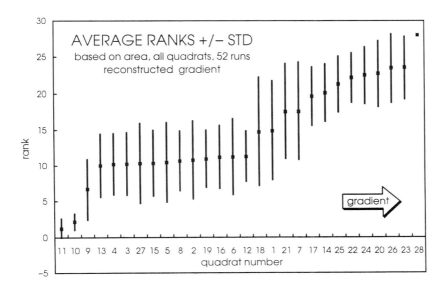

Fig. 12. Average ranks of quadrats on supposed gradient. Based on area occupied by individual taxa. Vertical bars represent one standard deviation in both directions. Average ranks and their standard deviations were obtained from 52 runs of the TSP algorithm. Quadrats are sorted according to their position on the gradient (ranks).

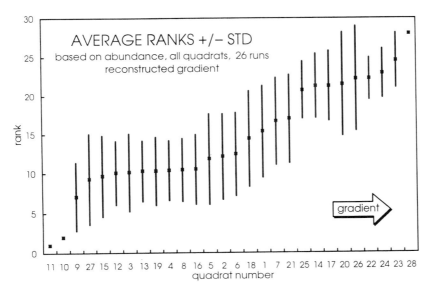

Fig. 13. Average ranks of quadrats on supposed gradient. Based on area occupied by individual taxa. Vertical bars represent one standard deviation in both directions. Average ranks and their standard deviations were obtained from 26 runs of the TSP algorithm. Quadrats are sorted according to their position on the gradient (ranks). Note high similarity of gradient to Fig. 12.

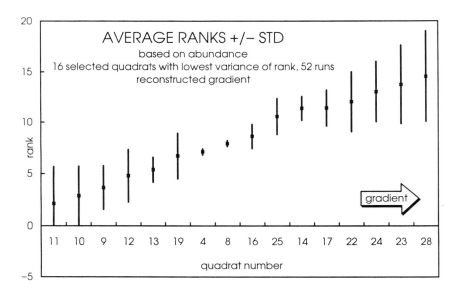

Fig. 14. Average ranks of selected 16 quadrats on supposed gradient. Based on area occupied by individual taxa. Vertical bars represent one standard deviation in both directions. Average ranks and their standard deviations were obtained from 52 runs of the TSP algorithm. Quadrats are sorted according to their position on the gradient (ranks). Compared to Fig. 13, pattern persists and standard deviations decreased.

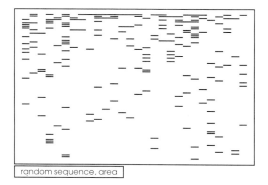

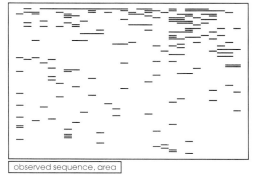

Fig. 15. Data matrix (reduced to presence/absence data for illustration) with quadrats sorted according to randomly generated sequence. Individual taxa are in rows, quadrats in columns. Taxa are sorted in descending order by number of quadrats in which they occur. Horizontal lines representing the taxa are interrupted by numerous and long gaps.

Fig. 16. Data matrix with quadrats sorted according to observed sequence, i.e. stratigraphically. Individual taxa are in rows, quadrats in columns. Time runs from left to right. Taxa are sorted in descending order by number of quadrats in which they occur. The gaps in horizontal lines representing taxa tend to close in comparison with random sequence (Fig. 15).

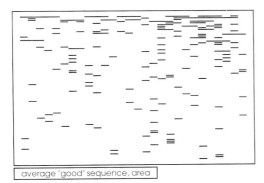

Fig. 17. Data matrix with quadrats sorted according to average sequence. Individual taxa are in rows, quadrats in columns. Taxa are again sorted in descending order by number of quadrats in which they occur. The gaps in horizontal lines representing taxa continue to close in comparison with previous sequences (Figs 15 and 16).

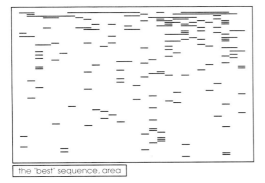

Fig. 18. Data matrix with quadrats sorted according to the best obtained sequence. Individual taxa are in rows, quadrats in columns. Taxa are again sorted in descending order by number of quadrats in which they occur. Relatively slight disturbance of horizontal lines representing taxa compared to previous sequences (Figs 15–17) is observable.

existence of the latent control must be deduced in another way. We believe that a review of several short enough paths can discriminate the presence of this control. When the short enough paths do not vary greatly and show some stable pattern of configuration, we assume that they are influenced by something 'behind' the data.

Unfortunately, the identification of a specified number of shortest Hamiltonian Paths is a very difficult task, even when compared with the TSP. There is no algorithm available, as far as we know, and a 'brute force' approach (e.g. a systematic search) is of no use here owing to the high number of quadrats. The same situation exists in the search for all Hamiltonian Paths shorter than some limit value. However, trapping in the local minima appeared to be useful for obtaining some very short sequences; we compared them with the best (shortest) calculated sequence for the purpose of assessing the robustness of the pattern.

The final task was to interpret the obtained gradient based on a short enough sequence. A simple statistical correlation has been used. We examined correlation between a calculated position of a quadrat on a gradient and the general characteristics of benthic communities, e.g. the Shannon index of diversity based on number of patches; percentage of biomass of retrocolonies relative to all colonies; percentage of skeletal mass relative to total biomass, maximum number of vertical levels; percentage of covered area; and total biomass. A correlation coefficient expressed the amount of dependence.

Results

A probabilistic approach to the question about randomness or non-randomness of the observed sequence clearly indicated that the observed sequence is too 'short' to be considered random. However, this conclusion is not based on all variables. Neither skeletal mass nor biomass led to the rejection of the null hypothesis giving a high risk of error (38% and 34%, respectively). On the other hand, abundance and occupied area gave highly significant results, which allowed rejection of the null hypothesis with an extremely low risk of error of 0.31% and 0.34%, respectively. The low significance of the skeletal mass might be expected, but the low significance of biomass, in contradiction to the very high significance of both abundance and occupied area, is surprising. The results are based on 10 000 random sequences for each variable.

We can conclude that the observed series possesses some kind of pattern. This pattern causes the sequence to be 'abnormally short' when distance is based on abundance or area occupied.

Extraction of some kind of a unidimensional gradient from the data was another objective of the analysis. The evidence that suggests that the analysis reflects a significant gradient is as follows: repeated runs of the TSP algorithm

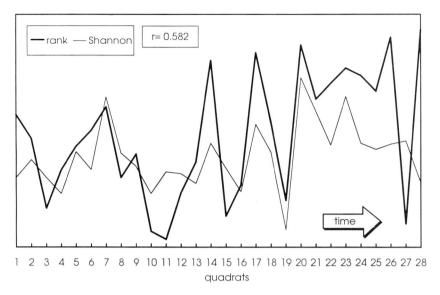

Fig. 19. Correlation of rank and Shannon diversity index in stratigraphic sequence. Correspondence of peaks and valleys of both curves is well visible, with minor exceptions especially for quadrats 11 and 27 (possible crises).

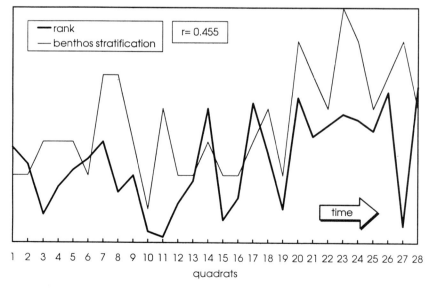

Fig. 20. Correlation of rank and stratification of benthic communities in stratigraphic sequence. Correspondence of peaks and valleys of both curves is good, with exception for both ends of the sequence (quadrats 1–6, and 27–28). Another discrepancy is at quadrat 11. However, the correspondence at other quadrats is excellent.

EXTINCTION/RECOVERY GRADIENTS, MORAVIA

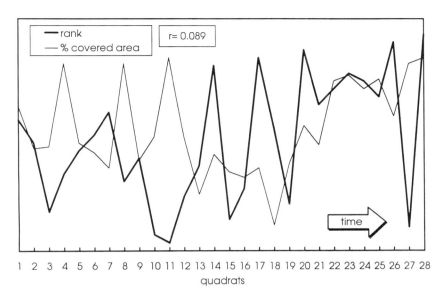

Fig. 21. Correlation of rank and percentage of covered area of quadrats in stratigraphic sequence. Peaks and valleys of both curves do not correspond in lower part of the sequence (left), but the correspondence improves higher, above the critical quadrats 10–11 (bed No. 5).

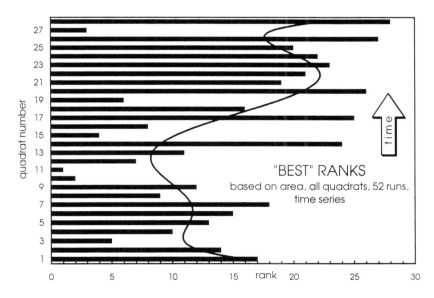

Fig. 22. Time series of quadrats with their positions on gradient (ranks) as received from the best sequence found in 52 runs of the TSP algorithm. Ranks are based on area occupied by individual taxa. Polynomial trendline of sixth degree smooths the data. Note low values (interpreted as crises) at quadrats 3, 10–11, 15, 19 and 27. The most expressive is at one third of the sequence. Gradual increase above this low is approximated as a recovery process.

provided several suboptimal sequences, by trapping in local minima. From these calculated sequences the average and modal quadrat ranks have been obtained, together with their standard deviations. Some quadrats showed stable rank, other quadrats used to vary moderately in rank. We tried then to exclude the quadrats with the highest standard deviations of rank and repeated the whole procedure with the reduced sequence. The pattern persisted and the standard deviations of rank decreased (Figs 12–14, see also Figs 23 and 24).

The existence of gradient was supported also by inspection of the sorted data matrices. We made a series of four data matrices that were sorted by quadrats according to the random rank of quadrats, observed rank of quadrats, average rank on the gradient (from several runs), and the 'best' rank, respectively. From the random rank matrix to the 'best' rank matrix, the values of variables for the taxa showed a clear tendency to cluster and to maintain high values inside of the cluster, and lower values on the margins. The number of 'holes' in the data (zero values corrupting the continuous series of non-zero values) also decreased remarkably (Figs 15–18). We consider the nearly unbroken series of taxa occurrences, with high values in the centre and tapering towards both ends, to be a typical response to various types of gradients.

Finally, we assessed the relationship or dependency between the rank and some general descriptive characteristics of quadrats. The highest correlation was established for rank and Shannon diversity index, with a correlation coefficient of 0.582. Although this is not a very high correlation (possibly because of the low number of quadrats compared, $n = 28$), inspection of the chart (Fig. 19) shows clearly that peaks and valleys of both rank and Shannon index agree well. Vertical stratification of benthic communities (number of bioconstructional floors) also correlated quite well with rank on the gradient, with a correlation coefficient of 0.455 (Fig. 20). In contrast, skeletization correlated poorly with rank (correlation coefficient 0.202), and the correlation of total biomass with rank was even worse (correlation coefficient 0.183). Total colonized area (Fig. 21) and percentage of retrocolonies did not correlate with rank at all, having correlation coefficients of 0.089 and 0.056, respectively. For other results see Table 3.

Interpretation and conclusions

Inspection of the raw data together with gradient ranks suggests that the detected gradient predominantly reflects colonization succession. One of the ends of the gradient embraces the pioneer, or impoverished assemblages, while the opposite end involves a variety of diversified and well-structured assemblages. When gradient rank values are assigned to quadrats in the section, the following picture is obtained (compare Figs 22 and 24). The time series starts with medium values interrupted by a moderate valley at quadrat 3 (bed 0). All of the lower part of the sequence (interval I, Figs 1, 2; see also Hladil *et al.* 1989) is characterized by irregularly placed, usually isolated patches of benthic organisms. The biotic cover of the whole surface is not as extensive as in the younger interval IV. The colonization of quadrat 3 is extensive and predominantly composed of *Kamaena* algae (sticks or felty covers) with several large patches of *Issinella* (seaweed). This quadrat lacks amphiporids, although they are very common in the profile. After a brief period of medium values, the rank reaches its minimum values at quadrats 10 and 11. Both quadrats lay within bed 5, consisting of several colonization levels intercalated with accumulations of early diagenetically cemented skeletal debris as well as clay and intraclasts. Their rank position is extremely low, regardless of the method used, suggesting that the beginning of the recovery after a significant crisis can be placed here. This agrees well with observed gradual colonization after the hardground formation between beds 4 and 5. Quadrat 10 is dominated by brachiopods; quadrat 11 is covered by large compound sheets of *Syringopora volkensis* (tabulate coral) and a stromatoporoid *Tienodictyon* sp. (Figs 5 and 6). Higher up, a rapid and continuous increase of the rank values occurs up to the level of quadrat 14 (bed 8). This trend does not agree with the previous assumption that interval II involves accidental assemblages of strongly impoverished faunas (Hladil *et al.* 1991). Another visible valley of ranks was detected at quadrat 15 (bed 11). This valley occurs within the darker beds of stratigraphic interval II (see Fig. 2), immediately above the breccia bed 10, which represents a possible tsunami deposit (Hladil *et al.* 1991).

A further slight increase of ranks was observed up to quadrat 18 (bed 14). The subsequent valley at quadrat 19 (bed 15) is not very pronounced. Above this level, a remarkable group of quadrats with very high rank values occurs up to quadrat 26 (bed 20A). This group represents temporary flourishing of coral–stromatoporoid assemblages. These assemblages (interval IV) differ in faunal composition from those that appear below the bed 4/5 boundary in interval I (see Fig. 2; compare Figs 3, 4).

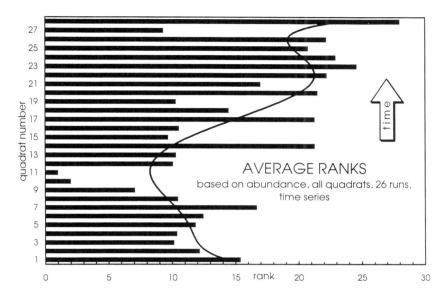

Fig. 23. Time series of quadrats with their average positions on gradient (ranks) based on abundance, as received from 26 runs of the TSP algorithm. In general, the pattern is almost the same as in case of area (see Fig. 22).

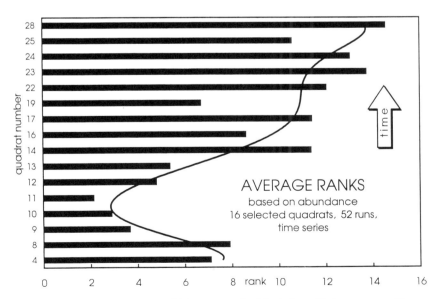

Fig. 24. Time series of 16 selected quadrats with lowest standard deviations. Their average positions on gradient (ranks) are based on abundance, received from 52 runs of the TSP algorithm. Generally, the pattern is again almost the same as in previous cases (compare Figs 22 and 23).

Quadrat 27 represents the last deep valley in rank. The rank values calculated for this quadrat fluctuated but they were always remarkably low. Moderately elevated hardgrounds on this palaeo-sea floor were occupied by crinoids surrounded by sponges. Other patches of benthic organisms (*Irregularina, Kamaena, Palaeoplysina*, see Fig. 10) are chaotically arranged and many of them look to be rudimentary or stressed. Both sediments and the pattern of colonization of the entire interval V differs from those of underlying beds. The lack of corals, as well as of other organisms with massive skeletons, is typical.

In the analysis, quadrat 28 showed high rank without exception. This quadrat possesses a high percentage value of covered surface (especially due to disseminated *Archaesphaera* tests affixed to the semi-lithified sediment surface). Sponges were distributed along moderate topographic highs on the otherwise flat sea floor, and were surrounded by *Rauserina*, and in turn by *Archaesphaera* foraminifers. The areas of moderately low relief seem to also have had concentrically arranged biotic elements. High biotic coverage and a concentric mosaic structure of colonization link this quadrat to older surface 25 of the bed 20A. However, the composition of benthic assemblage as well as the bulk of biomass differs strongly between the quadrat 28 and 25.

In summary, the deepest minimum (at quadrats 10, 11) can be correlated with a high-stress interval, whereas the succeeding quadrats (to quadrat 14 and eventually 18) show a vigorous increase in rank values, suggesting an extensive recovery of the benthic ecosystem. The intensity of the crisis at quadrats 10 and 11 supports its correlation with the Upper Kellwasser event although the presence of a vigorous recovery trend is unusual. Controversially, if the Kellwasser interval was placed at the bed 20A/20B boundary (interpretation C), then the gradient rank distribution would suggest the complete missing of the proper Kellwasser record.

The study was supported by The Grant Agency of the Czech Republic Project 0723, and the Geological Institute, Academy of Sciences of the Czech Republic, Project 5213. The authors are grateful for critical reading and corrections by Erle Kauffman (University of Colorado, Boulder), Douglas Erwin (Smithsonian Institution, Washington), Woody Hickcox (Emory University, Atlanta), Charles Holland (Trinity College, Dublin), Carl Stock (University of Alabama Tuscaloosa) and James Ebert (State University College, Oneonta).

Appendix: Terminology

Abundance. Number of individuals (for solitary organisms), or number of colonies (for clonal organisms) encountered in a quadrat.

Biomass production. Biomass production per year for short-lived organisms (e.g. *Archaesphaera* foraminifers) or those that rapidly increased their body size each year (majority of brachiopods). Corals and stromatoporoids episodically abandoned their underlying calices and either formed new bodies after rejuvenation or moved their older living bodies upward into the next gallery of calices. Both parts of soft bodies can hardly be distinguished. Thus we assigned all the last seasonal growth layer of the coral heads to 'biomass production'. This variable is related to one square metre of the palaeo-sea floor.

Gradient. A latent underlying unidimensional factor, influencing distribution of taxa and their abundance (or other biological variables). Gradients may be of a complex nature, involving environmental and biotic control as well as evolutionary, extinction, or depletion/recovery trends.

Peak. The local maximum value of the curve.

Quadrat. Herein, a reconstructed square, 2×2 m, of buried palaeo-sea floor, on which all co-existing benthic colonizers are documented. No strongly transported thanatocoenoses were included. Vagrant benthic, neritic and pelagic organisms were excluded as well. The quadrats have been reconstructed from individual fragments of surfaces on outcrop.

Rank. The position (order) of a quadrat on a reconstructed gradient. The rank is obtained in several ways: from the 'best' sequence found, or from several 'short enough' sequences as an average or median rank.

Retrocolonies. Some corals and stromatoporoids can occur in solitary as well as colonial form. A common situation is when a normally colonial organism stresses budding/dividing to favour the first settled individual. They are, in fact, solitary individuals of a normally clonal colony organism. This adaptation is partly fixed genetically but some remnant capability is retained to form rich colonies under the proper environmental conditions. Typical representatives of retrocolonies at the Mokrá quarries are among *Amphipora* and *Tabulophyllum*.

TSP. The Travelling Salesman Problem (TSP) is one of the classical problems of graph theory. The task is to find the shortest Hamiltonian Path through a graph (the Hamiltonian Path intersects every graph vertex exactly once). The number of all Hamiltonian Paths in a complete graph equals $n!$, where n is the number of vertices.

Valley. Term used for local minimum values of the curve.

References

DVOŘÁK, J., FRIÁKOVÁ, O., HLADIL, J., KALVODA, J. & KUKAL, Z. 1987. Geology of the Palaeozoic rocks in the vicinity of the Mokrá Cement Factory quarries (Moravian Karst). *Sborník Geologických Věd, Geologie*, **42**, 41–88.

FRIÁKOVÁ, O., GALLE, A., HLADIL, J. & KALVODA, J. 1985. A Lower Famennian from the top of the reefoid limestones at Mokrá (Moravia, Czechoslovakia). *Newsletters on Stratigraphy*, **15**, 43–56.

HLADIL, J., KALVODA, J., FRIÁKOVÁ, O., GALLE, A. & KREJČÍ, Z. 1989. Fauna from the limestones at the Frasnian/Famennian boundary at Mokrá (Devonian, Moravia, Czechoslovakia). *Sborník Geologických Věd, Palaeontologie*, 30, 61–84.

————, KREJČÍ, Z., KALVODA, J., GINTER, M., GALLE, A. & BEROUŠEK, P. 1991. Carbonate ramp environment of Kellwasser time-interval (Lesní lom, Moravia, Czechoslovakia). *Bulletin de la Société belge de Géologie*, **100/1–2**, 57–119.

————, RACKI, G., SZULCZEWSKI, M., SKOMPSKI, S. & MALEC, J. 1993. Field trip 1. Frasnian/Famennian and Devonian/Carboniferous boundary events in carbonate sequences of the Holy cross Mountains. *In:* NARKIEWICZ, M. (ed.) *Excursion guidebook – Global Boundary Events. An Interdisciplinary Conference, Kielce, Poland, September 27–29, 1993.* Warszawa.

Juvenile goniatite survival strategies following Devonian extinction events

MICHAEL R. HOUSE

Department of Geology, Southampton Oceanographic Centre, European Way, Southampton SO14 3ZH, UK

Abstract: During the Devonian there is a correlation between certain periods of sedimentary perturbations, often associated with hypoxia, and extinction events in the evolution of the goniatite and clymeniid ammonoids. A review is given of the survival and diversification strategies associated with these events. Survivors generally show novelty recognizable even in early stages. Characters involved include protoconch size and form, ornament and whorl form and degree of enrolling of the first whorl or so which completes the ammonitella. In post-ammonitella stages to the adult, changes include the nature of coiling, which is often polyphase, ornament and whorl form. Principally the character of new groups is defined by sutural changes following the proseptum, and new features of sutural ontogeny are usually recognizable in the first few millimetres of growth. Far from being a character merely indicating the plasticity of early stages, as is often claimed, it is suggested that larval strategy is crucial for both survival and subsequent radiation. Since, after an extinction event, radiation is only possible for survivors, the available biotic parameters for evolution are the stress-controlled factors at extinction events.

Ten years ago a review was given of extinction and radiation events in Devonian ammonoids which pointed out their association with widespread sedimentary perturbations, and event names were given linked with type sections showing typical lithological developments (House 1985a). A review of goniatite taxa has since been completed for a revision of the Devonian goniatite section of the *Treatise on Invertebrate Paleontology* and systematic collecting has been undertaken of successions around these events internationally. This has been partly in association with the work of the Subcommission on Devonian Stratigraphy and related to the establishment of boundary definitions for Devonian stages. Detailed reviews have been published on the Upper Devonian extinction and radiation events in the Frasnian and Famennian (Becker 1992, 1993a, b, c; Becker & House 1993, 1994a, b) and, for the Frasnian, precision has been aided by a new and much improved international zonation published for Western Australia (Becker et al. 1993) and New York (House & Kirchgasser 1993). A complete review of the Lower and Middle Devonian goniatite zonation has been published (Becker & House 1994a) based in part on new data from North Africa. Thus the time is appropriate to review the relationship between the extinction events established and the subsequent radiation of goniatite groups and to summarize views based on new data since the earlier account (House 1985a).

The recognition that certain Devonian sedimentary perturbations were widespread has a long history, for the names Kellwasser, Hangenberg and Annulata, used for three of the major events, date back to the last century. It has been the work of Ernst and colleagues on the Cretaceous of Germany that has led to a specific event terminology (Ernst et al. 1983). Given the relative homogeneity of chalk deposition, it is understandable that some events in the late Cretaceous are bedding plane events. Devonian event terminology was originally based on certain fossils thought to be characteristic (Walliser 1984, 1986), but was not primarily related to evolutionary extinctions on the large scale except in relation to the Kellwasser and Hangenberg events, the former related to the claims of McLaren (1970) and McLaren & Goodfellow (1990) that extinction there was due to bolide impact. Documentation of a relationship between environmental and eustatic changes and the evolution of ammonoid and other groups is also relatively recent (House 1985a; Becker 1992, 1993b, c, 1995b). This correlation was sharpened when event names were linked to lithological type sections, and the claim that there was a specific relation between evolutionary fluctuations and global sedimentary perturbations; the suggestion was made that there were many such events in the Devonian and that these often followed a phased pattern of several distinct successive small events (House 1985a, 1989). This was used to argue against a

From Hart, M. B. (ed.), 1996, *Biotic Recovery from Mass Extinction Events*,
Geological Society Special Publication No. 102, pp. 163–185

Fig. 1. Table showing Devonian Event terminology plotted against the conodont and ammonoid zonal scales.

	STAGES	CONODONT ZONES		AMMONOID GENOZONES	EVENTS	FAUNAL GUIDES	
D E V O N I A N	FAMMENIAN	praesulcata	VI	Wocklumeria	▬ HANGENBERG	Acutimitoceras	UPPER DEVONIAN
		expansa	V	Clymenia			
		postera	IV	Platyclymenia	═ ANNULATA	annulata	
		trachytera	III	Prolobites			
		marginifera			▬ ENKEBERG ═ CONDROZ		
		rhomboidea		Cheiloceras			
		crepida	II		▬ NEHDEN	Cheiloceras	
		triangularis					
	FRASNIAN	linguiformis			═ KELLWASSER	Crickites	
		rhenana					
		jamieae	I	Manticoceras	▬ RHINESTREET	Beloceras	
		hassi			═ MIDDLESEX	Sandbergeroceras	
		punct.+ trans.			FRASNE	Manticoceras	
		falsiovalvis					
	GIVETIAN	disparilis	III	Pharciceras			MIDDLE DEVONIAN
		hermanni			▬ TAGHANIC	Pharciceras	
		varcus	II	Maenioceras	═ PUMILIO	pumilio	
		hemiansatus			▬ KAČÁK	otomari	
	EIFELIAN	kockelianus			▭	plebeiforme	
		australis	I	Pinacites			
		costatus					
		partitus			CHOTEČ	jugleri	
	EMSIAN	patulus	IV	Anarcestes			LOWER DEVONIAN
		serotinus			▬ DALEJE	elegans	
		inversus					
		nothoperbonus	III	Anetoceras			
		gronberg			▬ ZLICHOV	Anetoceras	
		dehiscens					
	PRAGIAN	pireneae					
		kindlei	II				
		sulcatus			▭		
	LOCHKOVIAN	pesavis					
		delta	I				
		woschmidti / postwoschmidti			▭	uniformis	

single bolide as a cause. Several additions have been made to the event terminology based on typical facies developments, notably by Becker (1993*b*) and Chlupáč & Kukal (1987). The names in current use are illustrated in Fig. 1. One thing which should be noted is that although single names are often given to the events, their polyphase nature means that several steps are involved and that they operate over a significant period of time. It might be argued that the event terminology should be extended if such phases come to be recognized as almost universal. That would lead to a very complex terminology, and it is thought more appropriate that this should be done within the framework of the range of the main event as has been done, for example, by Schindler (1990) in the case of the Upper Kellwasser Event.

This paper will give some documentation of the recent literature on these events but is especially intended to comment on the nature of the ammonoid evolutionary radiation subsequent to them. The events were not of equal trauma for all ammonoid groups so emphasis will be placed on those events where new stocks evolved, especially at a family or higher taxonomic level, and on the nature of the novelty associated with them. A revised evolutionary diagram for Devonian ammonoids showing family ranges is given in Fig. 2 where the width of bars is related to generic diversity. The time scale is based on the new ammonoid zonation scheme for the Devonian (Becker 1993*b*; Becker *et al.* 1992; Becker & House 1994*a*, *b*).

Ammonoid early stages

The early Ammonoidea are characterized by a chambered shell with a marginal siphuncle, usually giving an associated ventral lobe to the suture: the growth lines show a ventral sinus, and there is an egg-shaped protoconch. Thus defined the group includes the Bactritida and probably starts at least in the Silurian (Ristedt 1981).

Tanabe *et al.* (1994) have argued that the ammonoid initial chamber 'has long been mistakenly called the protoconch' and correctly say that it is not homologous with the protoconch of gastropods or the prodissoconch of the Bivalvia.

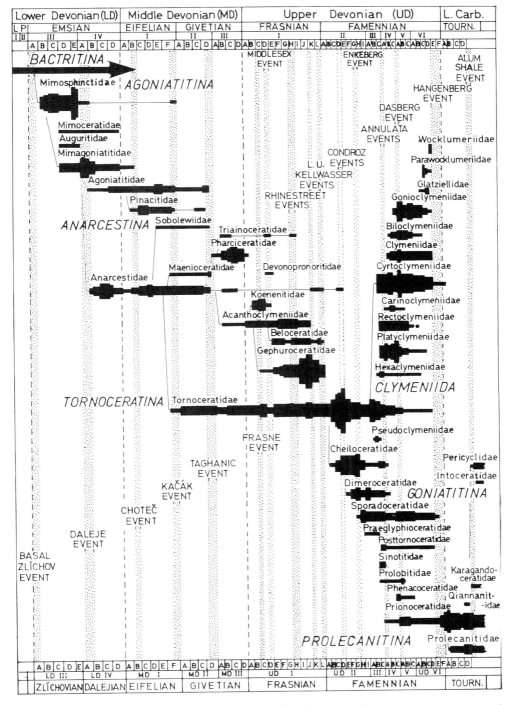

Fig. 2. Diagram showing range of Devonian ammonoid families. The width of bars corresponds to the number of genera or generic groups recognized. Updated from House (1979) mainly using data of Becker & House (1994 a, b) and Becker (1992, 1993a–c)

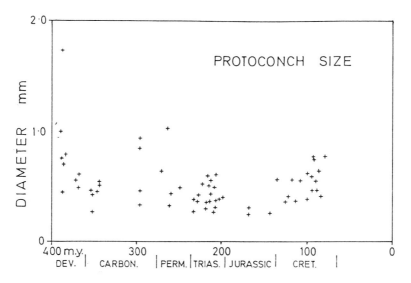

Fig. 3. Size of the protoconch in the Ammonoidea. Graph plotting diameter against time in thousand years. (Modified from House 1985b).

But 'protoconch' merely means 'initial chamber' and it is ridiculous to suggest that any embryological homology with other groups is even implied when the term is used, as it has been for ammonoids for well over a century. In textual use where there may be ambiguity, this is resolved by speaking of the 'ammonoid protoconch' and so on.

The protoconch, which is egg-shaped in early forms, soon develops a spiral axis in the Mimosphinctidae, and progressively through Devonian ammonoids a spindle-shaped form is developed, associated with even tighter coiling. A graph is given here of documented size in ammonoid protoconchs with data taken from many sources (Fig. 3; House 1985b). Excluding the largest known ammonoid protoconch, in *Agoniatites*, it would seem that there is little change in the range of diameter, although, given the tendency to spindle form from the mid-Devonian on, there is a slight trend towards smaller volume: there is insufficient data to attempt to assess this for many groups. The largest known protoconch in *Agoniatites* has an estimated volume of 1.5 mm^3. At the lower end is the Jurassic *Distochoceras* (Palframan 1967) with an estimated volume of 0.01 mm^3. Overall a slight tendency towards an r-type selection mode may be recognized. But it should be recalled that it is the ammonitella, and not the protoconch size, which is the best indirect measure of the size of the egg capsule.

At a variable distance from the protoconch there is often a varix or nepionic constriction. This is taken to indicate the normal hatching from the egg capsule. The early stage up to this constriction is referred to as the ammonitella (Drushchits & Khiami 1970), a name replacing the more ambiguous older term, the larval or nepionic stage. The ammonitella is commonly terminated by a thickening (welt) or constriction, sometimes showing on the outer shell, sometimes on the inner shell and as a groove on the adjacent internal moulds. These structures have commonly been called the nepionic constriction, primary constriction or primary varix. Landman & Waage (1982, p. 1293) suggest that the term constriction should be restricted to that on the ammonitella shell itself. As in the case of 'protoconch' above, it must be recognized that 'constriction' is merely a descriptive term, and can appropriately be used only where there is a constriction: if there is a welt-like swelling, that is something different. Thus a constriction on the outer shell, internal shell and internal mould can be encompassed with terms such as external shell constriction, internal shell constriction and internal mould constriction as appropriate. There is no way a common term can be commandeered for a specific technical use; such action is misleading and improper. Change of growth at the end of the ammonitella in some early Devonian goniatites is hardly shown by a constriction at all but sometimes by a slight change in shell direction, and sometimes by a slight change, or discordance in growth line pattern.

Subsequent to the common terminating constriction or welt of the ammonitella, in advanced forms, the growth lines change to suggest the presence of hyponomic and optic sinuses, and this has been taken to confirm that there is freedom from the egg capsule and that adult free-living life has begun (House 1965, p. 89). If this interpretation is correct, and it is widely accepted, then it is wrong to continue to term the ammonitella stage as larval, since larval refers in other groups to special post-egg stages. Here, the term post-ammonitella juvenile stage is used for the immediate, post-nepionic constriction whorls. In many Devonian goniatites the ammonitella may be ornamented with lirae, like the succeeding juveniles (Fig. 7f, g). This contrasts with the situation described in some later ammonoids by Bandel et al. (1982) and Landman (1988) which has given rise to suggestions that the shell of the ammonitella was endocochleate (Tanabe 1989). Other Devonian goniatites may have smooth ammonitellae (Fig. 12b, c) but there is no material well enough preserved to comment on the endocochleate hypothesis. Evidence from Mesozoic ammonites, in which original aragonite is preserved, and hence detailed shell structure can be elucidated, shows that the proseptum closing the protoconch is formed by the prismatic layer of the ammonitella beyond the protoconch (Birkelund 1981; Landman 1988) and a complete reorganization of shell deposition takes place with the terminating nepionic constriction of the ammonitella with an outer prismatic and inner nacreous layer forming subsequently. The quality of Devonian material is such that this has not been confirmed in early forms. Nevertheless, evidence of slight disruption or discontinuity, and often a constriction, occurs in an appropriate position which suggests that this curious character is part of the ancestral inheritance of the group.

Within the protoconch is a caecal apparatus, which either represents a swollen termination of the siphuncle or a discrete caecal sac and which may or may not touch the walls of the protoconch. The caecum is attached to the earlier protoconch wall by one or more prosiphonal strands (Fig. 4). Such structures are recorded in earlier orthoconic cephalopods which are referred to the Nautiloidea.

Nothing is known of the breeding and egg-laying behaviour of the ammonoids, but there are suggestions that some may share with many living cephalopods a habit of laying eggs in

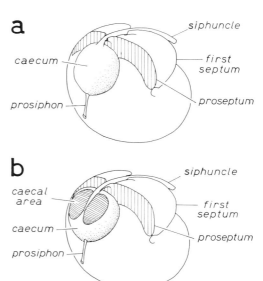

Fig. 4. The protoconch and caecal apparatus in *Tornoceras*, based on specimens from the Alden Marcasite, Givetian, New York. For details see House (1965). (**a**) Usual specimens with a caecum free from the protoconch wall; (**b**) occasional specimens with caecum touching the protoconch wall giving a caecal area. Diameters about 1.0 mm.

clusters benthonically, perhaps attached to sessile structures. The recent description of clusters of ammonitellae by Tanabe et al. (1993) may best be interpreted as such a cluster. S. M. Etches of Kimmeridge has kindly shown the writer a less ambiguous example from the Jurassic Kimmeridge Clay of Dorset. Then there are the examples known from the soles of turbidites where many scattered ammonitellae can occur, as in the Namurian *Reticuloceras* Zone of Pinhay Quarry, Exeter, Devon. The last can only satisfactorily be explained if the ammonitellae were benthonic, either recently hatched or detached from a mature egg cluster. The development of extreme dimorphism in some later ammonoid groups suggests that broods may be retained within the extremely large macroconch body chamber after fertilization: it is difficult otherwise to envisage a reason for such extremes of dimorphic size; dimorphism has been claimed in Devonian *Manticoceras*, *Tornoceras* and *Cheiloceras* by Makowski (1962). If this is common in ammonoids then the macroconch animal might deposit eggs in organized groups benthonically or according to its life style.

If egg-laying was often benthonic in ammonoids then there could be a specific advantage in the development of the caecal apparatus within the protoconch which characterizes the group. It may be that caecal apparatus functioned to extract embryonic liquid from the protoconch early and hence give· neutral to positive buoyancy to the ammonitella on hatching from a benthonic egg cluster, and hence an immediate mechanism for the young to migrate up and join the nekton, an idea suggested some time ago (House 1985b, p. 275). That such buoyant migration may lead to a planktonic or quasi-planktonic habit for some is suggested by fossil logs in, for example, the Rhinestreet Shale in the Devonian of New York State, where the altered woody surface is covered by very young stages of ammonoids. Such a habit may represent a temporary epiplanktonic stage although this is on the assumption that the logs were floating and sank during the life of the juvenile goniatites.

With such a range of shell form in ammonoids, and a probably related diversity of habit, it would be unwise to suggest that such patterns are universal among the Ammonoidea, either for a benthonic egg-laying habit, or a more planktonic stage in early post-ammonitella stage.

Zlíchov Event

The basal Emsian event was referred to by House (1985a p. 17) as the early Zlíchovian transgressive event but it was formalized by Chlupáč & Turek (1988, p. 119) in the form Basal Zlíchovian Event. It is best abbreviated as the Zlíchov Event (Fig. 1) since it corresponds to the lithological changes at the base of the Zlíchov Limestone of Bohemia and appears to correlate with a significant eustatic rise of sea level recognizable in many areas, and this is shown by subsequent faunal changes. The biotic changeover at this level is considerable but it has been long unrecognized because of international miscorrelation.

The origin and radiation of the coiled goniatites followed the Zlíchov Event, the earliest coiled goniatites of the Order Agoniatitida occurring early in the dacryoconarid *praecursor* Zone and late in the conodont *dehiscens* (= *kitabicus*) Zone, the base of which now defines the base of the Emsian Stage and of the regional Zlíchovian stage. What is extraordinary is the way in which these early forms achieved worldwide distribution, even within the *dehiscens* Zone and there are records from northern Canada, Europe, Asia and Australia (Chlupáč 1975;

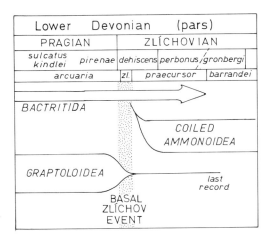

Fig. 5. Graptolite/Goniatite changeover. Diagram illustrating how the decline in the Graptoloidea, a dominant pelagic group of the earlier Palaeozoic, is replaced by the rise of the coiled Ammonoidea in the early Emsian, Lower Devonian.

Becker & House 1993). This spread is part of an international transgressive event.

The rise of the goniatites occurred *pari passu* with the extinction of graptolites in the early Emsian. The loss of most graptolites occurs by the Event, although remnants survive longer into the *gronbergi* Zone (Fig. 5). It is striking that the changeover between two of the dominant groups of the Palaeozoic nekton should take place so suddenly. A primary cause must be sought, perhaps, in plankton changes linked to ocean overturn accompanying the Basal Zlíchovian Event transgression.

The novelty in the earliest goniatites of the Mimosphinctidae is shown by the onset of coiling in the adult (Fig. 6). The morphological sequence shown is not documented in time within the early Zlíchovian. Indeed, the record suggests more tightly coiled stocks appear at the same time or before very loosely coiled forms in the late *dehiscens* Zone. But it is really only the sections in southwestern China which might allow clarification of this.

This coiling of the adult stages is reflected in the coiling progressively shown in the ammonitella (Fig. 6), with the protoconch initially being egg-shaped and later itself having a spiral axis. In none of these earliest goniatites of the Zlíchovian does the ammonitella coil wrap around the protoconch tightly, that is, there is always an umbilical perforation present. It would seem, therefore, that these are adaptive strategies not only of the adult, but of the early stages also.

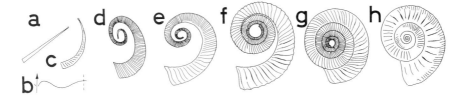

Fig. 6. Morphological sequence illustrating variation in shell morphology in the Emsian, Lower Devonian. (**a**) Bactritidae, *Lobobactrites*; (**b**) schematic suture-line for most forms; c–h, Mimosphinctidae; (**c**) *Kokenia obliquecostata* (Holzapfel); (**d**) *Anetoceras hunsrueckianum* Erben; (**e**) *Anetoceras arduennense* (Steininger); (**f**) *Erbenoceras advolvens* (Erben); (**g**) *Erbenoceras erbeni* House; (**h**) *Mimosphinctes erbeni* Bogoslovskiy.

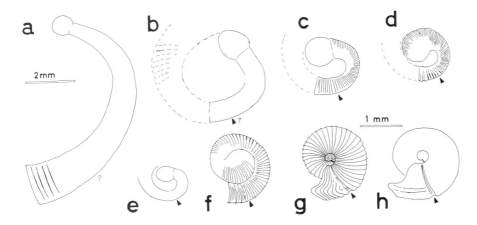

Fig. 7. Ammonitella and immediately following juvenile ornament in early ammonoids. Diagrams prepared from published photographs as indicated. (**a–e**) Scale bar, 2 mm; (**f–g**) scale bar 1 mm. (**a**) *Erbenoceras advolvens* (Erben). Emsian, northern Urals (Erben 1964). (**b**) *Mimosphinctes zlichovense* Chlupáč & Turek, Daleje Shales, Bohemia. (Chlupáč & Turek 1983). (**c, d**) *Mimagoniatites* aff. *fecundus* (Barrande): (**c**) *forme elliptique*; (**d**) *forme circulaire*. La Grange Limestone, Brittany (Erben 1964). (**e**) *Gyroceratites gracilis* (von Meyer). Wissenbacher Schiefer, Germany (Erben 1964). (**f**) *Agoniatites holzapfeli* Wedekind. Adorf, Givetian, Germany (Erben 1964). (**g**) *Tornoceras uniangulare* (Conrad), Leicester Marcasite, Givetian, New York State (House 1965). (**h**) *Truyolsoceras bicostatum* (Hall). Lower Gowanda Shale, Lower Famennian, New York State (House 1965).

Ornamentation of the protoconch in some early forms (Fig. 7d, f) is confluent with that of the ammonitella first whorl, and usually comprises raised slightly convex lirae. This continues until there is usually a slight constriction, the nepionic constriction (marked with a black arrow on Fig. 7), when the ammonitella stage ends. The length of this whorl is short in the Mimosphinctidae, sometimes less than 120° on the interpretation of Erben (1964), but rather variable, although whether variable from taxon to taxon or within a taxon is not clear. There is usually no break or discordance at the constriction and rather similar ornament may continue, or slight ribbing may begin (Erben 1964, pl. 7, fig. 3) which may become extremely well developed, mimicking later ammonites: *Mimosphinctes* was so named because of its similarity to Jurassic perisphinctids in this respect. In *Teicherticeras* and *Convoluticeras* (Teicherticeratinae) the whorls touch the protoconch within one volution of the ammonitella and early whorl.

The gyroconic *Kokenia* is the only member of the Mimosphinctidae to survive the subsequent Daleje Event if Becker & House (1994*a*) are correct in their taxonomic synonymy of *Metabactrites* and *Kokenia*. Neither are known between these two deepening events of the early Zlíchovian and early Dalejian and hence *Kokenia* may have occupied special, perhaps deeper-water, hypoxic niches or refugia.

Arising within the Zlíchovian, apparently after the initial radiation of the Mimosphinctidae, are the families Mimoceratidae, Mimagoniatitidae and Auguritidae, all sharing with the Mimosphinctidae a dominant lateral sutural lobe which is in that position from the earliest stages: this has been named an omnilateral lobe (O) in Russian terminology (Fig. 8). These late groups differ in the development of clear dorsal lobes to the suture. The Auguritidae are anomalous in forming adventitious lobes. They all still possess a perforate umbilicus. Much has still to be learnt on their derivation from the Mimosphinctidae. The start of the formation of true lobes is an important innovation although whether it is related to the development of cameral fluid hysteresis mechanisms has still to be demonstrated. An ocular growth line sinus is not shown in these early forms and it first appears in *Lenzites*.

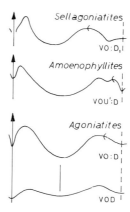

Fig. 9. Suture diagrams of members of the Agoniatitidae showing the ontogeny with a consistently lateral lobe (O) in the Givetian *Agoniatites*, and modest sutural modifications in the late Emsian *Amoenophyllites* and Givetian *Sellagoniatites*.

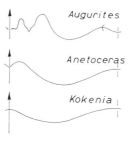

Fig. 8. Suture diagrams of some Emsian ammonoids of the Mimosphinctidae (*Kokenia* and *Anetoceras*) and Auguritidae (*Augurites*).

Daleje event

The Daleje Event was named by House (1985a p. 17) after the Daleje Shale of Bohemia, an early upper Emsian level which, until comparatively recently, was miscorrelated with the Eifelian shales of Germany and Belgium with great prejudice to evolutionary studies. The lower part corresponds closely with the '*gracilis*-Grenze' discussed in relation to the Lower/Middle Devonian boundary definition and with the entry of the dacryoconarid *Nowakia cancellata*; hence the names also applied of *gracilis* or *cancellata* Event (Walliser 1986, p. 405). The miscorrelation with the base of the Eifelian was resolved by Carls *et al.* (1972). In the type area for the event in Bohemia it is characterized by dark grey hypoxic shales and this level also marks an international transgression. Presumably perturbations resulting from the early stages of this event were responsible for the loss near the end of the Zlíchovian of almost all the Mimosphinctidae, many Mimagoniatitidae and the Auguritidae. However, the detailed history of the doubtless polyphase biostratigraphy still has to be elucidated.

The Agoniatitidae, with a completely closed umbilicus, but still with the omnilateral lobe (Fig. 9) radiates following the Daleje Event and is an extremely important group until extinguished at the Taghanic Event.

Novelty is also shown within the Daleje Shale interval (Becker & House 1994a) by the appearance of the Anarcestidae, only the early members of which, such as *Praewerneroceras* and *Latanarcestes*, have a small perforate umbilicus. Tight coiling in the ammonitella is now the rule. Also a subumbilical lobe appears in the early stages after the proseptum and this often migrates laterally to a lateral position towards the adult (Fig. 10), in contrast to the omnilateral lobe of the Zlíchovian Mimosphinctidae and their Dalejan to Middle Devonian derivatives in the Mimagoniatitidae and Agoniatitidae. The whorl form also changes, usually to more subinvolute form, but there are some Zlíchovian forms, such as *Archanarcestes*, which herald later conch styles. These changes appear as early stage adaptations in the Anarcestidae, a group

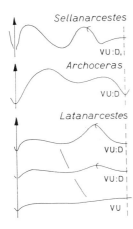

Fig. 10. Suture diagrams of members of the Anarcestidae showing the distinctive lateral migration of a near-umbilical lobe during the ontogeny of the Emsian *Latanarcestes*. Also shown is the late Frasnian *Archoceras*, and the late Emsian *Sellanarcestes* showing the development of a dorsal saddle.

thought to be the root stock for all subsequent Devonian and later ammonoids. The novelty in sutural ontogeny is recognizable in septa of the first whorl after the proseptum, that is, they usually occur in septa within the ammonitella. It would therefore seem likely that they had an early functional advantage.

Basal Chotec̆ Event

This event was named by Chlupác̆ & Kukal (1988 p. 125) to draw attention to the change between the Trebetov and Choteć Limestones of Bohemia which is characterized by the entry of dark coloured, hyopoxic limestones: the level is a little above the base of the Middle Devonian and Eifelian as now defined. It was termed by Walliser (1985) the *jugleri* Event, referring to the spread of *Pinacites* immediately above. Similar dark limestones, rich in dacryoconarids occur at this level in the Tafilalt (Walliser 1986, Becker & House 1994*a*).

By the early Eifelian Chotec̆ Event the mimagoniatitids and mimoceratids have been lost and several genera disappear before the level, such as *Gyroceratites*, *Latanarcestes*, *Mimagoniatites*, and *Paraphyllites*, and rather later, but before the main hypoxic level, several genera appear which become important later, such as *Foordites*, *Fidelites* and *Werneroceras* (Becker & House 1994*a*). Whether this is part of a phased pattern needs further documentation. Following the event, especially characteristic is the spread of *Pinacites*, a time-specific genus which is known in Alaska (House & Blodgett 1982), Europe and North Africa (reviewed by Becker & House 1994*a*), Turkey (Kullmann 1973), Asia (Bogoslovskiy 1982) and China (Zhong *et al.* 1992). This seems clearly to be an opportunistic spread, perhaps aided by higher sea levels. The actual characteristics of *Pinacites*, with acute ventrolateral and umbilicolateral saddles, are not demonstrated to be recognizable in the earliest whorls. Also the earliest *Cabrieroceras* of the *devians* Group enter and later a succession of new forms (documented by Becker & House 1994*a*).

Kac̆ák Event

This event was named by House (1985*a*) after the well-known black limestones of the Kac̆ák Member of Sbskro Formation in Bohemia. Several other names had been in use, such as the *rouvillei*, *crispiforme* or *otomari* Event (Walliser 1986). Truyöls-Massoni *et al.* (1990) pointed out that there was confusion resulting from the event showing several phases. Firstly, in the conodont *australis* Zone, *Cabrieroceras plebeiforme* enters and becomes widely spread; this is a probable senior synonym of *Cabr. rouvillei* and *Cabr. crispiforme*. Secondly there is the entry of *Parodiceras* and *Agoniatites vanuxemi* Group in the late *kockelianus* Zone and are associated with the onset of hypoxia. Thirdly, new forms of *Cabrieroceras* as well as *Nowakia otomari* characterize the later *kockelianus* Zone (former Lower *ensatus*) Zone. Fourthly, at about the acme of the event where pyritic levels are developed in the Tafilalt at Mech Irdane and Bou Tchrafine (Walliser 1991) and Jebel Amelane (Becker & House 1994*a*), the earliest of the Maenioceratidae occur. Fifthly, the base of the *hemiansatus* Zone marks the entry of the key zonal fossil and base of the Givetian as recommended by the Subcommission on Devonian Stratigraphy and ratified by IUGS. A sixth stage is characterized by entry of typical Tornoceratidae with very well developed lateral adventitious lobes and biconvex growth lines. If all these stages are embraced within the term Kac̆ák Event, then more detailed specification is required: much work has still to be done on this event.

It is instructive to compare the early stages of three groups of importance in the Givetian: the Agoniatitidae, Maenioceratidae and Tornoceratidae, only the last of which survived the subsequent Taghanic Event.

The Agoniatitidae includes forms with the largest ammonoid protoconch known (Fig. 11c–

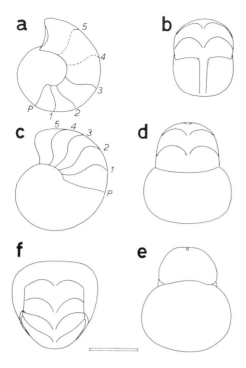

Fig. 11. Givetian protoconchs and early juvenile camerae. (a, b) *Maenioceras* aff. *terebratum* (G. F. Sandberger); based on British Geological Survey 1595CR from Portquin, North Cornwall. (c–f) *Agoniatites* cf. *costulatus* (D'Archiac & de Verneuil); based on author's Collection D.736, from Portquin, North Cornwall. Scale bar is 1 mm.

f) which reaches 1.7 mm in diameter, and is subglobular. Figure 12 illustrates an early stage with only 18 or so camerae which suggests that the ammonitella may have been about 3 mm in diameter. Where ornament has been illustrated for these stages (House & Becker 1994a, pl. 7, figs 15, 16), no sudden change in ornament from the ammonitella to first whorl is documented. The earliest whorl is evolute or slightly subevolute. The suture in this group remains 'omnilateral', but, as Fig. 11e shows, the earliest lateral lobe, which becomes that omnilateral lobe, is in an umbilical position for the first few septa. In the Agoniatidae the body chamber is short.

The Maenioceratidae, as far as is known, have a significantly smaller protoconch (Fig. 12a, b) and it is about 0.6 mm in diameter; the form is more rounded barrel-shaped than subglobular. At least in the specimen illustrated the siphuncle is large. The early whorls are cadiconic and subevolute, rather than evolute. The lateral lobe of the first few septa (Fig. 11a) shows it to be subumbilical in position, and it differs strongly from *Agoniatites* (Fig. 9) in that it migrates laterally during ontogeny (Fig. 13) which is why it is referred to the Anarcestina. But the chief novelty of this family is the addition of an adventitious lobe on ventrolateral saddle and this has developed in some forms by 8 mm diameter. Later the dorsal suture also becomes divided (Fig. 13). In the Maenioceratidae the body chamber is long, exceeding one whorl.

In the Tornoceratidae the protoconch is known to range from 0.55–1.0 mm in diameter (House 1965, p. 124) and the ammonitella may be up to 2.3 mm in diameter. The sharp change in ornament at the end of the ammonitella (Fig. 7g, h) has been described in detail elsewhere (House 1965). The first whorl is very tightly coiled around the protoconch and involution soon appears in *Tornoceras*. The sharp metamorphosis into the early stages is clear and the growth lines suggest well developed hyponomic and optic structures which show in the growth lines immediately following the ammonitella stage. The suture shows an umbilical lobe at the seam in the first few septa which remains in that position throughout ontogeny but a lobe soon develops on the arched lateral saddle (House 1965) giving the characteristic 'A-type' lobe of Schindewolf (1954) which is the characteristic of the Goniatitida (Fig. 14). Figure 15 illustrates how a more fundamental change may have occurred in the introduction of later juvenile growth being controlled by a change in characteristics of the controlling logarithmic spirals of the shell. This has not been described in earlier forms, but, truth to tell, little such detailed work has been done on early ammonoids so it cannot unequivocally be said to be an innovation. Similar patterns were described in Carboniferous goniatites by Kant & Kullmann (1973, 1980) and Kullman & Scheuch 1972).

The main novel characteristics of these three families, as compared with pre-Kačák Event types, is the tighter coiling of the ammonitella, and in the second two, the development of adventitious lobes rather later in the juvenile stages. At the Taghanic Event, the Agoniatitidae and Maenioceratidae are lost. By contrast, the Tornoceratidae survives all subsequent extinction and hypoxic events up to the Hangenberg Event when, whilst the family then becomes extinct, it is earlier derivatives from the Tornoceratidae which survive to give rise to all later ammonoids. Attention was drawn to this resilience of the family when it was suggested that the group had cold-water survival characteristics

Fig. 12. Ammonitellae of Devonian goniatites. (a) Protoconch and first whorl of *Agoniatites* cf. *costulatus* (D'Archiac & de Verneuil), illustrating the largest known ammonoid protoconch. Givetian, from Portquin, North Cornwall. British Geological Survey 95395, × 10. (b–d) *Sphaeromanticoceras rhynchostomum* (Clarke). Frasnian, from the Angola Shale in Hampton Brook, Erie Co., NY. New York State Museum (NYSM) 12097, × 20. (e–g) *Truyolsoceras bicostatum* (Hall). Lower Famennian, from the Gowanda Shale at Corell's Point, Lake Erie, NY. Author's Coll. D.1401, now in NYSM, × 20. (h–i) *Tornoceras (Tornoceras) uniangulare* (Hall) *aldenense* House. Givetian, from the Alden Marcasite, Alden, NY. Author's Coll. D. 1402, now in NYSM, × 20.

(House 1985a), this following ideas of Copper (1977). It now seems more appropriate to suggest survival strategies for hypoxic conditions are critical. In relation to the Kellwasser Event, Becker & House (1994b) have suggested that these may have been produced by ocean overturns.

Other groups which survived the Kačák Event, but which did not survive the Taghanic Event are the Sobolewiidae, which had developed convex growth lines, and the *Foordites* Group.

Pumilio Events

There are two levels of hypoxic-type limestones in the mid-Givetian of Germany known as the Lower and Upper *pumilio* Beds and these were named as events by Walliser (1986): they are also known in North Africa. Lottmann (1990) has reviewed the nature and distribution of these. They are characterized by small bivalves which were thought to be spat of *Stringocephalus* by early workers, but which are now assigned to new taxa. It appears that all the pre-existing

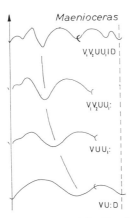

Fig. 13. Sutural ontogeny of the Givetian *Maenioceras* showing the anarcestid pattern and the distinctive adventitious ventrolateral lobe (V$_2$) of the Maenioceratidae and associated subdivision of the dorsal lobe.

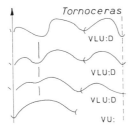

Fig. 14. Sutural ontogeny in the Tornoceratidae showing the development of an adventitious lateral lobe on the crest of the early lateral saddle in Givetian *Tornoceras*. This is the typical 'A-type' ontogeny of Schindewolf (1954).

was made to document this transgressive event internationally and analyse the faunal changes corresponding to it; these, it was argued, were significant (House 1975a, b). The main transgressive event seems to fall within the Middle *varcus* Zone of the conodont zonation and, unfortunately, conodont and ammonoid detailed biostratigraphy at this level raises a number of problems; the conodont *varcus* Zone is large and the faunas poorly descriminating or rare to absent in critical facies. It was Johnson (1970) who coined for North America the term Taghanic Onlap, and this was transposed by House (1985) to Taghanic Event, but both have their origin in the Taghanic Falls, one of the classic sections of the Tully Limestone of New York. Again it is clear that many details of phases have still to be elucidated for this event.

The extinctions preceding the acme of the event, probably in the late Middle *varcus* Zone, were considerable for ammonoids, and recently the significant effect on trilobites (Feist 1991) and corals (Oliver & Pedder 1994) has been documented. There was less faunal change for conodonts. The Agoniatitidae, Sobolewiidae, Pinacitidae and most Maenioceratidae became extinct together with many tornoceratids. The Maenioceratidae are thought to give rise to the main radiating group, the Pharciceratidae, and there may be transitional forms crossing the boundary. A derivative of the Anarcestidae, the Acanthoclymeniidae, arose somewhat later.

The Pharciceratidae arise during the Taghanic Event and are characterized by the proliferation of umbilical lobes (Fig. 16) which typically have rounded saddles and pinched terminations to the lobes. The radiation of this group formerly defined the basal Upper Devonian but now it is included in the late Givetian; only one species of *Petteroceras* is known to survive into the Upper Devonian, although similar but ribbed group, the Triainoceratidae (Fig. 2), may appear here and it reappears mainly at times of hypoxia. In the Pharciceratidae there is no evidence that the distinctive extra umbilical lobes are known earlier than the third or fourth whorl, that is, well after the ammonitella.

Another radiating group, the Acanthoclymeniidae, includes the simple-sutured *Pseudoprobeloceras* and *Ponticeras* as early members which shows a commonly developed feature in the Frasnian in several families, the formation of a large lateral saddle, hence the early name of von Buch for these forms 'magnosellarian'. It is also characterized by another new feature, the subdivision of the ventral lobe. Also appearing is the Eobeloceratidae with several pointed lobes and rounded saddles.

goniatite families survived this event, and it is curious in that it is not associated with a bloom in certain goniatite genera as in other similar Devonian events.

Taghanic Event

This event was in older times called the Upper Devonian transgression, especially following the correct correlation by James Clarke of the Tully Limestone of New York with levels then referred to the Upper Devonian in Europe. All that has changed with the redefinition of the base of the Upper Devonian to a much higher level (Klapper *et al.* 1987). Twenty years ago an attempt

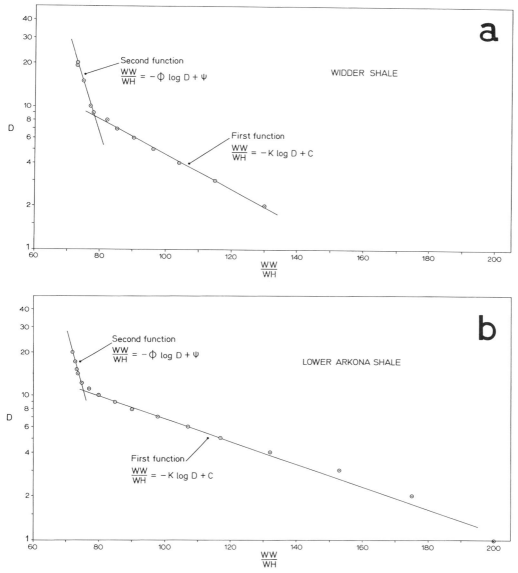

Fig. 15. Graphs showing the shell growth characteristics of forms of *Tornoceras* from the Givetian of New York State illustrating changes growth ratios during ontogeny. (a) *Tornoceras uniangulare widderi* House; (b) *Tornoceras uniangulare arkonense* House. Based on statistics given in House (1965).

Frasnes Event

There is a sequence of events embraced under names such as the Frasnes Event, the Genundewa Event and the *Manticoceras* Event. The former names refer to the transgressive pulse near the base of the Frasnian, the last to the genus which appears considerably after it and then dominates Frasnian faunas (Fig. 17). Much has still to be learnt concerning detailed evolution in these levels and also the juvenile stages of the groups involved. Early phases correspond to the development of *Koenenites*. One marker is *Timanites* a distinctive and very short-lived genus known in Europe, the Timan, Arctic Canada and Western Australia.

The loss of the multilobed Pharciceratidae is a mystery, as is why simpler-sutured stocks should

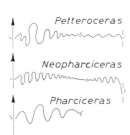

Fig. 16. Adult sutures in members of the Pharciceratidae from the late Givetian.

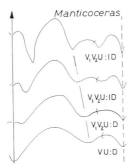

Fig. 17. Sutural ontogeny of the Frasnian *Manticoceras* showing the anarcestid lateral migration of an early subumbilical lobe and finally the adult suture with a large saddle on the mid-flanks.

replace them as dominant elements in the Frasnian faunas. These simpler-sutured forms are thought to have been derived from simple conservative stocks of the Anarcestidae through its derivatives in the early Acanthoclymeniidae. A possible ancestral group, *Ponticeras*, continues alongside the Gephurocerataceae in the Middle Frasnian.

Early stages of the Gephuroceratidae were documented by Clarke (1898) in barytic shell replacements from the Cashaqua Shales of New York which show incredible sheet-like flares given off frequently at progressive stages of the juvenile aperture (Fig. 12b–d). Otherwise most detail is from haematitic internal moulds of the Büdesheim fauna of the Eifel, but that is significantly younger than the originating forms (Clausen 1969). All evidence suggests that the family had small evolute early stages of the ancestral Acanthoclymeniidae. The typical suture recognizable as *Manticoceras* is seen in juveniles at less than 2 mm diameter (Clausen 1969, fig. 111b$_1$) so that it must develop in the first whorl: a similar situation is recorded by Bogoslovskiy (1969, fig. 78zh) in forms he named *Manticoceras sinuosum*.

Middlesex Event

This event was defined to correspond with the black Middlesex Shale of New York (House & Kirchgasser 1993) and was interpreted as a transgressive hypoxic event. In New York it is characterized by *Sandbergeroceras*. A similar level with *Sandbergeroceras* occurs in the Tafilalt of Morocco, and a transgressive horizon at the same level is dated in the Canning Basin Western Australia by *Manticoceras* cf. *evolutum* (Becker *et al.* 1993) which occurs just below the Middlesex Shale in New York. *Sandbergeroceras* is a multilobed and ribbed triainoceratid and is thought to be an opportunistic form: no records of such types are known in the earlier Frasnian and at that time it may have occupied special environments. Since typical gephuroceratids occur above this event, there is no evidence that it represents a significant extinction event although the Koenenitidae are lost a little earlier.

Rhinestreet Event

This event was defined to correspond with the Rhinestreet Shale of New York (House & Kirchgasser 1993), the thickest of the Frasnian black shales there, and probably occupying a very considerable time: several pulses of black shale within it extend well eastwards into the Catskill delta facies. Internationally this represents what may well be the maximum deeping in the whole Devonian, certainly of the Frasnian. In the Timan it corresponds to the bituminous Domanik Suite, further indicating anoxia. In the Canning Basin (Becker *et al.* 1993) a water deepening is indicated by the loss of earlier anoxic levels. As with the Middlesex Shale there is no particular evidence of extinction. But another multilobed group, the Beloceratidae (Figs 2, 18), appears, radiates widely and has wide international distribution. The ancestral *Probeloceras* occurs in the Cashaqua Shale, and the next stage, *Mesobeloceras*, probably in the boundary beds of the Rhinestreet Shale as part of a progressive deepening. New York does not then have the true *Beloceras*, indeed, until the late Rhinestreet, ammonoids are so rare that dissolution of their aragonite shells is suspected. But another multilobed form, *Wellsites*, appears which is referred to the Triainoceratidae. The striking occurrence and success of certain multilobed forms in the Middlesex and Rhinestreet

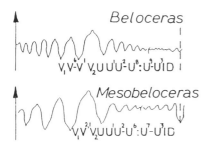

Fig. 18. Adult sutures of two members of the Frasnian Beloceratidae showing the multiplicity of umbilical and ventral lobes with sigmoidal shape and pinched terminations to the mature saddles.

Shales suggest that they had particular environmental advantages, and the association with hypoxic shales suggests that it was low oxygen tolerance.

Kellwasser Events

This is a major extinction period within the Devonian and one which has been said to be the fifth greatest in the Phanerozoic, but the statistics may include events before and after it. It is now recognized that the Lower and Upper Kellwasser Kalk horizons of Germany represent two stages of this, and both are black hypoxic type limestones. The main extinction level is associated with top of the Upper Kellwasser Kalk. The best documentation is by Schindler (1990) but there is an earlier classic study by Buggisch (1972). The literature on this boundary is formidable, with documentation for acritarchs (Vanguestaine et al. 1983), stromatoporoids (Stearn 1987), corals (Sorauf & Pedder 1986; Scrutton 1988; Oliver & Pedder 1994), ostracods (Lethiers & Feist 1991; Lethiers & Cassier 1995), trilobites (Feist 1991; Chlupáč 1994; Feist in Klapper et al. 1994), conodonts (Sandberg et al. 1988; Ziegler et al. 1990; Klapper et al. 1994) and other groups with increasing study on chemostratigraphy (McGhee et al. 1984; Grandjean et al., 1989, 1993; Joachimski & Buggisch 1992, 1993). For the contribution of meteorite or bolide impact as a cause (McLaren 1970; Goodfellow et al 1988; McLaren & Goodfellow 1990) there is little support (McGhee et al., 1984; Buggisch 1991; Girard et al. 1993; Becker & House 1994; Klapper et al. 1994).

Analysis of the goniatite decline at the Lower Kellwasser level has been given recently (Becker 1993; Becker & House 1993, 1994b). Mainly it is thought the Acanthoclymeniidae are lost, but this is probably only so if Enseites is assigned to the Gephuroceratidae. But many genera of the Gephuroceratidae do not survive the Lower Kellwasserkalk Event (Fig. 2).

However, it is the Upper Kellwasser Event and, for goniatites, the end of the Upper Kellwasser Event which marks the extinction of Gephuroceratacea and Beloceratacea, the groups which dominated the Frasnian. Now the Frasnian/Famennian boundary is defined (Klapper et al. 1994) immediatety above the Upper Kellwasser level, at a sudden regressive event.

The role of the tornoceratid *Phoenexites* as an opportunistic survivor has been documented by Becker (1993a, 1995b) but the protoconch and ammonitella are poorly known although by the post-ammonitella juvenile stages the distinctive falcate ribbing and ventro-lateral furrows have appeared. For *Truyolsoceras* in the early Famennian it has been shown that ventro-lateral furrows begin immediately after the nepionic constriction (Fig. 7h; House 1965). There is certainly a lull before the rise of the Cheiloceratidae which takes over as the dominant goniatite group by the conodont *crepida* Zone (Becker 1993a, c), but that is best considered with the next international transgressive event.

Nehden Event

This event was named by House (1985a) in relation to the widespread black shales of the early Nehdenian represented, for example, by the Nehdener Schiefer in Germany and the Dunkirk Shale in New York. It is rather above the base of the Famennian that the Cheiloceratidae are first known and they are shortly followed by a spectacular radiation of the family. This has been documented in detail by

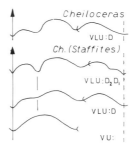

Fig. 19. Sutures in the Cheiloceratidae. The adult suture of *Cheiloceras (Cheiloceras)* and the ontogeny in *Cheiloceras (Staffites)*. Although very similar to the ancestral Tornoceratidae, the family differs in possessing convex, rather than biconvex growth lines and in the very long body chamber

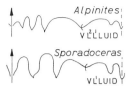

Fig. 20. Adult sutures in the Sporadoceratidae and Posttornoceratidae, groups derived from the Tornoceratidae and Cheiloceratidae respectively, by the development of ventrolateral adventitious lobes.

Becker (1993a, b). Despite the abundant haematized material, virtually nothing has been published on the early stages of *Cheiloceras* and related genera, and the range of size and form of protoconchs and ammonitella is not documented. Crude evidence of sutural ontogeny is known from earlier studies (Fig. 19) which show how closely the group is linked to the Tornoceratidae. The distinction lies in the adoption of convex growth lines, of a longer body chamber and the loss of the biconvexity of the aperture, which means the loss of the optic sinus and suggests the adoption of blindness, perhaps a result of a deeper-water nektonic or epi-benthonic habit. However, what, if any, the ammonitella modifications were, are not clear.

Condroz Event

This event was named by Becker (1993c) and is a two-phased international regressive event and not analogous with the transgressive events associated with anoxia. There are major extinctions of many Cheiloceratidae and the majority of tornoceratid genera (Becker 1995b). Radiations within this interval lead to the diversification of the Dimeroceratidae from advanced cheiloceratids, but whilst adult differentiation is clear, changes in early stages, if any, are unknown. Later there is the rise of the Sporadoceratidae and Posttornoceratidae (Fig. 20), groups arising in the late Nehdenian from the Cheiloceratidae and Tornoceratidae respectively (Becker 1995b), by the development of adventitious ventral lobes, but this appears to be a late stage feature and not characteristic of the juveniles.

Enkeberg Event

Although named by House (1985a), the precise events here have only been clarified by the work of Becker (1993a, c, 1995b), who has shown that extinctions in the Cheiloceratidae and Tornoceratidae are related to it, but nothing is documented on novelty in early stages of surviving forms. The chief subsequent event is the rise of the dorsally siphunculate Clymeniida (Fig. 2). Early stages of the clymeniids are very poorly known and it is only relatively recently that it has been recognized that the dorsally siphunculate condition appears in the first whorl, by the second to fourth septa, by the migration of the siphuncle (Bogoslovskiy 1976). It therefore must represent another adaptation of the immediate post-ammonitella stages. Although there has been wide discussion on the evolution of the Clymeniida, it is not based on any knowledge of the ammonitella or of the early stages, nor of their growth patterns. An origin of the Clymeniida from the Tornoceratidae is now generally conceded (House 1971; Korn 1992a; Becker 1995b).

Annulata Event

Whilst the Annulata Schiefer was named a long time ago, the event name was established by House (1985a). In Germany there are an upper and a lower Annulata Schiefer. These events have recently been thoroughly discussed by Becker (1992, 1993c). It represents another international event associated with transgression and hypoxia. The event is characterized by the wide international distribution of platyclymenids rather than by extinction or diversification. With respect to abundance within the hypoxic event, it is analogous to the abundance, for example, of *Crickites* within the Upper Kellwasser Kalk, or *Gyroceratites* within the Daleje Event. By the Annulata Event, however, the clymenids, of which *Platyclymenia* is an early representative, are already well diversified.

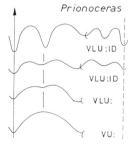

Fig. 21. Sutural development in the Prionoceratidae, differing from the Cheiloceratidae in the deeper ventral lobe and the dorsal suture. Carboniferous Goniatitina are developed from this stock by the subdivision of the mid-ventral lobe.

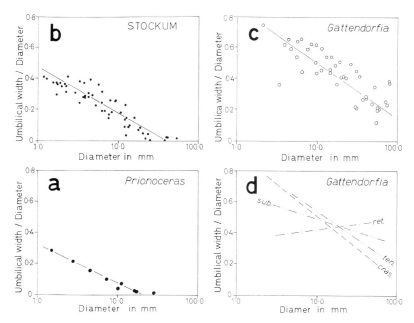

Fig. 22. Graphs showing the shell growth characteristics in a approximate evolutionary sequence from Devonian prionoceratids to early Carboniferous *Gattendorfia*. (a) *Prionoceras* Group with very involute shell form in early stages and throughout ontogeny. (b) Transitional 'Stockum fauna' of prionoceratids with open umbilicus in the early stages but later becoming less so in the adult (hence the increased negative slope). (c) *Gattendorfia* growth characteristics based on material from the early Tournaisian Hangenberg Kalk which are open umbilicate in both early and late stages. (d) Regression lines based on the separate species combined in (c) Key: *ret.*, *Gatt. reticulatum*; *sub*, *subinvoluta*; *ten*, *Gatt. tenuis*; *crass.*, *Gatt. crassa*. Data from Vöhringer (1960) and Price & House (1984).

Hangenberg Event

The classic expression of this event is the Hangenberg Schiefer in the Hönnetal valley, Germany, which is a sequence of grey shales with a level of black shales at the base which represents the main extinction level (Walliser 1986). Formerly the base of the Carboniferous was drawn at the entry of *Gattendorfia* a little higher. The importance of the extinction event has been recognized in detail since the work of Schindewolf (1937) and the subsequent radiation was elucidated by Vöhringer (1960) on the sequence in the Hangenberg Schichten above. The actual data were reviewed by Price & House (1984). Walliser (1986) documented the stages of extinction early in the Hangenberg Schiefer. Much work has since been done on these sections in relation to the redefinition of the Devonian/Carboniferous boundary (Becker 1988, 1993b, 1995b; Korn 1993). Subsequently precision has been given to the extinction of goniatites and the temporary continuance of certain clymenid genera especially by Korn and others (Clausen *et al.* 1994; Korn *et al.* 1994; Luppold *et al.* 1994). Within the Hangenberg Schiefer all goniatite groups apart from some prionoceratids, and all clymeniids become extinct and it is one of the major extinction events in the history of the Ammonoidea. It is now clear that there are several stages in the Hangenberg Event, one of the earliest being the chief extinction level of Walliser.

Recent proliferation of genera has made documentation of the prionoceratid history nomenclatorially complex (Fig. 21). An important survivor of the Hangenberg Event is *Mimimitoceras*, but the grounds for distinguishing it from *Prionoceras*, whilst clear in well preserved material, are difficult otherwise, making the distinction less than useful in practice. Common 'genera' currently assigned the Prionoceratidae include *Acutimitoceras*, *Nicimitoceras*, *Mimimitoceras* (Korn *et al.* 1994), also *Streeliceras* and *Rectimitoceras* (Becker 1995a). A study of ammonitellae from the Stockum Kalk in the track section (Clausen *et al.* 1994, p. 2) showed several different types represented (House 1993)

Fig. 23. Early stages of prionoceratids from the Stockum Limestone, a level of concretions almost exactly on the Devonian/Carboniferous boundary. (**a**) *Sulcimitoceras* aff. *yatskovi* (Kusina), × 25. (**b, c**) '*Prionoceras*' cf. *prorsum* (Schmidt), b × 20, c, × 0. (**d, e**) '*Prionoceras ornatissimum*' Schmidt; d, × 21; e, × 24.

including a form with a strangely sulcate venter (Fig. 23a) assigned to *Sulcimitoceras* (House 1993): it has been suggested (Becker 1995a) that such forms are pathological. This seems unlikely for the following reasons. Firstly, these sulcate ammonitellae are very numerous and easily distinguished from other members of the Stockum fauna, are a discrete group characterized by their sulcation, whorl form and weak ornament, and there are no intermediates or forms with poorly developed sulcate margin; all show the sulcate venter and such dominance of one discrete type would be quite unexpected as a pathological feature. Secondly, since they are ammonitellae, it is not clear how damage or infection could have occurred within the egg-capsule. Thirdly, the structures are perfectly symmetrical in all cases and that would be

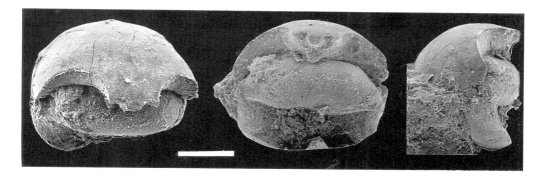

Fig. 24. Protoconch and first few septa of an early Carboniferous goniatite from the supposed Exshaw Shale of Alberta (House 1993) showing the characteristic division of the median ventral lobe. Scale bar 400 µm.

unusual in a pathological situation. Fourthly, it would be surprising that the young could survive so great a shell effect were it pathological. Fifthly, the occurrence of a similar feature in the type species of *Sulcimitoceras* shows that it was geographically widespread. Hence a genetic, rather than pathological interpretation accords better with the requirements of Occam's Razor.

Some of the statistical characteristics of forms close to the boundary are indicated in Fig. 22. Some of the early whorls are illustrated in Fig. 23. This fauna is within the interval of the Hangenberg Event perturbation yet it is instructive that the earliest whorls show so many discrete forms, suggesting niche selection even in earliest stages.

Recent work (Korn *et al.* 1994) has concluded from new collections that the Stockum Kalk is almost exactly on the boundary between the Devonian and Carboniferous and possible earlier, whereas Becker (1993c) takes the view that former collections were fractionally later. Since in any case this lies above what is the major extinction level of Walliser (1986), the diverse ammonitellae represented are part of the recovery pattern, and the variety in the early stages is remarkable.

Characteristics of the growth controls in the transition from Devonian to earliest Carboniferous prionoceratids are illustrated in Fig. 22. The ancestral prionoceratid types, with early and later whorls tightly coiled, are illustrated by the regression in Fig. 22a. With the Stockum fauna of Price & House (1984), the new characteristic of open-umbilicate early whorls is established (the diagram includes several different genera) but later whorls are more involute, hence the steeper slope. The earliest Carboniferous *Gattendorfia* faunas (Fig. 22c) show how the early whorls, even more evolute than in the Stockum fauna, are followed by generally more evolute outer whorls. Only *Gattendorfia reticulata* (Fig. 23d) has a growth pattern which crosses into the Stockum statistical field. Much more detailed work on successional ontogenies is needed, for this succession and others in the Devonian. The only really detailed study of heterochrony with a commencement of successional allometrical analysis for Devonian goniatites was by House (1965) for the Tornoceratidae. Studies which exclude thorough morphological and statistical analyses for successive populations through time are not correctly to be called studies of heterochrony (Korn 1992b).

It is clear that considerable sutural variety starts with a slightly later phase, in Bed 6 of the Hangenberg Schichten with *Gattendorfia*, characterized by very evolute inner whorls and an umbilical lobe which migrates ventrad. In Bed 5 (above), *Eocanites* enters, the first of the Prolecanitidae, in which additional umbilical lobes are thought to be produced; but much has still to be learnt about the ontogeny of early Prolecanitida. Following that, on the world scale, a significant radiation is seen with the rise of the Qiannanitidae, Karagandoceratidae, Pericyclidae, Prodromitidae and Intoceratidae (reviewed by Becker 1993c). With this radiation comes considerable variation in shell ornament and coiling patterns and several distinct sutural patterns including the typical divided ventral lobe of the Goniatitina and bizarre saddle subdivision (Prodromitidae). The distinctive characteristic of the true Carboniferous goniatites is the division of the median lobe, and this occurs in the first whorl beyond the protoconch (Fig. 24), representing another indication of how early stages of goniatites can be definitive.

Conclusions

After an extinction event it is self-evident that exploitation of subsequently available ecological niches can only be undertaken by survivors of the event. Yet survivors will have characteristics which are advantageous for the stress situation of the extinction event. In the Devonian cases considered here, the primary cause of the major events described appears to be hypoxic conditions almost always associated with transgression or with transgresion/regression couplets. That these provided anomalous environmental times is clear from their distinctive associated faunas, usually said to be epiplanktonic cryptodonts, such as *Buchiola*, or large benthonic forms such as *Panenka*, and often abundant pelagic dacryoconarids (before their late Frasnian near-extinction). Lithologically this is also indicated by sediments high in organic carbon and associated geochemical signatures, and often pyritization indicating sea-floor or sediment anoxia. Evolutionary survival in any group, and at any time, but especially at periods of such environmental stress, is dependent on the survival of the early stages to enable adult breeding populations to be established.

In the foregoing review of juvenile stages it has been emphasized that, following the major hypoxic environmental perturbations, new goniatite radiations are distinguished by characters which may often be recognized in the first few millimetres of conch size and within about one whorl of the termination of the ammonitella stage, which is thought to have been encapsulated in the egg sac. This is especially true of the Basal Zlíchov, Basal Daleje, Taghanic, Upper Kellwasser and Hangenberg Events. These particular radiations are of high taxon grade. There is no evidence of such radiations recapitulating successful characters of earlier radiations. It would appear to follow from this that at hypoxic events, there is often very high selection pressure operating on the early stages.

During other events, particularly the Choteč, Kačák, Lower Kellwasser and Annulata Events, opportunistic blooming of certain genera follow the acme of environmental perturbation represented by the events. In particular, the Upper Kellwasser and Annulata Events are associated with abundance within the record of the hypoxic level itself. High taxon changes are not usually an important factor.

There is increasing evidence that each 'event' represents a complex of environmental changes operating over a relatively long period of time, but probably within several of the upper band of Milankovitch oscillations, that is of Eccentricity frequencies of 100–400 ka. Some may indeed be climatically controlled.

Finally it should be stressed that an enormous amount of work is still needed on ammonoid early stages. It is the writer's experience that many genera can be recognized by the nature of their first whorl. Similarly the statistics of conch growth parameters are important. In practice the best material for such studies is pyritic since it is only then that simple breakdown of material to study the early stages is possible, but very little work of this type has been done.

This work has resulted from the support over many years of the Natural Environment Research Council. Special thanks are due to Dr R. T. Becker for his critical comments on an early draft and for the lively arguments with him on joint field work over the last decade, but that is not to imply his agreement with all views expressed.

References

BANDEL, K., LANDMAN, N. H. & WAAGE, K. M. 1982. Microornament on early whorls of Mesozoic ammonites; implications for early ontogeny. *Journal of Paleontology*, **56**, 386–391.

BECKER, R. T. 1988. Ammonoids from the Devonian–Carboniferous boundary in the Hasselbach Valley (Northern Rhenish Slate Mountains). *Courier Forschungsinstitut Senckenberg*, **100**, 193–213.

——— 1992. Zur Kenntnis von Hemberg-Stufe und *Annulata*-Schiefer im Nordsauerland (Oberdevon, Rheinisches Schiefergebirge, GK 4611 Hohenlimburg). *Berliner Geowissenschaftliche Abhandlungen*, *E*, **3(1)**, 3–41.

——— 1993a. Stratigraphische Gliederung und Ammonoideen-Faunen im Nehdenium (Oberdevon II) von Europa und Nord-Afrika. *Courier Forschungsinstitut Senckenberg*, **155**.

——— 1993b. Analysis of ammonoid palaeobiogeography in relation to the global Hangenberg (terminal Devonian) and Lower Alum Shale (Middle Tournaisian) Events. *Annales de la Société géologique de Belgique*, **115(2)**, 459–473.

——— 1993c. Anoxia, eustatic changes, and Upper Devonian to lowermost Carboniferous global ammonoid diversity. *In:* HOUSE, M. R. (ed.) *The Ammonoidea: Environment, Ecology and Evolutionary Change.* Systematics Association, Special Volume, **47**, 115–163.

——— 1995a. New faunal records and holostratigraphic correlation of the Hasselbachtal D/C – Boundary auxilliary stratotype. *Annales de la Société géologique de Belgique*, in press.

——— 1995b. Taxonomy and evolution of late Famennian Tornocerataceae (Ammonoidea). *Berliner geowissenschaftliche Abhandlungen, Reihe E*, **16**, in press.

——— & HOUSE, M. R. 1993. New early Upper Devonian (Frasnian) goniatite genera and the evolution of the 'Gephuroceratceae'. *Berliner*

Geowissenschaftliche Abhandlungen, Reihe E, **9**, 111–133.

—— & —— 1994*a*. International Devonian goniatite zonation, Emsian to Givetian, with new records from Morooco. *Courier Forschungsinstitut Senckenberg*, **169**, 79–135.

—— & —— 1994*b*. Kellwasser Events and goniatite successions in the Devonian of the Montagne Noire with comments on possible causations. *Courier Forschungsinstitut Senckenberg*, **169**, 45–77.

——, ——, KIRCHGASSER, W. T. & PLAYFORD, P. E. 1992. Sedimentary and faunal changes across the Frasnian/Famennian boundary in the Canning Basin of Western Australia. *Historical Biology*, **5**, 183–196.

——, —— & —— 1993. Devonian goniatite biostratigraphy and timing of facies movements in the Frasnian of the Canning Basin, Western Australia. *In:* HAILWOOD, E. A. & KIDD, R. B. (eds) *High Resolution Stratigraphy*. Geological Society, London, Special Publication, **70**, 293–321.

BOGOSLOVSKIY, B. I. 1969. [Devonian Ammonoidea. I. Agoniatitina]. *Trudy Paleontologicheskogo Instituta*, **124**, 1–341 [in Russian].

—— 1976. [Early ontogeny and clymeniid origin.] *Paleontologiskaya Zhurnal*, 41–50 [in Russian].

—— 1982. [Early Devonian and Eifelian ammonoids of the USSR, extend and zonation of the Eifelian]. *In: Biostratigrafia progranicnyck otloszenii nizhnegoi srednego devona]*. Akademia Nauk, SSSR, 23–26 [in Russian].

BUGGISCH, W. 1972. Zur Geologie und Geochemie der Kellwasserkalk und ihrer begleitenden Sedimente (Unteres Oberdevon). *Hessesches Landesamt für Bodenforschung, Abhandlungen*, **62**.

—— 1991. The global Frasnian–Famennian 'Kellwasser Event'. *Geologische Rundschau*, **80**, 49–72.

CARLS, P., GANDL, J., GROOS-UFFENORDE, H., JAHNKE, H. & WALLISER, O. H. 1972. Neue Daten zur Grenze Unter-/Mittel-Devon. *Newsletters on Stratigraphy*, **2(3)**, 115–147.

CHLUPÁČ, I. 1994. Devonian trilobites – evolution and events. *Geobios*, **27**, 487–505.

—— & KUKAL, Z. 1988. Possible global events and the stratigraphy of the Palaeozoic of the Barrandian (Cambrian–Middle Devonian, Czechoslovakia). *Sborník geologickiého věd, Geologie*, **43**, 83–146.

—— & TUREK, V. 1983. Devonian goniatites from the Barrandian area of Czechoslovakia. *Rozprravy Ústředniho ústavu geologickiého*, **46**, 1–159.

CLARKE, J. M. 1898. *The Naples Fauna: fauna with Manticoceras intumescens in Western New York, Part 1*. Albany.

CLAUSEN, C.-D. 1969. Oberdevonische Cephalopoden aus dem Rheinischen Schiefergebirge. II. Gephuroceratidae, Beloceratidae. *Palaeontographica, Abteilung A*, **132**, 1–178

——, KORN, D., FEIST, R., LEUSCHNER, K., GROOS-UFFENORDE, H., LUPPOLD, W., STOPPEL, D., HIGGS, K. & STREEL, M. 1994. Die Devon-Karbon-Grenze bei Stockum (Rheinisches Schiefergebirge). *Geologie und Paläontologie in Westfalen*, **29**, 71–95.

COPPER, P. 1977. Frasnian/Famennian mass extinction and cold-water oceans. *Geology*, **14**, 835–839.

DRUSHCHITS, V. V. & KHIAMI, N. 1970. Structure of the septa, protoconch walls and initial whorls in early Cretaceous ammonites. *Palaeontological Journal*, **1070**, 26–38.

ERBEN, H. K. 1964. Die Evolution der ältesten Ammonoidea (Lieferung I). *Neues Jahrbuch für Geologie und Paläontologie*, **120**, 107–202.

ERNST, G., SCHMID, F. & SEIBERTZ, E. 1983. Event stratigraphy im Cenoman und Turon. *Zitteliana*, **10**, 531–554.

FEIST, R. 1991. The late Devonian trilobite crises. *Historical Biology*, **5**, 197–214.

GIRARD, C., ROCCHIA, R., FEIST, R., FROGET, L., ROBIN, E. 1993. No evidence of impact at the Frasnian : Famennian boundary in the stratotype area, southern France. *Interdisciplinary Conference on Global Boundary Events, Kielce, 27–29 September 1993. Abstract.*

GOODFELLOW, W. D., GELDSETZER, H. H. J., McLAREN, D. J., ORCHARD, M. J. & KLAPPER, G. 1988. The Frasnian–Famennian extinction: current results and possible causes. *In:* McMILLAN, N. J., EMBRY, A. F. & GLASS, D. J. (eds) *Devonian of the World*. Canadian Society of Petroleum Geologists, **3**, 9–21.

GRANDJEAN, P., ALBARÈDE, F. & FEIST, R. 1989. REE variations across the Frasnian–Famennian boundary. *Terra Abstracts*, **1**, 184.

——, —— & —— 1993. Significance of rare earth elements in old biogenic apatites. *Geochimica et Cosmochimica Acta*, **57**, 2507–2514.

HOUSE, M. R. 1965. A study in the Tornoceratidae: the succession of *Tornoceras* and related genera in the North American Devonian. *Philosophical Transactions of the Royal Society of London*, **B250**, 79–130.

—— 1971. On the origin of the clymenid ammonoids. *Palaeontology*, **13**, 664–674.

—— 1975*a*. Facies and Time in Devonian tropical areas. *Proceedings of the Yorkshire Geological Society*, **40(2)**, 233–288.

—— 1975*b*. Faunas and Time in the Marine Devonian. *Proceedings of the Yorkshire Geological Society*, **40(4)**, 459–490.

—— 1978. Devonian ammonoids from the Appalachians and their bearing on international zonation and correlation. *Special Papers in Palaeontology*, **21**, 1–70.

—— 1979. Biostratigraphy of the early Ammonoidea. *Special Papers in Palaeontology*, **23**, 263–280.

—— 1985*a*. Correlation of mid-Palaeozoic ammonoid evolutionary events with global sedimentary perturbations. *Nature*, **313**, 17–22.

—— 1985*b*. The ammonoid time scale and ammonoid evolution. *In:* SNELLING, N. J. (ed.) *The Chronology of the Geological Record*. Geological Society, London, Memoir, **10**, 273–283.

—— 1989. Ammonoid extinction events. *Philosophical Transactions of the Royal Society of*

London, **B325**, 307–326.

—— 1993. Earliest Carboniferous goniatite recovery after the Hangenberg Event. *Annales de la Société géologique de Belgique*, **115**(2), 559–579.

—— & BLODGETT, R. B. 1982. The Devonian goniatite genera *Pinacites* and *Foordites* from Alaska. *Canadian Journal of Earth Sciences*, **19**, 1873–1876.

—— & KIRCHGASSER, W. T. 1993. Devonian goniatite biostratigraphy and timing of facies movements in the Frasnian of eastern North America. *In:* HAILWOOD, E. A. & KIDD, R. B. (eds) *High Resolution Stratigraphy*. Geological Society, London, Special Publication, **70**, 267–292.

JOACHIMSKI, M. M. & BUGGISCH, W. 1992. Carbon isotope shifts and the Frasnian/Famennian boundary: evidence for worldwide Kellwasser Events. *Abstract, Fifth International Conference on Global Bioevents, Göttingen, February 16–19*, 58, 59.

—— & —— 1993. Anoxic events in the late Frasnian – Causes of the Frasnian–Famennian faunal crisis. *Episodes*, **21**, 675–678.

JOHNSON, J. G. 1970. Taghanic Onlap and the end of North American Devonian provinciality. *Bulletin of the Geological Society of America*, **81**, 2077–2107.

KANT, R. & KULLMANN, J. 1973. 'Knickpunkte' im allometrischen Wachstum von Cephalopoden-Gehäusen. *Neues Jarhbuch für Geologie und Paläontologie, Monatshefte*, **142**, 97–114.

—— & —— 1980. Umstellung im Gehäusebau jungpaläozoischer Ammonoidea. Ein Arbeitskonzept. *Neues Jarhbuch für Geologie und Paläontologie, Monatshefte*. **11**, 673–685.

KLAPPER, G., FEIST, R. & HOUSE, M. R. 1987. Decision on the Boundary Stratotype for the Middle/Upper Devonian Series Boundary. *Episodes*, **10**, 97–101.

——, ——, BECKER, R. T. & HOUSE, M. R. 1994. Definition of the Frasnian/Famenian Stage boundary. *Episodes*, **16**, 433–441.

KORN, D. 1992a. Relationship between shell form, septal construction and suture line in clymeniid cephalopods (Ammonoidea; Upper Devonian). *Neues Jarhbuch für Geologie und Paläontologie, Abhandlungen*, **185**, 115–130.

—— 1992b. Heterochrony in the evolution of Late Devonian ammonoids. *Acta Palaeontologica Polonica*, **37**, 21–36.

—— 1993. The ammonoid fauna change near the Devonian–Carboniferous boundary. *Annales de la Société Geologique de Belgique*, **115** (2), 581–593.

——, CLAUSEN, C.-D., BELKA, Z., LEUTERITZ, K., GROOS-UFFENORDE, H., LUPPOLD, F. W, FEIST, R. & WEYER, D. 1994. Die Devon-Karbon-Grenze bei Drewer (Rheinisches Schiefergebirge). *Geologie und Paläontologie in Westfalen*, **29**, 97–147.

KULLMAN, J. 1973. Goniatite-coral associations from the Devonian of Istanbul, Turkey. *Ege Üniversitesi Fen Fakültesi Kitaplar Serisi*, **40**, 97–112.

—— & SCHEUCH, J. 1972. Absolutes und relatives Wachstum bei Ammonoideen. *Lethaia*, **5**, 129–146.

LANDMAN, N. H. 1988. Early ontogeny of Mesozoic ammonites and nautilids. *In:* WIEDMANN, J. & KULLMAN, J. (eds) *Cephalopods–Present and Past*. Schweizerbart'sche Verlagsbuchhandlung, 215–228.

—— & WAAGE, K. 1982. Terminology of structures in embryonic shells of Mesozoic ammonites. *Journal of Paleontology*, **56**, 1293–1295.

LETHIERS, F. & CASIER, J-F. 1995. Les Ostracodes du Frasnien terminal ('Kellwasser' Supérieur) de Coumiac (Montagne Noire, France). *Revue de Micropaléontologie*, **38**, 63–77.

—— & FEIST, R. 1991. La crise des ostracodes benthiques au passage Frasnien–Famennien de Coumiac (Montagne Noire, France méridionale). *Comptes Rendus de l'Academie des Sciences, Paris, Ser. II*, **312**, 1057–1063.

LOTTMANN, J. 1990. Die-*pumilio* Events (Mittel-Devon). *Göttinger Arbeiten zur Geologie und Paläontologie*, **46**.

LUPPOLD, F. W., CLAUSEN, C.-D., KORN, A. STOPPEL, D. 1994. Devon–Karbon-Grenzprofile im Bereich von Remscheid-Altenaer Sattel, Warsteiner Sattel, Briloner Sattel und Attendorn-Elsper Doppelmulde (Rheinisches Schiefergebirges. *Geologie und Paläontologie in Westfalen*, **29**, 7–69.

McCLAREN, D. J. 1970. Presidential address: time, life and boundaries. *Journal of Paleontology*, **44**, 801–815.

—— & GOODFELLOW, W. D. 1990. Geological and biological consequences of giant impacts. *Annual Review Earth and Planetary Sciences*, **18**, 123–171.

McGHEE, G. R. JR, GILMORE, J. S., ORTH, C. J. & OLSEN, E. 1984. No geochemical evidence for an asteroidal impact at late Devonian mass extinction horizon. *Nature*, **308**, 629–631.

MAKOWSKI, H. 1962. Problem of sexual dimorphism in ammonites. *Palaeontologica Polonica*, **12**, 1–92.

OLIVER, W. A. JR & PEDDER, A. E. H. 1994. Crises in the Devonian history of rugose corals. *Paleobiology*, **20**, 178–190.

PALFRAMAN, D. F. B. 1967. Variation and ontogeny in some Oxfordian ammonites: *Distichoceras bicostatum* (Stahl) and *Horioceras baugeri* (d'Orbigny) from England. *Palaeontology*, **10**, 60–94.

PRICE, J. D. & HOUSE, M. R. 1984. Ammonoids near the Devonian–Carboniferous boundary. *Courier Forschungsinstitut Senckenberg*, **67**, 15–22.

RISTEDT, H. 1981. Bactriten aus dem Obersilur Böhmens. *Mitteilungen der Geologisch-Paläontologischen Institut, Universitat Hannover*, **51**, 23–26.

SANDBERG, C. A., ZIEGLER, W., DREESEN, R. & BUTLER, J. L. 1988. Late Frasnian Mass Extinction: Conodont event stratigraphy, global changes, and possible causes. *Courier Forschungsinstitut Senckenberg*, **102**, 263–307.

SCHINDEWOLF, O. H. 1937. Zur stratigraphie und Paläontologie der Wocklumer Schichten. *Abhandlungen der Preussischen Geologischen Landesanstalt, Neue Folge*, **178**, 1–132.

—— 1954. On the development, evolution and terminology of the ammonoid suture line. *Bulletin of the Museum of Comparative Zoology at Harvard College*, **112(3)**, 217–237.

SCHINDLER, E. 1990. Die Kellwasser-Krise (hohe Frasne-Stufe, Oberdevon). *Göttinger Arbeiten zur Geologie und Paläontologie*, **46**, 1–115.

SCRUTTON, C. T. 1988. *Patterns of extinction and survival in Palaeozoic corals*. Systematics Association Special Volume, **34**, 65–88.

SORAUF, J. E. & PEDDER, A. E. H. 1986. Late Devonian rugose corals and the Frasnian–Famennian crisis. *Canadian Journal of Earth Science*, **23**, 1265–1287.

STEARN, C. W. 1987. Effect of the Frasnian-Famennian extinction event on stromatoporoids. *Geology*, **15**, 677–679.

TANABE, K. 1989. Endocochleate embryo model in the Mesozoic Ammonitida. *Historical Biology*, **2**, 183–196.

—— & OHTSUKA, Y. 1985. Ammonoid early shell internal structure: its bearing on early life history. *Lethaia*, **11**, 310–322.

——, LANDMAN, N. H. & MAPES, R. H. 1994. Early shell features of some late Paleozoic ammonoids and their systematic implications. *Transactions and Proceedings of the Palaeontological Society of Japan, New Series*, **173**, 384–400.

——, ——, —— & FAULKNER, C. J. 1993. Analysis of a Carboniferous embryonic ammonoid assemblage–implications for ammonoid embryology. *Lethaia*, **26**, 215–224.

TRUYÓLS-MASSONI, M., MONTESINOS, J. R., GARCIA-ALCALDE, J. L. & LEYVA, F. 1990. The Kacak–Otomari Event and its characterization in the Palentine Domain (Cantabrian Zone, N.W. Spain). *In:* KAUFFMAN, E. G. & WALLISER, O. H. (eds) *Extinction Events in Earth History*. Springer, Berlin, Lecture Notes in Earth History, **30**, 133–143.

VÖHRINGER, E. 1960. Die Goniatiten der unterkarbonischen *Gattendorfia*-Stufe im Hönnetal. *Fortschritte in der Geologie von Rheinland und Westfalen*, **3**, 107–196.

WALLISER, O. H. 1984. Geological processes and global events. *Terra Cognita*, **4**.

—— 1986. Natural boundaries and Commission boundaries in the Devonian. *Courier Forschungsinstitut Senckenberg*, **75**, 401–408 (dated 1985).

—— 1991. (ed.) *Morocco Field Meeting of the Subcommission on Devonian Stratigraphy*. International Union of Geological Sciences, Guidebook.

ZHONG, K., WU, Y. & YIN, B 1992. *Devonian of Guangxi–Stratigraphy of China, Part 1*. China University Geoscience Press.

ZIEGLER, W. & SANDBERG, C. A. 1990. The late Devonian conodont zonation. *Courier Forschungsinstitut Senckenberg*, **121**, 1–115.

The mid-Carboniferous rugose coral recovery

O. L. KOSSOVAYA

All-Russian Geological Research Institute (VSEGEI), Srednyi prospect 74, St Petersburg 199026, Russia

Abstract: The taxonomic diversity of the rugose corals following the mid-Carboniferous 'lesser mass extinction event' (Alekseev 1989) may be subdivided into two intervals. The survival interval includes long-ranging survivors, short-ranging survivors (*Dibunophyllum, Palaeosmilia, Diphyphyllum*), survivors with post-crisis ontogenetic changes (*Caninia, Bothrophyllum*), newly-appeared taxa with hypothetical ancestors (*Profischerina, Protodurhamina*) and those of cryptogenic origin (*Lytvophyllum*). The lower limit of the survival interval coincides with the *Eumorphoceras/Homoceras* Zone and can be identified in the Donetz Basin, Gornaya Bashkiria, Novaya Zemlja, North Timan, Cantabrian Mountains (Rodriguez *et al.* 1986) and the Western Interior of the USA (Sando 1989).
The recovery interval is established by the appearance of the immigrant genus *Petalaxis* and by increasing specific diversity in pre-existing lineages. Its lower limit corresponds to the *Pseudostaffella praegorsky–Profusulinella staffelliformis* boundary, the lower boundary of the Westphalian A in the Cantabrian Mountains and the lower boundary of the Atokan Series in the Western Interior of the USA.
The radiation interval was marked by the appearance of astreoid colonies in the Petalaxis–Ivanovia lineage at the *Fusulinella calaniae–F. vozhgalensis–F. kamensis* lower boundary.

The decrease in taxonomic diversity, including foraminifera, brachiopods, corals, cephalopods, crinoids, bryozoans and ammonoids, in the Late Visean–Early Serpukhovian interval has been identified by Sepkoski (1986). The same phenomenon has recently been established for conodonts (Nemirovskaya & Nigmadganov 1994).

Preliminary results of studies on the Late Palaeozoic Rugosa have shown the decrease of both the general number of genera and the number of newly appeared genera at this time (Fig. 1). Close to the beginning of the mid-Carboniferous about 90% of the Early Carboniferous genera became extinct. This event corresponds approximately to the *Eumorphoceras/Homoceras* zonal boundary and is classified as a 'lesser mass extinction' event, attributed to widespread regression and climatic cooling resulting from glaciation (Alekseev 1989). This analysis includes studies of the rugose corals from the complete Lower to Middle Carboniferous sections in Novaya Zemlja Archipelago and North Timan, as well as in the Donetz region and Gornaya Bashkiria; the latter being the type locality of the Bashkirian Stage. These successions were deposited in marginal epicontinental basins along the northeastern part of the Euro-American palaeocontinent. The sequence of deposits includes a range of shelf facies zones from detrital limestones of 4–5 standard facies zones to carbonate–terrigenous sediments of standard zones 2–3 of the outer shelf. This facies differentiation results in regional peculiarities of the rugose coral diversity.

The data on the rugose coral distribution from the Cantabrian Mountains (Rodriguez *et al.* 1986) and the Western Interior of the USA (Sando 1989) were used for comparative analysis of the general diversity dynamics in the evolution of this group at the extinction level followed by the survival–recovery interval. Post-crisis recovery coincides with transgression after the widespread regression at the end of Early Carboniferous time.

The dynamics of rugose corals is analysed separately for each of the four morphotypes identified by Fedorowski (1989), Sando (1989) and Rodriguez *et al.* (1986). These are:

I. morphoecotype – simple corals without dissepiments;
II. morphoecotype – simple corals with dissepiments;
III. morphoecotype – fasciculate corals;
IV. morphoecotype – massive colonial corals.

This investigation aims to define successive phases of rugose diversity after the mass extinction in the early *Homoceras* Zone.

From Hart, M. B. (ed.), 1996, *Biotic Recovery from Mass Extinction Events*, Geological Society Special Publication No. 102, pp. 187–199

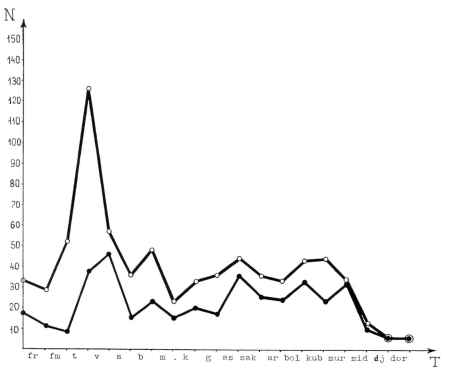

Fig. 1. Dynamics of rugose coral generic diversity during the Late Palaeozoic. The upper line is the total number of genera while the lower line represents the numbers of newly-appeared genera.

The mid-Carboniferous rugose coral extinction

The extinction level coincides with the base of the *Pseudostaffella bogdanovkensis* foraminiferal zone. Here the typical Early Carboniferous taxa, with complex axial structures, lonsdaleoid dissepiments, lamello-fibrous and pinnate fibrous microstructures, are eliminated. The general decline of diversity is characteristic for all morphotypes. In the North Island of the Novaya Zemlja Archipelago this level is identified by the disappearance of *Zaphrentites paralella* (morphoecotype I), *Siphonophyllia cyclindrica*, *S. spumosa*, *Palaeosilia* ex.gr. *stutchburyi*, *S. petrenkovi*, *Uralinia elegans*, *Sychnoelasma konincki*, *Cyathoclisia conisepta*, *Dibunophyllum turbinatum*, *Gangamophyllum boreale*, *Arachnolasma sinensis* (morphoecotype II), *Siphonodendron irregulare*, *S. caespitosum*, *Tschernoviphyllum podbotiense*, *Corwenia regularis*, *Lonsdaleia arctica*, *L. annulata*, *L. duplicata arctica* (morphoecotype III), *Actinocyathus longiseptata* (morphoecotype IV); see Gorsky (1951), Kropatcheva (pers. comm.), Figs 2 and 3. In the Cape Makarov section (Fig. 4) the uppermost Serpukhovian deposits are represented by thinly-bedded carbonate and terrigenous sediments, which contain *Palaeosmilia murchisoni* and *Carcinophyllum septentrionale*. Contrary to that, contemporaneous carbonate deposits, formed under the most favourable shallow-water conditions, contain a more diverse rugose coral fauna with a wide range of morphoecotypes. This is seen in the North Timan section (River Sula), where the extinction level is characterized by the disappearance of *Dibunophyllum gangamophylloides* (morphoecotype II), *Lonsdaleia arctica* (morphoecotype III) and *Actinocyathus ornata* (morphoecotype IV); see Fig. 5. The same diversity pattern was found in the Donetz region where about 90% of the taxa within all four morphoecotypes become extinct (Figs 6 and 7). The upper stratigraphic boundary of the eliminated taxa is determined by the appearance of *Homoceras* (limestone D5/10 of Aizenverg & Brazhnikova 1982) and coincides with the *Eumorphoceras/Homoceras* boundary.

The extinction event is identified in a much

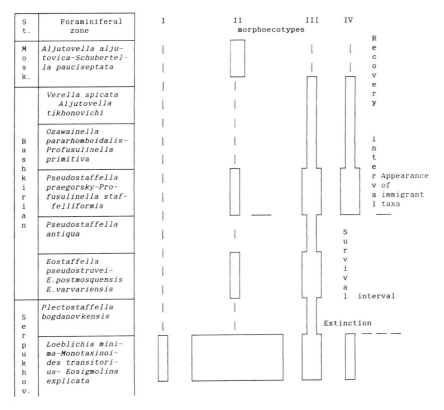

Fig. 2. Distribution of rugosan morphoecotypes in the Novaya Zemlja region in Late Serpukhovian and Bashkirian time.

wider geographical area. In the shallow-water facies of the Namurian in the Cantabrian Mountains (Rodriguez et al. 1986) this event is expressed in the abrupt decrease of coral diversity at the Namurian A and Namurian B boundary, which follows the Namurian A diversity maximum (Fig. 8). In the Western Interior of the USA the peak of rugose coral diversity is found in the lower part of the Chesterian Series (Sando 1989). This is followed by the disappearance of many taxa both in deep-water and shallow-water facies close to the lower *unicornis* Zone boundary.

Survival interval

The new taxonomic structure of the coral assemblages develops very slowly. The categories identified are long-ranging survivors belonging to morphoecotype I; short-ranging Early Carboniferous survivors belonging to morphoecotypes I and II; new genera belonging to morphoecotypes II and III and some transient Early Carboniferous genera showing important ontogenetic changes belonging to morphoecotype II.

Long-ranging taxa (of primitive morphology) belonging to morphoecotype I are poorly represented in the studied sections and they are not considered in this paper.

Rare, short-ranging, survivors are found only in some sections in the region studied. They are represented by specialized taxa possessing complex internal structures (e.g. *Dibunophyllum* from the *bogdanovkensis* Zone of the Donetz Basin) and those with non-trabecular fine structures (e.g. *Palaeosmilia* from the *bogdanovkensis* and lower part of the *pseudostruvei–postmosquensis varvariensis* Zones of Novaya Zemlja and the *pseudostruvei–postmosquensis–varvariensis* Zone of North Timan); see Figs 4 and 5. These lineages became extinct soon after the crisis without leaving any descendants (Fig. 9).

Stage	U. Ser.	Bashkirian		
Event, interval	E Survival		Recovery	Morpho-ecotype
Rugosa genera	Foraminiferal Zone			
	1 2 3 4	5 6 7		

Rugosa genera	1	2	3	4	5	6	7	Morpho-ecotype
Siphonophyllia	--							(II)
Palaeosmilia	----------							(II)
Zaphrentites	--							(I)
Sychnoelasma	--							(I)
Cyathoclisia	--							(II)
Dibunophyllum	--							(II)
Gangamophyllum	--							(II)
Carcinophyllum	-------							(II)
Arachnolasma	--							(II)
Siphonodendron	--							(III)
Tschernoviphyllum	--							(III)
Lonsdaleia	--							(III)
Actinocyathus	--							(IV)
Corwenia	--							(III)
Bothrophyllum			-------------------					(II)
Protodurhamina			---------					(III)
Profischerina			----------					(III)
Pseudokoninckophyllum			------					(II)
Donophyllum			----	---------				(III)
Lytvophyllum			------					(III)
Fomichevella				---------				(III)
Caninia				---------				(II)
Petalaxis				---------				(IV)
Cystolonsdaleia				---------				(IV)

Fig. 3. The distribution of rugose coral genera in the Novaya Zemlja region during the Late Serpukhovian to Bashkirian interval. The forminiferal zones (1–7) are as follows: 1. *Loeblichia minima–Monotaxinoides trasitorius–Eosigmolina explicata*; 2. *Plectostaffella bogdanovkensis*; 3. *Eostaffella pseudostruvei–E. postmosquensis–E. varvariensis*; 4. *Pseudostaffella antiqua*; 5. *Pseudostaffella praegorsky–Profusulinella staffelliformis*; 6. *Ozawainella pararhomboidalis–Profusulinella staffelliformis*; 7. *Verella spicata–Aljutovella tikonovichi*.

Survivors with post-crisis ontogenetic changes belong to two genera (*Caninia* and *Bothrophyllum*) with wide geographical and facies distributions. These changes are traceable at the specific level within *Caninia* lineages by the disappearance of the ephebic stage characteristic of the ancestor (*Caninia cornucopiae*). Its inner structure of the ephebic stage is characterized by a wide lonsdaleoid dissepimentarium (Poty, 1981). The earlier ontogenetic stages remain the same.

At the same time the fine structure of the descendant *Caninia* species becomes stable compared with its ancestor (Kossovaya & Kropatcheva 1993). Further evolution of the Cyathopsidae involves coenogenesis and heterochrony (Kossovaya 1989). The second genus, *Bothrophyllum*, derived from the Caninia lineage by the loss of the 'amplexoid' early neanic stage gives rise to a new family. The later development of the *Bothrophyllum* clade is based on a further complication of the middle ontogenetic stages resulting both in changes of septal arrangement and in the appearance (Kossovaya 1989) of axial structures (*Bothrophyllum pseudoconium, B. conicum*). Only a few species of these genera are known from the *Eumorphoceras/Homoceras* Zone; *Bothrophyllum simplex multiseptata* from the *pseudostruvei–postmosquensis–varvariensis* Zone of Cape Makarov, *B. berestovensis* from the *bogdanovskensis* Zone of the Donetz Basin and *Caninia* sp. from the *pseudostruvei–postmosquensis–varvariensis* Zone of North Timan.

Newly appeared genera consist of taxa with a hypothetical ancestor and those of cryptogenic origin (Fig. 9). The first group includes two genera of fasciculate corals (morphoecotype III). One of them, *Protodurhamina*, which possesses a rather complicated inner structure, appears abruptly at the very beginning of the survival interval (Fig. 4, loc. 797, bed 2) with foraminifera typical of the *bogdanovkensis* Zone. A possible ancestor can be suggested among fasciculate *Corwenia rugosa*, described from the Upper Serpukhovian of the Russian Platform (Hecker 1985). These ancestral *Corwenia* produce corallites with primitive early stages while reproducing under favourable conditions. They display variable axial dissepimentarium structures, as well as retardation of budding, resulting in a high potential for adaptivity and reproduc-

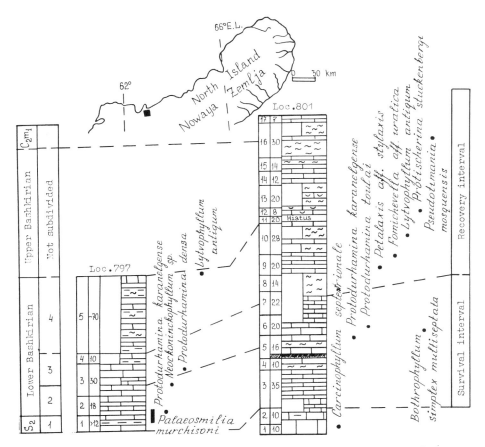

Fig. 4. Distribution of rugose corals in the uppermost Serpukhovian and Bashkirian sediments in Novaya Zemlja.

tion. Further phyletic evolution of *Protodurhamina* species is of a stepwise character and lasts up to the end of the survival interval.

Thus the few species of *Protodurhamina* (*P. densa, P. toulai, P. peculiare*) with different axial structures are characteristic of the *pseudostruvei–postmosquensis–varvariensis* Zone in both the Novaya Zemlja and North Timan sections (Figs 4 and 5). The second genus of this group, *Profischerina*, was probably derived from *Protodurhamina* by simplification of its inner structures. These genera are widespread in the contemporaneous deposits (Fig. 10) of Gornya Bashkiria (Ogar' 1985, 1990) and the Voronezh Anteclise (Kozyreva 1978, 1980), but the taxonomic composition is restricted to the above-mentioned species.

Among the cryptogenic genera, the fasciculate *Lytvophyllum*, *Darwasophyllum* (morphoecotype III) as well as the single *Neockoninckophyllum* and *Protokionophyllum* (morphoecotype II) are the first to appear after the extinction event.

The majority of the post-crisis genera form the ancestral stock for the Middle and Late Carboniferous families. Within the long survival interval (*bogdanovkensis–antiqua* Zones), two phases are distinguished. The first is determined by a few, widely distributed, taxa during *bogdanovskensis–postmosquensis–varvariensis* time. The second is characterized by a decline of diversity in the studied area (Figs 2, 6 and 11) and a stepwise replacement of species within the early lineages in the *antiqua* Zone.

The contemporaneous evolutionary law of rugose corals is observed in coral interval 10 of the Western Interior of the USA (Sando 1989). The equivalent interval of Namurian B in the Cantabrian Mountains shows the appearance of some new taxa within the impoverished assemblage (Rodriguez et al. 1986).

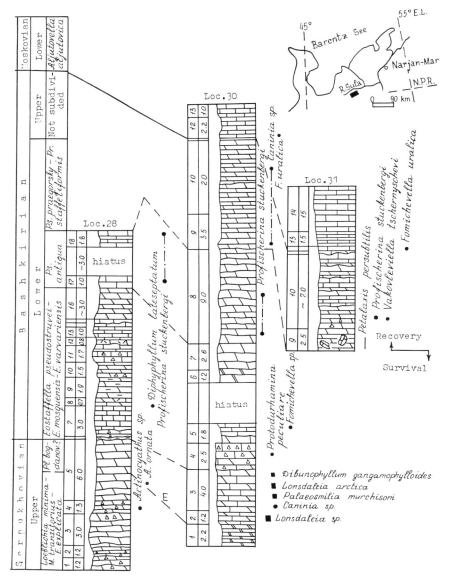

Fig. 5. Distribution of rugose corals in the uppermost Serpukhovian and Bashkirian sediments in the North Timan (River Sula) section.

Recovery interval of rugose coral diversity

The recovery interval can be readily documented in the Gornaya Bashkiria, Novaya Zemlja and North Timan sections. It begins approximately at the base of the *praegorsky–staffelliformis* fusulinid Zone with the sudden appearance and rapid expansion of massive, colonial, *Petalaxis* (morphoecotype IV) along the eastern margin of the Euro-American palaeocontinent (Figs 2, 5, 8 and 11). The other important feature is the diversification of the corals belonging to morphoecotypes II and III.

For the simple corals, derived from *Caninia*, the prolongation of early neanic stages results in the origin of new genera (*Yakovleviella, Pseudotimania*). In the conservative, long-lived, *Caninia* lineage the increase of species diversity through anaboly takes place. Amongst other survival taxa (e.g. *Bothrophyllum*) the long-ranging spe-

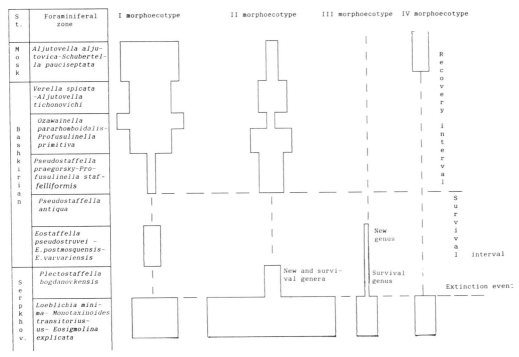

Fig. 6. Distribution of rugosan morphoecotypes in the Donetz Basin following the mid-Carboniferous extinction event. Taxonomic identifications based on Poljakova (1984), Vasiljuk in Aizenverg et al. (1983) and Fomitchev (1953).

cies *Bothrophyllum pseudoconicum* appears and becomes ancestral to many Late Carboniferous bothrophyllid genera. In the *Bothrophyllum* lineage further phylogeny is connected with the complication of ephebic stages (anaboly). Evolution of fasciculate corals (morphoecotype III) shows frequent replacement of taxonomically different corals within the same habitat. This is expressed in the simplification of the axial structures.

Another important component of the recovering assemblages are immigrant taxa, represented by massive colonies such as *Petalaxis* and *Cystolonsdaleia*. From the very beginning *Petalaxis* species evolve as two separate stocks. One of them, widely spread and long-ranging *(P. stylaxis)* is defined by the presence of one or two rows of dissepiments. The second stock includes species with more than two rows of dissepiments (Fig. 9). The appearance of *Petalaxis stylaxis* in the Cape Makarov and Gornaya Bashkiria successions coincides with the base of the *praegorsky–staffelliformis* Zone. In the former they were found together with *Pseudostaffella antiqua, P. grandis* and *P.* ex gr. *praegorsky*

(determination by V. Davydov, V. Matveev pers. comm.)

The beginning of the recovery interval can be established in the Western Interior of the USA by the appearance of exotic genera in coral interval 12 (Sando 1989), which corresponds with the *Profusulinella–Fusulinella* Zone (Atokan Series). In the Cantabrian Mountains the same event is expressed in the maximum diversity and appearance (Fig. 8) of *Petalaxis* at the base of Westphalian A (Rodriguez et al. 1986).

Radiation

The beginning of the radiation is easily identified in Novaya Zemlja, Moscow District, and the North Timan sections at the base of the *colaniae–vozhgalensis–kamensis* Zone. At present only some general features can be mentioned. The abrupt increase in diversity is established in all morphoecotypes, but it is better expressed in the evolution of massive colonial corals within the *Petalaxis* stock. It is based on the introduction of morphological

194   O. L. KOSSOVAYA

Stage	U. Serpukhovian			Bashkirian						Moskov.	
Event interval	E Survival				Recovery						
Rugosa coral genera	Foraminiferal Zone										Morpho-ecotype
	1		2	3	4	5	6	7	8	9	
Lm.	D3	D6	D7k	E1	E8	F1	G1-H4	I 2	K3	L1	
Amplexus	--------						----				(I)
Hapsiphyllum	--------										(I)
Zaphrentis	--------										(I)
Barithichisma	--------										(I)
Ufimia	--------										(I)
Leonardiphyllum	--------										(I)
Amandophyllum	.										(I)
Zaphriphyllum	--------										(I)
Palaeosmilia	--------										(II)
Caninia	.										(II)
Caninophyllum	--------										(II)
Dibunophyllum	--------------										(II)
Nevrophyllum	--------										(II)
Clisiophyllum	--------										(II)
Arachnolasma	--------										(II)
Eostrotion	.										(II)
Slimoniphyllum	--------										(II)
Zakowia	--------										(II)
Neockoninckophyllum	--------------		--	--	--	--	--	--	----------		(II)
Lithostrotion	--------										(IV)
Siphonodendron	--------										(III)
Aulina	--------										(IV)
Aulockoninckophyllum	--------										(II)
Actinocyathus	.										(IV)
Lonsdaleia	--------										(III)
Axophyllum	--------										(II)
Gangamophyllum	--------										(II)
Protokionophyllum			-----								(II)
Bothrophyllum			-----	--	--	--	--	--	------		(II)
Cyathaxonia						-----------------					(I)
Clinophyllum						----	-----				(I)
Monophyllum						----	-----				(I)
Stereolasma						------------					(I)
Lophophyllidium							----				(I)
Stereophrentis						-----------------					(I)
Pavastereophrentis						--------------					(I)
Kumpanophyllum						-----	----				(II)
Lophophyllum					-----		----				(II)
Yanophylloides							----				(II)
Cystilophophyllum						-----					(II)
Campophyllum							----				(II)
Orygmophyllum							----				(II)
Yakovleviella							----				(II)
Bothroclisia						-------------------					(II)
Axolithophyllum					-----						(II)
Carcinophyllum						-----					(II)
Lytvophyllum			----								(III)
Parastereophyllum			----								(I)
Homalophyllites			----								(I)
Bradyphyllum								----			(I)
Caninella								----			(I)
Sestrophyllum									----		(II)
Cystophora									?----		(IV)
Lonsdaleiastraea									?----		(IV)

Fig. 7. The distribution of rugose corals in the Donetz basin during Late Serpukhovian to Moskovian time. Foraminiferal zones as in Fig. 3 with the addition of: 8. *Aljutovella aljutovica–Schubertella pauciseptata*; 9. *Fusulina subpulchra–Aljutovella priscoidea*. Taxonomic identifications based on Poljakova (1984), Vasiljuk in Aizenverg *et al.* (1983) and Fomitchev (1953).

Russia***						Cantabrian Mountains Rodriguez et al. 1986			Western interior of USA. Sando. 1989				
Stage	Horizon	Ammonoids zones	Foraminiferal zones	Conodont zones	The main coral evolutionary event	Ser	St.	Corals	Sys.	Ser.	Con.	For.	Corals
Moskovian	Miachk-ovsky	Beds with *Pseudopa rallegoceras,Gla-phyrites*	*Fusulinella bocki. F.eopulchra, Fusulina cylindrica*	*Streptognathodus cancellosus-Neo-gnathodus roundyi*		Westphalian	D	**Peak of di-versity**	Pennsylvanian	Des Moines			Inter-val 13
	Podols-ky	*Pseudoparalego-ceras.Welleri-tes*	*Fusulinella colaniae-F.vozhgalensis-F kamensis*	*St.coninnus-Neo-gnathodus medexu-timus*	**Radiation** Appearance of astreoid colonies			max d					**Peak of diver-sity**
	Kashir-sky	*Paralegoceras, Eoewelwrites*	*Fusulinella subbul-chra - Aljutovella priscoidea*	*St. dissectum-N.medatultimus ---- St. transitivus.*									
	Vereis-ky	*Diaboloceras, Winslowoceras*	*Aljutovella alju-tovica-Schubertel-la pauciseptata*	*Neognathodus bothrops ---- Idiognathodus tuberculatum*	Gradual increase of diversity of		C	Increase					Gradual increase of diver-sity
Bashkirian	Asatau-sky	*Diaboloceras, Axinodosus*	*Verella spicata-Aljutovella tikhono-vichi*	*Declinognathodus marginadosus*	all morphocoty-types (replace-ment)		B	Decrease					Int 12
	Tasha-stynsky	*Brannoeras Gastrioceras*	*Ozawainella pararhomboidalis, Profusulinella pri-mitiva*										
	Askyn-bashsky	*Billinguites superbillinguites*	*Pseudostaffella praegorsky-Profu-sulinella staffelli-formis*	*St. expaneus - Idiognathodus sinuosus*	Appearance of **immi-grant taxa** (massive colonial *Petalaxis*) max d		A	Max diversi-ty. appeara-ns of *Peta-laxis* max d					Appea-rance of exotic genera
	Akavas-sky	*Verneilites verneuilli*	*Pseudostaffella antiqua*	*Idiognathodus sinuatus -Neognathodus simmetricus*	Increase of diversity	Namurian		Gradual inc-rease of diversity		Morrow		*Idiogna-thodus sinuatus*	Interval 11
	Sjuran-sky	*Reticuloceras Bashkortoceras*	*Eostaffella pseudostruvei - E.postmosquensis- E.varvariensis*										
Serpukhovian	Vozne-sensky	*Homoceras*	*Plectostaffella bog-danovkensis*	*Declinognathodus noduliferus s.l.* **	Appearance of I-III morphoecotypes n (*Protodurhamina* t. *Bothrophyllum*)		B					*Declino-gnathodus nodulife-rous*	Appear. of new nodulife-genera
	Zapal-tjubin-sky	*Eumorphoceras*	*Loeblichia minima-Monotaxinoides transitorius - Eosigmolina explica-ta*	*Gn.postbilineatus ---- Gn.bilineatus bo-landensis- Ad.un-icornis*	**Extinction** Max diversity of all morphoecotypes following by decline		A	Gradual dec-line diver-sity within I-IV morpho-ecotypes	Mis-sissi-pian	Chester		*Rh.primus A.unicor. --- A.navicu-la*	Interval Zone 6 ------ Disapp.

Fig. 8. Principal coral events in the Late Serpukhovian to Moskovian interval. **, zonation after Nemirovskaya & Nigmadganov (1993).

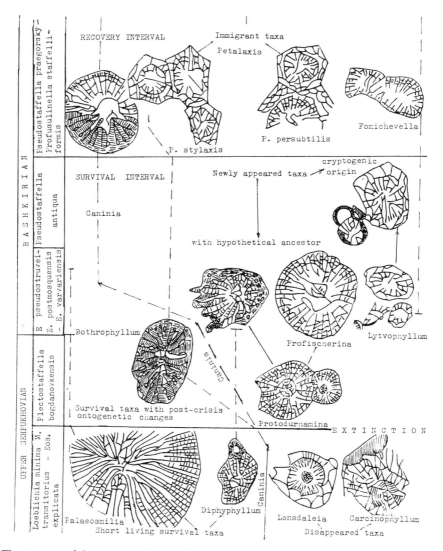

Fig. 9. The structure of the rugose coral assemblage during the extinction–recovery interval.

innovation such as minor septa (*P. vesiculosus, P. flexuosus*) and thamnasterioid and astreioid coralla (*Ivanovia*).

Two phases exist in the radiation interval. The first shows a maximum diversity and the appearance of new genera with new morphological features; the second shows a gradual decrease of specific diversity. The beginning of the radiation interval is well identified in the Cantabrian Mountains by a peak of diversity at the Westphalian C–Westphalian D boundary (Rodriguez *et al.* 1986). The same event allows the recognition of the radiation level at the lower boundary of the Desmonesian Series (Fig. 8) in the Western Interior of the USA (Sando 1989).

The upper level of the radiation interval seems to be restricted by the disappearance of massive colonial corals near to the Upper Moskovian boundary within a widespread area, including the Moscow Region, North Timan, Novaya Zemlja and Gornaya Bashkiria.

Conclusions

1. The disappearance of the typical rugose coral fauna coincides with the base of the *Homoceras*

CARBONIFEROUS RUGOSE CORAL RECOVERY

Stage	U.S	Bashkirian		
Event, interval		Survival	Recovery	Morpho-ecotype
Rugosa genera		Foraminiferal zone		
	1	2 3	4 5 6	
Zaphrentites	---			(I)
Darvasophyllu	---		-----	(III)
Protokionophyllum	-----------			(II)
Fomichevella		-------------------		(III)
Diphyphyllum		-----		(II)
Dibunophyllum		-----		(II)
Carcinophyllum		-----		(II)
Lytvophyllum		-----------		(II)
Protodurhamina		-----------		(III)
Pseudokonincko-phyllum		-----------------		(II)
Profischerina		----------------		(III)
Koninckophylloides			--------------	(II)
Stereolasma		-----		(I)
Bradyphyllum		-----		(I)
Petalaxis			-------	(IV)
Lophophyllidium			-----	(I)
Cravenia			-	(I)
Cyathaxonia			-----	(I)
Caninia			-----	(II)
Caninophyllum			-----	(II)
Yakovleviella			-----	(II)
Bothrophyllum			---------	(II)
Amygdalophylloides			---------	(II)
Eostrotion			---------	(II)

Fig. 10. The distribution of rugose corals in the Bashkirian Mountains (latest Serpukhovian to Bashkirian interval). The foraminiferal zones are those used in Fig. 3. The taxonomic identifications are based on Ogar' (1985, 1990).

Zone and seems to be synchronous.

2. The survival interval established at the lower *bogdanovskensis* Zone is characterized by a coral fauna of morphoecotypes I and III, including taxa as the result of ontogenetic change and new taxa with no clear origination.

3. Some evolutionary peculiarities of rugose corals are observed during the extinction–survival–recovery intervals. Thus, the pre-adaption of some rugose corals at mid-Carboniferous event is connected with early ontogenetic changes (coenogenesis), retardation and simplification of budding.

4. The recovery of rugose corals after the mid-Carboniferous 'lesser mass extinction' has an oscillating character with a few levels of high diversity. This type of diversification is a result of replacement, which is a more typical rugose coral evolutionary pattern within the survival–recovery interval.

This research was supported by the Russian Fundamental Research Foundation (Grant 94-05-17589), G. Soros Foundation and the Russian Academy of Natural Science (Program of Biological Diversity). I am very grateful to Prof. T. N. Koren for her valuable remarks and assistance with the English translation of the paper. I would also like to thank Dr V. Matveev who collected the rugose corals from Novaya Zemlja and Dr Kashik for advice on the lithological subdivision of the North Timan sections.

References

AIZENVERG, D. E. & BRAZHNIKOVA, N. E. 1982. The Upper Serpukhovian deposits of Donbass and their place in the stratigraphical scale USSR. *In:* [*The scale of the Carboniferous system from the modern data point of view.*] Nauka, Moscow, 58–73 [in Russian].

—— ASTAHOVA, T. V., BERCHENKO, O. I., BRAGNIKOVA, N. E., VDOVENKO, M. V., DUNAEVA, N. N., ZERNECKAYA, N. V., POLETAEV, V. I. & SERGEEVA, M. T. 1983. [*The Upper Serpukhovian substage of the Donetz basin.*] Kiev [in Russian].

ALEKSEEV, A. S. 1989. The global biotic crisis and mass extinction at Phanerozoic history of the Earth. *In:* [*The biotic events of the main Phanerozoic boundaries.*] Moscow State University, 22–47 [in Russian].

FEDOROWSKI, J. 1989. Extinction of Rugosa and Tabulata near the Permian/Triassic boundary. *Acta Palaeontologica Polonica*, **34**, 47–70.

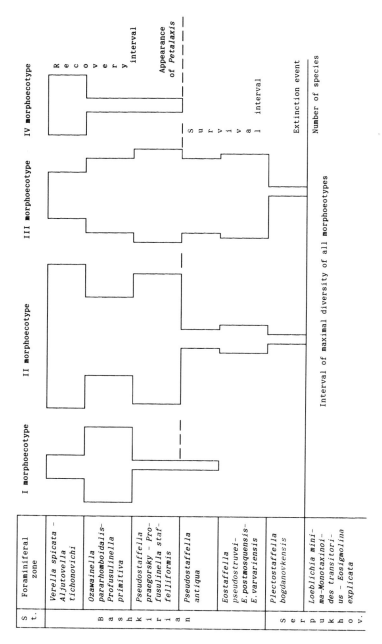

Fig. 11. Distribution of rugosan morphoecotypes during the Bashkirian in Gornaya Bashkiria. Taxonomic data after Ogar' (1985).

FOMITCHEV, V. D. 1953. [*Rugosa corals and stratigraphy of the Middle–Upper and Lower Permian deposits of the Donetz basin.*] Moscow [in Russian].

GORSKY, I. I. 1951. The Carboniferous and Permian corals of Novaya Zemlja. *Proceedings of the Scientific Research Institute of Arctic Geology*, XXXII, 1–165 [in Russian].

HECKER, M. R. 1985. The evolution of colony of some Carboniferous Rugosa. *Journal of Paleontology*, **4**, 12–20 [in Russian].

KOREN', T. N. (ed.) 1991. [*Zonal stratigraphy of Phanerozoic of USSR.*] Nedra, Moscow [in Russian].

KOSSOVAYA, O. 1989. New data on the morphogenesis and phylogeny of some Late Carboniferous and Early Permian rugose corals. *Memoir Association of Australasian Palaeontologists*, **8**, 109–113.

—— & KROPATCHEVA, G. 1993. New data on phylogeny on the Late Paleozoic tetracorals of the Suborder Caniniina. *In:* [*Phylogenetic aspects of palaeontology, Proceedings of the XXV VPO Session.*] St. Petersburg, 70–75 [in Russian].

KOZYREVA, T. A. 1978. The new Carboniferous genus–Protodurhamina (Rugosa) and its significance in the phylogeny of durhaminids. *Journal of Paleontology*, **1**, 20–24 [in Russian].

—— 1980. About the evolution of the colonial Middle Carboniferous corals. *In:* [*Corals and reefs of the phanerozoic of USSR.*] Nauka, Moscow, 130–136 [in Russian].

NEMIROVSKAYA, T. & NIGMADGANOV, I. 1993. The mid-Carboniferous Conodont Event. *Sonderdruck aus CFS-Courier*, **168**, 319–333.

OGAR', V. V. 1985. [*The Middle Carboniferous corals and detailed subdivisions of the Bashkirian and Moskovian stages of the Gornaja Bashkiria region.*] Summary of PhD Thesis, Kiev [in Russian].

—— 1990. The peculiarities of stratigraphical distribution of corals at the Middle Carboniferous in Bashkiria. *In:* [*The boundaries of biostratigraphical subdivisions of the Carboniferous in the Urals.*] 109–119 [in Russian].

POLJAKOVA, V. E. 1984. *The Late Serpukhovian corals of the Donetz basin and their stratigraphical significance.* Summary of PhD Thesis, Kiev [in Russian].

POTY, E. 1981. Recherches sur les tetracoralliaires et les heterocoralliaires du viséan de la Belgique. *Mededelingen Rijks Geologische Dienst*, **351**, 1–161.

RODRIGUEZ, S., SANDO, W. J. & KULLMAN, J. 1986. Utility of corals for biostratigraphic and zoogeographic analyses of the Carboniferous in the Cantabrian Mountains, Northern Spain. *Tribajos de Geologia, Universitade de Oviedi*, **16**, 37–60.

SANDO, W. J. 1989. Dymanics of Carboniferous coral distribution, western interior USA. *Memoir Association of Australasian Palaeontologists*, **8**, 10–30.

—— & BAMBER, E. W. 1984 Coral Zonation of the Mississippian System of Western North America. *In: Neuvième Congrès International de Stratigraphie et de Géologie du Carbonifère.* **2**, Southern Illinois University Press, 289–300.

SEPKOSKI, J. 1984. A kinetic model of Phanerozoic taxonomic diversity. III. Post-Paleozoic families and mass extinction. *Paleobiology*, **10**, 246–267.

Climate change, plant extinctions and vegetational recovery during the Middle–Late Pennsylvanian Transition: the Case of tropical peat-forming environments in North America

WILLIAM A. DIMICHELE[1] & TOM L. PHILLIPS[2]

[1]*Department of Paleobiology, National Museum of Natural History, Smithsonian Institution, Washington, DC 20560, USA*
[2]*Department of Plant Biology, University of Illinois, 505 So. Goodwin Ave., Urbana, IL 61801, USA*

Abstract: A major extinction of terrestrial plants occurred at the end of the Westphalian (Middle Pennsylvanian) in the lowland tropics of North America. Approximately 67% of the species in peat-forming mires, and at least half the species in clastic wetlands were eliminated by changing climatic conditions, probably protracted moisture deficits or exaggeration of seasonal dryness. Independent studies suggest that the end of the Westphalian was marked by deglaciation in Gondwana and increase in global temperature.
   In peat-forming habitats major floristic changes followed the extinction. Earliest Stephanian (Late Pennsylvanian) dominance-diversity patterns were highly variable temporally; several different plant groups that were of minor importance in Westphalian mires became major framework dominants, with high coal to coal variability. Ultimately opportunistic tree ferns, previously subdominant, became the dominant elements. Westphalian tree ferns were mostly small, cheaply constructed colonists with massive reproductive outputs and broad ecological amplitudes. Stephanian tree ferns were much larger and appear to have occupied sites for extended periods; they retained the cheap construction of earlier forms but added massive root mantles, which permitted greater height and girth. The 'marsh'-forming lycopsid *Chaloneria* also became common in Stephanian mires. These elements formed mire landscapes that had few analogues in the older Westphalian mire forests, dominated by tree lycopsids. Generic dominance patterns in the Stephanian became fairly consistent after the initial period of irregularity; however, the species composition of Stephanian mires was highly variable within generic themes.
   Ecological assemblages were persistent during most of the Westphalian and Stephanian. Patterns at the end-Westphalian suggest that high levels of species extinction disrupt self-regulatory properties of groups of species and create intervals of lottery-like ecological dynamics. Opportunists may have an advantage during these periods, but ultimately they must give rise to species capable of site occupation. Thus the size of the extinction, and the proximity of existing species that can recolonize vacated resource space, will dictate whether speciation or colonization will rebuild the new landscapes.

A major consequence of mass extinction is the vast change in ecological conditions. Disappearances of species or extreme reduction in their numbers or areal extent release previously sequestered resources, change the ecomorphic character of multispecies assemblages, and alter the numbers, kinds and strengths of interspecific interactions (Pimm 1991). The result is alteration of the ecological background through which evolutionary changes are filtered. Evidence from both the marine (Boucot 1983; Brett *et al.* 1990; Miller, 1993; Brett & Baird 1995) and terrestrial (Wing 1984; DiMichele & Phillips 1995) fossil records indicates that ecological assemblages may be long-lived with persistent (millions of years) dominance-diversity structure and resis-

tence to invasion, as long as extinction levels are low. Such patterns suggest that ecosystems can evolve properties of self-regulation and hierarchical structure (Miller 1990; Allen & Hoekstra 1992), which may have been important modes of organization on land and sea during much of the Phanerozoic. Separating periods of persistent dynamics are briefer intervals with elevated extinction levels and somewhat more chaotic dynamics. It is within these time periods that major evolutionary changes may be concentrated (Vermeij 1987; Brett & Baird 1995).
   Plant communities in the Pennsylvanian-age lowland tropics were characterized by long periods of vegetational persistence. Assemblages with similar species composition and quantita-

From Hart, M. B. (ed.), 1996, *Biotic Recovery from Mass Extinction Events*,
Geological Society Special Publication No. 102, pp. 201–221

tive dominance-diversity structure characterized particular habitats for intervals of millions of years before shifting abruptly to new equilibria. In both peat-forming mires (Phillips *et al.* 1985; DiMichele & Phillips 1995) and mineral-soil wetlands (Pfefferkorn & Thomson 1982) elevated levels of extinction and or origination accompanied major vegetational change. Regional climatic changes, particularly in rainfall, are the most likely factors causing vegetational change (Phillips & Peppers 1984; Cecil 1990; Winston 1990); like the Recent, the Pennsylvanian was a time of major polar glaciations (Frakes *et al.* 1992), and experienced extensive climatic and sea-level fluctuations (Ross & Ross 1987, 1988; Heckel, 1986, 1989; Cecil 1990).

The most profound vegetational change of the Pennsylvanian occurred during the Middle to Late Pennsylvanian (Westphalian–Stephanian) transition. At that time peat-forming mires were transformed from lycopsid dominance, which had persisted for more than nine million years, to tree-fern dominance. In addition, the hierarchical organization that characterized Westphalian mire landscapes (DiMichele & Phillips 1995) was replaced by vegetation with less spatio-temporal complexity (Willard & Phillips 1993). Westphalian landscapes had stereotypic patterns of resource partitioning largely defined by the tree lycopsids (Scott 1978; Collinson & Scott 1987*a, b*; DiMichele & Phillips 1985, 1994). This partitioning created a fabric that endowed ecosystems with a substantial degree of self-regulation of species turnover dynamics over geologically significant spans of time (DiMichele & Phillips 1994*b*). In contrast, the Early Stephanian wet tropics were dominated by opportunist tree-fern lineages or by seed plants that had survived the end-Westphalian extinctions (Pfefferkorn & Thomson 1982; Phillips *et al.* 1985). These extinctions, which affected both mires and clastic wetlands (Phillips *et al.* 1974; Gillespie & Pfefferkorn 1985; Peppers 1985), were at the root of the ecological differences. Such an extensive and temporally abrupt reorganization of plant communities suggests a threshold-like response of the system to extinction of the component species. As a consequence, the vegetation of the wet lowlands was taxonomically and structurally simplified, and the marked taxonomic differences between mires and clastic wetlands, which had characterized the Westphalian, were muted (DiMichele & Aronson 1992).

The study system: peat-forming mires

Peat formation was a major distinguishing feature of the Pennsylvanian tropics. Vast tracts of swampy lowlands were covered with mires, a landscape indicative of high rainfall and high groundwater tables distributed throughout most of the year (Clymo 1987). Mire floras were quantitatively and, to a large extent qualitatively distinct from those of surrounding clastic wetlands during the Westphalian (Middle Pennsylvanian). Lycopsids dominated peat substrates, on average, for the entire Westphalian with subdominant cordaitean gymnosperms from the Westphalian B through early D. In contrast, floodplains, levees, and other clastic lowland environments were dominated by pteridosperms, sphenopsids, and tree ferns (compare Phillips *et al.* 1985 and Pfefferkorn & Thomson 1982; Peppers & Pfefferkorn 1970). This distinction was dramatically reduced in the Late Pennsylvanian when tree ferns and pteridosperms became dominant elements in all lowland–wetland habitats (Pfefferkorn & Thomson 1982; Phillips *et al.* 1985; DiMichele & Aronson 1992), although species differences may have remained.

Peat substrates present major physiological challenges to plants (Schlesinger 1978). Generally low pH, flooding during substantial parts of the year, chelation of mineral nutrients and, especially in the case of domed peats, highly oligotrophic nutrient status, strongly select against most species. As such, mires can be visualized as semi-closed, edaphic islands. During the Pennsylvanian certain evolutionary lineages, most notably stigmarian lycopsids, were ecologically and evolutionarily centred in swamps and mires and had strongly partitioned these environments (Phillips & Peppers 1984; DiMichele & Phillips 1985, 1994*a*). Most groups, and the pteridosperms can serve as the best understood example, were centred in the broader lowland wetlands but had few or no species that grew uniquely or predominantly on peat substrates; generally only a small subset of species from these lineages could grow on peats (DiMichele *et al.* 1985; Beeler 1983; Schabilion & Reihman 1985). Sphenopsids, marattialean ferns, and cordaites had patterns similar to pteridosperms, although the cordaites, and to a lesser extent the marattialean ferns, may have evolved some mire-centered lineages (Costanza 1985; Trivett & Rothwell 1985; Lesnikowska 1989; Trivett 1992).

Sources of data

Data on mire vegetation come from two principal sources, coal-ball macrofossils and pollen-spore microfossils. Coal balls are carbonate concretions that contain structurally pre-

microfossils can be linked to parent plants (e.g. Mahaffy 1985; Willard 1993).

Most of our analysis and inference is based on coal-ball data of two types, profile and random sample. Profiles of coal balls are collected *in situ*; relative position of coal-ball layers within the coal bed is noted, which permits recovery of the original zonation of the plant litter (Fig. 1). Random samples of coal balls are collected from one locality without respect to the position of coal balls in the coal bed. Consequently, they represent the average composition of the permineralized peat at the collection site. A coal-bed summary can be obtained from a profile by averaging the composition of individual coal-ball zones. Palynological and compression fossil data are presented and discussed where appropriate, and in order to amplify coal-ball patterns. There are critical gaps in the coal-ball record, particularly in the early Stephanian, immediately following the extinctions at the end of the Westphalian, and palynological data are of particular importance in this interval. The coal-ball data base used in this analysis is described in Phillips *et al.* (1985), with some additions of coals from the Westphalian A and B of the United States (Winston & Phillips 1991; unpublished). Palynological data are mostly from Peppers (1985). Compression–impression species ranges are from Gillespie & Pfefferkorn (1979).

Fig. 1. Coal-balls *in situ*; Herrin No. 6 Coal Member, Carbondale Formation, late Westphalian D age; Old Ben Coal Company, No. 24 Mine, near Benton, Illinois. Coal balls are light-coloured material within the darker coal. Coal ball layers are extracted and analysed quantitatively, providing a partial record of vegetational change on the site through the time of peat accumulation.

served peat stages of the coal and that occur within coal seams (Phillips *et al.* 1976; DeMaris *et al.* 1983; Scott & Rex 1985). Coal balls preserve part of the original litter of the mire forest, often in exceptional anatomical detail. They occur in layers or aggregates (Fig. 1) and can replace much of the thickness of a coal seam, providing the basis for quantitative analysis of vegetational change during the history of peat accumulation (Phillips *et al.* 1977). Coals usually contain abundant pollen and spore microfossils, produced by plants of the mire vegetation. Incremental sampling of a coal seam can reveal fine-scale details of the history of peat accumulation, particularly when the more abundant

The pre-extinction system: late Westphalian mire landscapes

The late Westphalian was the zenith of coal-age tropical plant diversity and landscape complexity. The primordial Westphalian forests were a complex mixture of habitats each with distinct types of vegetation. Landscapes were strongly partitioned along taxonomic lines; each of several major clades had diversity and dominance peaks in particular types of physical settings (DiMichele & Phillips 1994).

A vegetational change occurred during the Westphalian–Stephanian transition that entailed not only extinction, but a fundamental alteration in the dynamics of wetland ecosystems. The Stephanian was a period of much greater vegetational uniformity among habitats. It is possible that the spectrum of environments colonized by plants was smaller than in the Westphalian, and that, overall, wetland vegetation encompassed significantly shorter physical gradients in substrate wetness, nutrient availability, and, perhaps, levels of physical disturbance.

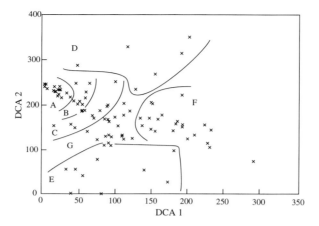

Fig. 2. Detrended correspondence analysis of coal-ball zones from late Westphalian D Herrin, Springfield, and Iron Post coals. Lines delimit groups of coal-ball zones with similar dominance patterns; letters refer to groups in Table 1 where dominance and life-history patterns are summarized.

Species assemblages in late Westphalian D mires

Peat-forming mires of the Westphalian D typically were composed of 40 to 50 species. Of the 50 species in our sample 31 were trees or subtrees, 19 were ground cover. As expected, this diversity was not randomly distributed among physical habitats, and several subenvironments can be recognized, each with a characteristic species assemblage (community-type) and dominance-diversity structure. The basic pattern is summarized in Fig. 2, an ordination of 96 zones from 7 profiles from three late Westphalian D coals, the Iron Post coal of Oklahoma, the Springfield coal of Indiana, and the Herrin coal of Illinois. This type of multicommunity landscape is typical of all late Westphalian D coals we have studied; more detailed analyses of the Springfield and Herrin coals are published (Phillips & DiMichele 1981; Eggert et al. 1983; DiMichele & Phillips 1988; Willard 1993).

The zones (vegetational stands) in Fig. 2 form three major groups: assemblages dominated by the monocarpic lycopsid tree *Lepidophloios hallii*, assemblages dominated by the polycarpic lycopsids *Diaphorodendron scleroticum* or *Sigillaria* sp., and assemblages dominated by *Medullosa* spp. and the polycarpic lycopsid *Paralycopodites brevifolius*. Average species richness among these assemblage-types ranges from 7.2–8.2; major differences reside in the kinds of life histories that dominate, the average percentages of ground cover, and the amounts of associated fusain and mineral matter. *Psaronius* tree ferns are components (5–15% biomass) of nearly all assemblages except those dominated by *Lepidophloios hallii* lycopsids. Calamites are minor elements and occur in greatest abundance in association with *Medullosa–Paralycopodites* and to a lesser extent with *Diaphorodendron* and *Sigillaria*.

Lepidophloios-dominated assemblages appear to have been best developed in habitats with extended periods of standing water. Species richness averages 7.2 in the 19 zones of the ordination, and includes low biomass and diversity of ground cover, vines, and understory trees and shrubs (Table 1). Quantitatively, lower vascular plants, particularly *Psaronius* tree ferns, are minor components and drop sharply in abundance as *Lepidophloios* proportions exceed 70% (DiMichele & Phillips 1988; Eble & Grady 1993). Ground cover and homosporous lower vascular plants both require exposed substrates, and their low abundances support the inference of flooded peat surfaces. In addition, the dominant *Lepidophloios* species produced seed-like 'aquacarps' (Phillips & DiMichele 1992), which had physical properties consistent with water dispersal (Phillips 1979).

Assemblages dominated by the polycarpic lycopsids, *Diaphorodendron* and/or *Sigillaria*, have variable composition. Species richness in these assemblages averages 7.7. Ground cover and vine richness averages 1.8 species per zone. However, quantitatively, ground cover and understory trees and shrubs were significant vegetational components, even if richness was low (Table 1). Levels of mineral matter and

EXTINCTION AND RECOVERY IN PENNSYLVANIAN COAL SWAMPS

Table 1. *Dominance and average abundance of ground cover and life history groups in ordinations of late Westphalian D coals and early Stephanian Calhoun coal.*

Group Dominant	n	GC	LVP	SP	HP	AQ
Late Westphalian D						
A *Lepidophloios* > 70%	7.2	1.7	2.7	1.7	0.3	2.5
B *Lepidophloios* > 50%	8.7	2.3	3.3	2.1	0.2	3.1
C *Lepidophloios* > 40%	7.1	1.7	3.2	1.6	0.1	2.2
D *Lepidodendron* > 40%	9.3	1.5	2.8	1.5	0.2	5.0
E *Diaphorodendron/Sigillaria* > 40%	7.7	1.8	2.6	1.9	0.8	2.5
F *Medullosa/Paralycopodites/Psaronius*	8.2	2.2	3.0	2.0	0.7	2.2
G Mixed Dominance	9.9	3.4	4.2	2.6	0.6	2.6
Average	8.3	2.1	3.1	1.9	0.5	2.8
Calhoun Coal–Early Stephanian						
A *Psaronius/Medullosa/Sigillaria*	8.0	3.5	3.5	2.5	2.0	000
B *Medullosa/Psaronius/Sigillaria*	10.0	5.5	5.5	3.2	1.0	000
C *Medullosa/Psaronius*	10.7	7.0	7.0	2.3	1.3	000
D *Psaronius*	10.0	7.0	7.5	2.5	000	000
Average	9.7	5.8	5.9	2.6	1.1	000

Group letters refer to coal-ball zones within areas marked by the same letters on the respective ordination (Figs 2 & 3). Numbers refer to average number of species within each category. n, species richness; GC, ground cover; LVP, homosporous lower vascular plants; SP, seed plants; HP, free-sporing heterosporous plants; AQ, aquacarpic plants (*sensu* Phillips and DiMichele 1992). Ground cover will also be listed in one of the other categories.

charcoal are not notably elevated in these assemblages. We suggest that they were characteristic of wet peats with occasional flooding and minor levels of physical disturbance.

Assemblages enriched in or dominated by medullosan pteridosperms and *Paralycopodites brevifolius* are associated with elevated levels of mineral matter and charcoal (Calder 1993; DiMichele & Phillips 1988; Johnson 1979). Average species richness is 8.2, with 2.2 species of ground cover or vines (Table 1). Subdominant species include a variety of life histories and growth forms, many of which are preferentially associated with mineral matter in coal (see literature summary in DiMichele & Phillips 1994) or that also occur in clastic swamp deposits. This vegetation can be characterized broadly as ecotonal, flourishing in habitats disturbed by flooding and possibly fire in areas often near the margins of peat bodies (Eble 1990; Eble *et al.* 1994).

Temporal patterns and hierarchical organization in the Westphalian

The type of landscape organization described above persisted within mires of western Euramerica for approximately two to three million years, based on the timescale of Hess & Lippolt (1986; Klein 1990). Landscapes partitioned among lycopsids, tree-ferns, and medullosans have been identified in all Late Westphalian D coals from which coal balls have been recovered: the Secor coal of Oklahoma (DiMichele *et al.* 1992), the Iron Post coal of Oklahoma (Phillips *et al.* 1985), the Springfield coal of Indiana (Eggert *et al.* 1983; Mahaffy 1988; Willard 1993), the Middle Kittaning coal of Pennsylvania (Feng 1989), the Herrin coal of Illinois (Phillips & DiMichele 1981; Mahaffy 1985; DiMichele & Phillips 1988), the Upper Freeport coal of Ohio (Phillips *et al.* 1985), the Baker coal of Kentucky (Phillips *et al.* 1985), and the Danville coal of Indiana. Palynological studies of other coals within this stratigraphic interval confirm a unity of composition and ecological structure (Peppers 1985; Kosanke 1988; Pierce *et al.* 1991; Eble *et al.* 1994).

When examined through the entire Westphalian, mire landscapes appear to be hierarchically organized.

1. Several levels of spatial organization can be identified, and each has certain temporal dynamics that operate only at that scale. In particular we recognize ecomorphs (guilds) within species assemblages (communities), and species assemblages within landscapes (DiMichele 1994).

2. The basic species assemblages (communities) identified in late Westphalian D mires occurred throughout the entire Westphalian. Distinct dominance-diversity structure, dominant species groups, and associated physical

attributes were conserved for > 9 Ma (DiMichele & Phillips 1995). Late Westphalian D mires were only one of several mire landscape types that existed sequentially during the Westphalian (Phillips & Peppers 1984; Phillips *et al.* 1985). The average species composition of earlier landscapes varied according to the proportions of the major species assemblages each contained. But the biotic characteristics of any particular species assemblage and the physical conditions with which it was associated remained distinct throughout the Westphalian. Thus, changes in the relative abundances of dominant species on a coal-bed average, and thus in the landscape composition of mires, was due mostly to changes in the proportions of different physical habitats.

3. Although there was substantial species turnover during the Westphalian, this turnover was largely on ecomorphic themes within habitats; species generally replaced others of the same clade, with similar growth architecture and life history, and remained within the habitat limits typical of those clades. Thus, in the broadest sense of ecomorphic patterns, intrahabitat organization persisted despite species turnover. Within habitat species turnover suggests an element of biotic control on species replacement, an emergent assemblage property beyond a strictly individualistic organization. We have attempted to document and detail these patterns elsewhere (DiMichele & Phillips 1994*a*).

Westphalian marattialean ferns

The marattialean ferns are a single, and most noteworthy, exception to the general pattern of Westphalian habitat restriction. Although broadly present in mires beginning in the Westphalian C (Peppers 1985), the first real abundances recorded by coal-ball macrofossils are not until near the Westphalian C–D transition in Iowa (Phillips *et al.* 1985). Marattialean ferns do not become generally abundant until the middle of the Westphalian D, and do so in both mires (Phillips & Peppers 1984) and clastic wetlands (Pfefferkorn & Thomson 1982).

Most *Psaronius* (*Pecopteris* in compression) species in the Westphalian were small trees, possibly a few were sprawling ground cover (Lesnikowska 1989). Large *Psaronius* trees began to appear in compression preservation during the Westphalian D (Pfefferkorn 1976), but truly large trees are not known from mires until the Stephanian. Unlike most other groups *Psaronius* species appear to have had very broad ecological amplitudes, permitting them to become interstitial opportunists within a variety of other vegetation types. Their free-sporing life histories apparently limited the colonizing ability of *Psaronius* species to environments with exposed substrates.

Extinction patterns and the Westphalian-Stephanian transition

The transition from the Westphalian (Middle Pennsylvanian) to the Stephanian (Late Pennsylvanian) was marked by a major extinction of tropical lowland plants, particularly in North America (western Euramerica). White & Thiessen (1913) first recognized the major floristic change, based on compression foliage. Later Kosanke (1947) detected the change in Illinois Basin coals using palynology. Recognition that this was a major extinction of lycopsid genera and included a major vegetational change became evident when coal-ball and palynological data were combined (Phillips *et al.* 1974). Monographic revisions of plant groups preserved in coal balls indicate a much broader extinction that includes most of the species of Westphalian tree ferns (Lesnikowska 1989) and pteridosperms (Taylor 1965). Palynological work in the U.S. (Peppers 1985; Kosanke 1988), and Europe (Stechesgolev 1975) demonstrates that the extinction was time transgressive from west to east across the Euramerican tropics, occurring in western Russia in the early part of the Stephanian. *Lycospora*-bearing lycopsids persisted in Europe during the Stephanian but at markedly reduced abundance (see Lorenzo 1979; Phillips *et al.* 1985). Stephanian lowland, wetland habitats were both more uniform and lower in species richness than those of the Westphalian, except perhaps in Cathaysia where floras typical of Euramerican wetlands persisted into the Late Permian (Phillips *et al.* 1985; Ziegler 1990; Guo 1992). Parts of the Cathaysian tropics apparently remained continuously wet throughout the Carboniferous, providing a refugium for the most moisture-sensitive floras.

Causation

Extinctions during the Westphalian–Stephanian transition appear to have been caused by changing climatic conditions in the tropics. Reduction in the amount of rainfall and increased seasonality in its distribution are the most likely factors to have affected the wetland biome. Peat-forming environments would have been the most sensitive landscapes within that biome. Because mires contained many edaphic-specialist taxa, climatic changes could have greatly affected migration routes between refugial safe sites. Lower levels of extinction are expected for clastic compression floras, which represent wetland plants with broader edaphic tolerances and thus access to a greater variety of migration routes.

Several studies have inferred regional drying in western Euramerica during the Westphalian–Stephanian transition. Phillips & Peppers (1984) used coal resource abundances from the Appalachians and Midcontinent coal basins as indicators of general environmental moisture availability. Their data suggest a marked period of drying coinciding with the transition and extinction. Cecil (1990) and Cecil et al. (1985) drew similar conclusions from the analysis of coal geochemistry, coal-body geometry, and analysis of clastic rocks associated with coals. Winston (1986, 1989) identified plant tissues in polished bocks of coal, and, based on current ecological understanding of the plants, inferred a climate signature like that of the other studies.

Recently Frakes et al. (1992) have re-evaluated the evidence for late Palaeozoic glaciations. They find that the Westphalian was a time of glacial maximum. This maximum ended abruptly at the Westphalian–Stephanian boundary, and the early Stephanian saw greatly diminished glaciation and possibly an increase in global temperature. Changes in the extent of polar glaciations directly affect the width and position of the intertropical convergence zone (see Parrish 1982; Ziegler et al. 1987), which in turn affect the amount and periodicity of annual rainfall. This joins with the other lines of evidence to suggest a global, allogenic cause for plant extinctions in the Westphalian tropical wetlands.

Turnover

Species origination and extinction were calculated on the basis of stratigraphic range data. Two types of calculation were performed using coal-ball data. The first considered the actual origination and extinction between coals through the later Westphalian D and early Stephanian. This method does not account for the Signor–Lipps effect, in which an extinction is 'smeared' backwards stratigraphically due to sampling effects. To account for this we also examined summary turnover between the latest Westphalian mires, a group uniformly dominated by lycopsids, tree ferns and pteridosperms, and early Stephanian mires, dominated by tree ferns and pteridosperms. In effect the comparison is restricted to two different types of

ecosystem, each internally uniform. The summary turnover also was broken down into two ecomorphic groups, trees and ground cover/vines/shrubs. Summary extinction and origination were calculated for compression–impression data reported by Gillespie & Pfefferkorn (1985), and for coal palynological data reported by Peppers (1985).

Table 2. *Species diversity, origination, and extinction during the late Westphalian and early Stephanian based on coal-ball data*

Coal	First occurrences	Last occurrences	Species richness
Stephanian			
Calhoun	3	—	31
Duquesne	5	7	36
Friendsville	5	0	30
Bristol Hill	5	2	27
Westphalian			
Baker	0	17	39
Lower Freeport	0	7	46
Herrin	7	2	48
Middle Kittaning	0	2	43
Springfield	3	4	47

Table 3. *Extinction, survivorship and origination based on spore-pollen data (from Peppers 1985)*

43 total species in late Westphalian
19 species terminate in late Westphalian D
    (44.2% extinction)
24 species range through W/S boundary
    (55.8% persistence)
26 total species in early Stephanian
2 species originate in early Stephanian
    (7.7% origination)

Species richness, origination, and extinction patterns are presented in Table 2 for nine stratigraphically successive coals from the late Westphalian and early Stephanian. A 'species' in this compilation is either a reconstructed whole plant for which isolated organs have been reassembled, or it is a diagnostic organ or group of organs. Seeds of pteridosperms, for example, are used as proxy for parental plant species, even though they may underrepresent true species diversity; at present there are too few reconstructions of most pteridosperm groups to use whole-plant data. *Psaronius* species are mostly presented as whole plants, based on the work of Lesnikowska (1989); reproductive organs were used as proxy for whole plants where their

Table 4. Extinction, survivorship and origination based on clastic compression data (from Gillespie & Pfefferkorn 1979)

18 total species in late Westphalian
9 species extinct in late Westphalian (50% extinction)
> *Lepidodendron aculeatum, Asolanus camptotaenia, Neuropteris heterophylla, N. rarinervis, Mariopteris nervosa, Alethopteris serlii, Annularia radiata, Annularia sphenophylloides, Sphenophyllum majus*

9 species range through W/S boundary (50% persistence)
> *Neuropteris scheuchzeri, N. ovata, Pecopteris miltonii, P. unita, P. hemitelioides, Asterophyllites equisetiformis, Annularia mucronata, A. stellata, Sphenophyllum emarginatum*

10 total species in early Stephanian
1 species originates in early Stephanian (10% origination)
> *Pseudomariopteris ribeyronii*

Table 5. Stratigraphic ranges of coal-ball plants in the late Westphalian and early Stephanian

	Stephanian				Westphalian					
	Ca	Du	Fr	Br	Da	Ba	LF	He	Ki	Sp
Bot. pseudoantiqua	x	=	=	=	=	=	=	x	=	x
B. forensis	x	x	x	x	x	x	=	x	=	x
B. cratis								x	=	x
Anachoropteris involuta (lateral)	x	x	x	x						
A. clavata	x	=	x							
A. gillotii	x	=	=	=	=	=	=	x	=	=
A. involuta (adaxial)					x	x	x	x	x	x
A. sp.								x		
Apotropteris minuta	x									
Sermaya biseriata	x	x	=	=	=	=	=	x	=	=
Z. berryvillensis (Biscalitheca)	x	x	x	x						
Z. illinoiensis (Corynepteris)	x	x	=	=	x	x	=	x	=	=
Rhabdoxylon americanum	x									
Ankyropteris brongniartii	x	x	=	=	=	=	=	x	=	x
Ankyropteris n. sp.	x	x	x	x						
Phillipopteris globoformis		x								
Doneggia compleura		x								
Calamite (Pendulostachys)	x									
Arthropitys (Calamocarpon)	x	x	x	=	x	x	=	x	=	x
Arthropitys (Calamostachys a.)	x	=	x							
Arthropitys (Palaeostachya d.)						x	=	=	=	=
Calamodendron						x	=	=	=	x
Sphenophyllum (Bowmanites sp.)					x	=	=	x		x
Sphenophyllum (Peltastrobus r.)					x	=	=	x		x
Callistophyton poroxyloides	x	x	x	x						
Callistophyton boysetii					x	x	x	x	x	x
Schopfiastrum decussatum	x	=	=	=	=	x	=	x	=	=
Heterangium (Conostoma quadrat.)	x	x	x							
Heterangium (C. villosum)	x	x								
Heterangium (C. platyspermum)	x	x	=	=	=	=	x	x	x	x
Heterangium (C. kestospermum)			x	=	=	=	=	x	=	x
Physostoma calcaratum								x	=	=
Cyathotheca ventilaria		x								
Stellastelara parvula		x	=	=	=	=	=	=	x	x
S. baxteri										x
Cordaixylon (Cardiocarpus ovi.)	x	x	x	=	x	x	=	x	=	x
Mesoxylon priapi (Mitro. vinc.)	x	x								
Psaronius (Scolecopteris majopsis)								x		
Psaronius (S. parkerensis)			x							
Psaronius (Ariangium pygmaem)	x									
Psaronius (S. calicifolia)										x
Psaronius (S. valumii)						x	=	x	=	x
Psaronius (S. mamayi)						x	=	x	=	x
Psaronius (S. gnoma)					x	x	=	=	=	x

EXTINCTION AND RECOVERY IN PENNSYLVANIAN COAL SWAMPS

	Ca	Du	Fr	Br	Da	Ba	LF	He	Ki	Sp
Psaronius (S. latifolia A)					x	x	=	x	x	x
Psaronius (S. minor)					x	=	=	x	=	x
Psaronius (S. altissima)						x	x	=	=	x
P. chasei (S. illinoensis)	x	x	x	x						
P. blicklei (S. monothrix)	x	=	x							
P. magnificus (S. latifolia B)	x	x	x	x						
Medullosa (Pachytesta hexang.)	x	x	x							
Medullosa (P. berryvillensis)	x	x	x							
Medullosa (P. incrassata)	x									
Medullosa (P. illinoense)	x	=	x	=	=	=	=	x	=	x
Medullosa (P. composita)										x
Medullosa (P. stewartii)		x?					x	=	=	x
Medullosa (P. saharasperma)						x	=	x	=	x
Medullosa (P. noei)										x
Medullosa (P. gigantea)					x	=	=	x	x	
Medullosa (P. vera)								x	=	=
Medullosa (P. hoskinsii)								x	=	=
Medullosa (Stephanospermum el.)	x	=	=	=	=	=	=	x	=	=
Medullosa (Stephanospermum sp.)			x	=	=	=	=	x		
Medullosa (Hexapterospermum sp.)		x								
Medullosa (H. delevoryii)								x		
Medullosa (Albertlongia incostata)								x		
Medullosa endocentrica	x	x	=	=	=	=	=	=	=	=
Sutcliffia insignis						x	=	x	x	x
Coronostoma quadrivasatum	x									
Lepidodendron hickii						x	x	x	x	x
Lepidophloios hallii					x	x	x	x	=	x
Lepidophloios johnsonii						x	=	x	=	x
Paralycopodites brevif.							x	x	x	x
Hizemodendron serratum						x	=	=	=	x
Diaphorodendron scleroticum					x	x	x	x	x	x
Synchysidendron resinosum					x	x	x	x	x	x
Sublepidophloios sp.									x	x
Chaloneria periodica					x	x	=	x	=	x
Chaloneria cormosa	x	x	x							
Sigillaria sp.					x	x	=	x	=	x
S. approximata (Mazocarpon oed.)	x	=	x	x						
Sigillaria (M. villosum)		x								
Sigillaria (M. bensonii)	x									
Paurodendron fraipontii	x	x	x	=	=	x	=	x	=	x

x, present in a coal; =, presence in a coal inferred by ocurrences in earlier and later coals. Ca, Calhoun coal (IL); Du, Duquesne coal (OH); Fr, Friendsville coal (IL); Br, Bristol Hill coal (IL); Da, Danville coal (IN); Ba, Baker coal (KY); LF, Lower Freeport coal (OH); He, Herrin coal (IL); Ki, Middle Kittaning coal (PA); Sp, Springfield coal (IL).

Table 6. *Tree extinction, survivorship and origination based on coal-ball data.* Psaronius *taxonomy from Lesnikowska (1989). Taxa of medullosans based on ovules, of* Arthropitys *on cones*

30 total species of trees in late Westphalian

26 species extinct in late Westphalian (86.7% extinction)

Calamodendron sp., *Palaeostachya decacnema, Lepidodendron hickii, Lepidophloios hallii, L. johnsonii, Diaphorodendron scleroticum, Synchysidendron resinosum, Hizemodendron serratum, Paralycopodites brevifolius, Sigillaria* sp., *Psaronius (Scolecopteris) majopsis, P. calicifolia, P. valumii, P. mamayi, P. gnoma, P. latifolia* (type A of Lesnikowska, 1979), *P. minor, P. altissima, Hexapterospermum delevoryii, Albertlongia incostata, Pachytesta composita, P. stewartii, P. saharasperma, P. noei, P. vera, P. hoskinsii*

 4 species range through W/S boundary (13.3% persistence)

Stephanospermum elongatum, Pachytesta illinoense, Pachytesta gigantea, Calamocarpon insignis

17 total species of trees in early Stephanian

13 species originate in early Stephanian (76.5% origination)

Psaronius chasei, P. blicklei, P. magnificus, P. parkerensis, P. (Araiangium) pygmaem, Hexapterospermum sp., *Pachytesta hexangulata, P. berryvillensis, P. incrassata, Sigillaria approximata, Sigillaria (Mazocarpon villosa), Sigillaria (Mazocarpon bensonii), Pendylostachys cingulatum*

Table 7. *Ground cover, vine and shrub extinction based on coal-ball data. Reproductive organs used as proxy for natural species where whole-plant reconstructions have not been completed*

18 total species of ground cover, vines and shrubs in late Westphalian
 6 species extinct in late Westphalian (33.3% extinction)
    Physostoma calcaratum, Callistophyton boysetii, Chaloneria periodica, Botryopteris cratis,
    Anachoropteris involuta (adaxial shoots), *A. cadyii*
12 species range through W/S boundary (66.6% persistence)
    Conostoma kestospermum, C. platyspermum, Schopfiastrum decussatum, Botryopteris 'pseudoantiqua',
    B. forensis, Anachoropteris gillotii, Sermaya biseriata, Zygopteris illinoiensis, Ankyropteris brongniartii,
    Medullosa endocentrica, Cordaixylon dumusum, Paurodendron fraipontii
25 total species of ground cover, vines and shrubs in early Stephanian
13 species originate in early Stephanian (52.0% origination)
    Callistophyton poroxyloides, Conostoma quadrivasatum, C. quadratum, Coronostoma villosum,
    Anachoropteris involuta (lateral shoots), *A. clavata, Zygopteris berryvillensis, Rhabdoxylon americanum,*
    Apotropteris minuta, Ankyropteris n. sp., *Phillipopteris globoformis, Norwoodia angustum,*
    Chaloneria cormosa

Table 8. *Summary extinction and origination patterns at Westphalian–Stephanian boundary based on coal-ball data*

	Extinctions	Originations
Trees	86.7%	76.5%
Ground cover	33.3%	52.0%
Total	66.7%	61.9%

associations with stems and leaves are unknown. Range-through data were used. This means that a species was assumed to be present in a given coal if it was found in both younger and older coals; range-through data compensate for differences in sampling intensities among coals. Coal balls have not been found in the last few coals of the Westphalian. However, palynological data indicate persistence of the same basic flora up to the end of the Middle Pennsylvanian (Peppers 1985; Kosanke 1988; Kosanke & Cecil 1989).

Tables 3 to 8 summarize the species extinction and origination patterns between the late Westphalian and early Stephanian; data in each table are from the same stratigraphic interval. Our ecological analyses indicate that late Westphalian and early Stephanian coals represent two distinct types of ecosystems, with ecological homogeneity characteristic of each interval.

Coal palynological data (Table 3, from Peppers 1985) sample mire ecosystems, the same environments as coal balls. Pollen and spore extinctions although elevated at the Westphalian–Stephanian boundary are lower than those found using coal balls. Many pollen and spore 'species' have been documented to come from more than one whole-plant species (e.g. Good 1975; Lesnikowska 1989; Willard 1989; DiMichele & Bateman 1992) and thus may

represent multi-species clades, which would reduce apparent extinction levels.

Table 4 summarizes turnover in compression–impression species listed in Gillespie & Pfefferkorn (1979); they used only species that occurred widely and abundantly in Euramerica, so the sample is not comprehensive. Summary extinction levels are near 50% for this sample and indicate that losses of dominant species were not simply confined to mires, but affected the entire tropical lowlands.

Ecomorphic extinction bias

Origination and extinction patterns for coal-ball plants were calculated from data in Table 5. Coal-ball data are divided into ecomorphic groups (Tables 6 and 7) and demonstrate that trees suffered proportionately nearly three times greater extinction than shrubs, ground cover, and vines. Despite such higher losses, tree origination in Stephanian mires was only 25% greater than that for the other groups. Thus the extinction was composed selectively of trees in all major phylogenetic groups. Nearly all the compression–impression taxa reported were trees. Coal-ball data are summarized in Table 8.

The post-extinction system: early Stephanian mire communities

Stephanian mires typically were composed of 25–35 species. Proportionally more of these species were ground-cover and vines than in Westphalian assemblages; 25 of 42 species in our Stephanian samples compared with 19 of 50 species in our Westphalian samples (Table 1). Trees continued to be the major biomass components of the mires, although tree dominance-diversity patterns were distinct from those of the Westphalian. In general, early Stephanian mires had stereotypic quantitative composition at the generic level; the most abundant peat producers were *Psaronius* tree ferns, medullosan pteridosperms, and sigillarian lycopsids. Data from both macrofossils (Lesnikowska 1989; Willard & Phillips 1993) and microfossils (Peppers 1985; Helfrich & Hower 1989; Willard & Phillips 1993) suggest, however, that species dominance patterns varied quite significantly through time from one coal seam to the next, but not in any directional fashion (Peppers 1985) or in association with clearly identifiable changes in the physical environment of the tropics.

Ordination of 11 zones from two coal-ball profiles from the Calhoun coal of Illinois summarizes a pattern typical of early Stephanian (Missourian) mires (Fig. 3). Coal-ball and palynological analyses have revealed similar patterns in the Friendsville and Bristol Hill coals of Illinois (Willard & Phillips 1993), the Otter Creek coals of Kentucky (Helfrich & Hower 1989), the Redstone coal of West Virginia (Grady & Eble 1990), and the Duquesne coal of Ohio (Pryor 1993). Overall, these analyses reveal significantly less taxonomic and dominance-diversity distinction among assemblages, and less clear ties to recurrent habitat factors than found in Westphalian mires. In a broad sense, Stephanian mire communities seem to be divisible into three or four principal assemblages. Most are heavily dominated by or enriched in tree-ferns. Of secondary importance are medullosans, and less commonly still are sigillarians, with variable amounts of tree ferns in either of these assemblages. *Sigillaria brardii* was a polycarpic lycopsid that typically occurred in clastic substrate habitats; in Stephanian mires sigillarian stands were most common during early phases of peat accumulation and thus enrich lower zones of some coal-ball and palynological profiles (Grady & Eble 1990; Grady *et al.* 1992; Willard & Phillips 1993; Helfrich & Hower 1989) in closest association with the clastic seat earth. Calamites can occur in association with any of the other assemblages, although they tend to be most abundant in the lower parts of the coal beds or in association with clastic partings (Grady & Eble 1990; Pryor 1993), a distribution similar to that found in the Westphalian.

There are additional elements of many Stephanian mires not found in the Calhoun coal samples from Berryville. Most importantly are assemblages enriched in the lycopsid *Chaloneria*, the source of *Endosporites* microspores. *Chalo-neria* was abundant in many coals, beginning in the earliest Stephanian, immediately after the extinctions of the latest Westphalian (Phillips *et al.* 1974; Peppers 1985). Megafossils or spores of *Chaloneria* are most common in the lower parts of coal beds or in association with clastic partings (Helfrich & Hower 1989; Grady & Eble 1990; Willard & Phillips 1993; Pryor 1993). Vegetation dominated by *Chaloneria* probably was physiognomically distinctive; the plant appears to have been short, maybe a metre or so in height, with a pole-like habit (Pigg & Rothwell 1983) and formed the Carboniferous equivalent of 'marshes' (DiMichele *et al.* 1979). Cordaites also have been found to be important components of some Stephanian mires (e.g., the Duquesne coal of Ohio, Rothwell 1988; Pryor 1993; the Friendsville coal of Illinois, Willard & Phillips 1993), where they occur most commonly in lower zones, or in proximity to clastic partings (Trivett & Rothwell 1985; Pryor 1993).

Using the Duquesne coal as an example, Pryor (1993) described recurrent assemblages as a series of 'successional' stages, beginning with cordaites and sigillarians, proceeding through marsh-like assemblages dominated by *Chaloneria*, and ending with progressively greater dominance of medullosans and ultimately *Psaronius* tree ferns. The progression appears to have been driven largely by temporal changes in edaphic conditions as an increasingly thick, nutrient-poor peat substrate developed. Similar 'successional' patterns have been described by Grady & Eble (1990) from palynological studies of the Redstone coal, where they correlate with directional changes in ash and inertinite content of the coal. The successional patterns retain the characteristic Stephanian pattern of generic rather than specific identity. The intermediate and ultimate tree-fern and medullosan species composition seems to be relatively unpredictable.

Short gradient lengths are typical of most Stephanian mires. The abundance of tree ferns imparts a similar ecomorphic aspect to most of the assemblages, and the generally uniformly high biomass of ground cover and vines suggests that substrates remained exposed for extended periods of time. The vagaries of density independent disturbance and subsequent colonization patterns appear to be major controls on local taxonomic composition.

When compared at the generic level, Stephanian coal-ball floras are similar to contemporaneous compression–impression floras from clastic, floodbasin habitats (compare data in Pfefferkorn & Thomson 1982; Phillips *et al.* 1985; DiMichele & Aronson 1992). This presents

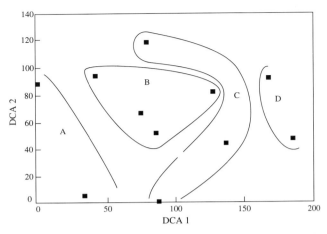

Fig. 3. Detrended correspondence analysis of coal-ball zones from Berryville VS 1 and VS 2, early Stephanian Calhoun coal. Lines delimit groups of coal-ball zones with similar dominance patterns; letters refer to groups in Table 1 where dominance and life-history patterns are summarized.

a sharp contrast to Westphalian patterns, and is indicative of a general homogenization of the tropical wetland vegetation in the early Stephanian.

Comparison of pre-extinction and postextinction landscapes

Life histories

The patterns of Westphalian and Stephanian mire communities cannot be compared in a one to one manner. Foremost, the biologies and life histories of the respective dominant taxa, lycopsids and tree ferns, were markedly different. The lycopsids were structurally diverse and had ecological strategies spanning the spectrum from weedy opportunists to site-occupying K-strategists (DiMichele & Phillips 1985; Phillips & DiMichele 1992), although direct comparisons to any extant forms are difficult. Tree ferns as a group, on the other hand, had broad ecological amplitudes, reflecting their massive reproduction and cheap construction; trees were non-woody, and were supported by adventitious root mantles composed mostly of air spaces (Ehret & Phillips 1977; Lesnikowska 1989). There is no evidence that any species of *Psaronius* were primary colonizers of the deep, standing-water habitats occupied by some species of Westphalian lycopsids. Palynological patterns from Westphalian coals indicate differential distributions of tree-fern species relative to contemporaneous channels, although the factors controlling these distributions are open to speculation at present (Willard 1993; Mahaffy 1988).

Tree ferns of Stephanian mires retained mass reproduction of small, highly dispersible spores, reflecting the opportunistic life histories of their ancestors. However, they also attained large sizes, estimated from organ diameters (stems, roots, petioles) and *in situ* trunk diameters of > 1 m; maximum organ dimensions are larger than any reported from Westphalian mires (Lesnikowska 1989). Clastic deposits of the late Westphalian do preserve large tree fern trunks (Pfefferkorn 1976; Lesnikowska 1989) suggesting that many Stephanian mire species may have descended from Westphalian clastic-swamp taxa. The dominance of tree ferns in mires mirrored their dominance in the clastic wetlands of the Stephanian, reducing the dominance-diversity disparity between mires and clastic wetlands that typified the Westphalian. Willard and Phillips (1993) report large *Psaronius* trunks that apparently persisted through most of the time of accumulation of a Stephanian coal-ball mass over 0.6 m thick that had replaced nearly the entire seam thickness. Mickle (1984) reports large diameter trunks in which the stem had rotted out at the base, implying that the stem could die back at the base as it grew from the top, supplied with water and nutrients by the living root mantle. These observations imply that large, site occupying *Psaronius* trees were components of Stephanian mires. Size and evidence of site occupation suggest that the more r-selected, opportunistic species of the Westphalian had been replaced by, or given rise to, more typically K-selected, site occupying forms.

Table 9. *Richness and abundance patterns of ground cover and vines by coal-ball zone in profiles of Westphalian Herrin coal and Stephanian Calhoun coal*

Coal	Age	Coal-ball Profiles	n/P	S/Z	%B/Z
Calhoun	Stephanian	Berryville VS 1 & 2	10	5.3	4.2
Herrin	Westphalian	Old Ben 3 & 5	9	1.8	3.0
Herrin	Westphalian	Sahara 4 & 5	13	3.3	4.3

n/P = total species richness of profiles, S/Z = average species richness per profile zone, %B/Z = average percent of total biomass per zone Ground cover and vines only.

Ecological gradients

Stephanian mires had markedly shorter ecological gradients than late Westphalian peat-forming ecosystems (Fig. 4). This is suggested by detrended correspondence analyses (Fig. 3) in which axis length reflects directly the extent of species turnover, four units representing approximately complete turnover between endpoints (Gauch 1982). Short Stephanian gradients are in part an artifact of lower species-level taxonomic resolution of tree fern and pteridosperm litter, a problem with coal-ball samples throughout the Pennsylvanian. This is mitigated, however, by the generally low species diversity of the dominant tree ferns and pteridosperms in any one deposit, determined by the relative abundances of diagnostic organs such as *Scolecopteris*, *Psaronius* stems, or pteridosperm ovules. Overall, the Stephanian gradient, the life histories of the plants (many requiring exposed substrates for life-cycle completion), and dominance-diversity patterns suggest that many of the physical habitats colonized and stabilized by Westphalian plants, particularly lycopsids, either did not occur or were unoccupied in Stephanian mires.

Ground cover and vines

Comparisons of ground cover and vine abundances are summarized in Table 9 for four coal-ball profiles from the Westphalian D Herrin coal and two profiles from the Stephanian Calhoun coal. Average biomass of ground cover is not significantly different between the Westphalian and Stephanian samples, nor is the total number of ground-cover species found within any of the profiles. However, the average number of species per coal-ball zone was markedly higher in the Stephanian samples. This difference reflects greater variability among Westphalian assemblages, where ground cover is lacking in many cases. This is consistent with longer ecological gradients in the Westphalian, and is reflected in greater spread of the data in ordinations of coal-ball profile zones. During the Stephanian the same species recur from one zone to the next in profiles, leading to a more uniform taxonomic signature in the groundcover across assemblages.

Disturbance

Disturbance levels within Pennsylvanian mires are difficult to quantify. There appear to have been several major disturbance agents: fire, indicated by fusain (mineral charcoal), severe floods, indicated by clastic partings or elevated levels of mineral matter in the coal, generally unfavorable climatic conditions, indicated indirectly by plant size, average thickness of coal seams. Disturbance levels may have differed between late Westphalian and early Stephanian mires, although evidence is indirect. Fusain (mineral charcoal) abundances are lower in the Stephanian based on coal-ball analyses (Phillips et al. 1985). Tree sizes are generally larger in the Stephanian, determined from maximum dimensions of isolated organs. 'Marshes' composed of small stature vegetation, such as *Chaloneria*, are proportionally more abundant in the Stephanian. Finally, Stephanian coals generally are considerably thinner than those of the late Westphalian. These factors present an ambiguous picture of disturbance. In general, however, they suggest that individual Stephanian mires may have had lower levels of serious disturbance than prevailed during the Westphalian, permitting the development of larger plants in virtually all lineages. However, the general thinness of coal seams, and the frequent occurrence of marsh-like vegetation suggests that climates and depositional settings conducive to peat-formation were shorter lived during the Stephanian

Stephanian gigantism

Plants of thick peat substrates, particularly those minimally influenced by the influx of clastics,

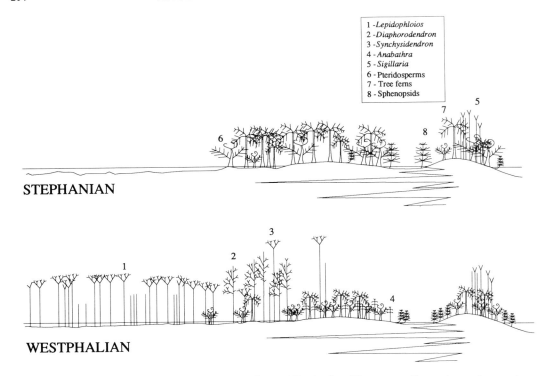

Fig. 4. Hypothetical intramire gradients in Westphalian and Stephanian. Wettest, standing water environments on left, intermittently flooded peat substrates in centre, clastic-peat transitional environments grade to 'levee' environments on left. Plants from similar lineages are shown to be larger in the Stephanian.

appear to have been smaller than their relatives on mineral substrates (Pfefferkorn 1976; DiMichele & DeMaris 1987; Gastaldo 1986), which is consistent with patterns in modern tropical mires (Anderson 1964, 1983). In the Westphalian, stem diameters of the dominant lycopsids recorded from coal balls are estimated to be 25–35 cm in *Diaphorodendron* and *Lepidophloios* respectively (DiMichele 1979, 1981), whereas trunks > 1 m diameter are encountered commonly in shales above or associated with coals. The only other Westphalian plants with moderate to large stem diameters were cordaites (*Pennsylvanioxylon* and *Mesoxylon*, up to 20 cm diameter) and calamites (*Arthropitys*, up to 12 cm diameter) from Westphalian B coals of the Appalachians. During the Westphalian, therefore, lycopsids may have been the only tall trees in what appear to have been stunted mire forests. Excluding the Westphalian B exceptions, maximum organ diameters (stems, roots, petioles) of subdominant taxa in mires were less than 10 cm, mostly less than 5 cm (e.g. Andrews & Agashe 1965; Costanza 1985; Lesnikowska 1989).

Stephanian coal balls reveal increases in the maximum sizes of tree ferns, pteridosperms, and sphenopsids, based on measurements of isolated organs. Large plant stature in Stephanian mires has been alluded to (Andrews & Agashe 1965; Galtier & Phillips 1985) but has not been examined systematically. Large size also has been documented for some Stephanian clastic compression taxa, such as *Calamites gigas* (Barthel & Kerp 1992).

Size increases within broad clades cannot be constrained more narrowly by phylogenetic analysis at the present time due to the incomparability of anatomically preserved and compression–impression based species. Lacking this framework, we suggest several possible causative mechanisms, perhaps interrelated. The simplest explanation is that trees of larger stature migrated into and occupied mire habitats following the extinction event, in the absence of competitors. Large size has been documented in several Westphalian groups dominant in clastic substrate habitats, including tree ferns (Pfefferkorn 1976) and pteridosperms (Wnuk & Pfefferkorn 1984; Pfefferkorn *et al.* 1984). In this case large stature would be a happenstance of the ancestry of the colonizing species. Suggestions that Stephanian mires were largely topogeneous (Grady & Eble 1990; Donaldson & Eble 1991) may have permitted the survival of species from, or derived from ancestors in, clastic substrate areas. Harvey & Dillon (1985) found higher levels of inertinites in Stephanian coals, suggesting higher nutrient levels either permitting attainment of larger size or the migration into mires of species requiring higher levels.

Because of the edaphic qualities of peat substrates, and the resultant 'island' nature of mires within the greater lowland wetlands, simple migration of species from clastic to peat substrates may have been inhibited by physical factors alone. In such a landscape mires may have presented opportunities much like modern islands, on which small to medium stature plants often undergo phyletic size increases. Lineages of ancestrally small plants show consistent trends to larger size in island environments where competition for resources is reduced (Carlquist 1974). In a highly disrupted post-extinction tropics, the edaphic-island nature of mires, in combination with greatly reduced species diversities, may have permitted the rapid evolution of larger size in many surviving clades. Given the extensive extinction of trees, the opportunity for island-type, low-competition, opportunistic dynamics should have been enhanced in the post-extinction period.

A final consideration is a general reduction in the amount of severe, intramire disturbance during the Stephanian. Under such conditions natural selection may have permitted trees to achieve larger sizes, ultimately driven by intrinsic developmental factors and evolutionary size increase.

Discussion

Extinction thresholds and ecological disequilibrium

The major vegetational change following the end-Westphalian extinctions suggests that high levels of species extinction, especially when concentrated in framework trees, leads to loss of ecosystem self-regulatory properties. Examination of the dynamics of the Westphalian indicates that persistence, both at the level of assemblages (communities) and multi-assemblage landscapes (DiMichele & Phillips 1995) was the dominant theme. Species turnover occurred between every coal bed, based on palynological (Peppers 1985) or coal-ball (Phil-

lips *et al.* 1985) data. Yet, the ecomorphic aspects of the major assemblages remained recognizably the same throughout the entire Westphalian. Species tended to be replaced ecologically by congeners with similar morphology and ecological preferences, although whether from migration into mires from an external source or by *in situ* speciation we cannot discern. These data indicate that as long as species turnover was low, the remaining species created a selective filter that only permitted certain forms to colonize vacated resource space. High levels of extinction, and associated major vegetational change suggest thresholds beyond which the remaining species cannot regulate subsequent colonization of available resources.

The largest extinction within the Westphalian is considerably smaller than that at the end. It occurred near the A–B boundary and is marked by the disappearance of *Lyginopteris* and *Neuralethopteris*, as well as other minor plants. However, the lycopsid trees, which were the major framework plants of mires, survived. Species turnover based on coal balls is approximately 30%; the coal-ball record of this interval is sparse, but there do appear to have been some changes in the vegetational organization of mires after the event. In particular, cordaites appeared as elements of 'ecotonal' assemblages, occupying heavily decayed peat surfaces, often in association with medullosans (e.g., Rock Springs coal of Tennessee, Phillips *et al.* 1985). Early cordaites may have extended the length of the Westphalian mire gradient by colonizing severely decayed peat surfaces, enlarging the scope of the 'ecotonal' parts of mires. Overall, though, the cordaites became a part of mires without disrupting the basic landscape structure. The Westphalian A–B extinction and its ecological consequences indicate that significant, but smaller extinctions, where the major tree forms survive, do not have the same catastrophic effect on ecosystem structure and composition as extinctions that remove the basic framework trees.

Coal-ball macrofossils do not occur in coals immediately above the Westphalian–Stephanian boundary. Consequently palynology provides the most complete record of the ecological events within mires that follow immediately after the major extinctions. With the loss of numerous species the environments of earliest Stephanian mires appear to have been open to colonization by new species or by expansion of surviving, previously subdominant lineages. Most notably, 'marshes' dominated by *Chaloneria* became major parts of some thin, boney coals. Through seven coals in the upper half of the Modesto Formation of Illinois coal-to-coal variability in spore-species dominance is high; included as occasional dominants on a whole-seam basis are *Chaloneria*, *Sigillaria*, and tree ferns producing *Punctatisporites obliquus* and *P. minutus* spores. The dominance patterns stabilize in the overlying Bond Formation, where *Sigillaria*, several tree fern species, and calamites begin to occur with greater consistency (Peppers 1985). The coal-ball record of the later Stephanian A consistently records tree-fern dominance with subdominant pteridosperms (Phillips *et al.* 1985; Willard & Phillips 1993; Pryor 1993).

It would appear from the palynological patterns during the early post-extinction interval that either climate or mire edaphic conditions were highly variable, and generally not suitable for the development of major peat bodies. Each successive mire was dominated by a different suite of species, including many that were subtrees, such as *Chaloneria*, which formed marsh-like vegetation. The mutualistic relationships that characterized the Westphalian had evolved over millions of years, beginning in the latest Devonian, permitting persistent vegetational patterns to appear. Disruption of this system apparently eliminated these kinds of interactions, and thus long-persistent vegetational patterns. As a result of low species diversity and little ecological structure (particularly evidence of strong resource partitioning), early Stephanian mire assemblages were highly invasible by species that could tolerate the rigours of low-nutrient, physically stressful, peat substrates. They offered opportunities for newly evolved species to establish under reduced resource competition and exclusion.

Rise to dominance of opportunists

The ultimate dominance of marattialen tree ferns reflects the rise of cheaply constructed plants with massive reproductive output of highly dispersible isospores (capable of establishing a new population from one spore under some circumstances). Westphalian marattialeans had what can be described in general terms as an opportunistic, fugitive life history strategy. Trees were generally of small stature; tree habit was made possible by a mantle of adventitious roots composed mostly of airspaces. The foliage bore large numbers of sporangia, in which were produced collectively billions of spores.

Ferns also recovered and dominated many landscapes after the Cretaceous–Tertiary extinctions, the now-famous 'fern-spike' (Tschudy *et al.* 1984). This brief period of dominance was not followed, however, by development of a fern-dominated vegetation. The K–T pattern may differ from the Westphalian–Stephanian pattern for several reasons. First, it is likely that the palynological fern-spike was produced by small, ground cover ferns rather than tree ferns. The potential of these plants to give rise to trees was limited. Second, it is clear that many arboreous lineages of seed plants survived the extinction as local populations. These populations ultimately fuelled the re-establishment of vertically stratified woodlands and forests (Wing & Tiffney 1987; Wolfe & Upchurch 1986, 1987; Hickey 1984; Johnson 1992; Johnson & Hickey 1990). Thus, sources of seed were available from which new ecosystems could begin to develop.

In contrast, Stephanian lowlands appear to have been regionally devastated by climate changes that, even if brief geologically, would have been long-term on an ecological timescale. Additionally, whole Westphalian clades were very narrowly distributed ecologically, particularly those centered in the lowland wetlands (rhizomorphic lycopsids, medullosan pteridosperms, and calamites) restricting the ability of many to survive as isolated populations. The heterosporous reproductive adaptations of many of the more advanced lycopsids strongly limited them to aquatic and semi-aquatic habitats (Phillips 1979), which would have constrained their ability to recolonize patchy open habitat spaces not connected by continuous wetlands. Finally, marattialians were trees, even if relatively small trees, which gave them the ability to spawn forms capable of canopy formation, and dominance of a stratified ecosystem. As disaster forms, marattialian tree ferns were thus capable of locating available resources rapidly and then dominating local landscapes in the face of other tree taxa through rapid growth, tolerance of low nutrient conditions, and abundant reproduction on site.

Implications for ecosystem dynamics

Vegetational dynamics of the Pennsylvanian tropics, and, in fact, of most times in the Earth's history that have been studied in detail (e.g. the Tertiary, Wing 1982; Burnham 1983), demonstrate a greater degree of congruence with the predictions of 'Clementsian' ecosystem theories than with 'Gleasonian' (individualistic) models. As a generality, neo-plant ecologists have embraced Gleasonian principles and rejected Clementsian concepts, particularly as originally formulated (Clements 1916). Pointing to Pleistocene palynological data (e.g. Webb 1988; Delcourt & Delcourt 1987; Overpeck *et al.* 1992), the 'fossil record' has been found to be in full support of the Gleasonian view. The differences between the Pleistocene patterns and those of deeper time do not reside in sampling scale; it is possible to resolve ecological time from palynological sampling nearly as finely in the Pennsylvanian as in the Pleistocene (e.g. Eble & Grady 1990; Willard 1993; Mahaffy 1985; Pierce *et al.* 1993). Such incompatibilities as exist may reflect real dynamical differences between the Pleistocene ice ages and the vast bulk of Earth history, differences between temperate and tropical dynamics, different scales of analysis, or different philosophies about what constitutes a vegetational unit of interest. Data from the pre-Pleistocene fossil record suggest a need for more serious consideration of modern attempts at Clementsian-type syntheses. The view of ecosystems as *systems*, with hierarchical organization and emergent properties above and beyond those of the component populations, has remained an integral part of ecology (Patton & Odom 1981; Salthe 1985; O'Neill 1989; O'Neill *et al.* 1986; Allen & Hoekstra 1992). However, due to the inherent complexity of ecosystems the theoretical elegance of Gleasonian individualism has yet to emerge. A complete theory of ecosystem organization must explain patterns on all scales of spatio-temporal observation; even if a general hierarchical model with the simplicity of the Gleasonian view has not come to fruition, the fossil record suggests a need for it, and for the rejection of individualistic models as all-powerful explanatory defaults.

The data from the Pennsylvanian represent an entirely different ecosystem from any that existed afterward. Dominance of substantial parts of the landscape by lower vascular plants, the development of ecological partitioning in concert with the modernization of the vascular plant flora, and the high degree to which the plants were structurally adapted to particular, narrow parts of the landscape, present ecologists with a natural test system for broader theories. To the degree that patterns of temporal behaviour in Pennsylvanian systems are congruent with those predicted for and documented in modern systems, our level of confidence in general principles is enlarged.

We thank Dan S. Chaney for assistance in preparation of the technical figures, and Mary Parrish for the artwork. Numerous people have contributed to this paper through discussions over the past 20 years; we particularly wish to thank Richard Bateman, Alicia Lesnikowska, James Mahaffy, Debra Willard and Richard Winston. TLP acknowledges the support of NSF grant EAR 8313094. This is contribution No. 24 from the Evolution of Terrestrial Ecosystems Consortium, which provided partial support for the research.

References

ALLEN, T. F. H. & HOEKSTRA, T. W. 1992. *Toward a Unified Ecology*. Columbia University Press, New York.

ANDERSON, J. A. R. 1964. The structure and development of the peat swamps of Sarawak and Brunei *Journal of Tropical Geography*, **18**, 7–16.

—— 1983. The tropical peat swamps of western Malesia. *In:* GORE, A. J. P. (ed.) *Ecosystems of the World, 4B; Mires: Swamp, Bog, Fen and Moor.* Elsevier, Amsterdam, 181–199.

ANDREWS, H. N. & AGASHE, S. N. 1965. Some exceptionally large calamite stems. *Phytomorphology*, **15**, 103–108.

BARTHEL, M. & KERP, H. 1992. *Calamites gigas* – an alternative development within Permian articulates. *Organisation Internationale de Paleobotanique, IVeme Conference Abstracts*, OFP Informations No. Spécial 16-B: 18.

BEELER, H. E. 1983. Anatomy and frond architecture of *Neuropteris ovata* and *Neuropteris scheuchzeri* from the Upper Pennsylvanian of the Appalachian Basin. *Canadian Journal of Botany*, **61**, 2352–2368.

BOUCOT, A. J. 1983. Does evolution take place in an ecological vacuum? *Journal of Paleontology*, **57**, 1–30.

—— 1990. Community evolution: its evolutionary and biostratigraphic significance. *In:* MILLER, W. III (ed.) *Paleocommunity Temporal Dynamics: the Long-term Development of Multispecies Assemblages.* Paleontological Society Special Publication, **5**, 48–70.

BRETT, C. E. & BAIRD, G. C. 1995. Co-ordinated stasis and evolutionary ecology of Silurian to Middle Devonian faunas in the Appalachian Basin. *In:* ERWIN, D. H. & ANSTEY, R. L. (eds) *New approaches to speciation in the fossil record.* Columbia University Press, New York, 283–315.

——, MILLER, K. B. & BAIRD, G. C. 1990. A temporal hierarchy of paleoecologic processes within a Middle Devonian epeiric sea. *In:* MILLER, W. III (ed.) *Paleocommunity Temporal Dynamics: the Long-term Development of Multispecies Assemblages.* Paleontological Society Special Publication, **5**, 178–209.

BURNHAM, R. J. 1983. Diversification and stable distribution with respect to temperature in early Tertiary Ulmoideae of western North America. *Abstracts, Botanical Society of America Annual Meeting*, 68.

CALDER, J. H. 1993. The evolution of a groundwater influenced (Westphalian B) peat-forming ecosystem in a piedmont setting: the No. 3 seam, Springhill coalfield, Cumberland Basin, Nova Scotia. *Geological Society of America, Special Paper*, **286**, 153–180.

CARLQUIST, S. 1974. *Island Biology*. Columbia University Press, New York.

CECIL, C. B. 1990. Paleoclimate controls on stratigraphic repetition of chemical and clastic rocks. *Geology*, **18**, 533–536.

——, STANTON, R. W., NEUZIL, S. G., DULONG, F. T., RUPPERT, L. F. & PIERCE, B. S. 1985. Paleoclimatic controls on late Paleozoic sedimentation and peat formation in the central Appalachian Basin (USA). *International Journal of Coal Geology*, **5**, 195–230.

CHANEY, R. W. 1947. Tertiary centres and migration routes. *Ecological Monographs*, **17**.

CLEMENTS, F. E. 1916. *Plant Succession, an Analysis of the Development of Vegetation*. Publication No. 242, Carnegie Institution, Washington, DC.

CLYMO, R. S. 1987. Rainwater-fed peat as a precursor of coal. *In:* SCOTT, A. C. (ed.) *Coal and Coal-Bearing Strata: Recent Advances*. Geological Society, London, Special Publication, **32**, 17–23.

COLLINSON, M. E. & SCOTT, A. C. 1987*a*. Factors controlling the organization and evolution of ancient plant communities, *In:* GEE, J. H. R. & GILLER, P. S. (eds) *Organization of Communities Past and Present*. Blackwell, Oxford, 399–420.

—— & —— 1987*b*. Implications of vegetational change through the geological record on models for coal-forming environments, *In:* SCOTT, A. C. (ed.) *Coal and Coal-Bearing Strata: Recent Advances*. Geological Society, London, Special Publication, **32**, 67–85.

COSTANZA, S. H., 1985. *Pennsylvanioxylon* of Middle and Upper Pennsylvanian coals from the Illinois Basin and its comparison with *Mesoxylon*. *Palaeontographica*, Abt. B, **197**, 81–121.

DELCOURT, P. A. & DELCOURT, H. R. 1987. *Long-term Dynamics of the Temperate Zone*. Springer, New York.

DEMARIS, P. J., BAUER, R. A. CAHILL, R. A. & DAMBERGER, H. H. 1983. *Prediction of Coal Balls in the Herrin Coal*. Illinois State Geological Survey, Contract-Grant Report 1982–3.

DIMICHELE, W. A. 1979. Arborescent lycopods of Pennsylvanian age coals: *Lepidophloios*. *Palaeontographica*, Abt. B, **171**, 57–77.

—— 1981. Arborescent lycopods of Pennsylvanian age coals: *Lepidodendron*, with description of a new species. *Palaeontographica*, Abt. B, **175**, 85–125.

—— 1994. Ecological patterns in time and space. *Paleobiology*, **20**, 89–92.

—— & ARONSON, R. B. 1992. The Pennsylvanian-Permian vegetational transition: a terrestrial analogue to the onshore-offshore hypothesis. *Evolution*, **46**, 807–824.

—— & BATEMAN, R. M. 1992. Diaphorodendraceae, fam. nov. (Lycopsida: Carboniferous): systematics and evolutionary relationships of *Diaphorodendron* and *Synchysidendron*, gen. nov. *American Journal of Botany*, **79**, 605–617.

—— & DeMARIS, P. J. 1987. Structure and dynamics of a Pennsylvanian-age *Lepidodendron* forest: colonizers of a disturbed swamp habitat in the Herrin (No. 6) coal of Illinois. *Palaios*, **2**, 146–157.

——, MAHAFFY, J. F. & PHILLIPS, T. L. 1979. Lycopods of Pennsylvanian age coals: *Polysporia*.

Canadian Journal of Botany, **57**, 1740–1753.

—— & PHILLIPS, T. L. 1985. Arborescent lycopod reproduction and paleoecology in a coal-swamp environment of late Middle Pennsylvanian age (Herrin Coal, Illinois, USA). *Review of Palaeobotany and Palynology*, **44**, 1–26.

—— & —— 1988. Paleoecology of the Middle Pennsylvanian-age Herrin coal swamp (Illinois) near a contemporaneous river system, the Walshville paleochannel. *Review of Palaeobotany and Palynology*, **56**, 151–176.

—— & —— 1994. Paleobotanical and paleoecological constraints on models of peat formation in the Late Carboniferous of Euramerica. *Palaeogeography, Palaeoclimatology, Palaeoecology*, **106**, 39–90.

—— & —— 1995. The response of hierarchically structured ecosystems to long-term climate change: a case study using tropical peat swamps of Pennsylvanian age. *In:* STANLEY, S. M., KNOLL, A. J. & KENNETT, J. P. (eds) *Effects of past global change on life*. National Research Council, Studies in Geophysics. 134–155.

——, —— & McBRINN, G. E. 1991. Quantitative analysis and paleoecology of the Secor coal and roof-shale floras (Middle Pennsylvanian, Oklahoma). *Palaios*, **6**, 390–409.

——, —— & PEPPERS, R. A. 1985. The influence of climate and depositional environment on the distribution and evolution of Pennsylvanian coal-swamp plants. *In:* TIFFNEY, B. H. (ed.) *Geological Factors and the Evolution of Plants*. Yale University Press, New Haven, 223–256.

DONALDSON, A. C. & EBLE, C. F. 1991. Pennsylvanian coals of central and eastern United States. *In:* GLUSKOTER, H. J., RICE, D. D. & TAYLOR, R. B. (eds) *Economic Geology*, U.S. Geological Society of America, The Geology of North America P-2, 523–546.

EBLE, C. F. 1990. A palynological transect, swamp interior to swamp margin, in the Mary Lee coal bed, Warrior Basin, Alabama. *In:* GASTALDO, R. A., DEMKO, T. M. & LIU YUEJIN (eds) *Carboniferous Coastal Environments and Paleocommunities of the Mary Lee Coal Zone, Marion and Walker Counties, Alabama*. Guidebook for Fieldtrip 6, 39th Annual Meeting, Southeastern Section, Geological Society of America. Alabama Geological Survey, Tuscaloosa, 65–81.

—— & GRADY, W. C. 1990. Paleoecological interpretation of a Middle Pennsylvanian coal bed from the Central Appalachian Basin, U.S.A. *International Journal of Coal Geology*, **16**, 255–286.

——, HOWER, J. C. & ANDREWS, W. M. 1994. Paleoecology of the Fire Clay coal bed in a portion of the Eastern Kentucky Coal Field. *Palaeogeography, Palaeoclimatology, Palaeoecology*, **106**, 287–307.

EGGERT, D. L., CHOU, C.-L., MAPLES, C. G., PEPPERS, R. A., PHILLIPS, T. L. & REXROAD, C. B. 1983. Origin and economic geology of the Springfield Coal Member in the Illinois Basin. Guidebook to Fieldtrip No. 9, Geological Society of America

Annual Meeting, 121–146.

EHRET, D. L. & PHILLIPS, T. L. 1977. *Psaronius* root systems – morphology and development. *Palaeontographica*, Abt. B., **161**, 147–164.

FENG, B.-C. 1989. Paleoecology of an upper Middle Pennsylvanian coal swamp from western Pennsylvania, U.S.A. *Review of Palaeobotany and Palynology*, **57**, 299–312.

FRAKES, L. E., FRANCIS, J. E. & SYKTUS, J. I. 1992. *Climate Modes of the Phanerozoic*. Cambridge University Press.

GASTALDO, R. A. 1986. An explanation for lycopod configuration, 'Fossil Grove', Victoria Park, Glasgow. *Scottish Journal of Geology*, **22**, 77–83.

GALTIER, J. & PHILLIPS, T. L. 1985. Swamp vegetation from Grand 'Croix (Stephanian) and Autun (Autunian), France and comparisons with coal ball peats of the Illinois Basin. *Compte Rendu, IX International Congress Carboniferous Stratigraphy and Geology*, **4**, 13–24.

GAUCH, H. G. 1982. *Multivariate Analysis in Community Ecology*. Cambridge University Press.

GILLESPIE, W. M. & PFEFFERKORN, H. W. 1979. Distribution of commonly occurring plant megafossils in the proposed Pennsylvanian System Stratotype. *In:* ENGLAND, K. J. *ET AL.* (eds) Proposed Pennsylvanian System Stratotype, Virginia and West Virginia. Ninth International Congress of Carboniferous Stratigraphy and Geology, Guidebook Field Trip No. 1, 87–96.

GOOD, C. W. 1975. Pennsylvanian-age calamitean cones, elater-bearing spores, and associated vegetative organs. *Palaeontographica*, Abt. B, **153**, 28–99.

GRADY, W. C. & EBLE, C. F. 1990. Relationships among macerals, minerals, miospores and paleoecology in a column of Redstone Coal (Upper Pennsylvanian) from north-central West Virginia (U.S.A.). *International Journal of Coal Geology*, **15**, 1–26.

——, FEDORKO, N. & EBLE, C. F. 1992. Paleoecology and paleoenvironments of peat formation in the Pittsburgh coal bed, northern Appalachian Basin, U.S.A. Abstracts, Geological Association of Canada Annual Meeting. *Geoscience Canada*, **17** (supplement), 42.

GUO, YINGTING. 1990. Palacoecology of flora from coal measures of Upper Permian in western Guizhou. *Journal of China Coal Society*, **15**, 48–54.

HARVEY, R. D. & DILLON, J. W. 1985. Maceral distributions in Illinois coals and their paleoenvironmental implications. *International Journal of Coal Geology*, **5**, 141–165.

HECKEL, P. H. 1986. Sea-level curve for Pennsylvanian eustatic marine transgressive-regressive depositional cycles along midcontinent outcrop belt, North America. *Geology*, **14**, 330–334.

—— 1989. Updated Middle-Upper Pennsylvanian eustatic sea level curve for midcontinent North America and preliminary biostratigraphic characterization. *11th International Congress of Carboniferous Stratigraphy and Geology, C.R.*, **4**, 160–185.

HELFRICH, C. T. & HOWER, J. C. 1989. Palynologic and petrographic variation in the Otter Creek coal beds (Stephanian, Upper Carboniferous), western Kentucky. *Review of Palaeobotany and Palynology*, **60**, 179–189.

HESS, J. C. & LIPPOLT, H. J. 1986. $^{40}Ar/^{39}Ar$ ages of tonstein and tuff anidines: New calibration points for improvement of the upper Carboniferous time scale. *Isotope Geoscience*, **59**, 143–154.

HICKEY, L. J. 1984. Changes in the angiosperm flora across the Cretaceous/Tertiary boundary. *In:* BERGGREN, W. A. & VAN COUVERING, J. A. (eds) *Catastrophies in Earth History*. Princeton University Press, 279–313.

JOHNSON, K. R. 1992. Leaf-fossil evidence for extensive extinction at the Cretaceous–Tertiary boundary, North Dakota, USA. *Cretaceous Research*, **13**, 91–117.

—— & HICKEY, L. J. 1990. Megafloral change across the Cretaceous/Tertiary boundary in the northern Great Plains and Rocky Mountains, U.S.A. *Geological Society of America, Special Paper*, **247**, 433–444.

JOHNSON, P. R. 1979. *Petrology and environments of deposition of the Herrin (No. 6) Coal Member, Carbondale Formation, at Old Ben Coal Company Mine No. 24, Franklin County, Illinois*. M.S. Thesis, University of Illinois, Urbana-Champaign.

KLEIN, G. DE V. 1990. Pennsylvanian time scales and cycle periods. *Geology*, **18**, 455–457.

KOSANKE, R. M. 1947. Plant microfossils in correlation of coal beds. *Journal of Geology*, **55**, 280–284.

—— 1988. Palynological studies of Middle Pennsylvanian coal beds of the proposed Pennsylvanian System Stratotype in West Virginia. *United States Geological Survey Professional Paper*, **1455**, 1–73.

—— & CECIL, C. B. 1989. Late Pennsylvanian climate changes and palynomorph extinctions. *Abstracts, 28th International Geological Congress*, **2**, 214–215.

LESNIKOWSKA, A. D. 1989. *Anatomically Preserved Marattiales from Coal Swamps of the Desmoinesian and Missourian of the Midcontinent United States: Systematics, Ecology, and Evolution*. Ph.D. Thesis, University of Illinois, Urbana-Champaign.

LORENZO, P. 1979. Les sporophylles de *Lepidodendron dissitum* Sauver, 1848. *Geobios*, **12**, 137–143.

MAHAFFY, J. F. 1985. Profile patterns of coal and peat palynology in the Herrin (No. 6) Coal Member, Carbondale Formation, Middle Pennsylvanian of southern Illinois. *Proceedings 9th International Congress of Carboniferous Stratigraphy and Geology*, **5**, 155–159.

—— 1988. Vegetational history of the Springfield coal (Middle Pennsylvanian of Illinois) and the distribution of the tree fern miospore, *Thymospora pseudothiessenii*, based on miospore profiles. *International Journal of Coal Geology*, **10**, 239–260.

MILLER, W. III. 1990. Hierarchy, individuality and paleoecosystems. *In:* MILLER, W. III (ed.) *Paleo-*

community *Temporal Dynamics: the Long-term Development of Multispecies Assemblages.* Paleontological Society Special Publication, **5**, 31–47.

—— 1993. Models of recurrent fossil assemblages. *Lethaia*, **26**, 182–183.

MICKLE, J. E. 1984. *Taxonomy of Specimens of the Pennsylvanian-age Marattialean Fern Psaronius From Ohio and Illinois.* Illinois State Museum, Scientific Paper, 19.

O'NEILL, R. V. 1989. Perspectives in hierarchy and scale. *In:* ROUGHGARDEN, J. A., MAY, R. M. & LEVIN, S. A. (eds) *Perspectives in Ecological Theory.* Princeton, Princeton University Press, 140–156.

—— DEANGELIS, D. L., WAIDE, J. B. & ALLEN, T. F. H. 1986. *A Hierarchical Concept of Ecosystems.* Princeton University Press, Monographs in Population Biology, 23.

OVERPECK, J. T., WEBB, R. S. & WEBB, T. III. 1992. Mapping eastern North American vegetation change of the past 18 ka: no analogues and the future. *Geology*, **20**, 1071–1074.

PARRISH, J. T. 1982. Upwelling and petroleum source beds, with reference to the Paleozoic. *AAPG Bulletin*, **66**, 750–774.

PATTON, B. C. & ODOM, E. P. 1981. The cybernetic nature of ecosystems. *American Naturalist*, **118**, 886–895.

PEPPERS, R. A. 1985. Comparison of miospore assemblages in the Pennsylvanian System of the Illinois Basin with those in the Upper Carboniferous of Western Europe. *9th International Congress of Carboniferous Stratigraphy and Geology, C.R.*, **2**, 483–502.

—— & PFEFFERKORN, H. W. 1970. A comparison of floras of the Colchester (No. 2) Coal and Francis Creek Shale. *In:* SMITH, W. H., NANCE, R. B., HOPKINS, M. E., JOHNSON, R. G. & SHABICA, C. W. (eds) *Depositional Environments in Parts of the Carbondale Formation – Western and Northern Illinois.* Illinois State Geological Survey Guidebook Series **8**, 61–74.

PFEFFERKORN, H. W. 1976. Pennsylvanian tree fern compressions *Caulopteris, Megaphyton,* and *Artisophyton* gen. nov. in Illinois. *Illinois State Geological Survey Circular 492.*

——, GILLESPIE, W. H., RESNICK, D. A. & SCHEIHING, M. H. 1984. Reconstruction and architecture of medullosan pteridosperms (Pennsylvanian). *The Mosasaur*, **2**, 1–8.

—— & THOMSON, M. 1982. Changes in dominance patterns in Upper Carboniferous plant-fossil assemblages. *Geology*, **10**, 641–644.

PHILLIPS, T. L. 1979. Reproduction of heterosporous arborescent lycopods in the Mississippian-Pennsylvanian of Euramerica. *Review of Palaeobotany and Palynology*, **27**, 239–289.

——, AVCIN, M. J. & BERGGREN, D. 1976. Fossil peat of the Illinois Basin. *Illinois State Geological Survey, Educational Series*, **11**.

—— & DIMICHELE, W. A. 1981. Paleoecology of Middle Pennsylvanian age coal swamps in southern Illinois – Herrin Coal Member at Sahara Mine No. 6. *In:* NIKLAS, K. J. (ed.) *Paleobotany,*

Paleoecology, and Evolution. Praeger, New York, **1**, 231–284.

—— & —— 1992. Comparative ecology and life-history biology of arborescent lycopsids in Late Carboniferous swamps of Euramerica. *Annals of the Missouri Botanical Garden*, **79**, 560–588.

——, KUNZ, A. B & MICKISH, D. J. 1977. Paleobotany of permineralized peat (coal balls) from the Herrin (No. 6) Coal Member of the Illinois Basin. *In:* GIVEN, P. N. & COHEN, A. D. (eds) *Interdisciplinary Studies of Peat and Coal Origins.* Geological Society of America, Microform Publication 7, 18–49.

—— & PEPPERS, R. A. 1984. Changing patterns of Pennsylvanian coal-swamp vegetation and implications of climatic control on coal occurrence. *International Journal of Coal Geology*, **3**, 205–255.

——, PEPPERS, R. A., AVCIN, M. J. & LAUGHNAN, P. F. 1974. Fossil plants and coal: patterns of change in Pennsylvanian coal swamps of the Illinois Basin. *Science*, **184**, 1367–1369.

——, —— & DIMICHELE, W. A. 1985. Stratigraphic and interregional changes in Pennsylvanian coal-swamp vegetation: environmental inferences. *International Journal of Coal Geology*, **5**, 43–109.

PIERCE, B. S., STANTON, R. W. & EBLE, C. F. 1991. Facies development in the Lower Freeport coal bed, west-central Pennsylvania, U.S.A. *International Journal of Coal Geology*, **18**, 17–43.

——, —— & —— 1993. Comparison of the petrography, palynology, and paleobotany of the Stockton coal bed, West Virginia and implications for paleoenvironmental interpretations. *Organic Geochemistry*, **20**, 149–166.

PIGG, K. B. & ROTHWELL, G. W. 1983. *Chaloneria* gen. nov.: heterosporous lycophytes from the Pennsylvanian of North America. *Botanical Gazette*, **144**, 132–147.

PIMM, S. L. 1991. *The Balance of Nature?* University of Chicago Press.

PRYOR, J. S. 1993. Patterns of ecological succession within the Upper Pennsylvanian Duquesne coal of Ohio (USA). *Evolutionary Trends in Plants*, **7**, 57–66.

ROSS, C. A. & ROSS, J. R. P. 1987. *Late Paleozoic Sea Levels and Depositional Sequences.* Cushman Foundation for Foraminiferal Research, Special Publication, **24**, 137–149.

—— & —— 1988. *Late Paleozoic transgressive-regressive deposition.* Society of Economic Paleontologists and Mineralogists Special Publication, **42**, 227–247.

ROTHWELL, G. W. 1988. Upper Pennsylvanian Steubenville coal-ball flora. *Ohio Journal of Science*, **88**, 61–65.

SALTHE, S. N. 1985. *Evolving Hierarchical Systems.* Columbia University Press, New York.

SCHABILION, J. T. & REIHMAN, M. A. 1985. Anatomy of petrified *Neuropteris scheuchzeri* pinnules from the Middle Pennsylvanian of Iowa: a paleoecological interpretation. *Proceedings 9th International Congress of Carboniferous Stratigraphy and Geology*, **5**, 3–12.

SCHLESINGER, W. H. 1978. Community structure, dynamics and nutrient cycling in the Okefenokee cypress swamp forest. *Ecological Monographs*, **48**, 43–65.

SCHUBERT, J. K. & BOTTJER, D. J. 1993. Recovery from the end-Permian mass extinction event: paleoecology of Lower Triassic marine invertebrate assemblages in the Great Basin. *Geological Society of America, Abstracts with Programs*, A-156.

SCOTT, A. C. 1978. Sedimentological and ecological control of Westphalian B plant assemblages from west Yorkshire. *Proceedings of the Yorkshire Geological Society*, **41**, 461–508.

—— & REX, G. 1985. The formation and significance of Carboniferous coal balls. *Philosophical Transactions of the Royal Society of London*, **311B**, 123–137.

STECHESGOLEV, A. K. 1975. Die entwicklung der Pflanzenbedeckung im Suden des Europaischen Teils der UdSSR, vom Ende des Mittelkarbons bis zum Perm. Umfang und Gliederung des oberen Karbons (Stefan). *C.R. 7th International Congress Carboniferous Stratigraphy and Geology (Krefeld)*, **4**, 275–280.

TAYLOR, T. N. 1965. Paleozoic seed studies: a monograph of the genus *Pachytesta*. *Palaeontographica*, Abt. B. **117**, 1–46.

TRIVETT, M. L. 1992. Growth architecture, structure, and relationships of *Cordaixylon iowensis* nov. comb. (Cordaitales). *International Journal of Plant Sciences*, **153**, 272–287.

—— & ROTHWELL, G. W. 1985. Morphology, systematics, and paleoecology of Paleozoic fossil plants: *Mesoxylon priapi* sp. nov. (Cordaitales). *Systematic Botany*, **10**, 205–223.

TSCHUDY, R. H., PILLMORE, C. L., ORTH, C. J., GILMORE, J. S. & KNIGHT, J. D. 1984. Disruption of the terrestrial plant ecosystem at the Cretaceous-Tertiary boundary, western interior. *Science*, **225**, 1030–1032.

VERMEIJ, G. J. 1987. *Evolution and Escalation*. Princeton University Press, Princeton.

WEBB, T. III. 1988. Eastern North America. *In:* HUNTLEY, B. & WEBB, T. III, (eds) *Vegetation History*. Kluwer, Dordrecht, 385–414.

WHITE, D. & THIESSEN, R. 1913. The origin of coal. *United States Bureau of Mines Bulletin*, **38**.

WILLARD, D. A. 1989. Source plants for Carboniferous microspores: *Lycospora* from permineralized *Lepidostrobus*. *American Journal of Botany*, **76**, 820–827.

—— 1993. Vegetational patterns in the Springfield coal (Middle Pennsylvanian, Illinois Basin): comparison of miospore and coal-ball records. *Geological Society of America*, Special Paper **286**, 139–152.

—— & PHILLIPS, T. L. 1993. Paleobotany and palynology of the Bristol Hill Coal Member (Bond Formation) and Friendsville Coal Member (Mattoon Formation) of the Illinois Basin (Upper Pennsylvanian). *Palaios*, **8**, 574–586.

WING, S. L. 1984. Relation of paleovegetation to geometry and cyclicity of some fluvial carbonaceous deposits. *Journal of Sedimentary Petrology*, **54**, 52–66.

—— & TIFFNEY, B. H. 1987. The reciprocal interaction of angiosperm evolution and tetrapod herbivory. *Review of Palaeobotany and Palynology* **50**, 179–210.

WINSTON, R. B. 1986. Characteristic features and compaction of plant tissues traced from permineralized peat to coal in Pennsylvanian coals (Desmoinesian) from the Illinois Basin. *International Journal of Coal Geology*, **6**, 21–41.

—— 1989. Identification of plant megafossils in Pennsylvanian-age coal. *Review of Palaeobotany and Palynology*, **57**, 265–276.

—— 1990. Implications of paleobotany of Pennsylvanian-age coal of the central Appalachian basin for climate and coal-bed development. *Geological Society of America Bulletin*, **102**, 1720–1726.

—— & PHILLIPS, T. L. 1991. *A Structurally Preserved, Lower Pennsylvanian Flora from the New Castle Coal Bed of Alabama*. Geological Survey of Alabama, Circular 157.

WNUK, C. & PFEFFERKORN, H. W. 1984. Life habits and paleoecology of Middle Pennsylvanian medullosan pteridosperms based on an *in situ* assemblage from the Bernice Basin (Sullivan County, Pennsylvania, U.S.A.). *Review of Palaeobotany and Palynology*, **41**, 329–351.

WOLFE, J. A. & UPCHURCH, G. R. JR. 1986. Vegetation and floral changes at the Cretaceous–Tertiary boundary. *Nature*, **324**, 148–152.

—— & —— 1987. Leaf assemblages across the Cretaceous–Tertiary boundary in the Raton Basin, New Mexico and Colorado. *Proceedings of the National Academy of Sciences, USA*, **84**, 5096–5100.

ZIEGLER, A. M. 1990. Phytogeographic patterns and continental configurations during the Permian Period. *In:* MCKERROW, W. S. & SCOTESE, C. R. (eds) *Palaeozoic Palaeogeography and Biogeography*. Geological Society, London, Memoir, **12**, 363–379.

——, RAYMOND, A., GIERLOWSKI, T. C., HORRELL, M. A., ROWLEY, D. B. & LOTTES, A. L. 1987. Coal, climate, and terrestrial productivity – the present and Early Cretaceous compared. *In:* SCOTT, A. C. (ed.) *Coal and Coal-Bearing Strata – Recent Advances*. Geological Society, London, Special Publication, **32**, 25–49.

Recoveries and radiations: gastropods after the Permo-Triassic mass extinction

DOUGLAS H. ERWIN & P. HUA-ZHANG

[1] *Department of Paleobiology, NHB-121, National Museum of Natural History, Washington, DC 20560, USA*

[2] *Nanjing Institute of Geology and Palaeontology, Academia Sinica, Nanjing, People's Republic of China*

Abstract: The biotic recovery following the end-Permian mass extinction begins with a long lag phase characterized by a low-provinciality, depauperate biota. Normal marine communities do not reappear until the latest Early Triassic (Spathian) and do not become common until the early Middle Triassic (Anisian). Three explanations have been offered for this pattern: first, that the end-Permian extinction so disrupted marine ecosystems that community assembly rules needed to be rewritten prior to the re-establishment of normal community structures; second, that the recovery was delayed by continuing harsh environmental conditions; and third, that the preservational and sampling bias is significant and the pattern more apparent than real.

Data on the stratigraphic, biogeographical and environmental distribution of Wordian–Anisian marine gastropod genera and species were used to evaluate these hypotheses. The results demonstrate (1) an initial removal of endemic (narrowly distributed) genera during the Wordian and Capitanian; (2) the presence of a low provinciality gastropod assemblage during the Djulfian and Changxingian; and (3) the persistence into the earliest Triassic of geographically widespread, environmentally tolerant genera, as well as the apparent (but unobserved), occurrence of numerous 'Lazarus' taxa. The data indicate that while preservational bias was significant, environmental dampening was also important.

The end-Permian mass extinction was the most widespread marine extinction of the Phanerozoic, eliminating perhaps 90% or more of all marine species (Erwin 1993, 1994) although with considerable variation in the magnitude of extinction between different groups. The lengthy lag phase between the apparent end of the extinction near the Permo-Triassic boundary and the reappearance of normal marine faunal assemblages in the Middle Triassic (Anisian) has long been one of the most puzzling aspects of this extinction, since recovery typically begins within a million years or so of a major mass extinction (Hallam 1991). Early Triassic faunas are typically depauperate, and are dominated by bivalves, ammonoids and some gastropods; in most faunal assemblages few species are present, but in very large numbers. Was the delayed recovery during the Early Triassic caused by the persistence of harsh environmental conditions which inhibited the reconstruction of normal marine communities (a form of environmental dampening), or did the magnitude of the extinction so disrupt normal communities that the recovery was delayed until new community assembly rules could form? P. Wignall has raised the additional possibility that preservational

bias played a significant role in producing the pattern.

These different possibilities are not mutually exclusive, and each may operate to some extent, or at different times during the Early Triassic. Nonetheless, they do suggest very different patterns of recovery. If the delayed recovery was largely mediated by environmental dampening, eurytopic organisms should reappear prior to more stenotopic taxa. Thus many gastropods and bivalves should appear before more stenotopic echinoderms. On the other hand, if recovery was delayed by ecological restructuring then the pattern of re-emergence should follow the palaeoecological equivalent of a successional phenomenon, with ecologically generalized opportunists reappearing first, followed by progressively more trophically complex taxa.

This rather simplistic view is complicated, however, by the acknowledged preservational biases across the Permo-Triassic boundary, which are likely to obscure the real pattern of recovery. If preservational biases are pervasive, the preserved pattern of recovery is likely to be randomized and the underlying pattern difficult to discern. In marked contrast to the Permian, Early Triassic gastropod assemblages are not

From Hart, M. B. (ed.), 1996, *Biotic Recovery from Mass Extinction Events*, Geological Society Special Publication No. 102, pp. 223–229

silicified, which dramatically reduces recovery (Erwin pers. comm.) Silicified faunas begin to return in the Anisian and may be partly responsible for the apparent re-emergence of normal marine communities. The significance of this problem remains unexamined, however. Sampling biases are likely to add additional complications.

Analysis of Early Triassic recovery has been confused by another phenomenon first noted by Batten (1973). Based on his compilation of gastropod distributions from the Guadalupian through Ladinian stages, Batten concluded that Djulfian and Scythian (Early Triassic) faunas were largely depauperate, with few species and large numbers of individuals. That some 32 Permian genera could have escaped the great end-Permian mass extinction by seemingly disappearing for perhaps 10 Ma seemed most remarkable. Jablonski (1986) christened these Lazarus taxa, in recognition of their disappearance and apparent rebirth. More importantly, however, Batten recognized that normal marine Anisian and Ladinian assemblages were more like those of Guadalupian than they were Jurassic gastropod assemblages. On this basis he argued that the major faunal turnover among gastropods occurred near the Triassic–Jurassic boundary.

Batten's (1973) data were worldwide, but, significantly, did not include data from South China. Late Permian faunas from South China were poorly known during the early 1970s, but have since proved to be one of the most important sources of information on the Permo-Triassic boundary (Erwin 1993). Chinese geologists have documented over 40 Permo-Triassic boundary sections and described numerous new taxa. Analysis of the data provides some important modifications to Batten's original conclusions, although the major thrust of his work remains. Additionally, the description of numerous new Early and Middle Triassic faunas further fleshes out his analysis.

The numerous Lazarus taxa demonstrate the pervasive nature of preservational problems during this mass extinction, raising the question of whether the recovery actually began early in the Triassic, but is obscured.

Data

This preliminary analysis is based on a compilation of 168 gastropod generic ranges and species occurrences from the Guadalupian (Wordian and Capitanian substages), Tatarian (Djulfian and Changxingian), Induan (Griesbachian and Dienerian), Olenekian (Smithian and Spathian) and Anisian stages. Many of the global data were developed by Erwin as part of his global Cambrian–Triassic gastropod generic database. This information was supplemented by Pan's compilation of stage-level gastropod generic occurrences in South China (see Pan & Erwin (1994) for additional details), and by detailed species occurrence information for the Guadalupian–Anisian interval. Some generic assignments were modified to reduce oversplitting. For example, Chinese records of *Porcellia*, a Devono-Carboniferous genus, were assigned to *Ambozone*, following Batten (1989). Similarly, *Rhaphistomella* appears to be the Triassic name for the more widely known *Glabrocingulum* (*Glabrocingulum*), and these genera were combined. Finally, 30 genera have been removed from the analysis. Fifteen of these represent questionable range extensions of genera to the Wujiapingian and Changxingian stages in China. A further 15 genera are new or undescribed, also recorded only from the Wujiapingian and Changxingian stages in China. The status of these genera remains unclear, and they were ignored pending further study. Despite this, these genera do suggest that large numbers of endemic species occurred in South China during the Changxingian.

Genera were characterized as widespread or restricted (endemic) based on their geographical distribution during the final two stages of their history. This dataset allowed the determination of the recorded global diversity (genera actually recorded from a substage) number of Lazarus genera per substage, total global diversity (recorded + Lazarus diversity) and the number of appearances and disappearances. Owing to uncertainties about the duration of these stages and substages, the data are simply reported by substage. The Permo-Triassic boundary is now dated at 251 Ma ago (Claoué-Long *et al.* 1991), some 4–6 Ma older than previously, but lack of temporal control has inhibited determination of the duration of the relevant stages.

Sampling problems

The occurrence of so many Lazarus taxa indicates the extent of sampling and preservational problems which plague analysis of this interval. Consequently, considerable caution should be used in interpreting diversity data. Seventeen of the 24 Lazarus genera first appear at the close of the Wordian. This largely reflects the loss of abundant silicified assemblages in the west Texas Permian Basin with the onset of the deposition of the massive Capitan Limestone. The reduced silicification in the Capitan Lime-

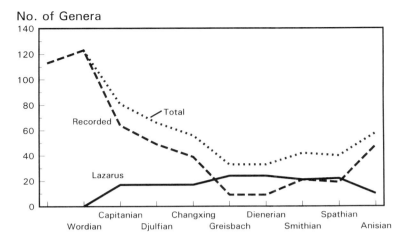

Fig. 1. Global gastropod diversity patterns from the Wordian through the Anisian, showing actual recorded diversity (genus is known from the interval), Lazarus diversity (genus is known from earlier and later stages) and total diversity. Note that the number of Lazarus taxa increases at the close of the Wordian, well before the Permo-Triassic boundary, and persists until the Anisian.

stone and equivalent formations produces an apparently less diverse gastropod fauna than in earlier units within the Permian Basin. Because of this preservational bias, it is difficult to determine how much of the signal accurately reflects a change in the biota and how much preservational failure.

Coincident with the appearance of numerous Lazarus taxa in the western United States, however, was an increase in the fidelity of preservation in South China. Pan & Erwin (1994) have demonstrated that Changxingian gastropod assemblages are far better preserved than earlier Permian assemblages in China. Thus, in contrast to most of the rest of the world, the fidelity of preservation of faunas from South China appears to increase toward the Permo-Triassic boundary, in part reflecting the relatively minor impact of Late Permian marine regression in South China. In west Texas, most normal marine deposits disappear at the end of the Capitanian as the basin begins to dry up. The only younger gastropod assemblage found in west Texas is a middle Djulfian fauna found in a limestone bed in the Rustler Formation (Walter 1953).

Extinction patterns

Global

Wordian–Anisian global diversity patterns are shown in Fig. 1. Generic diversity peaked at 123 during the Wordian and began a steady decline during the Capitanian. Thirty-eight genera (31%) disappear at the close of the Wordian, and 17 (14%) Lazarus genera develop, more than at the close of the Changxingian when only seven Lazarus taxa were added. This suggests the sampling and/or preservation biases responsible for the Lazarus effect date to the Wordian, rather than the end of the Permian. Only 14 genera (17% if Lazarus taxa are included, 22% without) disappeared at the end of the Capitanian, but 20 genera (30% and 41%) disappear during the Djulfian and 25 genera (45% and 64%) during the Changxingian.

Biogeographical

The global data indicate an early extinction peak during the Wordian, followed by increasing extinction during the Djulfian with a second peak during the Changxingian. Examination of the biogeographical data shows that this picture is somewhat misleading. Yet, as illustrated by Fig. 2, 56% of the generic extinctions during the Wordian were of endemic taxa. Of the 14 widespread genera which became extinct, most represent taxa which were widespread during the preceding (upper Leonardian) stage, but are only present in restricted geographical regions during the Wordian. Most of these endemics occur in west Texas and Malaysia. By the Djulfian and Changxingian very few endemic genera are present, except in South China.

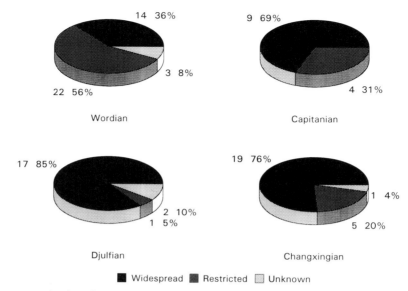

Fig. 2. Percentage extinction of geographically widespread and restricted genera during the final four substages of the Permian (Wordian–Changxingian). Both the number of genera and the percentage of the extinct genera are shown during each substage.

The effect of adding the problematic genera described previously would be to increase dramatically the number of endemic genera in South China. The latest Permian fauna in South China appears to have a large number of endemic forms in addition to the numerous widespread genera. Curiously, South China appears to be the only region which still has endemic genera by this time. This does not appear to be a result simply of the greater number of localities in South China, but rather reflects continuing occurrence of diverse and vibrant marine communities in South China. In contrast, other regions lost their endemics and are limited to geographically widespread genera.

Although these data are still somewhat preliminary, these results strongly suggest that the early phase of the end-Permian mass extinction involved the removal of endemic genera and the preferential persistence of widespread genera. By the Djulfian and Changxingian, gastropod provinciality was apparently reduced to a single, global province with continuing high endemicity in South China.

Triassic recovery

Phase 1: Induan

Few good Induan gastropod assemblages have been described, so most gastropods are recorded in faunal lists detailing earliest Triassic assemblages. Nine genera are found in deposits of this age (the bellerophontids *Bellerophon*, *Retispira*, *Euphemites* and ?*Statchella*, the pleurotomarids *Glabrocingulum* (*Glabrocingulum*) and *Worthenia*, the neritaceans *Naticopsis* and *Vernelia*, and the subulitid *Strobeus*. The life habits of Palaeozoic gastropods are notoriously difficult to interpret, but other than the predominance of bellerophontids in these assemblages, there are no apparent ecological similarities among these genera. These nine genera come from very different clades and evidently had very different life habits. All do share two characteristics, however: broad geographical distribution during the Late Permian and occupation of a broad variety of habitats, from offshore carbonate reefs to nearshore clastics and even the rigorous conditions of the Phosphoria Formation in the western United States. Additionally, all but three genera (*Statchella*, *Vernelia* and *Strobeus*) were species-rich during the Late Permian. The relative contribution of these characteristics awaits further study, but in conjunction with the information presented above, widespread geographical range at the genus level does appear to have enhanced survival during the end-Permian crisis and into the Triassic.

Phase 2: Olenekian

During the latter half of the Early Triassic, increased signs of recovery are evident. Twelve new genera have been described from this interval, mostly of neritaceans and loxonematids. They are joined by the nine genera from the Induan and three Lazarus taxa which reappear. Most gastropods from this interval seem to be found in nearshore clastics, as exemplified by the faunule described by Batten & Stokes (1986) from the Sinbad Member of the Moenkopi Formation in Southern Utah. Again, there is no obvious consistent pattern among these taxa.

Phase 3: Anisian

The Anisian is marked by the re-emergence of numerous Lazarus taxa and the apparent origination of many new genera. Reefs reappear and the Anisian faunas are the first truly normal marine assemblages of the Triassic. Among the gastropods, as Batten (1973) first observed, many of the Lazarus taxa have strong Guadalupian affinities and as a whole, the Anisian gastropod assemblage is remarkably similar to the Middle Permian. Nineteen new genera have been described from Anisian sediments and 12 more Lazarus genera re-emerge. Ten genera do not reappear until the Ladinian or, in a number of cases, the Carnian or Norian. As discussed below, the length of this gap raises questions of phylogenetic affinity. Many of these late appearing 'Lazarus taxa' may actually be homeomorphs (the 'Elvis' taxa of Erwin & Droser 1993).

The geographical affinities of the Lazarus genera are of some interest. Of the 17 genera with good data, 16 were present in west Texas during the Wordian; the only exception, *Magnicapulus*, is found in South China. In contrast, only 8 genera were found in South China prior to the Triassic. Of these eight, three, *Murchisonia*, *Dictyotomaria* and *Anomphalus*, do not appear in South China, although they are found in Europe or North America. Three Lazarus genera, *Glyptomaria*, *Eucycloscala* and *Natiria*, found in South China during the Smithian–Anisian are not recorded from South China during the Permian, suggesting they migrated to South China following the extinction. The Lazarus genera are a taxonomically varied group and share no obvious ecological or systematic characteristics.

There is a curious subset of the Lazarus genera: a total of eight genera which are traditionally believed to be Triassic, but which show up briefly in the Wordian (5), Capitanian (2) or Changxingian (1) and then disappear, to return in Middle Triassic. Each of these is known only from a restricted number of localities. Four are found in the displaced terranes of Japan, two in Malaysia, and one each in China and West Texas. Of the eight, six reappear in South China, two in the early Triassic of Utah (Batten & Stokes 1986). The significance of these eight genera is not completely clear, but they do reduce the significance of earlier conclusions (Erwin 1990) that the transition between Palaeozoic and Triassic gastropods was almost exclusively a function of the extinction and subsequent recovery.

Discussion

What does this analysis of gastropod diversity pattern reveal about the nature of the delayed early Triassic recovery, and specifically, about the causes of the Lazarus taxa? As noted in the introduction, the delayed recovery and the Lazarus effect are intimately related, but nonetheless distinct. The apparently delayed recovery may reflect (1) environmental dampening owing to the persistence of unfavourable environmental conditions into the Early Triassic; (2) disruption of community structure to such an extent that population sizes remain low and communities are poorly structured, with the re-emergence of 'normal' faunas tied to community restructuring; and finally (3) a persistent sampling or preservational bias. Similarly, the Lazarus phenomenon could reflect (1) limitation of a taxon to a biogeographically or environmentally restricted region which remains habitable (the refugia hypothesis); and (2) small population sizes which are below the limit of visibility to the fossil record due to community disruption. In either (1) or (2) the taxon may either lie within the former geographical or environmental range of the taxon or may represent movement to a new geographical or environmental area. Alternatively, (3) preservational bias may result when taxa were present in a particular area or environment but cannot be readily recovered from the rocks owing to taphonomic difficulties, and finally, (4) sampling problems may reflect a lack of sampling effort or effort expended in the wrong areas. Clearly, if either preservational or sampling biases are significant, there is not much left to explain and the first two possibilities need not be considered.

So what of the preservational and sampling problems? Preservational difficulties clearly account for many of the Lazarus taxa. Many of these lineages actually originated during the Wordian and Capitanian, well before the boundary. At least in West Texas, their appearance corresponds to a drop in the number and diversity of silicified assemblages and a change in carbonate environments. Thus many taxa may have been present but were not preserved. Silicified faunas are similarly rare during the Early Triassic, which is likely to reduce significantly faunal recovery. Taphonomic controls (see below) can provide some estimate of the significance of preservational bias.

The significance of sampling problems is less clear. At different levels, the diversity of the late Permian faunal assemblages from South China, as well as the Smithian gastropod fauna described by Batten & Stokes (1986) each demonstrate that faunas remain poorly sampled during the Late Permian–Early Triassic. What is particularly surprising, however, is that this new information has not changed diversity patterns as much as might be expected. Certainly the Chinese data include numerous range extensions which shift extinctions from the Wordian–Capitanian to the Changxingian, but the number of Lazarus taxa has dropped significantly during the Late Permian and not at all during the Induan. On the regional scale, Late Permian sampling problems do not seem a likely explanation for Lazarus taxa. Barring the discovery of another high-diversity region like South China, most Late Permian sequences have been sufficiently well studied to eliminate a significant sampling bias. Turning to the Triassic, Lower Triassic rocks were largely deposited during a global transgression. They are far more widespread than latest Permian rocks, and generally far more accessible. The difficulty is demonstrably not that they have not been sampled by palaeontologists and stratigraphers, but that the fauna is largely depauperate (and frankly, pretty boring!).

The use of taphonomic controls, pioneered by Bottjer & Jablonski (1988), provides an attractive avenue to explore to importance of preservational bias. Taphonomic controls are taxa with similar ecological, morphological and mineralogical characteristics as the group under study. The absence of the taphonomic control from a locality strongly suggests taphonomic bias as an explanation rather than another alternative. Examination of the nine gastropod genera found during the Induan indicates that most have robust shells and are readily preserved, even in the absence of silicification. Such is not the case for many of the Lazarus taxa, further demonstrating the importance of preservational bias. However, the Lazarus fauna also includes numerous robust genera which share similar morphological and mineralogical characteristics to the Induan fauna, and which were also geographically widespread during the Late Permian. Minimally, this includes *Warthia*, *Euomphalus*, *Straparollus*, *Ananias*, *Murchisonia* and perhaps *Marmolatella*. Several other genera might be included as well. The absence of these taxa cannot be explained simply as a result of preservational or sampling problems and demands consideration of the other possible explanations.

Small population sizes within the previous range of a species seem an unlikely explanation, assuming relatively normal sampling of the interval. Population sizes low enough to render a species invisible to the fossil record increase susceptibility to genetic drift and extinction. Thus population sizes must increase periodically in order to purge the population of deleterious alleles. Maintaining small population sizes over 8–10 Ma seems exceedingly implausible. There is an additional conundrum that most genera which survived were geographically and environmentally widespread prior to the extinction, and most were speciose as well. It is difficult to reconcile this pattern with the apparent invisibility of the Lazarus taxa.

The relative significance of environmental dampening v. delayed recovery for ecological reasons can only be determined through reference to the entire fauna, and thus requires a palaeoecological- rather than lineage-based approach. The data currently in hand for gastropods raise problems for either scenario. If environmental dampening had been effective, one would predict that eurytopic groups would be the first to return, followed by progressively more stenotopic groups. Lazarus taxa and new species should reappear gradually as environmental conditions moderate and without any taxonomic bias. In contrast, if the ecospace restructuring hypothesis was applicable, the duration of the lag phase would be controlled by the extent of community collapse. Recovery should vary between ecosystems based on the extent of the collapse (and thus the need for restructuring) and by trophic level, with detritivores and groups low on the food chain recovering first.

The occurrence of echinoids in mid-Griesbachian sediments suggests that normal marine conditions had returned by this time, at least in the Alps (R. Twitchett pers. comm.). Further studies are needed to confirm this, but it suggests that an environmental dampening may have been limited to the earliest part of the Triassic. However, there was no apparent increase in the rate of recovery during the late Early Triassic. In fact, on present evidence there is no evidence of either an initial recovery of eurytopic groups followed by stenotopic species, nor of gradual re-emergence of progressively more structured assemblages. Rather there are scattered assemblages like the Sinbad Member gastropods and then the return of full-fledged 'normal' marine communities during the Anisian.

This preliminary study demonstrates the need for future work along several different lines in order to understand better the nature of the post-Permian survival–recovery interval. First, the status of the 30 Late Permian problematic genera must be determined. Some may be new genera; others may represent stratigraphic and geographical range extensions for either older Permian genera or for Triassic taxa. Second, few phylogenetic studies have investigated clade relationships across the Permo-Triassic boundary, but those that have (for brachiopods, bryozoans and asteroids) all suggest that traditional systematics may exaggerate the magnitude of extinction at higher levels and miss phylogenetic links at lower levels. For example, traditional systematists viewed the echinoid genus *Miocidaris* as the only genus to survive the Permo-Triassic extinction and thus as the progenitor of all post-Palaeozoic echinoids (Paul 1988). Yet careful character studies and phylogenetic analysis reveal that at least one other genus must have survived the extention (Smith & Hollingsworth 1990). Preliminary analyses of articulate brachiopods (Carlson 1991) and bryozoans (Taylor & Larwood 1988) have reached similar conclusions. These results demonstrate the importance of detailed phylogenetic studies in reconstructing lineages, particularly during intervals of rapid morphological change when phylogenetic patterns may be obscure.

Finally, there is an ongoing need for more thorough investigations on the entire Early Triassic. Those studies which have been conducted largely focus on the immediate interval above the Permo-Triassic boundary. While such detailed stratigraphic and palaeontological studies must continue, preferably in combination with geochemical analyses, the later Early Triassic must be investigated as well. Particular attention should be given to the late phases with the re-emergence of the Lazarus taxa and the transition between the Olenekian and the Anisian.

References

BATTEN, R. L. 1973. The vicissitudes of the gastropods during the interval of Gaudalupian–Ladinian time. *In:* LOGAN, A. & HILLS, L. V. (eds) *The Permian and Triassic Systems and Their Mutual Boundary.* Canadian Society of Petroleum Geologists, Memoir, **2**, 596–607.

—— 1989. Permian gastropoda of the United States, 7, Pleurotomariacea: Eotomariidae, Lophospiridae, Glosseletinidae. *American Museum Novitates,* No. 2958, 1–64.

—— & STOKES, W. L. 1986. Early Triassic Gastropoda from the Sinbad Member of the Meenkopi Formation, San Rafael Swell, Utah. *American Museum Novitates,* No. 2864, 1–96.

BOTTJER, D. J. & JABLONSKI, D. 1988. Palaeoenvironmental patterns in the evolution of Post-Palaeozoic benthic marine invertebrates. *Palaios,* **3**, 540–560.

CARLSON, S. J. 1991. A phylogenetic perspective on articulate brachiopod diversity and the Permo-Triassic extinction. *In:* DUDLEY, E. (ed.) *The Unity of Evolutionary Biology. Proceedings of the 4th International Congress of Systematic and Evolutionary Biology.* Discorides Press, Portland, OR, 119–142.

CLAOUÉ-LONG, J. C., ZHANG, Z. C., MA, G. G. & DU, S. H. 1991. The age of the Permian–Triassic boundary. *Earth and Planetary Science Letters,* **105**, 182–190.

ERWIN, D. H. 1989. Regional paleoecology of Permian gastropod genera, southwestern United States and end-Permian mass extinction. *Palaios,* **4**, 424–438.

—— 1990. Carboniferous–Triassic gastropod diversity patterns and the Permo-Triassic mass extinction. *Paleobiology,* **16**, 187–203.

—— 1993. *The Great Paleozoic Crisis: Life and Death in the Permian.* Columbia University Press, New York.

—— 1994 The Permo-Triassic extinction. *Nature,* **367**, 231–236.

—— & DROSER, M. E. 1993. Elvis taxa. *Palaios,* **8**, 623–624.

HALLAM, A. 1991. Why was there a delayed radiation after the end-Paleozoic extinctions? *Historical Biology,* **5**, 257–262.

JABLONSKI, D. 1986. Causes and consequences of mass extinctions: a comparative approach. *In:* ELLICT, D. K. (ed.) *Dynamics of Extinction.* Wiley, New York, 189–229.

PAN, H. Z. & ERWIN, D. H. 1994. Gastropod diversity patterns in South China during the Chihsia–Ladinian and their mass extinction. *Palaeoworld,* **4**, 249–262.

PAUL, C. R. C. 1988. Extinction and survival in the Echinodermata. *In:* LARWOOD, G. P. (ed.) *Extinction and the Fossil Record,* Clarendon, Oxford, 155–170.

SMITH, A. B. & HOLLINGSWORTH, N. T. J. 1990. Tooth structure and phylogeny of the Upper Permian echinoid *Miocidaris keyserlingi. Proceedings of the Yorkshire Geological Society,* **48**, 47–60.

TAYLOR, P. D. & LARWOOD, G. P. 1988. Mass extinctions and the pattern of bryozoan evolution. *In:* LARWOOD, G. P. (ed.) *Extinction and the Fossil Record.* Clarendon, Oxford, 99–119.

WALTER, J. C. JR. 1953. Paleontology of Rustler Formation, Culbertson County, Texas. *Journal of Paleontology,* **27**, 679–702.

Recovery of the marine fauna in Europe after the end-Triassic and early Toarcian mass extinctions

A. HALLAM

School of Earth Sciences, University of Birmingham, Birmingham B15 2TT, UK

Abstract: A detailed analysis of species diversity increase, zone by zone up the Liassic section in northwest Europe, has been undertaken for the six fossil groups for which adequate data are available: bivalves, ammonites, rhynchonellid brachiopods, crinoids, foraminifera and ostracods. The general pattern is of a rapid increase through the Hettangian from a very low level after the end-Triassic mass extinction, followed by a slower rate of increase until the late Pliensbachian. Thereafter there was a drastic fall in the early Toarcian as a consequence of the mass extinction of that time, followed by a further rise continuing into the Middle Jurassic. A more general study takes into account data from southern Europe and extra-European localities, the best of which for the Liassic being in Argentina and Chile. The best data come from brachiopods and bivalves. The brachiopod pattern is similar to that described above for a more limited geographical region, with genera already occurring in the Hettangian and Sinemurian of southern Europe spreading into northwest Europe in the Pliensbachian. The bivalve data show that genera present in the Sinemurian and Pliensbachian of South America did not reach Europe until after the late Toarcian and early Middle Jurassic. Reef ecosystems, which were drastically affected by the end-Triassic mass extinction, did not re-establish themselves until the Pliensbachian; the early Toarcian event had a similarly deleterious effect.

There is a good correlation between the Hettangian to Pliensbachian diversity rise and rise of sea level, with the very low diversity values of the early Hettangian being associated with widespread dysoxic and anoxic conditions. The low diversities of the early Toarcian are also associated with anoxia correlated with a rapid sea-level rise, with the immigration into Europe of South American taxa taking place subsequently when environmental conditions ameliorated and the higher sea level permitted freer communication, most probably through the Hispanic Corridor across the present Central Atlantic region.

The end-Triassic mass extinction event in the marine realm has been established as one of the five biggest in the Phanerozoic (Raup & Sepkoski 1982; Sepkoski 1986), while a lesser but still important extinction event, wrongly ascribed by Sepkoski (1986) to the Pliensbachian, took place in the Early Toarcian (Hallam 1986, 1987). While both these events have received a fair amount of attention, this paper represents the first attempt to study specifically the biotic recovery that followed. By far the best data come from the comprehensively researched marine Lower Jurassic successions in Europe. Apart from Argentina and Chile, data from the rest of the world for this time interval are decidedly sketchy or limited at present, and are likely to remain so in many cases because of inadequacy of exposure or lack of appropriate facies.

Analysis of Liassic species diversity in Britain

A detailed analysis of taxonomic richness or diversity change through the Liassic (or Lower Jurassic) succession, recorded to the precision of ammonite zones, has proved possible for six fossil groups in Britain. For five of these, the bivalves, rhynchonellid brachiopods, crinoids, foraminifera and ostracods, the analysis is at species level. Because there is no consensus among specialists about species within the sixth group, the more taxonomically split ammonites, analysis is at generic level. Data of comparable quality from the European continent are not available at present, but judging from ammonites and bivalves at least (Hallam 1976) the pattern of change recognized in Britain is likely to be valid for the whole of northwest Europe.

The results are presented in Fig. 1. The bivalves contain the largest number of taxa and are hence least likely to show a pattern distorted by adventitious fluctuations. They show a progressive increase in species numbers up through the Hettangian, followed by a period of stability in the Sinemurian. Thereafter there is a further slight rise to a Liassic diversity peak in the Upper Pliensbachian (*margaritatus* Zone)

From Hart, M. B. (ed.), 1996, *Biotic Recovery from Mass Extinction Events*,
Geological Society Special Publication No. 102, pp. 231–236

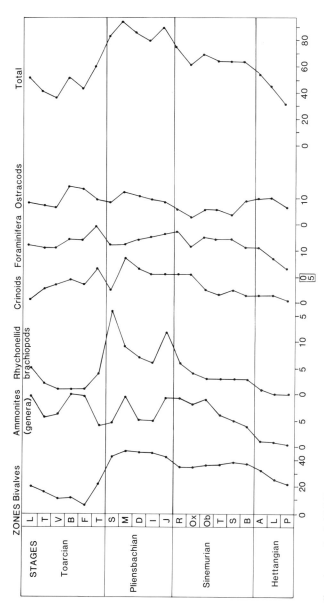

Fig. 1. Species diversity up the British Lower Jurassic succession. Hettangian: P, *planorbis*; L, *liasicus*; A, *angulata*. Sinemurian: B, *bucklandi*; S, *semicostatum*; T, *turneri*; Ob, *obstusum*; Ox, *oxynotum*; R, *raricostatum*. Pliensbachian: J, *jamesoni*; I, *ibex*; D, *davoei*; M, *margaritatus*; S, *spinatum*. Toarcian: T, *tenuicostatum*; F, *falciferum*; B, *bifrons*; V, *variabilis*; T, *thouarsense*; L, *levesquei*. Data sources: bivalves, Hallam (1976); ammonites, Dean *et al.* (1961); rhynchonellids, Ager (1956–1967); crinoids, Simms (1989); foraminifera, Jenkins & Murray (1989); ostracods, Bate & Robinson (1978).

followed by a sharp decline to a Liassic minimum in the Lower Toarcian (*falciferum* Zone), after which there is a renewed steady rise through to the highest Toarcian (*levesquei* Zone). The ammonites show a broadly similar pattern, the most notable difference being that the *falciferum* Zone does not show a taxonomic minimum. Allowing for the fluctuations due to smaller species numbers, the rhynchonellids exhibit a pattern closely resembling that of the bivalves.

All three other groups show a more or less progressive increase in diversity through the Hettangian, Sinemurian and Pliensbachian, with a tendency to decrease at the end of the Pliensbachian and in the Toarcian, but in no case is a *falciferum* Zone minimum followed by a younger Toarcian increase. The overall pattern displayed by all six groups is one of rapid diversity increase from a low value in the Hettangian, to slower increase progressively through the Sinemurian and Pliensbachian, followed by a sharp decline in the Early Toarcian and slight increase in the youngest two Toarcian zones. Accepting that the average duration of a Jurassic ammonite zone was about 1 Ma (Hallam *et al.* 1986), there was a more than three-fold increase in total taxa in about 13 Ma from the *planorbis* to the *margaritatus* Zone, followed by a more than two-fold fall to the *falciferum* Zone. The diversity almost doubled in about 2 Ma through the Hettangian, from the *planorbis* to the *angulata* Zone

The general pattern of change in Europe

An informative comparison can be made between Britain and southern Europe for the well documented brachiopods. This group is abundant and diverse in the Lower Liassic of a number of 'Tethyan' regions such as the Southern Alps (Gaetani 1970) and the Bakony Mountains of Hungary (Dulai 1992, 1993; Vörös 1993). A number of genera only reached Britain later in the Liassic, according to the monographs of Ager (1956–1967, 1990). This is shown in Table 1. Figure 2 shows a pattern of brachiopod diversity for the Lower and Middle Jurassic which, though described as global, is based largely on European data (Almeras 1964). This pattern is closely comparable to that of Fig. 1 in showing a progressive rise through the Hettangian, Sinemurian and Pliensbachian, followed by a sharp fall in the Toarcian. The further progressive increase through successive Middle Jurassic stages is broadly matched by bivalve species data for Europe (Hallam 1976).

Bivalve faunas occur only sparsely in most

Table 1. *First appearances of brachiopod genera in southern Europe and Britain*

Genus	Southern Europe	Britain
Cirpa	Hettangian	Pliensbachian
Cuneirhynchia	Sinemurian	Pliensbachian
Gibbirhynchia	Sinemurian	Pliensbachian
Homoeorhynchia	Sinemurian	Pliensbachian
Prionorhynchia	Hettangian	Pliensbachian
Stolmorhynchia	Pliensbachian	Toarcian
Lobothyris	Hettangian	Sinemurian

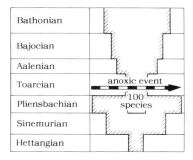

Fig. 2. Global diversity changes in brachiopods through the Early and Middle Jurassic. Based on data from Almeras (1964). Location of Toarcian anoxic event indicated.

of the Jurassic of southern Europe because of unsuitable facies (Hallam 1975a), but some interesting Hettangian faunas occur in the Southern Alps of Lombardy (Allasinaz 1992), the Northern Calcareous Alps of Austria (Golebiowski & Braunstein 1988) and the Mecsek Mountains of Hungary (Szente 1992). Unlike the brachiopods (and ammonites) these bivalve faunas are not evidently more diverse than those of northwest Europe, with the same species being recorded in both regions, but Szente tentatively records *Falcimytilus*, a genus that first appears in northwest Europe in the Bajocian (Hallam 1976). I personally have found a specimen of *Palmoxytoma* in the Hettangian of Kendelbach, Austria, a genus that first appears in Britain in the Pliensbachian. The important genus *Gryphaea*, which first appears in northwest Europe in the late Hettangian, occurs already in the late Triassic of the Arctic and the North American Cordillera, from Alaska to Nevada (McRoberts 1992). The species *G. arcuatiformis* is very similar to the earliest European species *G. arcuata*, except in its smaller maximum size, and is almost certainly

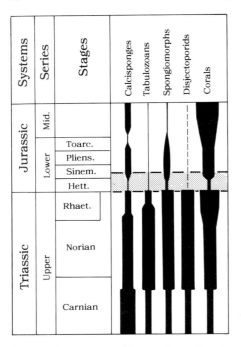

Fig. 3. Recovery of reef organisms after the end-Triassic mass extinction. Width of black bands proportional to diversity. Stippled zone is level at which reef communities are missing globally, according to Stanley (1988). Another such zone could have been added for the Toarcian. Adapted from Stanley (1986, fig. 3).

its direct ancestor.

The only microfaunal comparison between northern and southern Europe is that by El-Shaarawy (1982), who studied the foraminiferal and ostracod faunas of the Rhaetian Penarth Group and Hettangian–lower Sinemarian Blue Lias Formation of England and Wales and the Rhaetian Kössen Formation of the Northern Calcareous Alps of Austria. The Rhaetian deposits of northwest Europe were laid down in a low-salinity sea and hence the faunas are of restricted diversity, lacking in particular articulate brachiopods and ammonites, whereas the Kössen Formation was deposited in a warm, normal-salinity sea and hence has a high diversity of taxa including brachiopods and ammonites (Hallam & El-Shaarawy 1982). El-Shaarawy (1982) was able to establish that a number of species which entered the Blue Lias succession were already living in the Tethyan zone in the latest Triassic.

One of the most striking effects of the end-Triassic mass extinction was the loss of reef ecosystems, such as had flourished in the Northern Calcareous Alps and elsewhere until the end of the period. Nowhere in the world are reefs known in the Hettangian, and not until the Pliensbachian did they become fully re-established, as in Morocco (Agard & du Dresnay 1965). This pattern of change is indicated in Fig. 3, based upon a diagram of Stanley (1988). There was a further global disappearance of coral reefs in the Toarcian, though this is not indicated by Stanley, and a renewed re-establishment by the Bajocian, for which there is a good development in northern France (Hallam 1975b).

Continuing with the post-Toarcian recovery, much the best evidence comes from the rich Liassic bivalve faunas of Argentina and Chile (Damborenea 1987a, b; Riccardi et al. 1990). A number of genera that occur in the Pliensbachian of this region do not appear in Europe until much later, starting in the Late Toarcian (Table 2). If species or species groups are considered the number of taxa involved in migration to Europe is probably much greater. Thus bilobate *Gryphaea*, of the type that characterized the Middle Jurassic in Europe, have possible precursors in the Pliensbachian of Chile (Hallam 1982). It is a reasonable presumption that the South American faunas, and perhaps others from elsewhere along the margins of the Palaeo-Pacific, migrated to Europe to occupy ecological riches vacated after the Early Toarcian extinctions.

Table 2. *First appearances of bivalve genera in South America and Europe*

Genus	South America	Europe
Cucullaea	Pliensbachian	Bajocian
Lycettia	Pliensbachian	Upper Toarcian
Falcimytilus	Pliensbachian	Bajocian
Trichites	Pliensbachian	Bajocian
Pteroperna	Pliensbachian	Bathonian
Gervillaria	Pliensbachian	Upper Toarcian
Lopha	Pliensbachian	Aalenian
Myophorella	Pliensbachian	Upper Toarcian
?Neocrassinia ('Astarte' aureliae)	Pliensbachian	Upper Toarcian
Mesomiltha	Pliensbachian	Upper Toarcian

Discussion

Perhaps the most important question to be asked about the slow recovery after the end-Triassic mass extinction is: was it due primarily to a slow rate of evolution of survivors, or to unfavourable environmental conditions? The

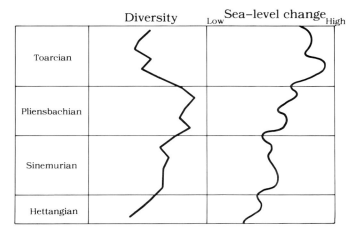

Fig. 4. Correlation of species diversity increase up the British Lower Jurassic succession, taken from Fig. 1, and eustatic sea-level rise, according to Hallam (1988).

evidence seems to favour the latter interpretation. Thus anoxic or dysoxic conditions were widespread at the bottom of the Early Hettangian sea in Britain (Wignall & Hallam 1991) and probably also over a much wider area in northwest Europe. Only gradually during the next few million years did these conditions ameliorate and allow the progressive influx of faunas from southern Europe, as particularly well exemplified by the brachiopods. There is a good correlation between the rise of species diversity during the Hettangian, Sinemurian and Pliensbachian and rise of sea-level (Fig. 4). The relatively rapid rise through the course of the Hettangian from a very low level at the start is likely to relate to increasing oxygenation of bottom waters. The replacement of *Liostrea* by *Gryphaea*, an immigrant from the Pacific or Arctic, in the late Hettangian could well be related to this. The migration into Britain in the Hettangian of foraminiferal and ostracod species that were already living in the Tethyan zone of Austria is associated with the marine transgression of that time, with normal-salinity waters replacing the low-salinity Rhaetian sea of northwest Europe.

The correlation of diversity with sea level breaks down in the Early Toarcian. A pulse of rapidly rising sea level was associated with the extensive spread of anoxic bottom waters, evidently a true oceanic anoxic event (Hallam 1987; Jenkyns 1988). This is the obvious cause of the well recorded mass extinction of benthic groups and explains the general fall in diversity. The subsequent recovery is evidently associated with amelioration of conditions and immigration from the eastern Pacific margins, presumably along the Hispanic Corridor in the central Atlantic region (Hallam 1983). This corridor is the likely result of tensional activity thinning continental crust prior to Middle Jurassic ocean opening, and continuing rise in sea level. No unequivocal Toarcian anoxic episode can be recognized in South America (Hallam 1986), which explains why many elements of the bivalve fauna, and no doubt much else, were able to survive and migrate to Europe in the Late Toarcian and Middle Jurassic. Both bivalve and brachiopod diversities continued to increase in the Middle Jurassic, when for bivalves they attained their Jurassic maximum (Almeras 1964; Hallam 1976) and reef habitats became re-established. These phenomena appear to be broadly related to an expansion of epicontinental seas associated with eustatic sea-level rise (Hallam 1994).

References

AGARD, J. & DU DRESNAY, R. 1964. La région minéralisée du J. Bon Dahar, près de Beni-Tajjite. *Notes et Memoires de la Service Géologique du Maroc*, **181**, 135–152.

AGER, D. V. 1956–1967. *The British Liassic Rhynchonellidae*. Monographs of the Palaeontographical Society.

—— 1990. *British Liassic Terebratulida (Brachiopoda). Part I*. Monographs of the Palaeontographical Society.

ALLASINAZ, A. 1992. The Late Triassic–Hettangian bivalves turnover in Lombardy (Southern Alps). *Rivista Italiana di Paleontologia e Stratigrafia*, **97**, 431–454.

ALMERAS, Y. 1964. Brachiopodes du Lias et du Dogger. *Documents des Laboratoires de Géologie de la Faculté des Sciences de Lyon*, **5**, 1–161.

BATE, R. H. & ROBINSON, E. 1978. *A stratigraphical index of British Ostracoda*. Seel House Press, Liverpool.

DAMBORENEA, S. E. 1987a. Early Jurassic Bivalvia of Argentian. Part I. Stratigraphical introduction and Superfamilies Nuculanacea, Mytilacea and Pinnacea. *Palaeontographica*, **A199**, 23–111.

—— 1987b. Early Jurassic Bivalvia of Argentina. Part II. Superfamilies Pteriacea, Buchiacea and part of Pectinacea. *Palaeontographica*, **A199**, 113–216.

DEAN, W. T., DONOVAN, D. T. & HOWARTH, M. K. 1961. The Liassic ammonite zones and subzones of the north-west European Province. *Bulletin of the British Museum (Natural History), Geology*, **4**, 438–505.

DULAI, A. 1992. The Early Sinemurian (Jurassic) brachiopod fauna of the Lokut Hill (Bakong Mts., Hungary). *Fragmenta Mineralogica et Palaeontologica*, **15**, 35–83.

—— 1993. Hettangian (Early Jurassic) megafauna and paleogeography of the Bakony Mts. (Hungary). *In:* PALFY, J. & VÖRÖS, A. (eds) *Mesozoic brachiopods of Alpine Europe*. Hungarian Geological Society, Budapest, 31–37.

EL-SHAARAWY, Z. 1982. *Foraminifera and Ostracoda of the topmost Triassic and basal Jurassic of England and Wales*. PhD Thesis, University of Birmingham.

GAETANI, M. 1970. Faune Hettangiane della parte orientale della Provincia di Bergamo. *Rivista Italiana di Paleontologia e Stratigrafia*, **76**, 355–442.

GOLEBIOWSKI, R. & BRAUNSTEIN, R. E. 1988. A Triassic–Jurassic boundary section in the Northern Calcareous Alps (Austria). *In: Excursion guide to the geologic sites of Rare Events in Geology*. IGCP Project 199. Bericht Geologisches Bundesanstalt, **15**, 39–46.

HALLAM, A. 1975a. *Jurassic environments*. Cambridge University Press.

—— 1975b. Coral patch reefs in the Bajocian (Middle Jurassic) of Lorraine. *Geological Magazine*, **112**, 383–392.

—— 1976. Stratigraphic distribution and ecology of European Jurassic bivalves. *Lethaia*, **9**, 245–259.

—— 1982. Patterns of speciation in Jurassic *Gryphaea*. *Paleobiology*, **8**, 354–366.

—— 1983. Early and mid Jurassic molluscan biogeography and the establishment of the Central Atlantic seaway. *Palaeogeography, Palaeoclimatology, Palaeoecology*, **43**, 181–193,

—— 1986. The Pliensbachian and Tithonian extinction events. *Nature*, **319**, 765–768.

—— 1987. Radiations and extinctions in relation to environmental change in the marine Lower Jurassic of north west Europe. *Paleobiology*, **13**, 152–168.

—— 1988. A re-evaluation of Jurassic eustasy in the light of new data and the revised Exxon curve. *In:* WILGUS, C. K., HASTINGS, B. S., ROSS, C. A., POSAMENTIER, H., VAN WAGONER, J. & KENDALL, C. G. St. C. (eds) *Sea-level changes: an integrated approach*. Society of Economic Paleontologists and Mineralogists, Special Publications, **42**, 261–273.

—— 1994. *An outline of Phanerozoic biogeography*. Oxford University Press.

—— & EL-SHAARAWY, Z. 1982. Salinity reduction of the end-Jurassic sea from the Alpine region into north western Europe. *Lethaia*, **15**, 169–178.

——, HANCOCK, J. M., LABRECQUE, J. L., LOWRIE, W. & CHANNELL, J. E. T. 1985. *In:* SNELLING, N. J. (ed.) *The chronology of the geological record*. Geological Society, London, Memoir, **10**, 118–140.

JENKINS, D. G. & MURRAY, J. W. 1989. *Stratigraphic atlas of fossil foraminifera*. Horwood, Chichester.

JENKYNS, H. C. 1988. The early Toarcian (Jurassic) anoxic event: stratigraphic, sedimentary, and geochemical evidence. *American Journal of Science*, **288**, 101–151.

MCROBERTS, C. A. 1992. Systematics and paleobiogeography of late Triassic *Gryphaea* (Bivalvia) from the North American Cordillera. *Journal of Paleontology*, **66**, 28–39.

RAUP, D. M. & SEPKOSKI, J. J. 1982. Mass extinctions in the marine fossil record. *Science*, **215**, 1501–1503.

RICCARDI, A. C., DAMBORENEA, S. E. & MANCEÑIDO, M. O. 1990. Lower Jurassic of South America and Antarctic Peninsula. *Newsletters in Stratigraphy*, **21**, 75–103.

SEPKOSKI, J. J. 1986. Phanerozoic overview of mass extinction. *In:* RAUP, D. M. & JABLONSKI, D. (eds) *Patterns and processes in the history of life*. Springer, Berlin, 277–295.

SIMMS, M. J. 1989. *British Lower Jurassic crinoids*. Monographs of the Palaeontographical Society.

STANLEY, G. D. 1988. The history of early Mesozoic reef communities, a three-step process. *Palaios*, **3**, 170–183.

SZENTE, I. 1992. Early Jurassic molluscs from the Mecsek Mountains (S. Hungary). A preliminary study. *Annales Universitatis Scientiarum Budapestiensis, Sectio Geologica*, **29**, 325–343.

VÖRÖS, A. 1993. Jurassic brachiopods of the Bakong Mountains (Hungary): global and local effects on changing diversity. *In:* PALFY, J. & VÖRÖS, A. (eds) *Mesozoic brachiopods of Alpine Europe*. Hungarian Geological Society, Budapest, 179–189.

WIGNALL, P. B. & HALLAM, A. 1991. Biofacies, stratigraphic distribution and depositional models of British onshore Jurassic black shales. *In:* TYSON, R. V. & PEARSON, T. H. (eds) *Modern and ancient continental shelf anoxia*. Geological Society, London, Special Publication, **58**, 291–309.

Foraminiferal recovery after the mid-Cretaceous oceanic anoxic events (OAEs) in the Cauvery Basin, southeast India

A. TEWARI, M. B. HART & M. P. WATKINSON

Department of Geological Sciences, University of Plymouth, Drake Circus, Plymouth PL4 8AA, UK

Abstract: The Cauvery Basin, SE India, is one of the best exposed late Mesozoic to Tertiary basins in India. The study of foraminiferal assemblages from the core samples obtained from two 120 m deep wells in the basin records the occurrence of two mid-Cretaceous anoxic events in the basin. Abrupt increases in planktonic:benthonic ratios and reductions in benthonic diversity are recorded in the late Albian and in the late Cenomanian–early Turonian. These events coincide with worldwide oceanic anoxic events (OAEs).

The OAEs had an impact on the microfauna of the basin, with approximately 35–45% of the benthonic species not surviving and major morphological changes occurring in the planktonic community. The late Cenomanian–early Turonian event was more significant than the late Albian event, with considerable readjustments occurring in the planktonic foraminifera. The genus *Rotalipora* disappeared and the genus *Hedbergella* was largely replaced by *Whiteinella*. *Marginotruncana* appeared for the first time and the *Dicarinella* population expanded considerably. *Praeglobotruncana* was the most tolerant genus, undergoing least change. The planktonic foraminifera evolved from small, weakly ornamented forms with poorly developed keels, into robust, well ornamented forms with well developed keels. The pattern of evolution of planktonic foraminifera suggests a recolonization of deeper water environments after the late Cenomanian–early Turonian anoxic event.

The break up of East Gondawanaland (India, Antarctica and Australia) in the late Mesozoic was responsible for the development of the Indian Ocean (Powell *et al*. 1988; Moores 1991). Major changes in the plate configuration took place which influenced the Cretaceous palaeogeography of the Indian sub-continent. New sedimentary basins were developed and two episodes of widespread flood-basalt extrusions occurred (Acharyya & Lahiri 1991). The Cretaceous basins of the east coast of India cut across the NW-trending Permian–Triassic Gondawana grabens (Sastri *et al*. 1981). They are often precursors of the Tertiary basins and present-day shelves. Others were intra-cratonic basins, occurring as narrow troughs which were later flooded by the Deccan basalts. Figure 1 shows the distribution of the Cretaceous rocks in the Indian sub-continent, together with the location of the Cauvery Basin.

The Cauvery Basin, SE India, is one of the best exposed late Mesozoic–Tertiary basins in India. It covers an area of about 25 000 km² on land, between latitude 12°15′N near Pondicherry and latitude 8°15′N to the south of Ramanathapuram (Fig. 1). The depth of the crystalline basement does not exceed more than 4.5 km in any part of the basin on land, but may reach nearly 6 km offshore. The basin is bounded to the west by the outcrops of the igneous and metamorphic rocks of the Indian Archaean shield. North of the latitude 10°N, the basin opens into the present offshore area of the Bay of Bengal. To the south, the basin is bounded to the east and southeast by the Archaean crystalline massifs of Sri Lanka. The basin is block-faulted, comprising horsts and grabens. It is a peri-cratonic shelf in the north and an intra-cratonic graben in the south (Sastri *et al*. 1973, 1981; Banerji 1983). Geophysical surveys and drilling by the Oil and Natural Gas Corporation of India have established that the basin comprises several depressions or sub-basins separated from each other by buried basement ridges (Ramanathan 1968; Sastri *et al*. 1977). The Ariyalur–Pondicherry depression is the largest and northern-most depression of the basin and contains three important outcrops, at Pondicherry, Vrindhanchalam and Ariyalur (Sastri *et al*. 1973, 1981; Banerji 1983).

In India, early Cretaceous organic rich black shales were first described from the Ariyalur–Pondicherry sub-basin by Subbaraman (1968). Later Bhatia & Jain (1969) reported similar Aptian–Albian black shales from the offshore parts of the Cauvery Basin. Govindan (1982,

From Hart, M. B. (ed.), 1996, *Biotic Recovery from Mass Extinction Events*, Geological Society Special Publication No. 102, pp. 237–244

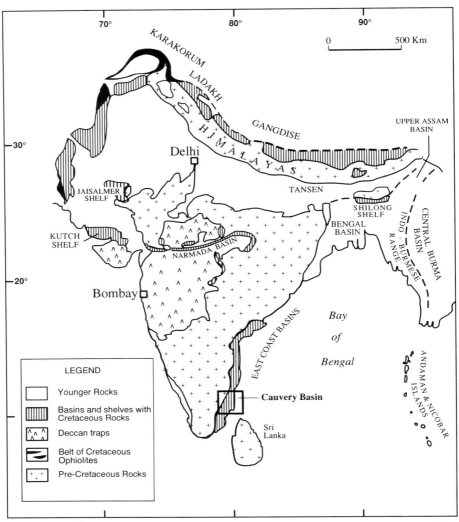

Fig. 1. Distribution of Cretaceous rocks on the Indian sub-continent, Cauvery basin in the enclosed box (after Acharyya & Lahiri 1991).

1993), for the first time, correlated the occurrence of these organic rich black shales in the Cauvery Basin with the global 'Cretaceous Anoxic Events'.

In the Cretaceous system, three intervals are reported as Oceanic Anoxic Events (Arthur & Schlanger 1979; Jenkyns 1980). These are (i) Late Barremian through Albian, (ii) Late Cenomanian to early Turonian, and (iii) Coniacian to Santonian. These Cretaceous OAEs record worldwide marine transgressions, when large amounts of organic carbon were accumulated because of the development of poorly oxygenated oceanic water masses and expanded oxygen minimum zones (Arthur & Schlanger 1979; Jenkyns 1980). Hart (1980a), Hart & Ball (1986) and Jarvis et al. (1988) suggest that the anoxic events had a major effect on the planktonic foraminifera and that the movement of the oxygen minimum zone in the water column has affected the evolution of the planktonic foraminiferal population.

This paper concentrates on the core samples obtained from two 120 m deep wells, Karai-4 and Karai-6, drilled between Karai and Kulakkalnattam villages, in the Ariyalur area (Fig. 2). The wells encounter grey–black shales throughout, with brown gypseous shales in the top 30 m from the surface. The gypsum is a post-depositional weathering phenomenon. Each sample

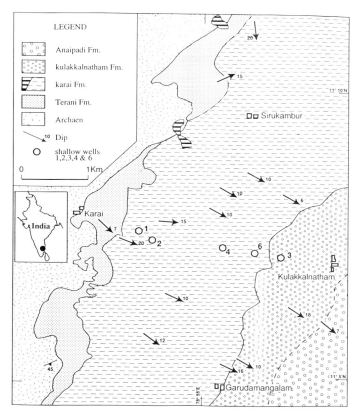

Fig. 2. Geological map of the Karai–Kulakkalnattam area showing positions of the Karai-4 and Karai-6 wells (after Venkataraman & Rangaraju 1965 in Ramanathan & Rao 1982).

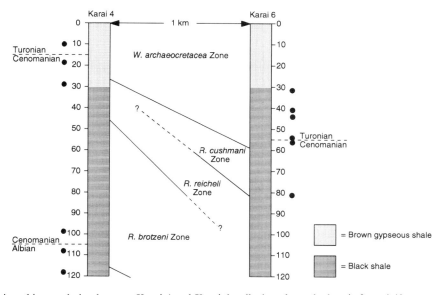

Fig. 3. Biostratigraphic correlation between Karai-4 and Karai-6 wells, based on planktonic foraminifera.

was divided into 500, 250, 125 and 63 micron grain size-fractions, and from each fraction size 301 complete foraminifera were picked. Where the 301 count was not met, the whole fraction was used. Foraminiferal assemblages of the *R. appenninica (Renz)* Interval Zone to *W. archaeocretacea* Pessagno Partial Range Zone are described from the two wells (Fig. 3). This study shows that the late Albian and the late Cenomanian–early Turonian, OAEs are recorded in the basin.

The Late Albian event

At a depth of about 100–110 m in the Karai-4 well, an abrupt increase occurs in the planktonic:benthonic ratio and a fall occurs in the benthonic diversity in the late Albian, within the *R. brotzeni* Interval Zone (Fig. 4). The Albian–Cenomanian boundary is placed at the first appearance of *Tritaxia pyramidata* Reuss 1862 in the Karai-4 well. A revised base of the Cenomanian is proposed just above the extinction of *Planomalina buxtorfi* (Gandolfi 1942) (Kennedy *et al.* 1995 pers. comm.). The event saw major morphological development amongst the *Rotalipora* Brotzen 1942 and *Praeglobotruncana* Bermudez 1952 (Fig. 5). During the event, the *Rotalipora* comprised small forms, each with a weakly developed keel. Soon after, the *Rotalipora* established themselves and expanded considerably, before disappearing in the late Cenomanian, when *Dicarinella* Porthault 1970 and *Whiteinella* Pessagno 1967 made their first appearance. Following the appearance of juvenile specimens of *R. appenninica* (Renz 1936) and *R. brotzeni* (Sigal 1948) during the anoxic event, well developed forms belonging to the species *appenninica, brotzeni, micheli* (Sacal & Debourle 1957), *montsalvensis* (Mornod 1950) and *reicheli* (Mornod 1950) appeared with the start of the recovery. The recovery of the *Rotalipora* saw a development in their morphological characters and an improvement in the quality of their preservation. Forms appear that possess well preserved primary and secondary apertures and a chamber profile marked by a well developed single keel continuing along the spiral side. Species of *R. reicheli* and *R. micheli* show well developed and very prominent adumbilical thickening. In fully grown adults of *R. reicheli* this periumbilical ridge is so well developed on all chambers that it forms a rampart along the umbilicus. The population was dominated by strongly asymmetrical *R. reicheli*, with a flat to gently concave spiral side and a strongly vaulted umbilical side. By the late Cenomanian, the *Rotalipora* started to reduce

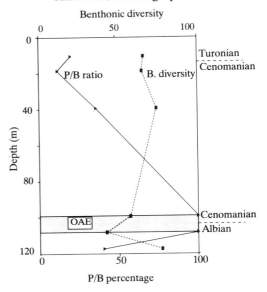

Fig. 4. Planktonic–benthonic ratio and benthonic diversity graphs showing the position of the late Albian–early Cenomanian OAE in the Karai-4 well.

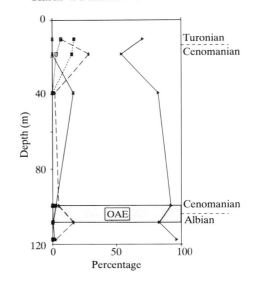

Fig. 5. Planktonic foraminifera percentage graph in the Karai-4 well, before and after the late Albian–early Cenomanian OAE.

in number before disappearing at the end of Cenomanian, when juvenile forms of *Dicarinella* and *Whiteinella* made their first appearance.

Praeglobotruncana appears to be the most tolerant genus during the event. Forms of *P. stephani* (Gandolfi 1942) and *P. gibba* Klaus 1960 appeared from before the crisis, survived the event and continued up. A major change of the ornamentation and spire height occurs, however, as the *Praeglobotruncana* population recovered from the event. *P. stephani* and *P. gibba* began with forms that have a low to moderate spire height and weakly calcified tests changing to a gentle beaded nature. These *stephani* and *gibba* have poorly preserved umbilical structures with chamber outlines marked by two very faint rows of pustules which becomes fainter and finally disappear completely. From a stock of such weakly developed morphological characters there evolved robust, highly calcified and beaded *stephani* and *gibba* with moderately high to very high trochospiral forms. The European Working Group on Planktonic Foraminifera (Robaszynski & Caron 1979) established that, with an increase in height of the spire, *stephani* evolves into *gibba*. The *stephani* and *gibba* in the Cauvery Basin follow the same *stephani* to *gibba* transformation rule, as seen in their European counterparts. There is, however, one exception. The forms reported from Europe (Robaszynski & Caron 1979) show a gradual increase in the height of the low to moderately high *stephani*, resulting in a mound-shaped high trochospiral *gibba*. In the Cauvery Basin examples, the increase in the height of the spire sees a more robust approach with inner whorls sitting like a lobe on the final whorl, thus giving a block-shaped high trochospiral *gibba*. These highly developed *gibba* and *stephani* also show very highly developed beaded ornamentation on their spiral sides, and very well developed rows of pustules. In some cases these rows of pustules could be referred to as two closely spaced keels in the final chambers of the last whorl.

In Alpine Europe the Albian records a number of distinctive anoxic/dysaerobic 'events'. In the Vocontian Trough, Breheret *et al.* (1986) record the foraminiferal changes across the Jacob and Paquier events but although they record the presence of dark, organic rich sediments in the lower Vraconian (uppermost Albian) they do not record the fauna from these horizons. This level is probably that known in other areas of Alpine Europe as the 'Briestroffer event'.

In the Sergipe Basin (Koutsoukos *et al.* 1991), organic-rich sediments are also well known from the Aptian–Albian interval, forming the primary source rock for the area. High total organic carbon (TOC) values are recorded in the uppermost Aptian (?Jacob) and the lowermost Albian (?Paquier) and these are, on various criteria, the maximum development of dysaerobic–anoxic conditions. The event is characterized by total absence of benthonic microfauna or very impoverished low-diversity assemblages of calcareous and agglutinated foraminiferas. At the end of the Albian, the southwestern region of the Sergipe Basin experienced shallowing of the water column, probably due to local uplift of the basement. Waning dysaerobic to oxic conditions were established which explains why the late Albian Breistroffer event is not recorded from the Sergipe Basin.

This study shows that the event in the Cauvery Basin appears to be much closer to the Albian–Cenomanian boundary being within, and at the end of, the *Planomalina buxtorfi* Zone. Some oxic conditions were present at this time. A low-diversity and impoverished benthonic population is present which suggests a waning dysaerobic–oxic condition was present in the basin, very similar to that of the late Albian in the Sergipe Basin. The late Albian anoxic event in the Cauvery Basin is not as strong as the late Cenomanian–early Turonian event. In the Cauvery Basin Govindan (1993) records two prominent peaks of TOC in Bhuvanagiri well-1 in the Middle–Late Albian (detailed zonation of well not published).

The Late Cenomanian–Early Turonian event

At a depth of 52–55 m in Karai-6 well, there is an increase in the planktonic : benthonic ratio and a fall in the benthonic diversity in the late Cenomanian–early Turonian, within the *W. archaeocretacea* Partial Range Zone (Fig. 6). The Cenomanian–Turonian boundary is placed at the last occurrences of *Orithostella indica* Scheibnerova 1972 subsp. *markisi* Narayanan 1975 and *Gavelinella baltica* Brotzen 1942 of late Cenomanian age, and at the first occurrence of *Lingulogavelinella turonica* Butt 1966 of early Turonian age. This event coincides with the worldwide Cenomanian–Turonian anoxic event which caused a major readjustment within the planktonic community (Fig. 7). *Rotalipora* disappeared completely and *Dicarinella*, *Whiteinella* and *Marginotruncana* Hofker 1956 came into existence. *Hedbergella* Bronnimann & Brown 1958, which was dominant before the event, is largely replaced by *Whiteinella*. Well preserved *W. aprica* (Loeblich & Tappan 1961), *W. archaeocretacea* Pessagno 1967, *W. baltica*

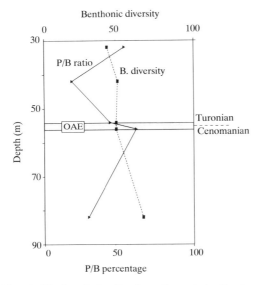

Fig. 6. Planktonic–benthonic ratio and benthonic diversity graphs showing the position of the late Cenomanian–early Turonian OAE in the Karai-6 well.

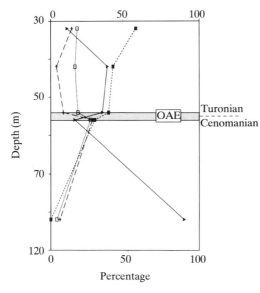

Fig. 7. Planktonic foraminifera percentage graph in the Karai-6 well, before and after the late Cenomanian–early Turonian OAE.

Douglas & Rankin 1969, and *W. brittonensis* (Loeblich & Tappan 1961) are rare before the event, but expanded considerably. *Praeglobotruncana* follows the same pattern of recovery as they did after the late Albian anoxic event, with a further enhancement in the spire height and beaded ornamentation.

The first *Dicarinella* to appear were juvenile *D. imbricata* (Mornod 1950) with low trochospiral forms and chamber outlines marked by a weak concentration of pustules. Soon *Dicarinella*, especially *D. hagni* (Scheibnerova 1962), dominated the population, along with *Praeglobotruncana* and *Whiteinella*. *D. hagni* reported from the Karai-6 well are different to those that have been reported as *D. hagni* from Europe. The European Working Group on Planktonic Foraminifera (Robaszynski & Caron 1979) report low trochospiral asymmetrical biconvex forms of *D. hagni* in Europe, differentiating them from the plano-convex '*concavata*' group. Under the '*concavata*' group (Robaszynski & Caron 1979) are included forms with clearly asymmetrical planoconvex to concavo-convex low trochospiral tests, a periumbilical ridge and early globigerina-like chambers elevated above the outer whorl as a little prominence in the centre of the test. The three forms belonging to the '*concavata*' group (*D. primitiva* (Dalbiez 1955), *D. concavata* (Brotzen 1934), and *D. asymetrica* (Sigal 1952)) occur one after the other in the Coniacian–Santonian interval. The first *hagni* which appear in the Karai-6 well have all the characters of a true *D. hagni* as described from Europe (Robaszynski & Caron 1979). They have a low trochospiral-asymmetrical biconvex test with two closely spaced keels, often disappearing on the last or penultimate chamber. Soon after the OAE, *D. hagni* shows major morphological changes with a tendency for the development of forms very similar to the '*concavata*' group. From the typical *D. hagni* develop morphotypes having a moderate–asymmetrical trochospiral test with a more convex spiral side in comparison to the typical *hagni*, although the umbilical side still is more convex. These morphotypes have less closely spaced double keels with the spiral keel diverging from the umbilical keel to give a slight imbricated pattern on the dorsal side. At the same time, within the *hagni* population, appeared planoconvex to concavo-convex morphotypes with the inner whorls sitting like a knob in the final whorl. Both these morphotypes show a development of periumbilical ridge at least in the final one or two chambers. The full recovery of *D. hagni* saw well developed planoconvex to concavo-convex *hagni* with spiral

MID-CRETACEOUS OAEs, CAUVERY BASIN

side flat to gently concave and a periumbilical ridge in the last one or two chambers. The only ways these plano-convex *hagni* are different from the 'concavata' group are that they occur with a foraminiferal assemblage belonging to *W. archaeocretacea* Zone, of late Cenomanian–early Turonian age and they do not have early globigerina-shaped chambers.

Jacob & Sastry (1950) report a new species, *Globotruncana indica*, along with a Cenomanian fauna, from the Cauvery Basin, having a plano-convex test with two keels. The European Working Group on Planktonic Foraminifera (Robaszynski & Caron 1979) consider *G. indica* to show a morphological similarity with *D. hagni*. In the absence of topotypes or any duplicate material available for the study during the preparation of the Atlas of Mid-Cretaceous Planktonic Foraminifera (Robaszynski & Caron 1979), the working group could not fully clarify the taxonomic status of *G. indica*. *D. hagni* was therefore retained as a more adequate taxon to identify the European forms. Lamolda (1977) describes a new plano-convex species, *D. elata*, from northern Spain, ranging from the late Cenomanian(?)–early Turonian to the early Coniacian. Lamolda differentiates *D. elata* from *G. indica* by the presence of a single keel instead of a double keel on the outer whorl of *elata*. The plano-convex *hagni* in the Karai-6 well are different from *D. elata* because they have a periumbilical ridge in the final one or two chambers and two well developed keels. In the absence of clear taxonomic status for *G. indica* and *D. elata*, *D. hagni* is used here as the valid name, following the European Working Group (Robaszynski & Caron 1979).

In the genus *Marginotruncana*, juveniles of *M. marginata* (Reuss 1845) were the first to appear. The full expansion of the *Marginotruncana* population was dominated by *M. schneegansi* (Sigal 1952). The *M. schneegansi* in the Karai-6 well includes forms with a very high trochospiral test. These are different from the *M. schneegansi* reported from Europe (Robaszynski & Caron 1979), which have a low trochospiral test.

Conclusions

It has been suggested (Hart & Bailey 1979; Hart 1980*b*) that the planktonic : benthonic ratio and the evolution of the planktonic foraminifera can be used, with caution, to indicate major changes in water depth. The late Albian and the late Cenomanian–early Turonian events in the Cauvery Basin are represented by an increase in the planktonic : benthonic ratio, with a fall by approximately 35–45% of the benthonic popu-

lation. This coincides with the sea-level maxima and eustatic rises in sea-levels, which caused the development of an expanded oxygen minimum zone. The planktonic foraminiferal growth pattern suggests a late Albian and a late Cenomanian–early Turonian sea-level maxima with intermittent periods of fluctuating sea-level in the basin. The recovery of the fauna from the late Albian OAE was dominated by shallow water *Hedbergella* and intermediate water *Rotalipora* and *Praeglobotruncana*. By the mid-Cenomanian, relatively deeper-water *R. reicheli* and *R. micheli* were established and expanded considerably, but disappeared completely at the end of the Cenomanian. At the Cenomanian–Turonian boundary, there was a second influx of deeper water fauna. The development of high spired and highly calcified *Praeglobotruncana* and *Marginotruncana* and of plano-convex *Dicarinella* were a significant development towards the return of deeper water fauna after the Cenomanian–Turonian event.

The study of foraminiferal assemblage in the Cauvery Basin suggests the following:

1. The mid-Cretaceous anoxic events OAE-1 and OAE-2, which may have resulted from world wide marine transgressions, are also present in the Cauvery Basin.
2. The Late Albian and Cenomanian–Turonian boundary intervals were the times of sea-level maxima, with intermittent periods of fluctuating sea-level.
3. These anoxic events are responsible for major changes in the planktonic foraminiferal population.
4. The planktonic foraminifera indicate that the late Cenomanian–early Turonian OAE was followed by recolonization of deeper water environments.
5. A comparison of the planktonic fauna from the Cauvery Basin with other basins in the world gives an impression that the fauna from SE India is composed of more robust individuals. After these mid-Cretaceous OAEs, a more ornamented and more prominently keeled fauna dominated the Cauvery Basin.

We thank D. J. van der Zwaan for kindly providing the Karai well samples. David Peacock is thanked for his help in the preparation of Figs 3–7 and for reading the manuscript. John Abraham is thanked for drafting Figs 1 and 2.

References

ACHARYYA, S. K. & LAHIRI, T. C. 1991. Cretaceous palaeogeography of the Indian subcontinent; a review. *Cretaceous Research*, **12**, 3–26.

ARTHUR, M. A. & SCHLANGER, S. O. 1979. Cretaceous 'oceanic anoxic events' as causal factors in development of reef-reservoired giant oil fields. *AAPG Bulletin*, **63**, 870–885.

BANERJI, R. K. 1983. Evolution of the Cauvery Basin during Cretaceous. *In:* MAHESHWARI, H. K. (ed.) *Cretaceous of India. Proceedings of the symposium on 'Cretaceous of India: palaeoecology, palaeogeography and time boundaries'.* Lucknow, India, 22–39.

BHATIA, S. B. & JAIN, S. P. 1969. Dalmiapuram formation: A new lower Cretaceous horizon in South India. *Indian Geological Association Bulletin*, **2**, 105–108.

BREHERT, J. G., CARRON, M. & DELAMETTE, M. 1986. Niveaux Riches en Matière Organique dans L'Albien Vocontien; Quelques Caractères du Paleoenvironnement; Essai d'interpretation Genetique. *Document de la Bureau recherches geologie et minieres*, **110**, 141–191.

GOVINDAN, A. 1982. Imprints of Global 'Cretaceous Anoxic Events' in East Coast Basins of India and their implications. *Oil and Natural Gas Commission Bulletin*, **19**, 257–270.

—— 1993. Cretaceous Anoxic Events, Sea Level Changes and Microfauna in Cauvery Basin, India. *In:* BISWAS, S. K. *et al.* (eds) *Proceedings of Second Seminar on Petroliferous Basins of India*, **1**, 161–176.

HART, M. B. 1980*a*. A water depth model for the evolution of the planktonic Foraminiferida. *Nature*, **286**, 252–254.

—— 1980*b*. The recognition of mid-Cretaceous sea-level changes by means of Foraminifera. *Cretaceous Research*, **1**, 289–297.

—— & BAILEY, H. W. 1979. The distribution of planktonic Foraminiferida in the mid-Cretaceous of N.W. Europe. *Aspekte der Kreide Europas*, IUGS Series, **A6**, 527–542.

—— & BALL, K. C. 1986. Late Cretaceous anoxic events, sea-level changes and the evolution of the planktonic foraminifera. *In:* SUMMERHAYES, C. P. & SHACKLETON, N. J. (eds) *North Atlantic Palaeoceanography*. Geological Society, London, Special Publication, **21**, 67–78.

JACOB, K. & SASTRY, M. V. A. 1950. On the occurrence of *Globotruncana* in Uttatur stage of the trichinopoly Cretaceous, South India. *Current Science*, **16**, 266–268.

JARVIS, I., CARSON, G. A., COOPER, M. K. E., HART, M. B., LEARY, P. N., TOCHER, B. A., HORNE, D. & ROSENFELD, A. 1988. Microfossil Assemblages and the Cenomanian–Turonian (late Cretaceous) Oceanic Anoxic Event. *Cretaceous Research*, **9**, 3–103.

JENKYNS, H. C. 1980. Cretaceous anoxic events: from continents to oceans. *Journal of the Geological Society, London*, **137**, 171–188.

KOUTSOUKOS, E. A. M., MELLO, M. R., DE AZAMBUJA FILHO, N. C., HART, M. B. & MAXWELL, J. R. 1991. The Upper Aptian–Albian Succession of the Sergipe Basin, Brazil: An Integrated Paleoenvironmental Assessment. *AAPG Bulletin*, **35**, 479–497.

LAMOLDA, M. A. 1977. Three new species of planktonic foraminifera from the Turonian of northern Spain. *Micropaleontology*, **23**, 470–477.

MOORES, E. M. 1991. Southwest U.S.–East Antarctic (SWEAT) connection: A hypothesis. *Geology*, **19**, 425–428.

NARAYANAN, B. & SCHEIBNEROVA, V. 1975. Lingulogavelinella and Orithostella (Foraminifera) from the Uttatur Group of the Trichinopoly Cretaceous, South (Peninsular) India. *Revista Espanola De Micropaleontologia*, **7**, 25–36.

POWELL, C. McA., ROOTS, S. R. & VEEVERS, J. J. 1988. Pre-breakup continental extension in East Gondawanaland and the early opening of the eastern Indian Ocean. *Tectonophysics*, **155**, 261–283.

RAMANATHAN, Rm. & RAO, V. R. 1982. Lower Cretaceous foraminifera from the subsurface sediments of 'Kallakkudi embayment', Cauvery Basin, India. *Oil and Natural Gas Commission Bulletin*, **19**, 39–80.

RAMANATHAN, S. 1968. *Stratigraphy of Cauvery Basin with respect to its oil prospects.* Memoir Geological Society India, **2**, 153–167.

ROBASZYNSKI, F. & CARON, M. 1979. *Atlas of Mid-Cretaceous planktonic foraminifera (Boreal Sea and Tethys)*. Cahiers de Micropaleontologie, Editions du Centre National de la Recherche Scientifique, 2 parts.

SASTRI, V. V., RAJU, A. T. R., SINHA, R. N., VENKATACHALA, B. S. & BANERJI, R. K. 1977. Biostratigraphy and Evolution of the Cauvery Basin, India. *Journal of the Geological Society of India*, **18**, 355–377.

——, SINHA, R. N., GURCHARAN SINGH & MURTI, V. S. 1973. Stratigraphy and Tectonics of Sedimentary Basins on East Coast of Peninsular India. *AAPG Bulletin*, **57**, 655–679.

——, VENKATACHALA, B. S. & NARAYANAN, V. 1981. The evolution of the East Coast of India. *Palaeogeography, Palaeoclimatology, Palaeoecology*, **36**, 23–54.

SUBBARAMAN, J. V. 1968. Surface and subsurface geology of the area round Dalmiapuram, Trichinopoly District. *In: Cretaceous–Tertiary formations of South India.* Memoir Geological Society of India (for 1966), **2**, 92–98.

Benthonic foraminiferal mass extinction and survival assemblages from the Cenomanian–Turonian Boundary Event in the Menoyo section, northern Spain

DANUTA PERYT[1] & MARCOS LAMOLDA[2]

[1] *Institute of Paleobiology, Polish Academy of Sciences, Al. Zwirki i Wigury 93, 02-089 Warsaw, Poland*
[2] *Facultad de Ciencias, Universidad del Pais Vasco, Apdo 644, 48080 Bilbao, Spain*

Abstract: Foraminiferal response to the Cenomanian–Turonian Boundary Event was studied from a 50 m thick section in Menoyo, northern Spain, representing the uppermost *Rotalipora cushmani* and *Whiteinella archaeocretacea* Zones. Taxonomic and stratigraphic studies on benthonic foraminiferal assemblages indicate that the studied section represents mass extinction and survival intervals, with the mass extinction boundary in the lowermost part of the *Whiteinella archaeocretacea* Zone.
 Stepped extinction within benthonic foraminifers was observed in the uppermost *Rotalipora cushmani* Zone. In the late phase of mass extinction several species became extinct (e.g. *Gavelinella intermedia–cenomanica–baltica* group, *Tritaxia pyramidata* (Reuss)), some others temporarily disappeared; Lazarus taxa (e.g. *Tritaxia tricarinata* (Reuss), nodosariids) and progenitor (e.g. *Globorotalites* sp. 1) taxa appeared. Disaster (e.g. *Praebulimina elata*) species along with opportunistic taxa (e.g. *Gyroidinoides praestans* (Magniez-Jannin), *Ammobaculites parvispira* Ten Dam) colonized vacated ecospace in the middle part of the survival interval, i.e. in the topmost part of the *Rotalipora cushmani* Zone. Opportunistic taxa dominated assemblages in the *Whiteinella archaeocretacea* Zone, i.e. in the higher part of the survival interval.
 The recorded changes in benthonic foraminiferal assemblages most likely reflect the decline in oxygenation level of the bottom waters at the end of the *Rotalipora cushmani* Zone and the persistence of these unfavourable conditions in the *Whiteinella archaeocretacea* Zone.

A major palaeoceanographic event often referred to as the Oceanic Anoxic Event II (OAE II) or the Cenomanian–Turonian Boundary Event (CTBE) took place in the latest Cenomanian–earliest Turonian (Schlanger & Jenkyns 1976; Arthur *et al.* 1987). CTBE involved among others a significant faunal turnover (Raup & Sepkoski 1984). Raup & Sepkoski (1984) esti-

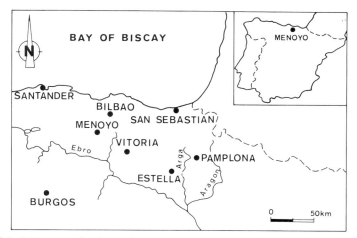

Fig. 1. Location of the Menoyo section in northern Spain.

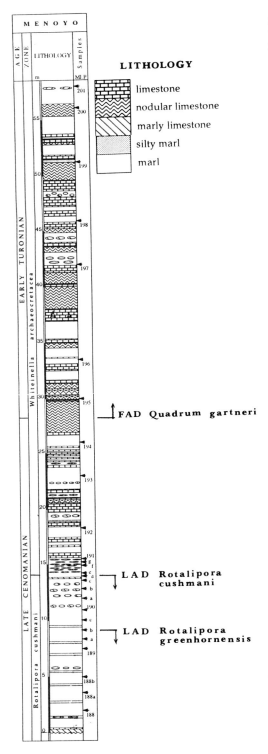

mated that 70–85% of species did not survive the C–T boundary. The C–T mass extinction is recognized as stepped mass extinction (Koch 1980; Kauffman 1984, 1986, 1988). According to this theory mass extinction episodes occur as series of steps spread over 1–4 Ma, associated with abrupt major marine palaebiogeographical changes and environmental perturbations of unusual magnitude and character. Patterns of survival and recovery from the Cenomanian–Turonian mass extinction for different groups have been proposed already (e.g. Harries & Kauffman 1990; Koutsoukos et al. 1990; Kauffman & Harries 1993).

The aim of this work is to trace benthonic foraminiferal changes across the Cenomanian–Turonian boundary in the Menoyo section, N Spain, in order to estimate the response of deeper-water benthonic foraminifera to the CTBE and to establish a pattern of recovery of this group from the C–T mass extinction.

The studied section (Fig. 1) is located in N Spain. Palaeogeographically it belongs to the Navarro–Cantabrian line, which corresponded to an outer platform/bathyal setting.

During the Cenomanian and Turonian the platform subsided significantly so that more than 2000 m of sediment accumulated in the area of study (Lamolda 1978; Lamolda et al. 1987). The Menoyo section represents probably one of the most expanded Cenomanian–Turonian boundary intervals. It is about four times thicker than that at Eastbourne (S England), five times that at Pueblo Colorado (USA), and seven times that at Dover (cf. Gale et al. 1993). The section at Menoyo offers therefore a particularly complete representation of the highest Cenomanian and lowest Turonian succession and an uninterrupted sequence across the Cenomanian–Turonian boundary in carbonate facies.

The studied interval is composed generally of two lithological complexes: lower (marly/silty) and upper (calcareous) (Fig. 2). At the bottom there is a 20 m-thick sequence of siltstones with marly limestone intercalations. The sequence is overlain by a 35 m-thick calcareous series composed mainly of limestones and nodular limestones interbedded with thin marlstones and siltstones. Paul et al. (1994) have shown that limestone and marl sediments in the Menoyo section represent Milankovitch rhythms. The

Fig. 2. Lithostratigraphy and bioevents in the Cenomanian–Turonian boundary interval in the Menoyo section.

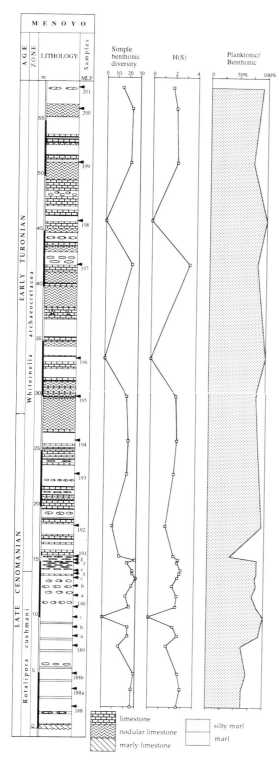

rhythms manifest themselves there as thick marly intervals, sometimes with scattered nodular limestones, which pass up into increasingly crowded thin limestone beds, or more rarely into a thick limestone (Paul *et al.* 1994). In the studied interval the bioturbation was low; three thin marly beds were laminated and without any signs of bioturbations (samples MLP 189c, MLP 196, MLP 198).

Material and methods

Twenty six samples from a 55 m-thick sequence in the Menoyo section were studied. 400–500 specimens from the >100 microns size fraction were picked up and all species in each sample were identified. P/B ratio (relative abundances of planktonic and benthonic foraminifers in the assemblages) and relative abundances on benthonic species within assemblages were calculated and H(S), the Shannon–Weaver heterogeneity index of benthonic foraminiferal assemblages, was computed. The results are presented on Fig. 3.

Palaeoenvironment

The Cenomanian–Turonian boundary interval in the Menoyo section yields quite abundant foraminifera, both planktonic and benthonic (Fig. 3) Planktonic forms are recorded in all studied samples, benthonic ones in all but three (MLP 189c, MLP 196, MLP 198).

The number of benthonic foraminiferal species is moderately high. The most diverse benthonic foraminiferal fauna is recorded in the topmost part of the *Rotalipora cushmani* Zone. In that interval the number of benthonic species varies between 20 and 28. In the *Whiteinella archaeocretacea* Zone diversity of benthonic foraminiferal fauna is moderate and in most of samples is comprised between 19 and 22.

H(S), the Shannon–Weaver heterogeneity index, is generally high: >2.0, with maximum value 2.8. Values of H(S) between 2.0 and 2.8, equivalent to those found in the studied section, are prominent on the Recent continental shelf off the eastern United States at depths of 100–2000 m (Buzas & Gibson 1969).

Fig. 3. Results of quantitation analyses of benthonic foraminiferal assemblages in the Cenomanian–Turonian boundary interval in the Menoyo section: simple benthonic diversity; H(S), the Shannon–Weaver heterogeneity index of benthonic foraminiferal assemblages; P/B ratio.

P/B ratio values vary from 60–100%. The lowest values (60–70%) were recorded in the lower part of the section, within a brief interval shortly below the last appearance of planktonic foraminifer *Rotalipora greenhornensis* (Morrow); in the upper part of the succession P/B ratio is high and in most samples is higher than 80%.

P/B ratio indicates outer shelf/upper bathyal for the Menoyo section (cf. Murray 1991).

Depth distribution of recorded benthonic foraminiferal genera (Fig. 4) (cf. Murray 1991) combined with quantitative data on studied fauna indicate an outer shelf to upper bathyal water depth during latest Cenomanian/early Turonian time at the Menoyo section. Similar conclusions were obtained by Colin *et al.* (1982) on the basis of ostracods.

Intervals in which benthonic foraminifera were not recorded, representing black, laminated marls, are interpreted as accumulated under anoxic bottom conditions.

Biostratigraphy and foraminiferal analyses

The studied interval represents the Cenomanian–Turonian boundary interval and comprises the upper part of the *Rotalipora cushmani* and the *Whiteinella archaeocretacea* planktonic foraminiferal Zones. The boundary between the Cenomanian and Turonian is placed at the first appearance of the coccolith species *Quadrum gartneri* Prins and Perch-Nielsen (Birkelund *et al.* 1984; Gorostidi 1993) within the *Whiteinella archaeocretacea* Zone (Fig. 2).

Benthonic foraminiferal turnover

Figure 4 shows benthonic species ranges in the Cenomanian–Turonian boundary interval and Fig. 5 abundance fluctuations of dominant species. Figures 6 and 7 illustrate dominant species.

A measure of relative success of species within the total foraminiferal population is its abundance fluctuations in time. There is a direct relationship between the abundance of species within community and environment. Abundance fluctuations of benthonic foraminifera are, therefore, sensitive palaeoceanographic indicators responding to changing palaeotemperature, salinity, nutrient and oxygen conditions. Extinction of several rare species will have a minor impact on the total population as their combined species abundances are not likely to exceed 2–10%. Extinction of one or more of the dominant species, however, will have a major impact as they may comprise 50% or more of the total population. The total number of benthonic species in the studied interval is 74 (Fig. 4), however, the number of species which contributes significantly to the total foraminiferal population is only 15 (Fig. 5).

The major faunal turnover occurred in Late Cenomanian, in the late *Rotalipora cushmani* chron. In the lower part of the section accelerated stepwise disappearance within benthonic foraminiferal fauna is clearly visible. In the topmost part of the *Rotalipora cushmani* Zone 33 species disappeared; some of them became extinct, others disappeared only temporarily. In the same interval one can observe the stepped appearance of many species. Altogether in this interval 42 species appeared.

Within dominant species very rapid disappearance of *Tritaxia* spp. is recorded in the lowermost part of the studied interval, followed by a brief interval of domination of *Gubkinella graysonensis* (Tappan) which was soon replaced by *Gyroidinoides praestans* Magniez-Jannin. This sequence of events was terminated by complete removal of benthonic foraminifera from the area for some time. This event corresponds with the extinction of planktonic foraminifer *Rotalipora greenhornensis* (Morrow). Soon after, most of the previously dominating species recovered, except *Tritaxia pyramidata* (Reuss) and *Gavelinella intermedia–cenomanica–baltica* group which became extinct. Some other species profited from vacant niches and became significant contributors to the assemblages, e.g. *Tappanina eouvigeriniformis* (Keller), *Praebulimina elata* Magniez-Jannin, *Reophax*? sp., *Lingulogavelinella globosa* (Brotzen). After some time again an important change in the benthonic foraminiferal assemblages occurred. *Textularia chapmani* Lalicker and *Reophax*? sp. became extinct. *Tappanina eouvigeriniformis* (Keller), *Praebulimina elata* Magniez-Jannin became very rare while *Lingulogavelinella globosa* (Brotzen) and *Ammobaculites* spp. bloomed. This change in the structure of benthonic foraminiferal assemblages correlates with the extinction of *Rotalipora cushmani* (Morrow). In the following interval (samples MLP 192 to MLP 195), the main components of assemblages were *Lingulogavelinella globosa* (Brotzen), *Gyroidinoides praestans* Magniez-Jannin and *Ammobaculites* spp. Sample MLP 196 was devoid of benthonic foraminifera. The topmost part of the studied interval was dominated by three species: *Lenticulina rotulata* Lamarck, *Ammobaculites parvispira* Ten Dam and *Valvulineria lenticula* (Reuss), except sample MLP 198, which again was barren of a benthonic group.

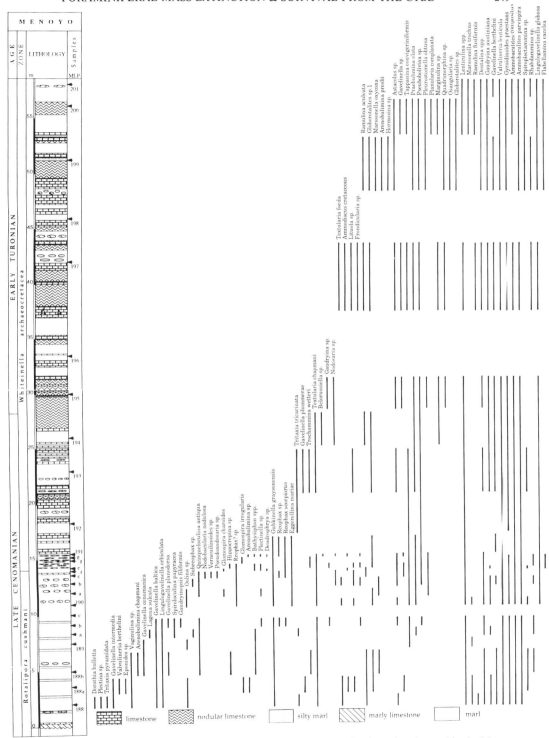

Fig. 4. Ranges of benthonic foraminifera species in the Cenomanian–Turonian boundary interval in the Menoyo section.

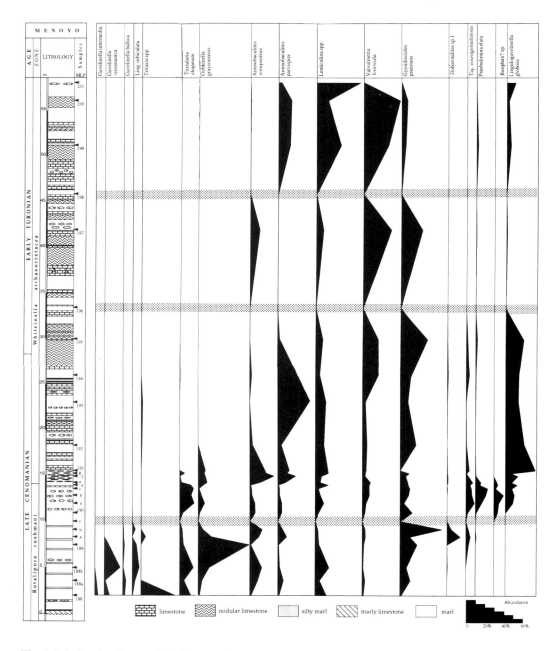

Fig. 5. Relative abundances of dominant species in the Cenomanian–Turonian boundary interval in the Menoyo section. Horizontal bands, intervals barren of benthonic foraminifers.

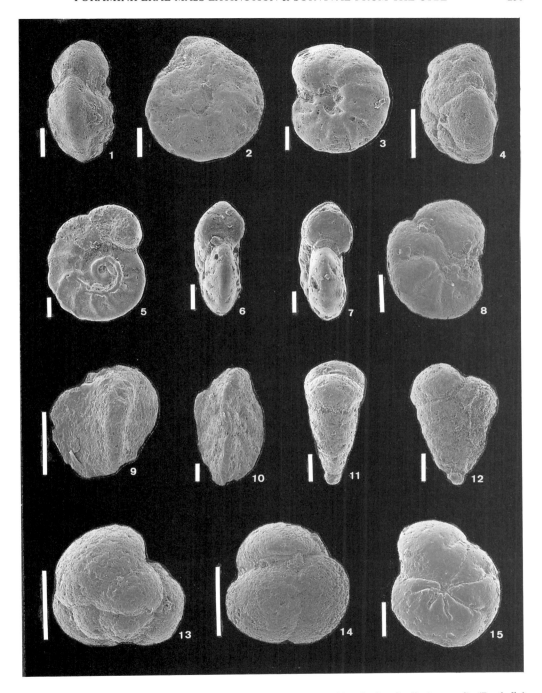

Fig. 6. 1. *Lingulogavelinella orbiculata* (Kusnezova), sample MLP 189a; 2. *Gavelinella intermedia* (Berthelin), sample MLP 189a; 3, 7. *Gavelinella baltica* Brotzen, sample MLP 188; 5, 6. *Gavelinella cenomanica* Brotzen, sample MLP 188; 4. *Lingulogavelinella orbiculata* (Kusnezova), sample MLP 188; 8, 15. *Lingulogavelinella orbiculata* (Kusnezova), sample MLP 189; 9. *Tritaxia pyramidata* (Reuss), sample MLP 189a; 10. *Tritaxia tricarinata* (Reuss), sample MLP 188; 11, 12. *Textularia chapmani* Lalicker, sample MLP 190; 13, 14. *Gubkinella graysonensis* (Tappan), sample MLP 189. Scale bar = 100 μm.

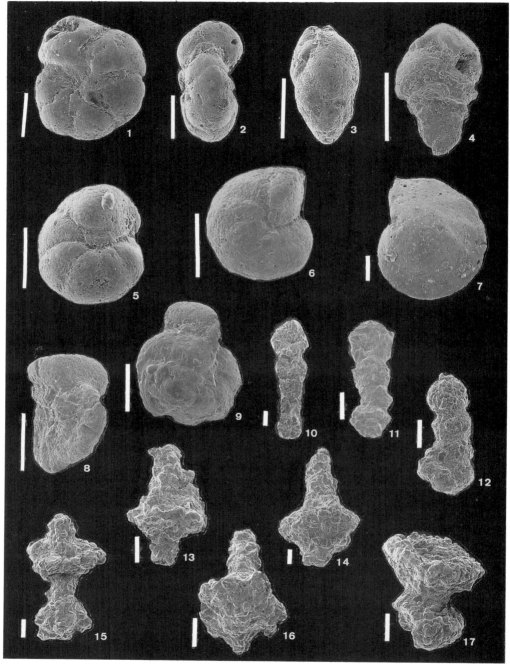

Fig. 7. 1. *Lingulogavelinella globosa* (Brotzen), sample MLP 193; 2. *Lingulogavelinella globosa* (Brotzen), sample MLP 190; 3. *Praebulimina elata* Magniez-Jannin, sample MLP 190; 4. *Tappanina eouvigeriniformis* (Keller), sample MLP 189; 5. *Valvulineria lenticula* (Reuss), sample MLP 190; 6. *Gyroidinoides praestans* Magniez-Jannin, sample MLP 188; 7. *Lenticulina rotulata* Lamarck, sample MLP 189; 8, 9. *Globorotalites* sp. 1, sample MLP 189a; 10. *Ammobaculites parvispira* Ten Dam, sample MLP 190d; 11. *Ammobaculites parvispira* Ten Dam, sample MLP 189a; 12. *Ammobaculites compositus* Magniez-Jannin, sample MLP 190; 13, 16, 17. *Reophax*? sp., sample MLP 190b; 14. *Reophax*? sp., sample MLP 190c; 15. *Reophax*? sp., sample MLP 190a. Scale bar = 100 μm.

Palaeoecology

Paul *et al.* (1994) postulated enhanced preservation of organic matter during the Cenomanian–Turonian boundary interval in the Menoyo section. Enhanced preservation of organic matter may impact benthonic biota in two ways: by causing oxygen deficiency in bottom waters, and by changing the type of food available.

Do changes in benthonic foraminiferal assemblages in the Menoyo section reflect changes in oxygenation of bottom waters and/or changes in character of food supply? Recent studies documented that assemblages of modern benthonic foraminifers from sediments accumulated under dysaerobic conditions differ both in faunal composition and morphology from those of oxic deposits (Phleger & Soutar 1973; Bernard 1986; Perez-Cruz & Machain-Castillo 1990; Sen Gupta & Machain Castillo 1993; Kaiho 1994). The faunas from the oxygen minimum zone are characterized by high abundance, low diversity, small size and high dominance, typically with 2 or 3 calcareous species constituting up to 80% of the total assemblage; some agglutinated taxa (*Textularia*, *Reophax*) are also present in oxygen-deficient environments as well (Koutsoukos *et al.* 1990; Murray 1991).

However, oxygen deficiency alone almost never acts as a limiting agent, hampering the metabolism. The greatest control comes from the changes in the amount of food available as well as from possible changes in the quality of food (Van der Zwaan 1982). Infaunal or potentially infaunal taxa generally are more tolerant of low oxygen concentrations and are therefore the first to profit from the high food availability. The ultimate downward organic flux rate controls both the food availability and the oxygen concentrations at the bottom (Barmavidjaja *et al.* 1992; Jorissen *et al.* 1992). Deposit feeders with the infaunal or semi-infaunal life position as living within the sediment are thought to be more tolerant of oxygen deficiency and these groups might survive and even grow in abundance under these environmentally stressed conditions.

Environmental requirements of Recent benthonic foraminifera have been the subject of many studies, and broad classifications of sediment type/depth morphotypic associations have been proposed (Corliss 1985; Jones & Charnock 1985; Corliss & Chen 1988; Kaiho 1994). Simple classification of benthonic foraminifera into two groups – those with epifaunal and infaunal modes of life – is often disordered by motile species which, living free of the muddy sediment, may migrate down to penetrate unconsolidated soft mud in search of food and/or may move within the sediment because the microhabitat preference of individual taxa may change as a function of the depth of the redox front (Jorissen 1988; Corliss & Emerson 1990; Koutsoukos *et al.* 1990; Barmavidjaja *et al.* 1992). Free-living species may be found in different microhabitats in response to changing environmental conditions and food supply. This group of species, by changing habitats from epifaunal to infaunal, is regarded as highly adaptable and tolerant (Linke & Lutze 1993). Infaunal species dominate assemblages associated with relatively high organic-carbon values (Corliss & Chen 1988). Latest Cenomanian low trochospiral inflated and laterally compressed *Lingulogavelinella globosa* Brotzen and lenticular planispiral biconvex *Lenticulina rotulata* Lamark are interpreted as forms of inferred infaunal/semi-infaunal modes of life (Koutsoukos *et al.* 1990; Leary & Peryt 1991).

Figure 8 shows relative abundances of infaunal/semi-infaunal and epifaunal morphogroups in the assemblages.

In the lowermost part of the studied interval (samples MLP 188 to MLP 189b) epifaunal morphogroups make 20–55% of the total assemblages. In the following sample, MLP 189c, benthonic foraminifera are absent. A considerable decrease in the contribution of epifaunal morphogroups to the benthonic assemblages is recorded in the short interval (samples MLP 190a–g, 191) just following the interval barren of benthonic fauna. Epifauna there makes up only 4–30% of the total assemblages.

Starting from the beginning of the *Whiteineila archaeocretacea* Zone, the contribution of the foraminiferal epifauna to the assemblages gradually increases upwards (samples MLP 192 to MLP 201). In this interval, however, benthonic foraminifera were twice excluded from the area. They were not recorded in samples MLP 196 and MLP 198.

Recorded changes in the morphotypic composition of benthonic foraminiferal assemblages indicate fluctuation of dysaerobic to anoxic conditions at Menoyo during late Cenomanian–early Turonian time. In oxygen-depleted waters and probably with higher organic content, infauna and semi-infauna made up 70–96% of the population. When oxygen conditions ameliorated and less organic matter was available, epifauna began to expand. Complete anoxia occurred when benthonic foraminifera were not present in the sediment.

Late Cenomanian low oxygen tolerant benthonic foraminiferal faunas have been recorded from several regions (North America:

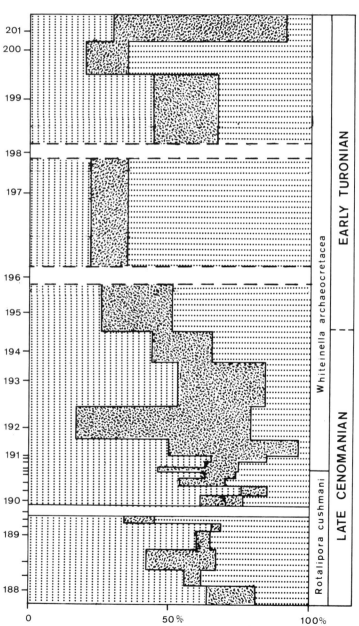

Fig. 8. The infaunal:epifaunal ratio in the Cenomanian–Turonian boundary interval in the Menoyo section.

FORAMINIFERAL MASS EXTINCTION & SURVIVAL FROM THE CTBE

Leckie 1985; Bernard 1986; NW Pacific: Kaiho *et al.* 1993; South Atlantic: Koutsoukos *et al.* 1990; Lincolnshire, UK: Hart & Bigg 1981; Anglo-Paris Basin: Jarvis *et al.* 1988; Danish–Polish Trough: Peryt & Wyrwicka 1991, 1993; Peryt *et al.* 1994; Bohemian Basin: Ulicny *et al.* 1993; Vocontian Basin: Tronchetti & Grosheny 1990). These regions show very similar morphotypic associations; the faunal differences which nevertheless exist most probably reflect provincialism within benthonic foraminifers.

Pattern of benthonic foraminiferal survival from Cenomanian–Turonian mass extinction at Menoyo

Kauffman & Harries (1993) proposed a model for survival and recovery after mass extinction. To these intervals are related such terms as extinct taxa, survivors, opportunistic species, disaster species, progenitor taxa and Lazarus species.

Kauffman & Harries (1993) divide the period between mass extinction and restructuring of communities and ecosystems into three intervals.

1. The late phases of the mass extinction, during which time Lazarus taxa disappear; taxa that will survive the extinction and persist through to the recovery interval show significant changes in population structure, progenitor taxa arise.
2. A short survival interval with (a) an early crisis phase characterized by rare survivors and blooms of disaster and (b) a later population expansion phase characterized by population blooms among ecological opportunists, expansion of populations among surviving resident and immigrant species, and the early return of some Lazarus taxa.
3. The recovery interval with (a) an early phase characterized by widespread return of Lazarus taxa and (b) a later phase of recovery characterized by more rapid, continuous increase in diversity as new lineages arise and radiate, and speciation continues among surviving lineages.

The recovery phase eventually results in basic restructuring of communities and eco-systems.

Abundance fluctuations of dominant species, infaunal/semi-infaunal to epifaunal ratio and distribution of benthonic foraminifera in the section, allow characterization of the lowermost part of the section as representing the late phase of mass extinction (samples MLP 188 to MLP 189b); the higher part of the section represents the survival interval.

Stepped extinction of dominant species char-acterizes the late phase of mass extinction and early phase of survival interval (Fig. 9). *Gavelinella intermedia–cenomanica–baltica* group, *Tritaxia pyramidata* (Reuss), *Textularia chapmani* Lalicker and *Gubkinella graysonensis* (Tappan) became extinct during this period. *Gyroidinoides praestans* Magniez-Jannin, *Ammobaculites* spp., *Lenticulina rotulata* Lamarck are interpreted as opportunistic taxa. Included in this group are species which normally occurred in relatively low numbers in complex pre-extinction communities, but which underwent significant population expansion during the survival interval in the absence of significant space/resource competition.

None of the dominant species was classified as representing Lazarus taxa, i.e. species which temporarily disappeared at or near the peak of mass extinction and reappeared during the survival or recovery intervals as environmental conditions normalized. However, most of the rare or common species which disappeared from the section during the late phase of mass extinction belong to this group, e.g. nodosariids, *Trochammina wetteri* Stelck and Wall, *Textularia foeda* Reuss, *Tritaxia tricarinata* (Reuss).

To the late phase of mass extinction are also related progenitor species, i.e. species which evolved during the late, most stressful phases of the mass extinction interval, and which persist during the survival and recovery intervals, with or without giving rise to new taxa. *Globorotalites* sp. 1 and *Tappanina eouvigeriniformis* (Keller) are included to this group.

The survival interval is distinctly tripartite.
(a) A very short interval, completly devoid of benthonic foraminifera, corresponding to a 'dead' zone of Kauffman & Harries.
(b) Also a short interval, in which along with opportunistic taxa and species which are going to be extinct by the end of this interval, disaster species appeared, i.e. species which are normally rare and widely dispersed under pre-extinction environments, but which are specifically adapted to stressed environmental conditions associated with peak mass extinction and survival intervals. Dominant species in this group are *Praebulimina elata* Magniez-Jannin, *Reophax*? sp., *Lingulogavelinella globosa* (Brotzen); rare and common species in this group include *Quinqueloculina antiqua* Franke, *Spiroloculina papyracea* Burrows, Sherborn and Bailey, *Nodobacularia nodulosa* (Chapman), *Glomospira charoides* (Jones and Parker).
(c) A relatively long interval dominated mainly by opportunistic species, i.e. *Gyroidinoides praestans* Magniez-Jannin, *Valvulineria lenticula* (Reuss). *Lenticulina* spp., *Ammobaculites* spp.

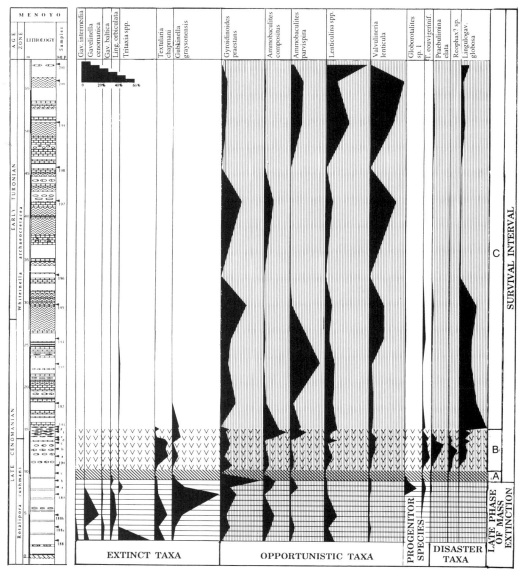

Fig. 9. Mass extinction and survival intervals in the Cenomanian–Turonian boundary interval in the Menoyo section. Diagram shows the relationships of the repopulation pattern to groups of taxa representing extinct, opportunistic, progenitor and disaster species which contribute to the assemblages in each repopulation phase. Horizontal lines, late phase of mass extinction interval (disaster species are not present); oblique lines (a), lowermost part of the survival interval devoid of benthonic foraminifera; chevrons (b), middle part of the survival interval where assemblages are dominated mainly by disaster and opportunistic species; white space (c), upper part of the survival interval where assemblages are dominated mainly by opportunistic taxa.

along with *Lingulogavelinella globosa* (Brotzen) in the lower part of this interval.

Conclusions

The studied interval represents the late phase of mass extinction and survival intervals. The recovery interval is beyond the studied part of the section. In the survival interval may be distinguished a brief interval immediately following the final major extinction event, in which benthonic biota are absent: the dead zone; followed also by a relatively brief interval characterized by the population bloom of disaster taxa and finally an interval which is characterized by the domination of benthonic population by a few opportunistic taxa.

The recorded changes in benthonic foraminiferal assemblages including the stepped extinction of several foraminiferal species, temporary disappearance of many rare and common species, bloom of disaster species in the lower part of the survival interval and opportunistic species in the upper part of the survival interval, most likely reflect the decline of the oxygenation level of the bottom waters at the end of the *Rotalipora cushmani* Zone and the persistence of these unfavourable conditions in the *Whiteinella archaeocretacea* Zone.

The interval in which faunal turnover was observed represents the most stressful environmental conditions; increasing oxygen depletion caused stepped extinction or temporary disappearance from the area of several species and in the peak of the event when complete anoxia occurred, temporary disappearance of all benthic species. When oxygen conditions ameliorated a little, benthonic foraminifera reappeared. However, twice in early Turonian, bottom waters became anoxic. Because at that time benthonic populations were dominated by opportunistic species well adapted to low oxygen conditions, these brief anoxic periods did not cause extinctions within communities but only temporary disappearances of benthonic biota.

The low content of benthonic foraminiferal epifauna in the uppermost Cenomanian confirms strong oxygen deficiency in bottom waters and probably an increased amount of organic matter. Improvement of oxygen conditions in bottom waters is reflected by expansion of epifauna which characterize areas with less food availability.

We thank Michele Caron and Malcolm Hart for their helpful comments on the manuscript. D. Peryt gratefully acknowledges partial support from the Basque government's 'Programa de Perfeccionamiento y Morilidad del Personal Investigaor' and from the European Community Science, Research and Development Division, granted by project ERB-CIPA-CT-92-0132. This is a contribution to the DGICYT project no. PS90/91.

References

ARTHUR, M. A., SCHLANGER, S. O. & JENKYNS, H. C. 1987. The Cenomanian–Turonian Anoxic Event, II: palaeoceanographic controls on organic matter production and preservation. *In:* BROOKS, J. & FLEET, A. (eds) *Marine Petroleum Source Rocks.* Geological Society, London, Special Publication, **26**, 401–420.

BARMAVIDJAJA, D. M., JORISSEN F. J., PUSKARIC, S & VAN DER ZWAAN, G. J. 1992. Microhabitat selection by benthic foraminifera in the northern Adriatic Sea. *Journal of Foraminiferal Research,* **16**, 297–317.

BERNARD, J. M. 1986. Characteristic assemblages and morphologies of benthic foraminifera from anoxic organic-rich deposits: Jurassic through Holocene. *Journal of Foraminiferal Research,* **16**, 207–215.

BIRKELUND, T., HANCOCK, J. M., HART, M. B., RAWSON, J., ROBASZYNSKI, F., SCHMID, F. & SURLYK, F. 1984. Cretaceous stage boundary proposals. *Bulletin of the Geological Society of Denmark,* **33**, 3–20.

BUZAS, H. A. & GIBSON, T. G. 1969. Species diversity: Benthonic Foraminifera in Western North Atlantic. *Science,* **163 (3862)**, 72–75.

COLIN, J-P., LAMOLDA, M. A. & RODRIGUEZ-LAZARO, J. 1982. Los ostracodos del Cenomaniense superior y Turoniense de la Cuenca Vasco-Cantabrica. *Revista Española de Micropaleontologia,* **14**, 187–220.

CORLISS, B. H. 1985. Micro-habitats of benthic foraminifera within deep-sea sediment. *Nature,* **314**, 435–438.

—— & CHEN, C. 1988. Morphotype patterns of Norwegian Sea deep-sea benthic foraminifera and ecological implications. *Geology,* **16**, 716–719.

—— & EMERSON, S. 1990. Distribution of Rose Bengal stained deep-sea benthic foraminifera from the Nova Scotian continental margin and Gulf of Maine. *Deep-Sea Research,* **37**, 381–400.

GALE, A. S., JENKYNS, H. C., KENNEDY, W. J. & CORFIELD, R. M. 1993. Chemostratigraphy versus biostratigraphy: data from around the Cenomanian–Turonian boundary. *Journal of the Geological Society, London,* **150**, 29–32.

GOROSTIDI, A. 1993. *Nanofósiles calcáreos y eventos del Cretácico Medio-Superior de la Región Vasco-Cantábrica.* PhD thesis, Universidad del Pais Vasco.

HARRIES, P. J. & KAUFFMAN, E. G. 1990. Patterns of survival and recovery following Cenomanian–Turonian (Late Cretaceous) mass extinction in the Western Interior Basin, United States. *In:* KAUFFMAN, E. G. & WALLISER, O. (eds) *Extinction Events in Earth History.* Springer, Berlin, Heidelberg, Lecture Notes in Earth Sciences, **30**, 277–298.

HART, M. B. & BIGG, P. J. 1981. Anoxic events in the Cretaceous chalk seas of north-west Europe. *In:* NEALE, W. J. & BRASIER, M. B. (eds) *Microfossils from Recent and Fossil Shelf Seas.* Ellis Horwood for the British Micropalaeontological Society, 177–185.

JARVIS, J., CARSON, G. A., HART, M. B., LEARY, P. N., TOCHER, B. A., HORNE, D. & ROSENFELD, A. 1988. Microfossil assemblages and the Cenomanian–Turonian (late Cretaceous) Oceanic Anoxic Event. *Cretaceous Research,* **9**, 3–103.

JONES, R. W. & CHARNOCK, M. A. 1985. 'Morphogroups' of agglutinating foraminifera, their life positions and feeding habits and potential applicability in (paleo)ecological studies. *Revue de*

Paleobiologie, **4**, 311–320.

JORISSEN, F. J. 1988. Benthic Foraminifera from the Adriatic Sea: Principles of Phenotypic variation. *Utrecht Micropaleontological Bulletin*, **37**, 1–176.

——, BARMAVIDJAJA, D. M., PUSKARIC, S. & VAN DER ZWAAN, G. J. 1992. Vertical distribution of benthic foraminifera in the northern Adriatic Sea: The relation with the organic flux. *Marine Micropaleontology*, **19**, 131–146.

KAIHO, K. 1994. Benthic foraminiferal dissolved-oxygen index and dissolved-oxygen levels in the modern ocean. *Geology*, **22**, 719–722.

——, FUJIWARA, O. & MOTOYAMA, I. 1993. Mid-Cretaceous faunal turnover of intermediate-water benthic foraminifera in the northwestern Pacific Ocean margin. *Marine Micropaleontology*, **23**, 13–49.

KAUFFMAN, E. G. 1984. Toward a synthetic theory of mass extinction. Abstracts with programs. *Geological Society of America, 97th Annual Meeting, Reno*, **16**, 555–556.

—— 1986. High resolution event stratigraphy: regional and global Cretaceous Bio-events. *In:* WALLISER, O. H. (ed.) *Global Bio-events*. Springer, Berlin-Heidelberg, Lecture Notes in Earth Sciences, **8**, 279–335.

—— 1988. The dynamics of marine stepwise mass extinction. *In:* LAMOLDA, M. A., KAUFFMAN, E. G. & WALLISER, O. H. (eds) *Paleontology and Evolution, Extinction events*. Revista Espagnola de Paleontologia, No. Extraordinario, 57–71.

—— & HARRIES, P. J. 1993. A model for survival and recovery after mass extinction. *In: Global Boundary Events. Abstracts of an Interdisciplinary Conference in Kielce, Warszawa*. 27.

KOCH, C. F. 1980. Bivalve species duration, aerial extent and population size in a Cretaceous sea. *Paleobiology*, **6**, 184–192.

KOUTSOUKOS, E. A. M., LEARY, P. N. & HART, M. B. 1990. Latest Cenomanian–earliest Turonian low oxygen tolerant benthonic foraminifera: a case study from the Sergipe basin (N. E. Brazil) and the western Anglo-Paris basin (southern England). *Palaeogeography, Palaeoclimatology, Palaeoecology*, **77**, 143–177.

LAMOLDA, M. A. 1978. Le passage Cenomanien–Turonien dans la coupe de Menoyo (Alaya, Alava). *Cahiers de Micropaleontologie*, **4**, 21–27.

——, LOPEZ, G. & MARTINEZ, R. 1987. Turonian integrated Biostratigraphy in the Estella Basin (Navarra, Spain). *In:* WIEDMANN, J. (ed.) *Cretaceous in the Western Tethys. Proceedings 3rd International Cretaceous Symposium, Tübingen 1987*. Stuttgart, 145–159.

LEARY, P. N. & PERYT, D. 1991. The Late Cenomanian oceanic anoxic event in the western Anglo-Paris Basin and southeast Danish–Polish Trough: survival strategies of and recolonization by benthonic foraminifera. *Historical Biology*, **5**, 321–335.

LECKIE, R. M. 1985. Foraminifera of the Cenomanian–Turonian boundary interval, Greenhorn Formation, Rock Canyon Anticline, Pueblo, Colorado. *In:* PRATT, L. M., KAUFFMAN, E. G. & ZELT, F. B. (eds) *Fine-grained deposits and biofacies of the Western Interior Seaway: Evidence of cyclic sedimentary processes*. Society of Economic Paleontologists and Mineralogists, 139–150.

LINKE, P. & LUTZ, G. F. 1993. Microhabitat preferences of benthic foraminifera – a static concept or a dynamic adaptation to optimize food acquisition? *Marine Micropaleontology*, **20**, 215–234.

MURRAY, J. W. 1976. A method of determining proximity of marginal seas to an ocean. *Marine Geology*, **22**, 103–119.

—— 1991. *Ecology and Palaeoecology of Benthic Foraminifera*. Longman Scientific and Technical, London.

PAUL, C. R. C., MITCHELL, S., LAMOLDA, M. A. & GOROSTIDI, A. 1994. The Cenomanian–Turonian Boundary Event in northern Spain. *Geological Magazine*, **131**, 801–817.

PEREZ-CRUZ, L. L. & MACHAIN-CASTILLO, M. L. 1990. Benthic foraminifera of the oxygen minimum zone, continental shelf of the Gulf of Tehuantepec, Mexico. *Journal of Foraminiferal Research*, **20**, 312–325.

PERYT, D. & WYRWICKA, K. 1991. The Cenomanian–Turonian Oceanic Anoxic Event in SE Poland. *Cretaceous Research*, **12**, 65–80.

—— & —— 1993. The Cenomanian–Turonian boundary event in Central Poland. *Palaeogeography, Palaeoclimatology, Palaeo-ecology*, **104**, 185–197.

——, ——, ORTH, C., ATTREP, M. JR & QUINTANA, L. 1994. Foraminiferal changes and geochemical profiles across the Cenomanian/Turonian boundary in Central and southeast Poland. *Terra Nova*, **6**, 158–165.

PHLEGER, F. B. & SOUTAR, A. 1973. Production of benthic foraminifera in three East Pacific oxygen minima. *Micropaleontology*, **19**, 110–115.

RAUP, D. M. & SEPKOSKI, J. J. 1984. Periodic extinctions of families and genera. *Science*, **231**, 833–836.

SCHLANGER, S. O. & JENKYNS, H. C. 1976. Cretaceous oceanic anoxic events: causes and consequences. *Geologie en Mijnbouw*, **55**, 179–184.

SEN GUPTA, B. K. & MACHAIN-CASTILLO, M. L. 1993. Benthic foraminifera in oxygen-poor conditions. *Marine Micropaleontology*, **20**, 183–202.

TRONCHETTI, G. & GROSHENY, D. 1991. Les assemblages de foraminifères benthiques au passage Cenomanien–Turonien à Vergons, S-E France. *Geobios*, **24**, 13–31.

ULIČNÝ, D., HLADIKOVÁ, J. & HRADECKÁ, L. 1993. Record of sea-level changes, oxygen depletion and the $\delta^{13}C$ anomaly across the Cenomanian–Turonian boundary, Bohemian Cretaceous Basin. *Cretaceous Research*, **34**, 211–234.

VAN DER ZWAAN, G. J. 1982. Paleoecology of Late Miocene Foraminifera. *Utrecht Micropaleontological Bulletins*, **25**, 5–201.

Planktonic foraminifera recovery from the Cenomanian–Turonian mass extinction event, northeastern Caucasus

NATALIYA A. TUR

All-Russian Geological Research Institute (VSEGEI), Sredny pr. 74, 199026 St. Petersburg, Russia

Abstract: The changing structure of the planktonic foraminiferal assemblages is examined from the Upper Cenomanian–Lower Coniacian succession of 'Carbonate Dagestan'.

The analysis of the morphological and taxonomical diversity shows that the deep-water forms are most severely affected by the mass extinction event. The repopulation of their post-crisis assemblages is rapid. The composition of the shallow-water population remains more or less stable.

The foraminiferal dynamics suggests the main developmental phases. The Extinction Phase occurs during the *R. cushmani* Zone, when the bathypelagic rotaliporids became extinct as a result of an expansion of the oxygen-minimum conditions. During the *D. hagni* Zone the survival community is characterized by the primitive forms and forms with an incipient keeled structure. The Early Recovery Phase, occurring during the *H. helvetica, M. coronata* and early *M. tarfayaensis* Zones, is characterized by the appearance of the first marginotruncanids. The Later Recovery Phase, beginning in the later *M. tarfayaensis* Zone, involves a maximum of morphological diversity of marginotruncanids in the *M. concavata* Zone.

The main trend among the Cretaceous planktonic foraminifera is the evolution of keeled forms with a complex umbilical system from primitive unkeeled globular ancestors (Caron & Homewood 1983). It has been suggested that this process may be related to their colonization of deeper water environments (Hart 1980; Caron 1983). Thus foraminifera speciated by invading different levels in the water column. The thin shelled, spinose forms inhabited the surface waters and the more robust keeled forms with longer life cycles and lower reproductive rates occupied deeper environments. Environmental stresses caused the disappearance of deeper-water well evolved taxa and simultaneously stimulated an expansion of the primitive forms adapted to a fluctuating environment. The most critical stresses were connected with the periods of time when the oxygen-minimum zone interfered with the life habits of the selected taxa (Hart & Ball 1986).

The Cenomanian–Turonian Oceanic Anoxic Event (OAE) and global changes in the ocean-climate system related to the eustatic highstand in the sea-level caused the dramatic changes in the population structure of foraminifera. The Late Cenomanian extinction is defined as one of the three major extinctions of Cretaceous foraminifera (Hart & Leary 1990). The distinct extinction peak in the Late Cenomanian is marked in the Sepkoski histogram of percent extinction of the foraminifera (Sepkoski 1990). The strong effect of this event on the evolution of planktonic foraminifera has already been repeatedly described and is well known (Jarvis *et al.* 1988; Leary *et al.* 1989). An expansion and intensification of the oxygen-minimum zone in the water column initially caused the decline and extinction of bathypelagic keeled foraminifera, particularly rotaliporids.

Gradual planktonic foraminiferal recolonization of the vacated and newly formed deep-water niches began in the Turonian. Diversity analysed at the species level increases in the Turonian–Coniacian interval (Hart & Leary 1990).

This paper attempts to show the nature of regional repopulation of planktonic foraminiferal assemblages in the northeastern Caucasus. The quantity of the appearing, disappearing and key planktonic and benthonic taxa (species and genera) for the Late Cretaceous in this area was calculated by Botvinnik (1978). According to these data, the following phases were distinguished for the post-Cenomanian foraminiferal evolution: extinction (Early Turonian), establishment (Late Turonian–Early Campanian) and the following bloom (Late Campanian–Early Maastrichtian). Our study aims at a better understanding of the dynamics of the post-crisis globotruncanid assemblages, based on analyses of both qualitative and quantitative data, in accordance with the revision of the Ceno-

From Hart, M. B. (ed.), 1996, *Biotic Recovery from Mass Extinction Events*,
Geological Society Special Publication No. 102, pp. 259–264

manian–Turonian and Turonian–Coniacian boundary positions and the newly proposed foraminiferal subdivision for the section studied.

Lithology and microbiostratigraphy

The Bass section is located in the southern highlands of Chechnya, which is a part of the large tectonic unit known as 'Carbonate Dagestan' (Fig. 1). This unit is characterized by the development of typical, box-like folds. The Upper Cretaceous highly fossiliferous sediments are exposed in cliff-like, extensive outcrops up to 1300 m thick. The Cenomanian–Coniacian sequence, represented by rhythmically bedded marly limestones and marls, are overlain by Santonian–Maastrichtian limestones.

The interval studied embraces the Karanaiskya and Dzhengutaiskaya (partly) Formations (see Fig. 2). Within the Karanaiskaya Formation sequence four Members are distinguished. The grey marly limestones with bentonites (Mmbr 1) are overlain by light-coloured limestones with numerous layers of black marl and flint concretions (Mmbr 2). In the overlying limestones with flints the dark layers disappear (Mmbr 3). The uppermost part of the Karanaiskaya formation consists of stylolitic chalk-like limestones (Mmbr 4). The overlying sediments of the lower part of the Dzhengutaiskaya Formation are represented by grey marly limestones. A full sedimentological description of these formations may be found in Smirnoff et al. (1986).

The Cenomanian–Turonian (C–T) boundary in 'Carbonate Dagestan' was drawn at the base of the Karanaiskaya Formation, which corresponds to the base of the *Praeglobotruncana imbricata* Zone (Orel 1970) or the *Hedbergella hoelzli* Zone (Samyschkina 1983). However, in the assemblages diagnostic for the above-mentioned foraminiferal zones, these authors indicated the presence of both Cenomanian rotaliporids and typical Turonian forms, although vertical ranges of the species were not shown. Thus, in the present author's opinion, the C–T boundary position was not proved by the vertical distribution of foraminiferal taxa. The base of the Formation is also correlated with the base of the *Inoceramus labiatus* Zone (Pergament & Smirnoff 1978), but the inoceramid bivalves determined before as the Early Turonian species are rare in the lower part of the Karanaiskaya Formation. They have not been described well enough and need further research and taxonomical revision. The position of the C–T boundary was not, therefore, precisely located.

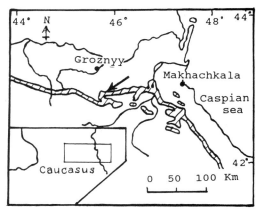

Fig. 1. Sketch map of studied area. The outcrop of Upper Cretaceous deposits is shaded. The arrow indicates the location of the Bass section.

Samples from limestones and marls in the Bass section yielded abundant foraminiferal assemblages. For the first time in this region, foraminifera were isolated from the hard rocks by the use of acetic acid which has allowed us to get new, and more precise, biostratigraphic data. The principal internationally recognized horizons based on globotruncanids are all recognized in the studied section. It is possible to distinguish six foraminiferal zones (Fig. 2), which correspond to the zonal sequence established in many complete sections within Europe (Robaszynski et al. 1980; Hilbrecht 1986; Jarvis et al. 1988). In another section of the Crimea–Caucasian region (the Aksu–Dere section) a similar microfaunal succession around the C–T boundary has also been described (Kopaevich & Walaszczyk 1990).

In the Bass section the proposed position of the C–T boundary has been defined by the disappearance of the planktonic foraminiferal genus *Rotalipora*. The extinction of this genus is a widespread biostratigraphic event, and it is strongly recommended by Marks (1984) and Salaj (1986) that 'the boundary be made to coincide with this level' (Marks 1984, p. 167). This recommendation was taken as the basis for drawing the boundary in the Bass section since we examined only the planktonic foraminifera. The boundary has been drawn at the base of a layer, 3 m thick, of laminated dark marl. In this calcareous fossils are absent and the C org. content reaches up to 15.1% (Naidin & Kiyashko 1994). It can be taken as representative of the C–T Boundary Event (Jenkyns 1980, 1985). Thus the lower part of the Karanaiskaya Formation (mbrs 1–2), previously referred to

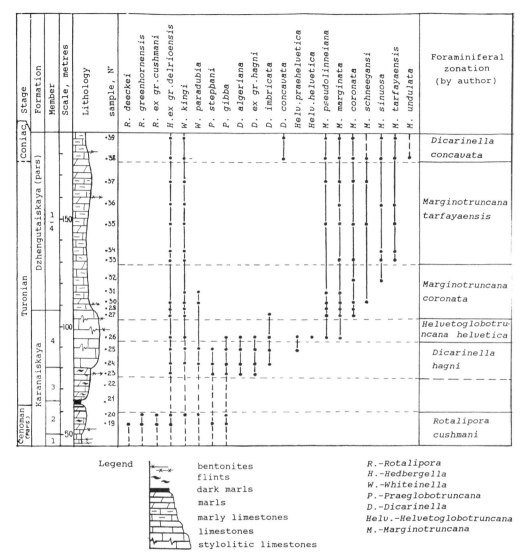

Fig. 2. Biostratigraphical distribution of the selected planktonic species in the Bass section. Stratigraphy after Smirnoff & Lopatkin (pers. comm.). The positions of the Cenomanian–Turonian and Turonian–Coniacian boundaries are adapted by the present author.

the Lower Turonian, is assigned to the upper part of the *Rotalipora cushmani* Zone of Cenomanian age.

The position of the Turonian–Coniacian boundary within the Dzhengutaiskaya Formation of this section has also been revised. In 'Carbonate Dagestan' the boundary was drawn at the base of the *Inoceramus schloenbachi* Zone (Smirnoff et al. 1986). It must be noted that the macrofaunal position of the Turonian–Coniacian boundary in 'Carbonate Dagestan' is not discussed in this paper. The macrofauna of the Bass section was not examined in detail. The boundary in this section was drawn on the basis of some discoveries of inoceramids identified by Pergament (1977 pers. comm.) as species of *I. wandereri* and *I. koeneni* at the level of 155 m. The data from this investigation of the planktonic foraminifera have indicated the possibility of drawing this boundary at a higher level (see Fig. 2). Its position has been defined by the appearance of *Dicarinella concavata* and that

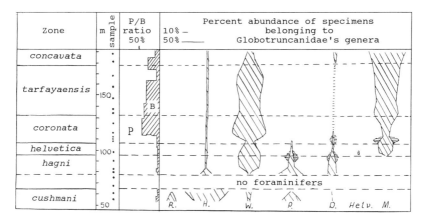

Fig. 3. Changes in the planktonic assemblages. For legend see Fig. 2. The horizontal shading indicates a percentage of intermediate forms from Praeglobotruncana to Dicarinella and from Dicarinella to Marginotruncana.

Zone	Dynamics of morphological diversity		Dynamics of taxonomical diversity
concavata		Maximum of morphological diversity of abundant marginotruncanids	Recovery: Early Phase
tarfayaensis		Appearence of more specialized marginotruncanids	Recovery: Later Phase
coronata, *helvetica*		The first typical marginotruncanids; the intermediate forms from *Dicarinella* to *Marginotruncana*	Recovery: Later Phase
hagni		Survivals among praeglobotruncanids; large morphotypes of new *Dicarinella hagni* lineage	Survival Phase
	no foraminifers		
cushmani		The last single-keeled rotaliporids	Extinction Phase

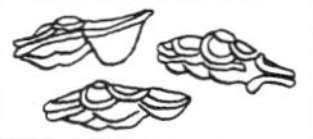

Fig. 4. Dynamics of diversification of foraminiferal assemblages.

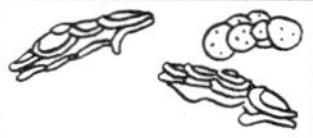

seems to be in agreement with the traditional understanding of this boundary (Marks 1984; Salaj 1986). Further research towards correlation with foraminiferal fauna from other Dagestanian sections to confirm the boundary position in the 'Carbonate Dagestan' is necessary.

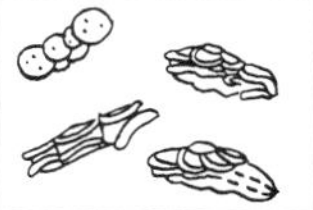

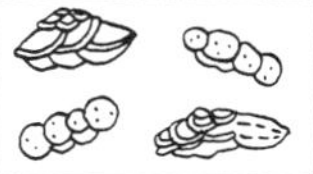

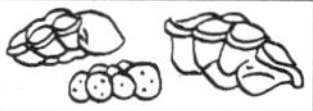

Foraminiferal dynamics: extinction and repopulation

The foraminiferal assemblages are characterized by planktonic/benthonic (P/B) ratios of between 55% and 95% (Fig. 3). The high ratio and the presence of numerous keeled globotruncanids are considered to be indicative of a relatively deep depositional environment in the basin.

In the rich planktonic assemblages, besides the individuals with the distinct specific feature, there are numerous intermediate forms bearing the features of the different species and even genera. At present, all the forms are attributed to traditionally recognized Turonian and Coniacian species although in most cases these species appeared to be polymorphic groups. The systematic work is in progress and the results will be given elsewhere.

For reconstruction of the post-crisis repopulation of planktonic foraminifera in this basin, the percentage of the specimens of the various genera has been calculated. Varieties within a species possessing characters of two or more genera from a morphological point of view have also been counted (see Fig. 3). Similar intraspecific varieties have been described by Robaszynski *et al.* (1990) who considered that their appearance was especially typical during periods of evolutionary radiation.

The extinction, survival and recovery phases have been identified and are shown in Fig. 4.

Extinction

This phase corresponds to the upper part of the *cushmani* Zone. Planktonic taxa represent over 95% of the foraminiferal assemblages in which primitive morphotypes of shallow-water hedbergellids dominate. The number of keeled rotaliporids and praeglobotruncanids gradually decreases upwards towards the 'black marls'. Deep-water keeled rotaliporids become extinct probably as a result of the expansion of the oxygen-depleted water masses.

Foraminiferal remains have not been found in these black marls. Rare, badly preserved specimens occur in the overlying beds. These may have been caused by primary factors (such as another step in oxygen depletion) or by diagenetic dissolution.

Survival

The early repopulation or survival phase may correspond to the *hagni* Zone. It is characterized by long-ranging primitive opportunists (*Hedbergella* spp. and *Whiteinella* spp.) and the pre-adapted praeglobotruncanids with an incipient keeled structure. The species diversity of the non-keeled taxa is more or less stable within the interval studied. In contrast, the keeled taxa gradually begin to dominate (up to 55% of the planktonic assemblages) being represented by new species and varieties. Among the keeled forms there are specimens showing features similar both to *Praeglobotruncana* and *Dicarinella* genera. Newly appeared, large double-keeled morphotypes of the *Dicarinella hagni* group inhabit vacant deep-water niches and become important elements in the assemblages (up to 35%).

Recovery

One can assume that the early recovery phase takes place during the *helvetica* Zone, *coronata* Zone and early part of the *tarfayaensis* Zone. This phase starts with the appearance of the well developed double-keeled marginotruncanids with sigmoidal umbilical sutures. Gradual upward increase in their diversity to 8 species is characteristic. In many regions the appearance of *Marginotruncana* is related to the main transgressive peak in the Turonian. This genus appears to be adapted to the deeper-water niches. Besides typical marginotruncanids there are forms transitional from *Dicarinella* to *Marginotruncana*. It should be noted that the abundance and diversity of typical marginotruncanids increased, in contrast to the slowdown in the development of less evolved dicarinellids.

The late recovery phase may be identified within the uppermost part of the *tarfayaensis* Zone and the *concavata* Zone. In this section studied there is a renewed influx of *Marginotruncana* varieties although the specific composition is the same as in the underlying beds. Their numerical abundance reaches 80%. At the same time a few typical dicarinellids appear once again. They are represented by deep-water plano-convex morphotypes belonging to the newly-evolving *concavata* lineage. More favourable and stable deep-water environments may be responsible for this development.

Conclusions

The changing structure of the foraminiferal faunas during the latest Cenomanian and the earliest Turonian in the Dagestan Basin was controlled, as everywhere, by the Oceanic Anoxic Event.

Deep-water planktonic forms were affected the most severely and the repopulation of their post-crisis assemblages was relatively rapid. Their development was paralleled by the shallow-water forms, which did not show such strong reductions and subsequent rapid diversification.

The re-occupation of the vacated deep-water niches by the new keeled species and numerous intermediate forms took place during the recovery phase.

Upper Cretaceous sections in the 'Carbonate Dagestan' have a great potential for testing and improving the proposed repopulation model.

The author thanks Yu. P. Smirnoff for considerable help in collecting samples during the field work on the 'Carbonate Dagestan' sections, T. N. Koren and A A. Atabekyan for critical reading of the manuscript, and

References

BOTVINNIK, P. V. 1978. [Some regularities of the Later Cretaceous Foraminifera development in the North-East Caucasus]. *Trudy of XVIII Session of the All-Union Palaeontological Society*, 147–154. Nauka, Leningrad [in Russian].

CARON, M. 1983. La spéciation chez les foraminifères planctiques: une réponse adaptée aux contraintes de l'environnement. *Zitteliana*, **10**, 671–676.

—— & HOMEWOOD, P. 1983. Evolution of early planktic foraminifers. *Marine Micropaleontology*, **7**, 453–462.

HART, M. B. 1980. A water depth model for the evolution of the planktonic foraminifera. *Nature*, **286**, 252–254.

—— & BALL, K. C. 1986. Late Cretaceous anoxic events, sea-level changes and the evolution of the planktonic foraminifera. *In:* SUMMERHAYES, C. P. & SHACKLETON, N. J. (eds) *North Atlantic Palaeoceanography*. Geological Society, London, Special Publication, **21**, 67–78.

—— & LEARY, P. N. 1990. Periodic bioevents in the evolution of the planktonic Foraminifera. *Lecture Notes in Earth Sciences*, **30**, Extinction Events in Earth History, 325–331.

HILBRECHT, H. 1986. Die Turon-Basis im Regensburger Raum: Inoceramen, Foraminiferen und "events" der Eibrunner Mergel bei Bad Abbach. *Neues Jahrbuch fur Geologie und palaeontologie, Abhandlungen*, **172**, 71–82.

JARVIC, I., CARSON, G. A., COOPER, M. K. E., HART, M. B., LEARY, P. N., TOCHER, B. A., HORNE, D. & ROSENFELD, A. 1988. Microfossil assemblages and the Cenomanian–Turonian (Late Cretaceous) oceanic anoxic event. *Cretaceous Research*, **9**, 3–103.

JENKYNS, H. C. 1980. Cretaceous anoxic events: from continents to oceans. *Journal of the Geological Society, London*, **137**, 171–188.

—— 1985. The early Toarcian and Cenomanian–Turonian anoxic events in Europe: comparisons and contrasts. *Geologische Rundschau*, **74**, 505–518.

KOPAEVICH, L. F. & WALASZCZYK, I. 1990. An integrated inoceramid–foraminiferal biostratigraphy of the Turonian and Coniacian strata in south-western Crimea, Soviet Union. *Acta Geologica Polonica*, **40**, 83–96.

LEARY, P. N., CARSON, G. A., COOPER, M. K. E., HART, M. B., HORNE, D., JARVIS, I., ROSENFELD, A. & TOCHER, B. A. 1989. The biotic response to the Late Cenomanian oceanic anoxic event; evidence from Dover, SE England. *Journal of the Geological Society, London*, **146**, 311–317.

MARKS, P. 1984. Proposals for the recognition of boundaries between Cretaceous stages by means of planktonic foraminiferal biostratigraphy. *Bulletin of the Geological Society of Denmark*, **33**, 163–170.

NAIDIN, D. P. & KIYASHKO, S. I. 1994. [Geochemical description of the Cenomanian–Turonian boundary sediments of the Crimea. Article 1. Lithological composition, substance of organic carbon and some elements.] *Bulletin of Moscow Society of Natural Testers*, **69**, 28–42, [in Russian].

OREL, G. V. 1970. [*Biostratigraphy of the Upper Cretaceous deposits of Dagestan and paleoecology of foraminifers.*] Doctoral Thesis (Summary), Rostovsky University [in Russian].

PERGAMENT, M. A. & SMIRNOFF, YU. P. 1978. [The vertical distribution and stratigraphical significance of the inoceramids in the Upper Cretaceous sequence of Dagestan.] *Trudy of All-Union Colloquium on the inoceramids*, **1**, 94–113. Nauka, Moscow [in Russian].

ROBASZYNSKI, F., AMEDRO, F., FOUCHER, J.-C., GASPARD, D., MAGNIEZ-JANNIN, F., MANIVIT, H. & SORNAY, J. 1980. Synthèse biostratigraphique de l'Aptien au Santonien du Boulonnais à partir de sept groupes paléontologiques: foraminifères, nannoplancton, dinoflagellés et macrofaunes. *Revue de Micropaléontologie*, **22**, 195–321.

——, CARON, M., DUPUIS, Chr., AMEDRO, F., GONZALEXZ DONOSO, J.-M., LINARES, D., HARDENBOL, J., GARTNER, S., GALANDRA, F. & DELOFRE, R. 1990. A tentative integrated stratigraphy in the Turonian of Central Tunisia: formations, zones and sequential stratigraphy in the Kalaat Senan Area. *Bulletin des Centres de Recherches Exploration–Production Elf-Aquitaine*, **6**, 119–225.

SALAJ, J. 1986. Proposition of Turonian boundaries of the Tethyan realm on the basis of foraminifers. *Geologický Zborník. Geologica Carpathica*, **37**, 4, 483–500.

SAMYSCHIKINA, K. G. 1983. [*Foraminifera and stratigraphy of the Cretaceous sediments of the East Caucasus.*] Nauka, Moscow [in Russian].

SEPKOSKI, J. J. JR. 1990. The taxonomic structure of periodic extinction. *In:* SHARPTON, V. L. & WARD, P. D. (eds) *Global catastrophes in Earth history; An interdisciplinary conference on impacts, volcanism, and mass mortality*. Geological Society of America, Special Paper, **247**, 33–44.

SMIRNOFF, JU. P., MOSCVIN, M. M. & TKACHUK, G. A. 1986. [North Caucasus and Pre-Caucasus.] *In:* ALIEV, M. M., KRYLOV, N. A., PAVLOVA, M. M. (eds) [*Upper Cretaceous of the South of USSR.*] Nauka, Moscow, 22–104 [in Russian].

Recovery of the food chain after the Late Cenomanian extinction event

M. B. HART

Department of Geological Sciences, University of Plymouth, Drake Circus, Plymouth PL4 8AA, UK

Abstract: The Late Cenomanian bio-event is accepted as a globally synchronous extinction event that is characterized by significant biological changes which are coincident with isotopic and geochemical anomalies. The extinction event is characterized by changes on the macro-, meio-, micro- and nanno-scale and, in many examples, it has been suggested that these are 'step-wise'. In the United Kingdom successions the steps are emphasized by depositional non-sequences, but in all cases the biological changes are in the same order. In microfaunal terms this involves a marked reduction in the benthonic foraminifera followed closely by a restriction of the planktonic foraminifera to surface-dwelling morphotypes. The event is also characterized by short-lived floods of *Heterohelix* sp. (small, biserial planktonic foraminiferids), *Bulimina* sp. (small, benthonic foraminiferids often characteristic of low-oxygen environments, calcispheres (?calcified dinoflagellates) and radiolaria. Following the extinction events, in the very latest Cenomanian, the recovery phase begins. Almost immediately the calcareous nannoplankton flora is restored but the dinoflagellates do not recover until much later in the Turonian. The benthonic foraminifera recover slowly with the fauna of the Early Turonian being of low diversity, with long-ranging taxa – including a large (?deep water), internally complex, agglutinated genus (*Labyrinthidoma*). The planktonic foraminifera recovered quickly with the *Praeglobotruncana*, *Dicarinella*, and *Marginotruncana* faunas appearing in succession within 100 000–200 000 years of the end of the extinction event. Using a model of the normal food chain it is possible to identify the order in which the basic building blocks needed for ecosystem recovery are put in place following the extinction event.

IGCP 335 is specifically charged with the study of biotic recovery following mass extinction events. What is meant by recovery? As there is a wide range of interpretations of what actually constitutes recovery, it is first of all necessary to establish some ground rules. In a range of English dictionaries the following definitions can be found.

To recover: regain possession or use or control of, acquire or find (out) again, reclaim . . . ; bring or come back to life, consciousness, health, or normal state or position . . . ; retrieve, make up for, get over, cease to feel effects of . . .

Recovery: act or process of recovering or being recovered.

In these definitions there are many phrases commonly used by palaeontologists to describe recovery from 'events' of any kind of magnitude. If one lifts out some of the key words the recovery should mean the process involved in attaining a previous (healthy) position or as clear a derivative of that position as is possible. This last phrase is necessary as taxa, once extinct, cannot normally reappear in the succession although closely related, ecologically similar, taxa may indeed replace those that have disappeared at the event (see Jablonski 1986; Erwin & Droser 1993 for a discussion of Lazarus and Elvis taxa). It would be reassuring to regard every global bio-event as a chain that leads from normality through the event or events back to normality. Figure 1 represents a very idealized model of such an extinction bio-event. In any geological succession there is a certain level of extinction, but where this is concentrated either at one level, or in a step-wise pattern, it can be properly described as a mass extinction event. The problem for palaeontologists is that of defining where normality ends and normality begins again.

What is the normal biological situation in the fossil record? As the greater part of the biological world cannot be preserved in the sedimentary record it is almost impossible for palaeontologists to identify recovery to an undefinable point. In any healthy (mentioned in one of the definitions) community, the whole of the food chain must be in place and functioning properly. There are many models of the food chain available, and that presented in Fig. 2 is a modified version of one (based on Ducklow & Taylor 1991) used by Laybourn-Parry (1992, fig. 6.18). The overwhelming majority of the individuals mentioned in that

From Hart, M. B. (ed.), 1996, *Biotic Recovery from Mass Extinction Events*,
Geological Society Special Publication No. 102, pp. 265–277

265

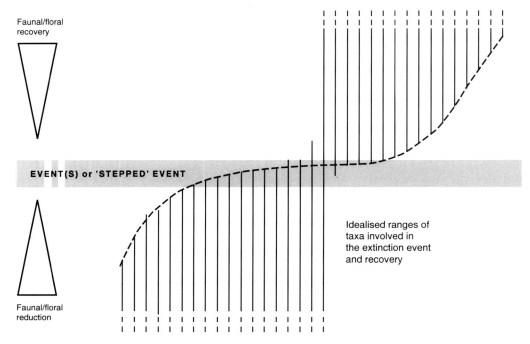

Fig. 1. Schematic representation of the reduction and recovery of a fauna and flora across an extinction event.

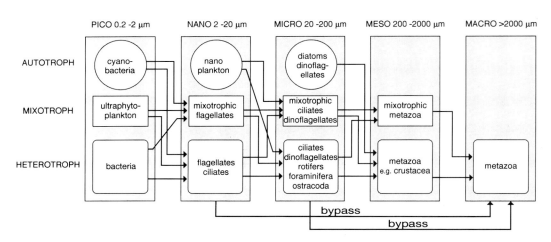

Fig. 2. Possible model of a marine food chain of the sort that might be applicable to the Cretaceous (modified from a model provided by Laybourn-Parry 1992).

figure leave little or no fossil record. Rare records of bacteria isolated from Precambrian materials should not be taken as a sign that bacteria, naked amoebae, cilliates, rotifers, etc., are all routinely known from sediment samples. This is quite clearly not the case. It is very rare, for example, for subsets of the same sample to be used to identify populations of foraminiferids, ostracods, dinoflagellates, haptophytes, diatoms, radiolarians, ebridians, calpionellids and silicoflagellates. Even this quite comprehensive list does not even approach the full list of pico-, nanno- and micro-plankton that makes up the marine food chain.

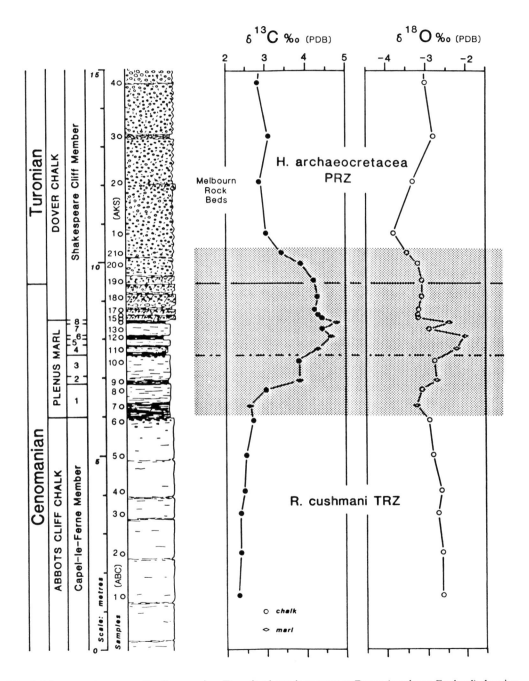

Fig. 3. The succession across the Cenomanian–Turonian boundary event at Dover (southeast England) showing the lithostratigraphy, sample locations and the carbon isotope curve (after Jarvis et al. 1988).

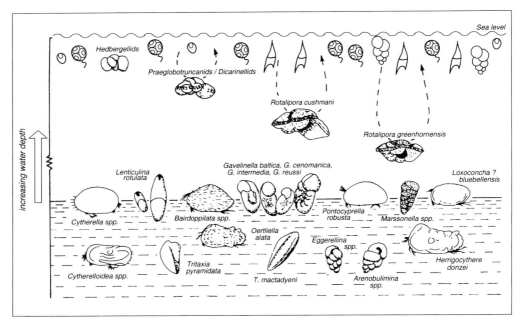

Fig. 4. Reconstruction of the ecosystem that may have existed during the deposition of the upper part of the Abbots Cliff Chalk Formation and Bed 1 of the Plenus Marl succession. The symbols are explained in Fig. 5.

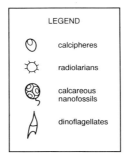

Fig. 5. Explanation of some of the symbols used in Figs 4, 6–8 and 10–13.

Palaeontologists must, therefore, assume that provided representatives of all the groups that can be fossilized are present, then a healthy food chain is in place and that those components not found as fossils must be there in order that the system, as a whole, can operate.

The Late Cenomanian event

One of the best locations at which to study the Late Cenomanian extinction event is at Shakespeare Cliff (Dover), although other good sections are available at Eastbourne (Sussex), Culver Cliff (Isle of Wight), Betchworth (Surrey), Shillingstone (Dorset) and at a number of locations in northeast England (see Hart et al. 1991, 1993). The Dover succession was used as the basis for an integrated investigation of the Late Cenomanian extinction event by Jarvis et al. (1988), although Eastbourne has also received some attention (Gale et al. 1993) as it is a slightly expanded succession. Figure 3 shows the Dover succession, lithostratigraphy, sample horizons and the carbon isotope curve based on the data of Carson in Jarvis et al. (1988).

At the top of the Lower Chalk (Abbots Cliff Chalk Formation), in the *Calycoceras guerangi* Zone (see Fig. 4), there would appear to be a normal assemblage of foraminiferids, ostracods, dinoflagellates, calcispheres, calcareous nannofossils and their dependent ammonites, brachiopods, echinoderms, bivalves, etc. Icthyoliths, fish scales and sharks teeth recorded from micropalaeontological residues also show that there was a reasonable vertebrate fauna in the Late Cenomanian marine ecosystem. Jarvis et al. (1988) provide details of the fauna and flora present in the Abbots Cliff Chalk Formation, based on the analysis of subsets of precisely located samples. Figure 4 gives an indication of how this fauna and flora might appear in this

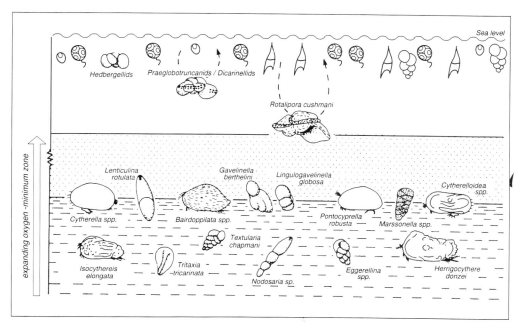

Fig. 6. Reconstruction of the ecosystem that may have existed during the deposition of Bed 2 of the Plenus Marl succession.

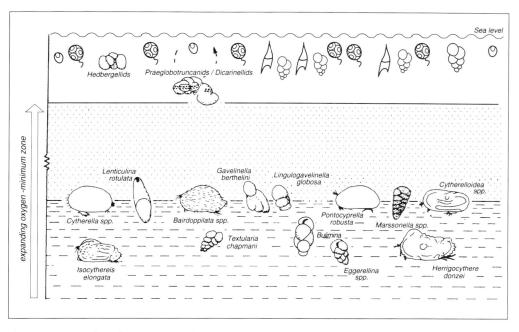

Fig. 7. Reconstruction of the ecosystem that may have existed during the deposition of Bed 4 of the Plenus Marl succession.

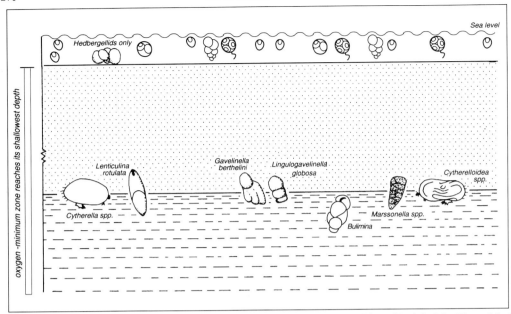

Fig. 8. Reconstruction of the ecosystem that may have existed during the deposition of Bed 6 of the Plenus Marl succession.

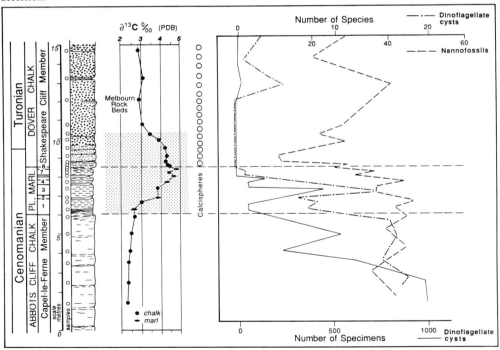

Fig. 9. The Dover succession across the Late Cenomanian extinction event using data from Jarvis et al. (1988). Note in particular the appearance of the floods of calcispheres immediately following the maximum level of the carbon isotope excursion. The calcareous nannofossil diversity recovers very quickly after the event while the diversity of the dinoflagellate cysts does not recover at all until well into the Turonian. Note also that the abundance (number of specimens in a set volume of sample) of dinoflagellate cysts also remains at a low level throughout this interval.

FOOD CHAIN RECOVERY AFTER THE LATE CENOMANIAN EVENT 271

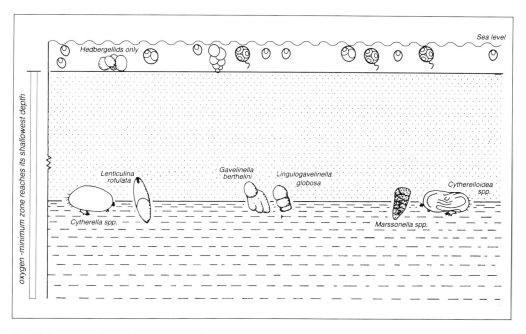

Fig. 10. Reconstruction of the ecosystem that may have existed during the deposition of Bed 8 of the Plenus Marl succession.

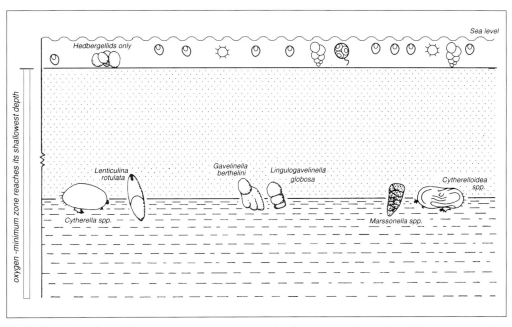

Fig. 11. Reconstruction of the ecosystem that may have existed during the deposition of the basal part of the Melbourn Rock succession.

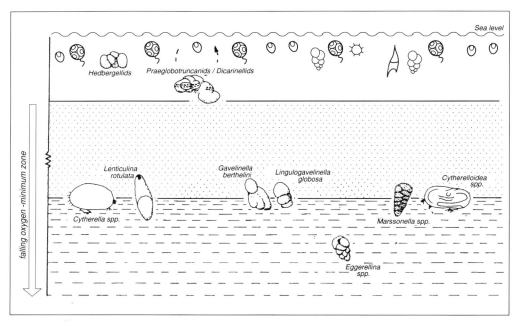

Fig. 12. Reconstruction of the ecosystem that may have existed during the deposition of the middle of the Melbourn Rock succession.

Late Cenomanian ecosystem. Figure 5 provides a key to some of the symbols used in all the reconstructions.

A similar set of four diagrams is given in Jarvis et al. (1988), in which an outline explanation is also provided. The reconstruction in Fig. 4 shows some of the more important ostracods and benthonic foraminiferids together with four representative species of planktonic foraminifera. These are shown in their postulated life position, in which juveniles occupy the eutrophic surface waters while the more mature individuals migrate downwards into more oligotrophic surface waters until they reach their optimum depth for reproduction. This follows the models of Hart & Bailey (1979), Hart (1980), Caron (1983), Caron & Homewood (1983) and Jarvis et al. (1988). The surface water plankton is represented by symbols that depict calcispheres, dinoflagellates, calcareous nannofossils (sensu lato) and the biserial planktonic foraminiferid, *Heterohelix* spp. Individual taxa are not represented within this fauna and flora. Across the Abbots Cliff Chalk Formation/Bed 1 Plenus Marl boundary Leary & Hart (1989) have proposed an increase in water depth. This conclusion was based on a detailed biometric analysis of planktonic foraminiferids, especially *Rotalipora cushmani* (Morrow). A water depth increase is, however, in direct opposition to Jeans et al. (1991) who have proposed a glacial control for the changes associated with the Plenus Marls (and the Cenomanian/Turonian boundary event).

At the Bed 1/Bed 2 boundary there were quite marked changes in the benthonic fauna and the expansion of the oxygen minimum zone is thought to have limited the water column available to the planktonic foraminifera. As a result *Rotalipora greenhornensis* (Morrow) has been removed (Fig. 6). If the migration of the oxygen minimum zone onto the shelf had simply been caused by a rise in sea-level then there would not have been any effect on the microfauna living in the water column as there would still have been the same column of water available to them. This was explained and illustrated in Hart (1993).

In Bed 4 (Fig. 7) the oxygen minimum zone has ascended further in the water column and the benthonic foraminiferal population is now limited to very few taxa. This fauna has been joined by *Bulimina* sp., a genus quite commonly found in modern dysaerobic environments. Modern *Globobulimina* live deep in the sediment, feeding on degraded organic matter and/or the

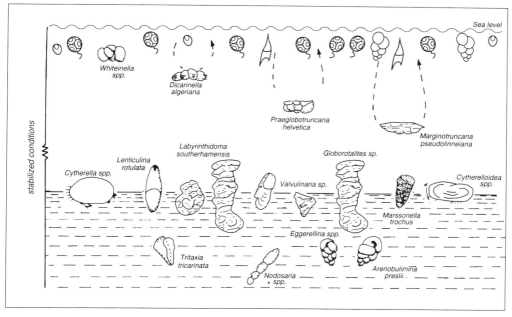

Fig. 13. Reconstruction of the ecosystem that may have existed during the deposition of the Southerham Marls and Chalks in the *Terebratulina lata* Zone of the mid-Turonian.

associated bacteria under near oxic conditions (Corliss 1985; Buzas *et al.* 1993; Jorissen *et al.* 1994). A flood of buliminids is also recorded in the K/T boundary clay (Speijer & Van der Zwaan 1996). This assemblage of benthonic foraminiferids is quite typical of this stage in the Late Cenomanian event (see Koutsoukos *et al.* 1990; Peryt & Lamolda 1996). Within the planktonic foraminiferids, *R. cushmani* has disappeared and despite a small 'bloom' of early dicarinellids (see Jarvis *et al.* 1988) the fauna is restricted to *Hedbergella* spp. and *Praeglobotruncana* spp. These have been joined by what appears to be a flood of small, biserial, taxon *Heterohelix* spp. (including *H. moremani* Cushman and *H. globulosa* (Ehrenberg)). In the overlying Bed 6 (Fig. 8) there are even further faunal and floral reductions. The ostracod fauna is reduced to almost two taxa and the benthonic foraminiferids are reduced to *Lenticulina rotulata* (s.l.) *Gavelinella berthelini* (Ten Dam) – which is often described at this level as *Gavelinella tourainensis* Butt – *Marssonella oxycona* (Reuss), *Lingulogavelinella globosa* (Brotzen) and *Bulimina* sp.

In the plankton there are, however, major changes. The planktonic foraminiferids are reduced to *Hedbergella* spp., although many individuals have begun to show the typical overhanging last chamber which gives the aperture a more umbilical aspect. This effectively changes the generic determination from *Hedbergella* to *Whiteinella*. While the number of calcareous nannofossil species is reducing quite sharply (see Fig. 9) the biggest change is in the dinoflagellate flora. Figure 9 shows counts based on the work of Tocher (in Jarvis *et al.* 1988) which show that the dinoflagellate flora has almost totally disappeared. At this point in the succession the calcispheres, which have been present throughout the Cenomanian (Banner 1972), suddenly become very abundant. Sample residues (especially the 125–63 μm size fraction) are dominated by these tiny calcareous spheres. Calcareous spherical bodies were first described as *Lagena ovalis* and *Lagena sphaerica* by Kauffman in Heer (1865) and these, together with many related forms, have become taxonomically 'lumped' as *Pithonella* Lorenz (1902). Calcispheres have been variously described as protozoans (Adams *et al.* 1967; Banner 1972), algal cysts (Bonet 1956; Rupp 1968; Villain 1975) and calcified dinoflagellate cysts (Keupp 1978; Willems 1985, 1988, 1990, 1992). The work of Willems is particularly compelling as he has illustrated clear evidence of what appears to be

Table 1. *The reduction and recovery of the preservable elements of the food chain across the Late Cenomanian event*

Stratigraphic Position	B.F.	B	O	P.F. SW	IW	DW	H	D	C	N	R	State of F.C.
S. Marls	●	·	●	●	●	●	·	●	●	●		$\frac{2}{3}$
M. Rock	·	·	·	●	·		·	·	●	●	●	3
b. M. Rock	·	·	·	●			●	·	●	·	●	4
P.M. Bed 8	·	·	·	●			●	·	●	●		3
P.M. Bed 7	·	·	·	●			●	·	●	●		3
P.M. Bed 6	·	●	·	●			●	·	●	●		3
P.M. Bed 5	·	●	·	●	·		●	●	·	●		3
P.M. Bed 4	·	●	●	●	●		●	●	·	●		$\frac{2}{3}$
P.M. Bed 3	●	·	●	●	●	·	·	●	·	●		$\frac{2}{3}$
P.M. Bed 2	●	·	●	●	●	·	·	●	·	●		$\frac{2}{3}$
P.M. Bed 1	●	·	●	●	●	●	·	●	·	●		1
A.C.C.F.	●	·	●	●	●	●	·	●	·	●		1

B.F. - benthonic foraminiferids
B. - *Bulimina* sp.
O. - ostracods
P.F. - planktonic foraminiferids [SW -shallow water; IW -intermediate water; DW -deep water]
H. - *Heterohelix* sp.
D. - dinoflagellate cysts
C. - calcispheres
N. - calcareous nannofossils (*sensu lato*)
R. - radiolarians

State of F.C. -state of food-chain [1 -all in place; 2 -slight disturbance; 3 -major disturbance; 4 -almost collapsed]

dinoflagellate-like tabulation inside calcispheres. Futterer (1976) has shown that *Thoracosphaera albatrosina* Kamptner has an opening strongly resembling a dinoflagellate abundance (Speijer & Van der Zwaan 1996) immediately above the K/T boundary. Hart (1991) concluded that there was considerable circumstantial evidence (Fig. 9) to indicate that the calcispheres are most abundant in the interval where dinoflagellate cysts are either missing or drastically reduced. Either they are some form of dinoflagellate response to a stressed environment or they are suddenly colonizing a niche left vacant by the collapse of the dinoflagellate population. It must be stressed that internal structures (e.g. tabulation) have not been observed (by the author) in broken specimens from this level in the succession.

It is not just in the UK where this abundance of calcispheres has been recorded. In northern Spain (Caus pers. comm.), Bauges Massif (Haute-Savoie, France), Poland, Oman and Brazil the same relationships have been observed. In the case of the sections in the French Alps (Flaine and Col de la Colombière), the basal Seewen Limestone is flooded with calcispheres (Hart 1991, fig. 1) in association with *Whiteinella archaeocretacea* Pessagno, *W. aprica* (Loeblich & Tappan), *Praeglobotruncana praehelvetica* (Trujillo) and very early *Marginotruncana* spp. This assemblage is so distinctive that it can be readily identified as very latest Cenomanian or earliest Turonian wherever it is found. This can be confirmed using macrofaunal data (see Fig. 3) for the location of the Cenomanian–Turonian boundary. The calcispheres may,

FOOD CHAIN RECOVERY AFTER THE LATE CENOMANIAN EVENT

therefore, be fulfilling the criteria for a 'disaster' taxon (or taxa).

In the overlying Plenus Marl Bed 8 (Fig. 10) the surface water plankton hardly changes, except that the *Heterohelix* fauna returns to more normal proportions. The calcispheres still dominate the very finest grain size fraction (125–63 μm). In the benthonic community *Bulimina* sp. is very rare to absent but this may be due to the diagenesis of the nodular chalks at this level and in the overlying Melbourn Rock. Plenus Marl Bed 8 and the immediately overlying sample (basal Melbourn Rock, Fig. 11) mark the maximum levels of the carbon isotope excursion and the greatest effect on the fauna and flora. The basal Melbourn Rock does, however, mark one other change. Figure 9 shows the diversity of calcareous nannofossils (*sensu lato*) and this drops quite suddenly at this level. Recovery, unlike that of the organic-walled dinoflagellate cysts, is rapid, but in the association shown in Fig. 11 there is an influx of radiolarians. This is one of the few horizons in the onshore chalk succession of the UK (known to the author) where radiolaria can be found. The specimens are very badly preserved and, as they are calcified, they are almost impossible to extract. Most of the time they are seen only in thin section. The fauna was first described by Hill & Jukes Browne (1895) and, to date, these authors have provided the only illustrations. The fauna is composed entirely of spumellarians and has been described from the Melbourn Rock (or equivalent horizons) of Royston (Cambridgeshire), Leagrave (Bedfordshire), Pitstone (Buckinghamshire), Tring (Hertfordshire), Wallington (Oxfordshire), Dover (Kent), Bincombe (Dorset), Axmouth (Devon) and other localities in Lincolnshire and Humberside. Spumellarians have also been found associated with the Late Cenomanian extinction event in many localities including Brazil (Koutsoukos & Hart 1990).

In the higher levels of the Melbourn Rock (Fig. 12) the plankton is beginning to recover. The calcareous nannofossils are more abundant again, a few dinoflagellate cysts are present and only isolated radiolarians have been recorded. The calcispheres are still the most obvious feature of micropalaeontological residues, with *Heterohelix* spp. reduced to normal levels of abundance. Within the larger elements of the plankton *Praeglobotruncana* spp. (*P. stephani* (Gandolfi) and *P. gibba* Klaus) and *Dicarinella* spp. (*D. imbricata* (Mornod) and *D. hagni* (Scheibnerova)) have appeared. If the models (Hart & Bailey 1979; Hart 1980; Caron 1983; Caron & Homewood 1983) are correct, this would indicate that the planktonic foraminifer-

ids are again exploiting the deeper-water, more oligotrophic, environments. In the Southerham Marls and adjacent chalks (mid-Turonian) the fauna and flora is almost totally recovered (Fig. 13). The calcareous nannofossils (*sensu lato*) are present in large numbers and relatively high diversity. The organic-walled microplankton are still relatively reduced (see FitzPatrick 1996) although they are present throughout the Turonian. The planktonic foraminiferids appear to be fully recovered with an abundant fauna that includes species such as *P. helvetica* (Bolli), *Marginotruncana pseudolinneiana* Pessagno and *M. sigali* (Reichel). The benthonic foraminiferids have also recovered, although the fauna is not as diverse, or as abundant, as that recorded in the Cenomanian (Hart & Carter 1975). At this level in the succession there is a rather large species, recently described (Hart 1995) as *Labyrinthidoma southerhamensis*. The species is known from the Turonian succession between the New Pit Marls and the Southerham Marls (see Jenkyns *et al.* 1994, for the stratigraphy of this part of the succession). The sea-floor community at this level appears to be relatively healthy, although the presence of occasional dark-coloured marl bands with reduced benthonic faunas and chondritiform burrow systems may indicate repeated, short-lived, dysaerobic events. As far as can be ascertained most of the food chain is back in place.

Summary

Using a somewhat crude ranking system (Table 1) it is possible to attempt a summary of the changes recorded in the lowest (detectable) elements of the food chain. The maximum disturbance (=food chain almost collapsed) appears to be in the basal Melbourn Rock. This followed a period of 'major disturbance' in Beds 5–8 of the Plenus Marl succession. Nearly all the food chain is back in place by the Early–mid-Turonian (*P. helvetica* Zone), although the lack of a fully recovered organic-walled dinoflagellate cyst population may be a slight anomaly. Although sampling has been undertaken in the interval between the Melbourn Rock and the Southerham Marls more detailed work on the fauna/flora is required. Preliminary data suggest that immediately above the Melbourn Rock the fauna/flora are almost identical to that recorded in the Southerham Marls (i.e. recovery is almost completed).

It was suggested at the outset that a study of the preservable elements of the food chain might act as a guide to the recovery observed above the Late Cenomanian extinction event. Table 1 shows the reduction, minimum and recovery phases, and it would be interesting to see how these relate to the distribution of the meio- and macro-fauna.

The author thanks his many colleagues for a number of stimulating discussions on the food chain during the writing of this paper. Special thanks go to Dr John Green (Plymouth Marine Laboratory) and Dr Tony Matthews (Department of Biological Sciences, University of Plymouth), both of whom have a much wider experience of biological systems. John Abraham is thanked for the final drawing of the diagrams.

References

ADAMS, T. D., KHALILI, M. & SAID, A. K. 1967. Stratigraphic significance of some oligosteginid assemblages from Lurestan Province, northwest Iran. *Micropaleontology*, 13, 55–67.

BANNER, F. T. 1972. *Pithonella ovalis* from the early Cenomanian of England. *Micropaleontology*, 18, 278–284.

BONET, F. 1956. Zonificacion microfaunistica de las calizas cretacicias del este de Mexico. *Boletin de la Asociacion del geologia petroleo de Mexico*, 8, 389–488.

BUZAS, M. A., CULVER, S. J. & JORISSEN, F. J. 1993. A statistical evaluation of the microhabitats of living (stained) infaunal benthic foraminifera. *In:* LANGER, M. R. (ed.) *Foraminiferal Microhabitats.* Marine Micropalaeontology, 20, 311–320.

CARON, M. 1983. La spéciation chez les foraminifères planctiques: une réponse adaptée aux contraintes de l'environnement. *Zitteliana*, 10, 671–676.

—— & HOMEWOOD, P. 1983. Evolution of early planktic foraminifera. *Marine Micropalaeontology*, 7, 453–462.

CORLISS, B. H. 1985. Microhabitats of benthic foraminifera within deep sea sediments. *Nature*, 314, 435–438.

DUCKLOW, H. W. & TAYLOR, A. H. 1991. Modelling-Session Summary. *In:* REID, P. C., TURLEY, C. M. & BURKILL, P. H. (eds) *Protozoa and Their Role in Marine Processes.* NATO ASI Series G: Ecological Sciences, 25, Springer, Heidelberg & Berlin, 431–442.

ERWIN, D. H. & DROSER, M. L. 1993. Elvis Taxa. *Palaios*, 8, 623–624.

FITZPATRICK, M. E. J. 1996. Recovery of Turonian dinoflagellate cyst assemblages from the effects of the oceanic anoxic event at the end of the Cenomanian in southern England. *This volume.*

FUTTERER, D. 1976. Kalkige Dinoflagellaten ('Calciodinelloidea') und die systematische Stellung der Thoracosphaeroideae. *Neues Jahrbuch der Geologisch–Paläontologisches Abhandlungen*, 151, 119–141.

GALE, A. S., JENKYNS, H. C., KENNEDY, W. J. & CORFIELD, R. M. 1993. Chemostratigraphy versus biostratigraphy: data from around the Cenoma-nian–Turonian boundary. *Journal of the Geological Society, London*, 150, 29–32.

HART, M. B. 1980. A water depth model for the evolution of the planktonic foraminifera. *Nature*, 286, 252–254.

—— 1991. The Late Cenomanian calcisphere global bioevent. *Proceedings of the Ussher Society*, 7, 413–417.

—— 1993. Cretaceous foraminiferal events. *In:* HAILWOOD, E. A. & KIDD, R. B. (eds) *High Resolution Stratigraphy.* Geological Society, London, Special Publication, 70, 227–240.

—— 1995. *Labyrinthidoma* Adams, Knight and Hodgkinson; an unusually large foraminiferal genus from the chalk facies (Upper Cretaceous) of S. England and N. France. *In:* KAMINSKI, M. A., GEROCH, S. & GASINSKI, M. A. (eds) *Proceedings of the Fourth International Workshop on Agglutinated Foraminifera, Krakow, Poland.* Grzybowski Foundation, Special Publication, 3, 123–130.

—— & BAILEY, H. W. 1979. The distribution of planktonic Foraminiferida in the mid-Cretaceous of N.W. Europe. *Aspekte der Kreide Europas, IUGS Series A*, 6, 527–542.

—— & CARTER, D. J. 1975. Some observations on the Cretaceous Foraminiferida of SE England. *Journal of Foraminiferal Research*, 5, 114–126.

—— & LEARY, P. N. 1991. Stepwise mass extinctions: the case for the Late Cenomanian event. *Terra Nova*, 3, 142–147.

——, DODSWORTH, P. & DUANE, A. M. 1993. The Late Cenomanian Event in Eastern England. *Cretaceous Research*, 14, 495–508.

——, ——, DITCHFIELD, P. W., DUANE, A. M. & ORTH, C. J. 1991. The Late Cenomanian event in Eastern England. *Historical Biology*, 5, 339–354.

HEER, O. 1865. *Urwelt der Schweiz.* Friedrich Schulthess, Zurich.

HILL, W. & JUKES-BROWNE, A. J. 1895. On the occurrence of Radiolaria in Chalk. *Quarterly Journal of the Geological Society, London*, 51, 600–609.

JABLONSKI, D. 1986. Causes and consequences of mass extinctions. *In:* ELLIOT, D. K. (ed.) *Dynamics of Extinction.* Wiley, New York, 183–229.

JARVIS, I., CARSON, G., HART, M., LEARY, P., TOCHER, B. A., HORNE, D. & ROSENFELD, A. 1988. Microfossil assemblages and the Cenomanian–Turonian (late Cretaceous) oceanic anoxic event. *Cretaceous Research*, 9, 3–103.

JEANS, C. V., LONG, D., HALL, M. A., BLAND, D. J. & CORNFORD, C. 1991. The geochemistry of the Plenus Marls at Dover, England: evidence of fluctuating oceanographic conditions and of global control during the development of the Cenomanian–Turonian δ^{13}C anomaly. *Geological Magazine*, 128, 603–632.

JENKYNS, H. C., GALE, A. S. & CORFIELD, R. M. 1994. Carbon- and oxygen-isotope stratigraphy of the English Chalk and Italian Scaglia and its palaeoclimatic significance. *Geological Magazine*, 131, 1–34.

JORISSEN, F. J., BUZAS, M. A., CULVER, S. J. &

KUEHL, S. A. 1994. Vertical distribution of living benthic foraminifera in submarine canyons of New Jersey. *Journal of Foraminiferal Research*, **24**, 28–36.

KEUPP, H. 1978. Calcisphaeren des Untertithon der sudlichen Frankenalb und die systematische Stellung von *Pithonella* Lorenz 1901. *Neues Jahrbuch der Geologisch–Paläontologisches Monatshefte*, **2**, 87–98.

KOUTSOUKOS, E. A. M. & HART, M. B. 1990. Radiolarians and Diatoms from the mid-Cretaceous successions of the Sergipe Basin, Northeastern Brazil; palaeoceanographic assessment. *Journal of Micropalaeontology*, **9**, 45–64.

——, LEARY, P. N. & HART, M. B. 1990. Latest Cenomanian–Earliest Turonian low-oxygen tolerant benthonic Foraminifera: a case study from the Sergipe Basin (NE Brazil) and the Western Anglo-Paris Basin (Southern England). *Palaeogeography, Palaeoclimatology, Palaeoecology*, **77**, 145–147.

LAYBOURN-PARRY, J. 1992. *Protozoan Plankton Ecology*. Chapman & Hall, London.

LEARY, P. N. & HART, M. B. 1989. The use of the ontogeny of deep water dwelling planktonic foraminifera to assess basin morphology, the development of water masses, eustacy and the position of the oxygen minimum zone in the water column. *Mesozoic Research*, **2**, 67–74.

LORENZ, T. 1902. Geologische Studien im Grenzgebiete helvetischer und ostalpiner Fazies II. Der sudliche Rhatikon. *Bericht der Naturforchenden Gesellschaft zu Freiburg*, **12**, 34–95.

PERYT, D. & LAMOLDA, M. 1996. Benthonic foraminiferal mass extinction and survival assemblages from the Cenomanian–Turonian Boundary Event in the Menoyo section, N Spain. *This volume.*

RUPP, A. W. 1968. Origin, structure and environmental significance of Recent and fossil calcispheres. *Geological Society of America Special Paper*, **101** 186 (Abstract).

SPEIJER, R. P. & VAN DER ZWAAN, G. J. 1996. Extinction and survivorship of southern Tethyan benthic foraminifera across the Cretaceous/Palaeogene boundary. *This volume.*

VILLAIN, J.-M. 1975. 'Calcisphaerulidae' (Incertae Cedia) du Cretace superieur du Limbourg (Pays-Bas), et d'autres regions. *Palaeontographica (A)*, **149**, 193–242.

WILLEMS, H. 1985. *Tetramerosphaera lacrimula*, eine intern gefacherte Calcisphaere aus der Ober-Kreide. *Senckenbergiana Lethaea*, **66**, 177–201.

—— 1988. Kalkige Dinoflagellaten-Zysten aus der oberkretazischen Schreibkreide-Fazies N-Deutschlands. *Senckenbergiana Lethaea*, **68**, 433–477.

—— 1990. *Tetratropis*, eine neue Kalkdinoflagellaten-Gattung (Pithonelliodeae) aus der Ober-Kreide von Lagerdorf (N-Deutschland). *Senckenbergiana Lethaea*, **70**, 239–257.

—— 1992. Kalk-Dinoflagellaten aus dem Unter-Maastricht der Insel Rugen. *Wissenschaftliche Zeitschrift fur Geologie*, **20**, 155–178.

Recovery of Turonian dinoflagellate cyst assemblages from the effects of the oceanic anoxic event at the end of the Cenomanian in southern England

M. E. J. FITZPATRICK

Department of Geological Sciences, University of Plymouth, Drake Circus, Plymouth PL4 8AA, UK

Abstract: The oceanic anoxic event during the latest Cenomanian caused the decline and temporary extinction of many species of marine micro-flora and -fauna. Many of these species reappeared during the Turonian. The recovery of the marine phytoplankton population after the oceanic anoxic event is examined. Detailed sampling of Turonian sediments from Kent, Sussex and the Isle of Wight (southern England) was undertaken to analyse the changes in the dinoflagellate cyst assemblages through time. Chalk, marl and flint samples were collected with no bias. Abundance and diversity of the assemblages varies greatly between sections and samples. Assemblages at the base of the Turonian (the Melbourn Rock Beds) are poor to barren. This is attributed to the lasting effect of the latest Cenomanian oceanic anoxic event. There is a general increase in diversity and abundance upwards through the succession, although assemblages never reach the richness that is seen in other Cenomanian sections.

Analysis of the latest Cenomanian interval, lithologically known as the Plenus Marl Formation, has been carried out by Jarvis *et al.* (1988) and by Leary *et al.* (1989). Jarvis *et al.* (1988) proposed that the dark marls were deposited in a period of increased upwelling and that an expansion of the marine oxygen minimum zone caused a worldwide oceanic anoxic event (OAE).

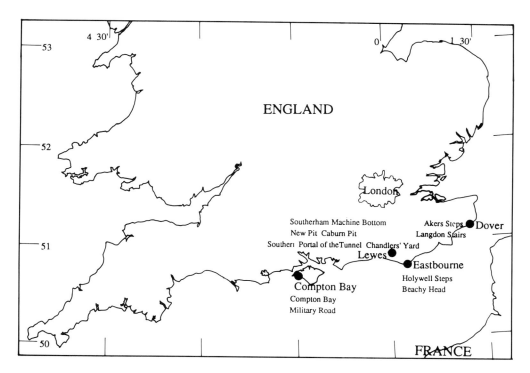

Fig. 1. Map of southern England showing the locations of the areas studied.

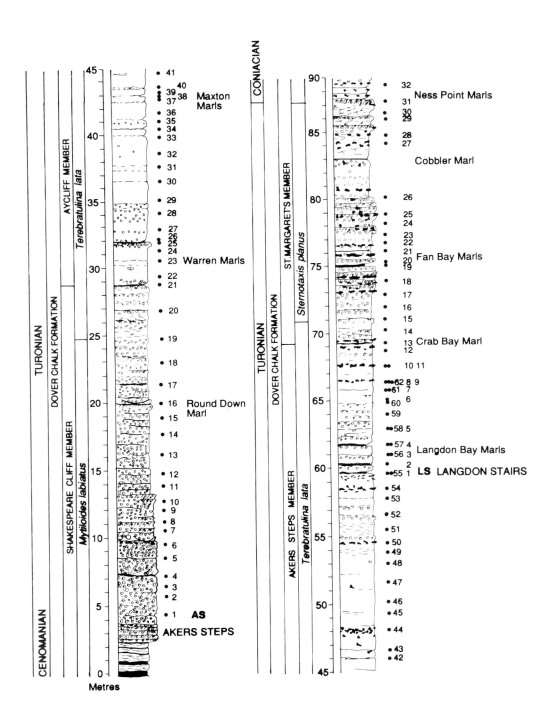

Fig. 2. Composite lithological log of the Turonian sections sampled from Dover, Kent.

RECOVERY OF TURONIAN DINOFLAGELLATE CYST ASSEMBLAGES

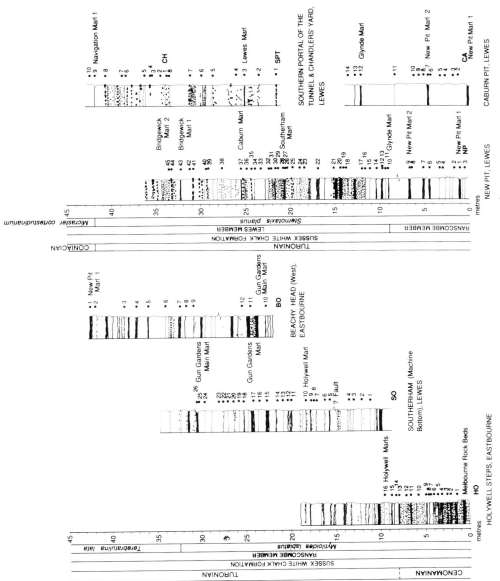

Fig. 3. Composite lithological log of the Turonian sections sampled from Eastbourne and Lewes, Sussex.

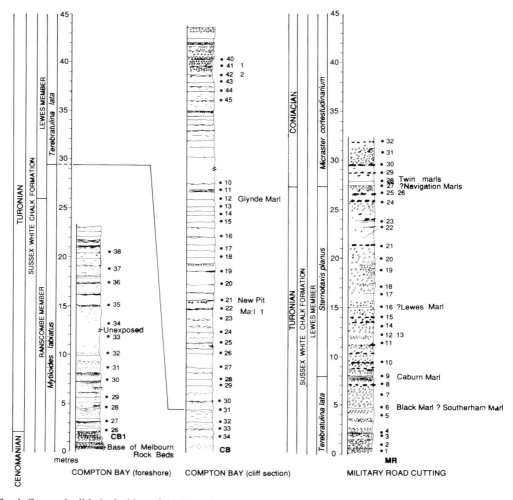

Fig. 4. Composite lithological log of the Turonian sections sampled from the Isle of Wight.

The disappearance of various micro-faunas and -floras at the end of the Cenomanian from Dover were analysed by Jarvis et al. (1988), who briefly describe their recovery. This paper gives a detailed analysis of the recovery of the dinoflagellate cyst assemblages during the period that followed the Cenomanian, the Turonian, at Dover (Kent), Sussex and the Isle of Wight.

Sampling and methods

Samples were collected from composite sections of the Turonian from Kent, Sussex and the Isle of Wight (Fig. 1). The sections were chosen for their completeness, lithological and palaeontological characteristics and for their accessibility. The lithologies of all the sections are described in detail by Woodroof (1981), Mortimore (1986) and by Robinson (1986). Detailed lithological and palaeontological logs of the sections from Kent, Sussex and the Isle of Wight are shown in Figs 2 to 4 respectively.

Sampling was carried out to collect specimens from all important horizons and notable fossil occurrences. The aim was to sample the whole Turonian at relatively closely spaced intervals, usually of about 1 m vertically, unless access was difficult or hazardous. 250–500 g samples of fresh rock were collected from as little vertical extent as was practical (50–100 mm). Chalk, marl and flint samples were collected without bias. Samples have been designated letters and numbers depending on which section they are taken from (Table 1). The succession is composed of nodular chalks at the base (Melbourn Rock Beds), which becomes less nodular upwards, with marl and flint bands becoming more common. The samples were prepared using standard palynological methods (Barss & Williams 1973; Phipps & Playford 1984), using hydrochloric (HCl) and hydroflouric (HF) acids. Samples from Dover were processed using HCl only, at the BP Research Centre, Sunbury (J. Chitolie pers. comm.).

Table 1. *Labels given to the different sections sampled*

Dover	Akers Steps (AS), Langdon Stairs (LS).
Sussex	Holywell Steps (HO), Southerham Machine Bottom (SO), Beachy Head (BO), New Pit (NP), Caburn Pit (CA), Southern Portal of the Tunnel (SPT), Chandler's Yard (CH)
Isle of Wight	Compton Bay (CB1, CB), Military Road (MR).

Taxonomic palynology and morphogroups

The dinoflagellate cyst species encountered are listed in Table 2. See FitzPatrick (1992, 1995) for more detailed discussion of these species. Those marked with an asterisk in Table 2 are first appearances and are not reported previously from the Cenomanian. The species described in this study have been assigned to the morphological groupings of Evitt (1985). Four of these groups (Gc, Gv, Gn & Gi) dominate the assemblages in each of the regions studied. Table 3 notes the characteristics of these groups and the commonly occurring species.

Results

The samples were analysed for their dinoflagellate cyst content using normal counting procedures. Relative occurrence data were used based upon a count of 300 palynomorphs, which is statistically reliable (Buzas 1979). When fewer than 300 specimens occur in a sample, all specimens are counted and the total is shown. 'Abundance' is the total number of specimens counted per slide and 'diversity' is the number of species in a sample. Measurements of dominance were also made. This considers the nature of the assemblage and whether it is dominated by one or two species. The results from the Isle of Wight are not discussed in detail, as the abundances and diversities are too low to produce reliable results.

Abundance (total number of specimens)

The total number of specimens in each sample from all regions studied varies from 0 to >300 (Figs 5 to 7). Chalks, marls and flints each yield barren and rich dinoflagellate cyst assemblages (Figs 8 to 13). Abundances and diversities do not appear to be related to any sedimentological features, such as hardgrounds, chalks, marls or flints. In some cases, flints seem preferentially to preserve a more abundant and diverse assemblage than the chalk. The type of flint may have some control on this. Clayton (1986) suggests that nodular flints represent further stages of silicification, which may alter the organic fossils whereas *Thalassinoides* flints preserve a more diverse and abundant assemblage. Chalk and marl samples have similarly variable diversities and abundances (Figs 8, 9, 11, 12). The diversities are generally low to moderate (10–30) for chalk, marl and flint. These diversities may be related to a stressed environment, such as in a restricted basin. The high sea-levels that are thought to have prevailed during parts of the Turonian (Haq *et al.* 1988) would have connected the seaways allowing the introduction of species to most parts of the Anglo-Paris basin. Parts of the basin may have remained cut off by topographic features which may have restricted the diversity of the floras. This is demonstrated in the atypical floras recovered from Dover, when compared with other areas in the Anglo-Paris basin (Tocher & Jarvis 1987; FitzPatrick 1995).

Samples at the base of the Turonian are often barren or have lower than average abundances (0–100). This is particularly noticeable at Dover, where samples (AS1 to 16) from Akers Steps have a maximum of 100 specimens in any one sample. Samples from the same stratigraphic level (within the Melbourn Rock Beds) at Holywell Steps, Sussex, also show lower abundances than occur higher in the succession. None is barren, but abundances as low as 50 occur. The lowermost Turonian samples on the Isle of Wight also show variable abundances with some barren samples, although two with >300 specimens occur. The abundances generally increase through the *Terebratulina lata* zone, but they remain variable in all areas.

Diversity

Diversity also varies greatly (Figs 14 to 16). At the base of the Turonian (within the Melbourn Rock Beds), the diversities vary from 0–13 at Dover and from 4–24 in Sussex. Maximum diversities occur in the middle Turonian, with

284   M. E. J. FITZPATRICK

Table 2. *List of all dinoflagellate cyst species encountered during this study*

Species are listed in their morphological groups and an asterisk denotes their first appearance during the Turonian.

D Morphogroup
Dinogymnium sp. cf. *D. westralium* Cookson & Eisenack 1958, emend. May 1977

Pp Morphogroup
Eurydinium saxoniensis Marshall & Batten 1988
Isabelidinium belfastense (Cookson & Eisenack 1961) Lentin & Williams 1977

Px Morphogroup
Palaeohystrichophora infusorioides Deflandre 1935
Subtilisphaera pontis-mariae (Deflandre 1936) Lentin & Williams 1976

Gv Morphogroup
Canningia collivieri Cookson & Eisenack 1960
Canningia reticulata Cookson & Eisenack 1960, emend. Helby 1987
Canningia sp.B
Cyclonephelium compactum – membraniphorum complex Marshall & Batten 1988
Cyclonephelium distinctum Deflandre & Cookson 1955
Cyclonephelium membraniphorum Cookson & Eisenack 1962
Senoniasphaera rotundata Clarke & Verdier 1967*
Senoniasphaera sp. A*

Gc Morphogroup
Odontochitina costata Alberti 1961, emend. Clarke & Verdier 1967
Odontochitina operculata forma A (Wetzel 1933) Deflandre & Cookson 1955
Odontochitina operculata forma B (Wetzel 1933) Deflandre & Cookson 1955*
Xenascus ceratioides (Deflandre 1937) Lentin & Williams 1973

Gp Morphogroup
Microdinium ornatum Cookson & Eisenack 1960
Microdinium sp. A*
Rhiptocorys veligera Lejeune-Carpentier & Sarjeant 1983

Gq Morphogroup
Dinopterygium cladoides Deflandre 1935
Dinopterygium medusoides (Cookson & Eisenack 1960) Stover & Evitt 1978

Gs Morphogroup
Achomosphaera ramulifera (Deflandre 1937) Evitt 1963
Achomosphaera reginensis Corradini 1973
Achomosphaera sagena Davey & Williams 1966
Cribroperidinium spp.
Endoscrinium campanula (Gocht 1959) Vozzhennikova 1967
Gonyaulacysta cassidata (Eisenack & Cookson 1960) Sarjeant 1966
Hystrichodinium pulchrum Deflandre 1935
Hystrichodinium voigtii (Alberti 1961, emend. Sarjeant 1966) Davey 1974
Hystrichostrogylon membraniphorum Agelopoulos 1964
Pterodinium cingulatum (Wetzel 1933) Below 1981 var. cingulatum (1973)
Pterodinium ?cornutum Cookson & Eisenack 1962
Spiniferites ?dentatus (Gocht 1959) Lentin & Williams 1973, emend. Duxbury 1977
Spiniferites multibrevis (Davey & Williams 1966) Below 1982
Spiniferites ramosus (Ehrenberg) Loeblich & Loeblich 1966 var. gracilis (Davey & Williams 1966) Lentin & Williams 1973
Spiniferites ramosus (Ehrenberg) Loeblich & Loeblich 1966 var. ramosus (1973)
Spiniferites ramosus (Ehrenberg) Loeblich & Loeblich 1966 var. reticulata (Davey & Williams 1966) Lentin & Williams 1973
Spiniferites spp.
Stephodinium coronatum Deflandre 1936
Xiphophoridium alatum (Cookson & Eisenack 1962) Sarjeant 1966

Gi Morphogroup
Callaiosphaeridium asymmetricum (Deflandre & Courteville 1939) Davey & Williams 1966
Florentinia buspina (Davey & Verdier 1976) Duxbury 1980*

RECOVERY OF TURONIAN DINOFLAGELLATE CYST ASSEMBLAGES

Florentinia deanei (Davey & Williams 1966) Davey & Verdier 1973
Florentinia ferox (Deflandre 1937) Duxbury 1980
Florentinia torulosa (Davey & Verdier 1976) Lentin & Williams 1981*
Florentinia spp.
Hystrichosphaeridium bowerbankii Davey & Williams 1966
Hystrichosphaeridium conispiniferum Yun 1981
Hystrichosphaeridium tubiferum (Ehrenberg 1838) var. *brevispinum* (Davey & Williams 1966) Lentin & Williams 1973
Hystrichosphaeridium tubiferum (Ehrenberg 1838) Deflandre 1937 var. *tubiferum* emend. Davey & Williams 1966
Kleithriasphaeridium readei (Davey & Williams 1966) Davey & Verdier 1976
Oligosphaeridium complex (White 1842) Davey & Williams 1966
Oligosphaeridium sp. cf. *O. complex* (White 1842) Davey & Williams 1966
Oligosphaeridium poculum Jain 1977*
Oligosphaeridium prolixispinosum Davey & Williams 1966

Gn Morphogroup
Cleistosphaeridium sp. cf. *C.armatum* (Deflandre 1937) Davey 1969
Cleistosphaeridium clavulum (Davey 1969) Below 1982
Cleistosphaeridium multispinosum (Singh 1964) Brideaux 1971
Coronifera oceanica Cookson & Eisenack 1958, emend. May 1980
Dapsilidinium laminaspinosum (Davey & Williams 1966) Lentin & Williams 1981
Heterosphaeridium difficile (Manum & Cookson 1964) Ioannides 1986*
Heterosphaeridium heteracanthum (Deflandre & Cookson 1955) Eisenack & Kjellstrom 1971*
Kiokansium polypes (Cookson & Eisenack 1962) Below 1982, emend. Duxbury 1983
Litosphaeridium sp. A (Marshall & Batten 1988)*
Litosphaeridium sp. cf. *L. siphoniphorum* Davey & Williams 1966; emend. Lucas-Clark 1984 var. *siphoniphorum* Lucas-Clark 1984
Pervosphaeridium monasteriense Yun 1981
Pervosphaeridium sp. cf. *P. pseudhystrichodinium* (Deflandre 1937) Yun 1981

Gx Morphogroup
Apteodinium deflandrei (Clarke & Verdier 1967) emend. Lucas-Clark 1987
Batiacasphaera euteiches (Davey 1969) Davey 1979*
Cassiculosphaeridia reticulata Davey 1969
Ellipsodinium rugulosum Clarke & Verdier 1967
Exochosphaeridium sp. cf. *E.arnace* Davey & Verdier 1973
Exochosphaeridium phragmites Davey et al. 1966
Leberidocysta chlamydata (Cookson & Eisenack 1962) Stover & Evitt 1978
Leberidocysta defloccata (Davey & Verdier 1973) Stover & Evitt 1978
Prolixosphaeridium conulum Davey 1969
Tanyosphaeridium variecalamus Davey & Williams 1966
Trichodinium castanea (Deflandre 1935) Clarke & Verdier 1967
Wallodinium anglicum (Cookson & Hughes 1964) Lentin & Williams 1973

Table 3. *Characteristics of morphological groupings above and the commonly occurring species*

Gc: Ceratioid outline with apical, antapical and postcingular horns. *Odontchitina costata* Alberti 1961, emend. Clarke & Verdier 1967, *O. operculata* (Wetzel 1933) Deflandre & Cookson 1955.

Gv: Dorsoventrally compressed cysts. *Cyclonephelium distinctum* Deflandre & Cookson 1955, *C. membraniphorum* Cookson & Eisenack 1962, *Senoniasphaera rotundata* Clarke & Verdier 1967.

Gn: Skolochorate cysts with a non-tabular arrangement of spines or processes. *Heterosphaeridium difficile* (Manum & Cookson 1964) Ioannides, *Odontchitina costata* 1986, *Pervosphaeridium* sp. cf. *P. pseudhystricho-dinium* (Deflandre 1937) Yun 1981.

Gi: Paratabulation indicated by the arrangement of the processes and/or other ornament. *Hystrichosphaeridium bowerbankii* Davey & Williams 1966, *Oligosphaeridium prolixisphaeridium*, *Oligosphaeridium complex* (White 1842) Davey & Williams 1966, *Florentinia buspina* (Davey & Verdier 1976) Duxbury 1980, *Florentinia torulosa* (Davey & Verdier 1976) Lentin & Williams 1981, *Litosphaeridium* sp. A Marshall & Batten (1988).

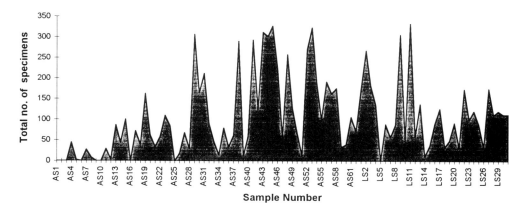

Fig. 5. Abundance (total number of specimens) per sample from Dover, Kent.

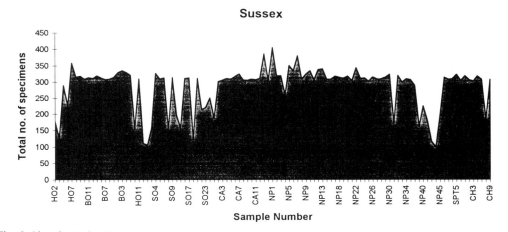

Fig. 6. Abundance (total number of specimens) per sample from Sussex.

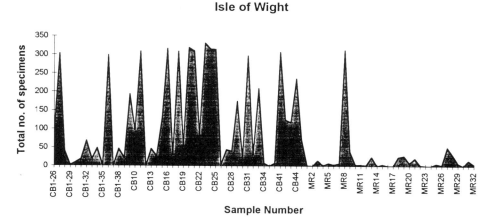

Fig. 7. Abundance (total number of specimens) per sample from the Isle of Wight.

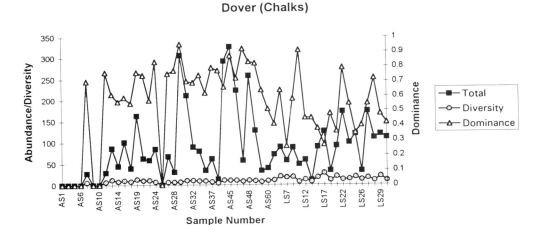

Fig. 8. Abundance, diversity and dominance of Chalk samples from Dover, Kent.

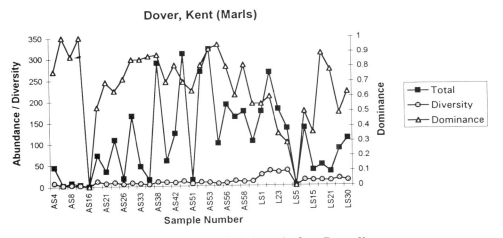

Fig. 9. Abundance, diversity and dominance of Marl samples from Dover, Kent.

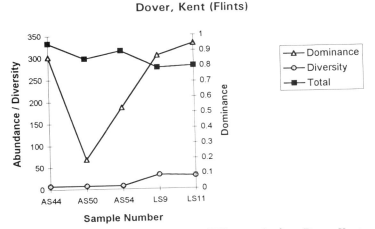

Fig. 10. Abundance, diversity and dominance of Flint samples from Dover, Kent.

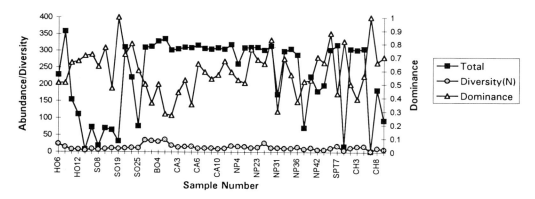

Fig. 11. Abundance, diversity and dominance of Chalk samples from Sussex.

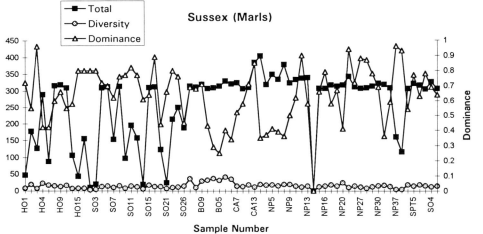

Fig. 12. Abundance, diversity and dominance of Marl samples from Sussex.

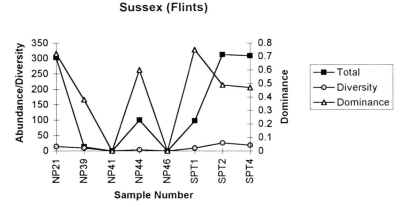

Fig. 13. Abundance, diversity and dominance of Flint samples from Sussex.

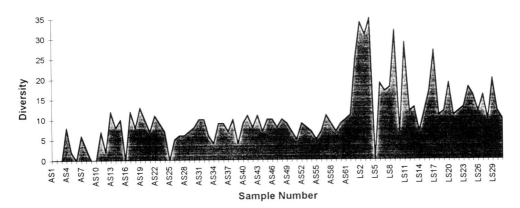

Fig. 14. Diversity of samples from Dover, Kent.

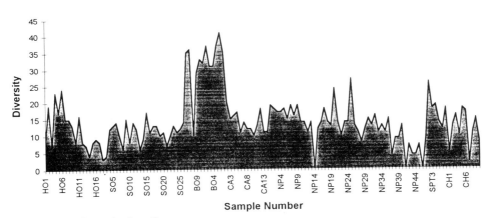

Fig. 15. Diversity of samples from Sussex.

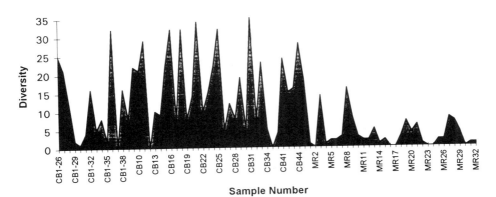

Fig. 16. Diversity of samples from the Isle of Wight.

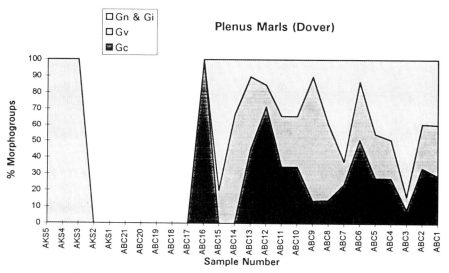

Fig. 17. Gc, Gv, Gn & Gi morphological groups from the Plenus Marls, Dover (Kent). Data taken from Jarvis *et al.* (1988).

diversities up to 42 in samples from Beachy Head. Samples from the base of the sequence at Compton Bay, the Isle of Wight show greater diversities (0–32) than those in Kent or Sussex. In Sussex, where the recovery and the abundances are good (300 or greater), the diversities still vary (0–42).

Dominance

The dominance (D) of samples was calculated using the following equation from Goodman (1979).

Dominance (D) =
$$\frac{\text{Total number of the two most common species}}{\text{Total number of specimens counted}}$$

Highest dominance = 1; no dominance = 0.5; lowest dominance = 0.

The dominance varies in all areas (Figs 8 to 13). The highest dominances were found in samples of low abundances and diversities (< 30), which might be expected.

Statistical analyses

Statistical measurements were made to analyse the relationship between the dinoflagellate cyst assemblages and their stratigraphic occurrences. Various clustering methods were used on the raw data and the morphological groups formulated in this study. These do not exhibit any significant clustering, except for the R-mode analysis (morphogroups) which show some trends. The D, Pp, Gp and Gq morphogroups are always closely grouped. This is interpreted as being due to their rare occurrence, which is their only similarity. Other groups which have unrelated occurrences include Gc, Gv, Gi and Gn. The variability of the assemblages between samples is therefore unexplained, because the statistical analyses did not produce any confident results.

Comparison with results from the Cenomanian

A large percentage of species (91%) recur after the oceanic anoxic event and have been called Lazarus taxa (Jablonski 1986). Only 9% of the total 80 species described in this study occur for the first time in the Turonian. These are marked with an asterisk (*) on Table 2. It appears that the OAE did not have a long lasting effect on the dinoflagellates. This may have been achieved by the avoidance of stressed conditions through the non-motile resting stages (cysts) of their life cycle (Dale 1976, 1983; Evitt 1985).

Much more diverse and abundant dinoflagellate cyst assemblages have been documented from the Cenomanian of Dover (Jarvis *et al.* 1988) and Humberside (Duane 1992), before the

RECOVERY OF TURONIAN DINOFLAGELLATE CYST ASSEMBLAGES 291

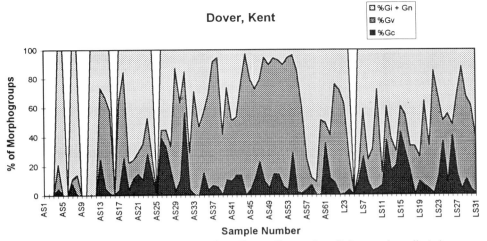

Fig. 18. Gc, Gv, Gn & Gi morphological groups from Dover, Kent, using all the samples collected.

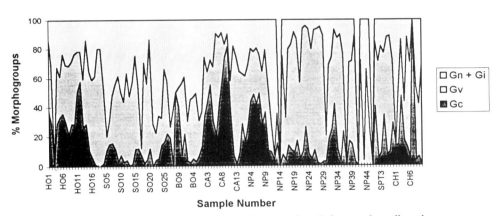

Fig. 19. Gc, Gv, Gn & Gi morphological groups from Sussex, using all the samples collected.

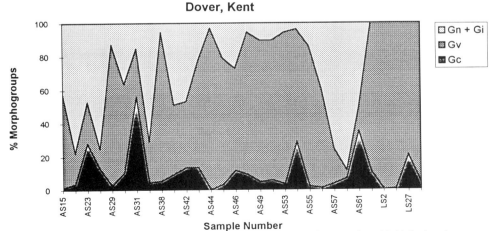

Fig. 20. Gc, Gv, Gn & Gi morphological groups from Dover, Kent, using samples with high abundance.

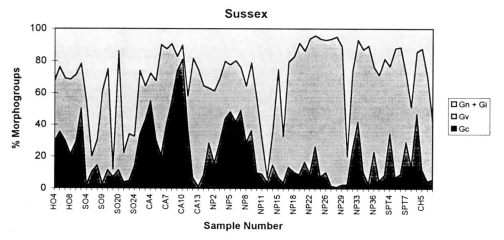

Fig. 21. Gc, Gv, Gn & Gi morphological groups from Sussex, using samples with high abundance.

onset of the OAE. The richest, most diverse samples from the Turonian do not match those from the Cenomanian. Duane (1992) records diversities of between 39 and 66 from the middle Cenomanian of Kent, and between 0 and 32 from the Late Cenomanian of Humberside. The dinoflagellate cysts exhibit low to moderate diversities (Plenus Marls Beds 1 to 7 have a diversity of 4–14; Plenus marls Beds 7 and 8 have a diversity of 0–1; Post Plenus Marls have zero diversity) but the abundances reach up to 1000 s per sample (Jarvis et al. 1988). The species occurrences from the Plenus Marls at Dover (Jarvis et al. 1988) are grouped into their dominant morphogroups as for the Turonian (Fig. 17). The four most dominant morphogroups are the same as in the Turonian, i.e. Gc, Gv, Gn and Gi.

Palaeoenvironmental significance of the morphological groupings and their recovery

Dinoflagellate cysts presently occur in both marine and fresh-water sediments. These cysts form in response to environmental stresses, usually producing 'temporary' cysts which often remain in the sediment. In this way the cell can over-winter or become dormant during periods of low nutrient levels, or of abnormal temperatures, salinities or oxygen levels, until it receives a stimulus to excyst and return to its thecate stage. Most, if not all, cysts are accepted as hypnozygotic bodies of the sexual life-cycle (Evitt 1985).

The palaeoenvironmental significance of recent and fossil dinoflagellate cyst assemblages has been discussed in many papers (e.g. Scull et al. 1966; Vozzhennikova 1967; Wall & Dale 1968; Harland 1973; Dale 1976, 1983; Harker et al. 1990). The ecological factors affecting these assemblages are, however, only beginning to be more fully understood. Salinity, water temperature, water depth (McKee et al. 1959; Reid 1975; Koethe 1990) and distance from the shoreline (Wall et al. 1977) are known to affect the distribution of dinoflagellate cysts. Several trends are recognized in the occurrence of genera, species and informal morphogroupings or families and these have been substantiated by other workers.

Four major groups of dinoflagellate cysts with different affinities were noted by Harker et al. (1990), these being of gonyaulacoid, peridinioid, gymnodinioid and uncertain origins. Fossil gonyaulacoids are divided by Harker et al. (1990) into those related to the modern genus Gonyaulax Diesing 1866, those related to the modern genus Ceratium Schrank 1793 and those represented by the genus Cyclonephelium Deflandre & Cookson 1955; emend. Sarjeant & Stover 1978. The first two appear to be most common in stable marine sediments, and the third is more common in near-shore, unstable environments. The three groups are represented in this investigation by the following morphogroups. Morphogroups Gn and Gi (Florentinia spp., Hystrichosphaeridium spp. and Oligosphaeridium spp.) are representative of the group Gonyaulax. Morphogroup Gc (Odontochitina spp.) represents the Ceratium group and morphogroup Gv

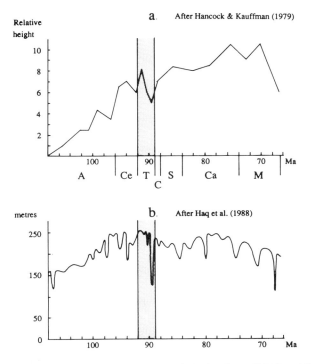

Fig. 22. Comparison of sea-level curves produced by Hancock & Kauffman (1979) and Haq *et al.* (1988).

(*Cyclonephelium* spp. and *Senoniasphaera* spp.) is representative of the *Odontochitina* group.

Gonyaulax (Gn & Gi), *Cyclonephelium* (Gv) and *Odontochitina* (Gc) occurrences are shown in Figs 18 and 19 as a percentage of the three groups for each sample. These results were modified and barren samples, or those with a low abundance, omitted to show the general trends (Figs 20 and 21). There are many controls on dinoflagellate cyst assemblages, and variations in assemblages from the Anglo-Paris Basin indicate that conditions varied within the basin, despite the high sea-levels and uniform shelf sea. Such parameters as proximity to shoreline and water depth have an important association with specific dinoflagellate cyst assemblages. The water depth may have been determined by the underlying geological features, with topographical highs providing shallower regions and perhaps restricting the circulation of water masses. Such restricted areas may have been short-lived, but others may have developed increased salinities and thus a stressed environment. Such variables are difficult to quantify and there is no single dinoflagellate cyst species which can be considered specific to a particular palaeoenvironmental setting. A series of assemblages and predominances of certain morphogroups can, however, be strongly indicative of certain conditions.

The palaeoenvironmental interpretations that can be drawn from these results are limited by the sample interval (approximately 1 m). The sample coverage of the Turonian as a whole does, however, provide a strong basis for the following general comments.

Dover

There appears to have been little terrestrial influence during the Turonian as few bisaccate pollen and spores are recorded. The Family Gymnodiniaceae (D) and Peridiniaceae (Pp & Px) are rarely present and the assemblages are dominated by Gonyaulacoid (Gn, Gi, Gp & Gs) and Ceratioid (Gv) genera and species. This indicates that the palaeoenvironment fluctuated between near-shore, shallow and deeper water marine conditions, as suggested by Harker *et al.* (1990). The variation in water depth was probably caused by a series of transgressive and regressive sequences which may have been locally modified by tectonics. Figures 18 and 20 exhibit the dominance of *Cyclonephelium*

(Gv) in relation to the two other groups throughout much of the succession, interpreted as reflecting an unstable near-shore environment. During the early and latest parts of the Turonian, the *Odontochitina* (Gc) and *Gonyaulax* (Gn & Gi) groups became dominant, reflecting deeper water and a more stable environment. This broadly agrees with the sea-level curves (Fig. 22) of Hancock & Kauffman (1979) and of Haq *et al.* (1988), who recognize a period of transgressive conditions during the early and latest parts of the Turonian. This dominance of the Gc and Gn & Gi groups also reflects the recovery of the dinoflagellate cyst assemblages during the Turonian and may have been related to the changing water depths.

Sussex

The three representative groups of Harker *et al.* (1990) are also plotted for the assemblages from Sussex in Figs 19 and 21. The succession is thicker in Sussex than at Dover, and the interpreted pattern of sea-level fluctuations is slightly different. The lower and uppermost parts of the succession are dominated by the *Gonyaulax* (Gn & Gi) and *Odontochitina* (Gc) groups, indicating deeper and more stable marine conditions. There was also a period of more stable, deeper water conditions during the latest early and mid-Turonian, indicated by the dominance of these two groups. These transgressive periods alternate with intervals of nearer shore conditions, as indicated by the dominance of the *Cyclonephelium* (Gv) group. This expanded sequence records more fluctuations in sea-level during the Turonian than recorded at Dover, indicating it was in a shallower part of the basin. The intervals when deeper water conditions dominated generally correspond with the transgressive peaks exhibited by the sea-level curve of Haq *et al.* (1988). The sea-level curve of Hancock & Kauffman (1979) corresponds with information from the dinoflagellate cysts for the early and late Turonian transgressive phases, but Hancock & Kauffman (1979) do not record a mid-Turonian transgressive period. As this sea-level curve is based partly on information from NW Europe, it is surprising that it does not correspond more closely with the depth changes recorded by the changing dinoflagellate cyst assemblages from southern England. This indicates that local influences (e.g. tectonics) may have played an important role in locally modifying sea-level changes.

It is possible to recognize changes in the cyst assemblages that may be linked to changing environmental conditions (following Harker *et al.* 1990). The palynological morphogroups (Figs 19 and 21) therefore reflect the marine conditions under which they were deposited and agree with the transgressive peaks described by Hancock (1989) for the Turonian. Stable marine conditions or more unstable nearer-shore situations can be invoked, which generally correspond to changes in sea-level documented from other evidence by Hancock & Kauffman (1979), Haq *et al.* (1988) and by Hancock (1989). The dinoflagellate cyst assemblages at Sussex record a third fluctuation of marine conditions during the latest early and mid-Turonian. These variations may be accounted for by local tectonics, whose deformation is described by Bergerat & Vandycke (1994). Alternatively it may have been missed in the extant global record. The causal mechanism for these sea-level changes is not suggested on the basis of these new observations.

Conclusions

The dinoflagellate cyst assemblages were one of the first groups to recover from the oceanic anoxic event (OAE) at the close of the Cenomanian (Jarvis *et al.* 1988), with their abundances and diversities reaching a maximum during the mid-Turonian. The diversities are, however, lower than the diverse assemblages in the Cenomanian noted by other workers. Many species present in the Turonian recur after their apparent demise at the end of the Cenomanian, hence the term Lazarus taxa. Only 9% of the species described occur for the first time in the Turonian. Diversity and abundance indices vary greatly throughout the succession and are not lithologically controlled. The assemblages are dominated by four morphological groupings and these can be used to infer variation in the depositional environment.

The work for this paper was carried out at the University of Plymouth and was supported by a University Research Assistantship. The author would like to thank M. Hart and D. Peacock for their critical comments and advice in the preparation of this manuscript. Thanks also to D. Peacock for help with the diagrams. Lithological detail, and assistance in examining the sections was provided by R. N. Mortimore and I. Jarvis.

References

AGELOUPOULOS, J. 1964. *Hystrichostrogylon membraniphorum* n.g. n.s.p. aus dem Heiligenhafner Kieselton (Eozan). *Neues Jarhbuch fur Geologie und Palaontologie, Monastshefte*, 673–675.

ALBERTI, G. 1961. Zur kenntnis mesozoischer und altterierer Dinoflagellaten und Hystrichosphaerideen von Nord- und Mitteldeutschland sowie einigen anderen europaischen Gebieten. *Palaeontographica*, Abt.A, **116**, 1–58.

BARSS, M. S. & WILLIAMS, G. L. 1973. Palynology and Nannofossil processing techniques. *Geological Survey of Canada*, Paper, **73-26**, 22.

BELOW, R. 1981. Dinoflagellaten-Zysten aus dem oberen Hauterive bis unterer Cenoman Sud West-Marokkos. *Palaeontographica* Abt. B, **176**, 1–145.

—— 1982. Dinoflagellate cysts from Valanginian to Lower Hauterivian sections near Ait Hamouch, Morocco. *Revista Espanola de Micropalaeontologia*, **14**, 23–52.

BERGERAT, F. & VANDYCKE, S. 1994. Palaeostress analysis and geodynamical implications of Cretaceous–Tertiary faulting in Kent and the Boulonnais. *Journal of the Geological Society, London*, **151**, 439–448.

BRIDEAUX, W. W. 1971. Palynology of the Lower Colorado Group, Central Alberta, Canada. I. Introductory remarks, geology and microplankton studies. *Palaeontographica* Abt. B, **135**, 53–114.

BUZAS, M. A. 1979. The measurement of species diversity. *In:* LIPPS, J. G., BERGER, W. H., BUZAS, M. A., DOUGLAS, R. G. & ROSS, C. A. (eds) *Foraminiferal Ecology and Palaeoecology*. Society of Economic Palaeontologists and Mineralogists, Short Course Notes, **6**, 3–9.

CLARKE, R. F. A. & VERDIER, J. P. 1967. An investigation of microplankton assemblages from the Chalk of the Isle of Wight, England. *Verhandelingen der Koninklijke Nederlandsche Akademie van Wetenschappen, Afdeeling Natuurkunde, Eerste Reeks*, **24**, pls 1–17.

CLAYTON, C. J. 1986. The chemical environment of flint formation in Upper Cretaceous chalks *In:* SIEVEKING, G. DE C. & HART, M. B. *The scientific study of flint and chert*. Cambridge University Press, 43–54.

COOKSON, I. C. & EISENACK, A. 1958. Microplankton from Australian and New Guinea Upper Mesozoic sediments. *Proceedings of the Royal Society of Victoria*, **70**, 17–79.

—— & —— 1960. Microplankton from Australian Cretaceous sediments. *Micropalaeontology*, **6**, 1–18.

—— & —— 1961. Upper Cretaceous microplankton from the Belfast No. 4 bore, southwestern Victoria. *Proceedings of the Royal Society of Victoria*, **74**, 69–76.

—— & —— 1962. Additional microplankton from Australian Cretaceous sediments. *Micropaleontology*, **8**, 485–507.

CORRADINI, D. 1973. Non-calcareous microplankton from the Upper Cretaceous of the Northern Apennines. *Bollettino della Societa Paleontologia Italiana*, **11**, 119–197.

DALE, B. 1976. Cyst formation, sedimentation and preservation: Factors affecting dinoflagellate assemblages from Trondheimsfjord, Norway. *Review of Palaeobotany & Palynology*, **22**, 36–60.

—— 1983. Dinoflagellate resting cysts: 'benthic plankton'. *In:* FRYXELL, G. A. (ed.) *Survival strategies of the Algae*. Cambridge University Press, 69–136.

DAVEY, R. J. 1969. Non-calcareous microplankton from the Cenomanian of England, Northern France and North America, Part I. *Bulletin of the British Museum (Natural History) Geology*, **17**, 103–180.

—— 1974. Dinoflagellate cysts from the Barremian of the Speeton Clay, England. *Symposium on Stratigraphic Palynology, Birbal Sahni Institute of Palaeobotany, Special Publication*, **3**, 41–75.

—— 1979. Marine Apto-Albian palynomorphs from Holes 400A and 402A, IPOD leg 48, Northern Bay of Biscay. *Initial Reports of the Deep Sea Drilling Project*, **48**, 547–577.

—— & VERDIER, J. P. 1973. An investigation of microplankton assemblages from latest Albian (Vraconian) sediments. *Revista Espanola Micropalaeontologia*, **5**, 173–212.

—— & —— 1976. A review of certain nontabulate Cretaceous Hystrichosphaerid dinocysts. *Review of Palaeobotany and Palynology*, **22**, 307–335.

—— & WILLIAMS, G. L. 1966. The genus *Hystrichosphaeridium* and its allies. *In:* DAVEY, R. J., DOWNIE, C., SARJEANT, W. A. S. & WILLIAMS, G. L. *Studies on Mesozoic and Cainozoic dinoflagellate cysts. Bulletin of the British Museum (Natural History) Geology*, Supplement 3, 53–106.

——, DOWNIE, C., SARJEANT, W. A. S. & WILLIAMS, G. L. 1966. Fossil dinoflagellate cysts attributed to *Baltisphaeridium*. Studies on Mesozoic and Cainozoic dinoflagellate cysts; *Bulletin of the British Museum (Natural History) Geology*, Supplement 3, 157–175.

——, ——, & —— 1969. Generic reallocations. Appendix to *Studies on Mesozoic and Cainozoic dinoflagellate cysts. Bulletin of the British Museum (Natural History) Geology*, Appendix to supplement 3, 15–17.

DEFLANDRE, G. 1935. Considerations biologiques sur les microorganismes d'origine planctonique conserves dans les silex de la Craie. *Bulletin biologique de la France et de la Belgique*, **69**, 213–244.

—— 1936. Microfossiles des silex Cretaces. Premiere partie. Generalites flagelles. *Annales de Paleontologie*, **25**, 151–191.

—— 1937. Microfossiles des silex Cretaces. Deuxieme partie. Flagelles incertae sedis. Hystrichosphaerides, Sarcodines. Organismes divers. *Annales de Paleontologie*, **26**, 51–103.

—— & COOKSON, I. C. 1955. Fossil microplankton from Australian late Mesozoic and Tertiary sediments. *Australian Journal of Marine and Freshwater Research*, **6**, 242–313.

DIESING, K. M. 1866. Revision de Prothelminthen. *S - B. Akad. Wissensch. Wien, math.-nat.*, **287**

DUANE, A. M. 1992. *Palynological investigations of Cenomanian chalks and marls from England*. PhD Thesis, University of Plymouth.

DUXBURY, S. 1977. A palynostratigraphy of the Berriasian to Barremian of the Speeton Clay of Speeton, England. *Palaeontographica, Abt. B,* **160**, 17–67.

—— 1980. Barremian phytoplankton from Speeton, East Yorkshire. *Palaeontographica, Abt. B,* **173**, 107–146.

—— 1983. A study of dinoflagellate cysts and acritarchs from the Lower Greensand (Aptian to Lower Albian) of the Isle of Wight, Southern England. *Palaeontographica, Abt.B,* **186**, 18–80.

EHRENBERG, C. G. 1838. Tber das Massenverhaltniss der jetzt lebenden kiesel-infusorien und uber ein neues Infusorien-Conglomerat als Polirschiefer von Jastraba in Ungarn. *Abhandlungen der Preussischen Akadamie der Wissenschaften,* **1936**, 108–135.

EISENACK, A & KJELLSTROM, G. 1971. Katalog der fossilen Dinoflagellaten, Hystrichospharen und verwandten Mikrofossilien. Band III. Acritarcha. *E. Schweizerbart'sche Verlagsbuchhandlung, Monatshefte.*

EVITT, W. R. 1963. A discussion and proposals concerning fossil dinoflagellates, Hystrichospheres and Acritarchs. *Proceedings of National Academy of Sciences, Washington,* **49**, 158–164.

—— 1985. *Sporopollenin dinoflagellate cysts: their morphology and interpretation.* American Association of Stratigraphic Palynologists Foundation.

FITZPATRICK, M. E. J. 1992. *Turonian dinoflagellate cyst assemblages from southern England.* PhD thesis, University of Plymouth.

—— 1995. Dinoflagellate cyst biostratigraphy of the Turonian from southern England. *Cretaceous Research,* (in press).

GOCHT, H. 1959. Mikroplankton aus dem Nordwestdeutschen Neokom (Teil II). *PalaeoZeitschrift,* **33**, 50–89.

GOODMAN, D. K. 1979. Dinoflagellate 'communities' from the Lower Eocene Nanjemoy Formation of Maryland, U.S.A. *Palynology,* **3**, 169–190

HANCOCK, J. M. 1976 for 1975. The Petrology of the Chalk. *Proceedings of the Geologists' Association,* **86**, 499–535.

—— 1989. Sea-level changes in the British region during the Late Cretaceous. *Proceedings of the Geologists' Association,* **100**, 565–594.

—— & KAUFFMAN, E. G. 1979. The great transgressions of the Late Cretaceous. *Journal of the Geological Society of London,* **136**, 175–186.

HAQ, B. U., HARDENBOL, J. & VAIL, P. R. 1987. Chronology of fluctuating sea levels since the Triassic. *Science,* **235**, 1156–1167.

——, —— & —— 1988. Mesozoic and Cenozoic chronostratigraphy and cycles of sea-level change. *In: Sea-level changes – an integrated approach.* Society of Economic and Petroleum Mineralogists, Special Publication, **42**, 71–108.

HARKER, S. D., SARJEANT, W. A. S. & CALDWELL, W. G. E. 1990. Late Cretaceous (Campanian) organic-walled microplankton from the Interior Plains of Canada, Wyoming & Texas: biostratigraphy, palaeontology and environmental interpretation. *Palaeontographica, Abt. B,* **219**, 1–243.

HARLAND, R. 1973. Dinoflagellate cysts and acritarchs from the Bearpaw Formation (Upper Campanian) of southern Alberta, Canada. *Palaeontology,* **16**, 665–706.

——, ARMSTRONG, R. L., COX, A. V., CRAIG, L. E., SMITH, A. G. & SMITH, D. G. 1990. *A geologic time scale.* Cambridge University Press.

HELBY, R. 1987. *Muderongia* and related dinoflagellates of the latest Jurassic to earliest Cretaceous of Australia. *In:* JELL, P. A. (ed.) *Australian Mesozoic Palynology.* Association of Australian Palaeontologists Memoir, **4**, 135–141.

IOANNIDES, N. S. 1986. Dinoflagellate cysts from Upper Cretaceous–Lower Tertiary sections, Bylot and Devon Islands, Arctic Archipelago. *Geological Survey of Canada: Bulletin,* **371**, 1–49.

JABLONSKI, D. 1986. Causes and consequences of mass extinctions. *In:* ELLIOT, D. K. (ed.) *Dynamics of Extinction,* Wiley, New York, 183–229.

JAIN, K. P. 1977. Additional dinoflagellates and acritarchs from the Grey Shale Member of the Dalmaipuram Formation, South India. *Palaeobotanist,* **24**, 107–194.

JARVIS, I., CARSON, G. A., COOPER, M. K. E., HART, M. B., LEARY, P. N., TOCHER, B. A., HORNE, D. & ROSENFELD, A. 1988. Microfossil assemblages and the Cenomanian–Turonian (late Cretaceous) oceanic anoxic event. *Cretaceous Research,* **9**, 3–103.

KOETHE, A. 1990. Paleogene dinoflagellates from Northwest Germany – biostratigraphy and paleoenvironment. *Geologisches Jahrbuch,* **118**, 3–111.

LEARY, P. N., CARSON, G. A., COOPER, M. K. E., HART, M. B., HORNE, D., JARVIS, I., ROSENFELD, & TOCHER, B. A. 1989. The biotic response to the late Cenomanian oceanic anoxic event; evidence from Dover, SE England. *Journal of the Geological Society London,* **146**, 311–317.

LEJEUNE-CARPENTIER, M. & SARJEANT, W. A. S. 1983. Restudy of some smaller dinoflagellate cysts from the Upper Cretaceous of Belgium. *Annales de la Societe Geologique de Belgique,* **106**, 1–17.

LENTIN, J. K. & WILLIAMS, G. L. 1973. Fossil dinoflagellates: index to genera and species. *Geological Survey of Canada,* Paper **73–42**.

—— & —— 1976. A Monograph of fossil peridinoid dinoflagellate cysts. *Bedford Institute of Oceanography,* Report series **BI-R-75-16**.

—— & —— 1977. Fossil dinoflagelllates: index to genera and species. *Bedford Institute of Oceanography,* Report series **BI-R-77-8**.

—— & —— 1981. *Fossil dinoflagellates: index to genera and species.* Bedford Institute of Oceanography, Report Series **BI-R81-12**, 1–345.

LOEBLICH, A. R. Jr. & LOEBLICH, A. R. III 1966. Index to the genera, subgenera and sections of the Pyrrhophyta. *Studies in Tropical Oceanography, Miami,* **3**.

LUCAS-CLARKE, J. 1984. Morphology of species of Litosphaeridium (Cretaceous Dinophyceae). *Palynology,* **8**, 165–193.

—— 1987. *Wigginsella* n. gen. *Spongodinium* and *Apteodinium* as members of the *Aptiana-Ventrio-*

sum complex (fossil Dinophyceae). *Palynology*, 11, 155–184.

McKee, E. D., Chronic, J. & Leopold, E. B. 1959. Sedimentary belts in Lagoon of Kapingamarangi atoll. *AAPG Bulletin*, 43, 501–562.

Manum, S. & Cookson, I. C. 1964. Cretaceous microplankton in a sample from Graham Island, Arctic Canada, collected during the second 'Fram' expedition (1898–1902) with notes on microplankton from the Hassel Fmn., Ellef Ringhes Island. *Schrifter utgitt av det Norske Videnskaps-Akademie; Oslo, I Mat-Naturv. Klasse, Ny series*, 17, 1–35.

Marshall, K. & Batten, D. J. 1988. Dinoflagellate cyst associations in Cenomanian–Turonian 'Black shale' sequences of Northern Europe. *Review of Palaeobotany and Palynology*, 54, 85–103.

May, F. E. 1977. Functional morphology, palaeoecology and systematics of *Dinogymnium* tests. *Palynology*, 1, 103–121.

——— 1980. Dinoflagellate cyst of the Gymnodiniaceae, Peridiniaceae and Gonyaulacaceae from the Upper Cretaceous, Monmouth Group, Atlantic Highlands, New Jersey. *Palaeontographica Abt.B*, 172, 10–116.

Mortimore, R. N. 1986. Stratigraphy of the Upper Cretaceous White Chalk of Sussex. *Proceedings of the Geologists' Association*, 97, 97–139.

Phipps, D. & Playford, G. 1984. *Laboratory techniques for extraction of palynomorphs from sediments*. Papers, Department of Geology, University of Queensland, 11, 1–23.

Reid, P. C. 1975. A regional subdivision of dinoflagellate cyst around the British Isles. *New Phytologist*, 75, 586–603.

Robinson, N. D. 1986. Lithostratigraphy of the Chalk Group of the North Downs, Southeast England. *Proceedings of the Geologists' Association*, 97, 141–170.

Sarjeant, W. A. S. 1966. Dinoflagellate cysts with *Gonyaulax*-type tabulation. *In:* Davey, R. J., Downie, C., Sarjeant, W. A. S. & Williams, G. L. Studies on Mesozoic and Cainozoic dinoflagellate cysts. *Bulletin of the British Museum (Natural History) Geology*, Supplement, 3, 107–156.

——— & Stover, L. E. 1978. Cyclonephelium and Tenua: a problem in dinoflagellate cyst taxonomy. *Grana*, 17, 47–54.

Schrank, F. von P. 1793. Mikroskopische Wahrnemungen, *Der Naturforscher*, 27, 26–37.

Scull, B. J., Felix, C. J., McCaleb, S. B. & Shaw, W. G. 1966. The interdisciplinary approach to palaeoenvironmental interpretations. *Transactions of the Gulf Coast Association Geological Societies*, 16, 81–117.

Singh, C. 1964. Microflora of the Lower Cretaceous Mannville Group, East-Central Alberta. *Research Council of Alberta*, Bulletin 44, 1–322.

Stover, L. E. & Evitt, W. R. 1978. *Analyses of pre-Pleistocene organic-walled dinoflagellates*. Stanford University Publications in the Geological Sciences, 15.

Tocher, B. A. & Jarvis, I. 1987. Dinoflagellate cysts and stratigraphy of the Turonian (Upper Cretaceous) near Beer, southeast Devon, England. *In:* Hart, M. B. (ed.) *Micropalaeontology of Carbonate Environments*. Ellis Horwood, Chichester, 138–175.

Vozzhennikova, T. F. 1967. *Fossilized Peridinid Algae in the Jurassic, Cretaceous and Palaeogene deposits of the USSR*. National Lending Library for Science and Technology. Translation from Russian.

Wall, D. & Dale, B. 1968. Modern dinoflagellate cysts and evolution of the peridiniales. *Micropalaeontology*, 14, 265–304.

———, ———, Lohmann, G. P. & Smith, W. K. 1977. The environmental and climatic distribution of dinoflagellate cysts in modern marine sediments from regions in the North and South Atlantic. *Marine Micropalaeontology*, 2, 121–200.

Wetzel, O. 1933. Die in organischer Substanz erhaltenen Mikrofossilen des baltischen Kreide-Feuersteins mit einem sediment–petrographischen und stratigraphischen. Anhang. *Palaeontographica Abt. A*, 77, 141–188.

White, H. H. 1842. VI. On fossil Xanthidia. *Microscopical Journal*, 11, 35–40.

Woodroof, P. B. 1981. *Faunal and stratigraphic studies in the Turonian of the Anglo-Paris Basin*. PhD Thesis, Oxford University.

Yun, H. S. 1981. Dinoflagellaten aus der Oberkreide (Santon) von Westfalen. *Palaeontographica, Abt. B*, 177, 1–89.

Post-crisis recovery of Campanian desmoceratacean ammonites from Sakhalin, far east Russia

ELENA A. YAZYKOVA

VSEGEI, Srednii pr. 74, 199026 St Petersburg, Russia

Abstract: The Santonian–Campanian boundary in Sakhalin marks a major Cretaceous mass extinction event. In the late Santonian there was a decrease in the taxonomic diversity of ammonoids as well as significant changes in shell morphology which were probably related to a global regression event. However, the fauna of the Pacific region is endemic and precise correlation between the Pacific and European regions is impossible. In this paper the stratigraphic position of the boundary between the Santonian and Campanian in Sakhalin is revised and the post-crisis recovery of the ammonoid diversity is discussed. Detailed sampling of the Santonian–Campanian sequence of Sakhalin (Naiba section, Orlovka section and others) allows identification of the phyllogenetic links between the different ammonite morphotypes in the pre-crisis communities. Comparison with adjacent regions in northeast Russia and Japan is also attempted. The ammonite extinctions at the Santonian–Campanian boundary are a good marker for stratigraphic correlation. During the crisis event ammonites of the superfamilies Acanthocerataceae and Desmocerataceae were affected most strongly whereas phylloceratids and lytoceratids were not significantly changed at that time. Heteromorphic ammonites are not discussed here. The pre-crisis community is characterized by a predominance of Acanthocerataceae, while the Desmocerataceae had a subordinate role. Representatives of these two groups form a community with large-size shells and strong ornamentation and this is not seen in the ammonites of the post-crisis community. The maximum occurrence of the Superfamily Desmocerataceae is recorded in the Middle Campanian, when the first diversification occurred (Family Pachydiscidae). Zonal correlation of the Santonian to Campanian succession of northeast Russia and Japan is discussed.

Ammonoid evolution was interrupted by several mass extinction events during the Late Cretaceous. One of the most important crises occurred at the Santonian–Campanian boundary and strongly affected the studied groups. The Late Santonian sea-level changes mark a worldwide regression (Matsumoto 1977; Hancock & Kauffman 1979; Hancock 1993) and it was probably those changes in the environment that were the main reason for the reductions of taxonomic diversity and the changes in the shell morphology of the ammonites.

The Sakhalin's Santonian to Campanian marine terrigenous succession with ammonite fossils consists mostly of sandstones and black shales with a total thickness of up to 2000 m. Typical facies are shales with beds and lensoid beds of sandstones up to 1 m thick, tuffaceous sandstones and siltstones. In the Santonian the sandstone facies was widespread while the mudstone facies is more common in the Campanian. Numerous fossils have been found, mainly in marly nodules that form 'concretion beds' in some sections. The ammonite distribution in the Sakhalin succession is quite variable,

ranging from single specimens to great abundance in the different sedimentary units. Discussion of the evolution of the Campanian ammonites and their population dynamics is based on detailed sampling of the Bykovskaya and Krasnoyarkovskaya Formations (Fig. 1) as well as three sections of the Zhonkierskaya Formation located along the Orlovka and Onora Rivers and in the Cape of Zhonkier (Fig. 2).

The Naiba section is the reference stratigraphic section for Sakhalin Island (Poyarkova 1987). Three Campanian ammonite zones (*Anapachydiscus* (Neopachydiscus) *naumanni*, *Pachydiscus (P.) egertoni* and *Canadoceras multicostatum*) have been recognized in this section (Zonova *et al.* 1993). The underlying Santonian and Maastrichtian deposits contain a continuous sequence of ammonite zones (Zonova *et al.* 1993).

Ranges of the ammonite species have been precisely established in the Sakhalin section and this allows the identification of regionally useful ammonite zones (Fig. 3). In addition to this, specimens from synchronous deposits of Kam-

From Hart, M. B. (ed.), 1996, Biotic Recovery from Mass Extinction Events,
Geological Society Special Publication No. 102, pp. 299–307

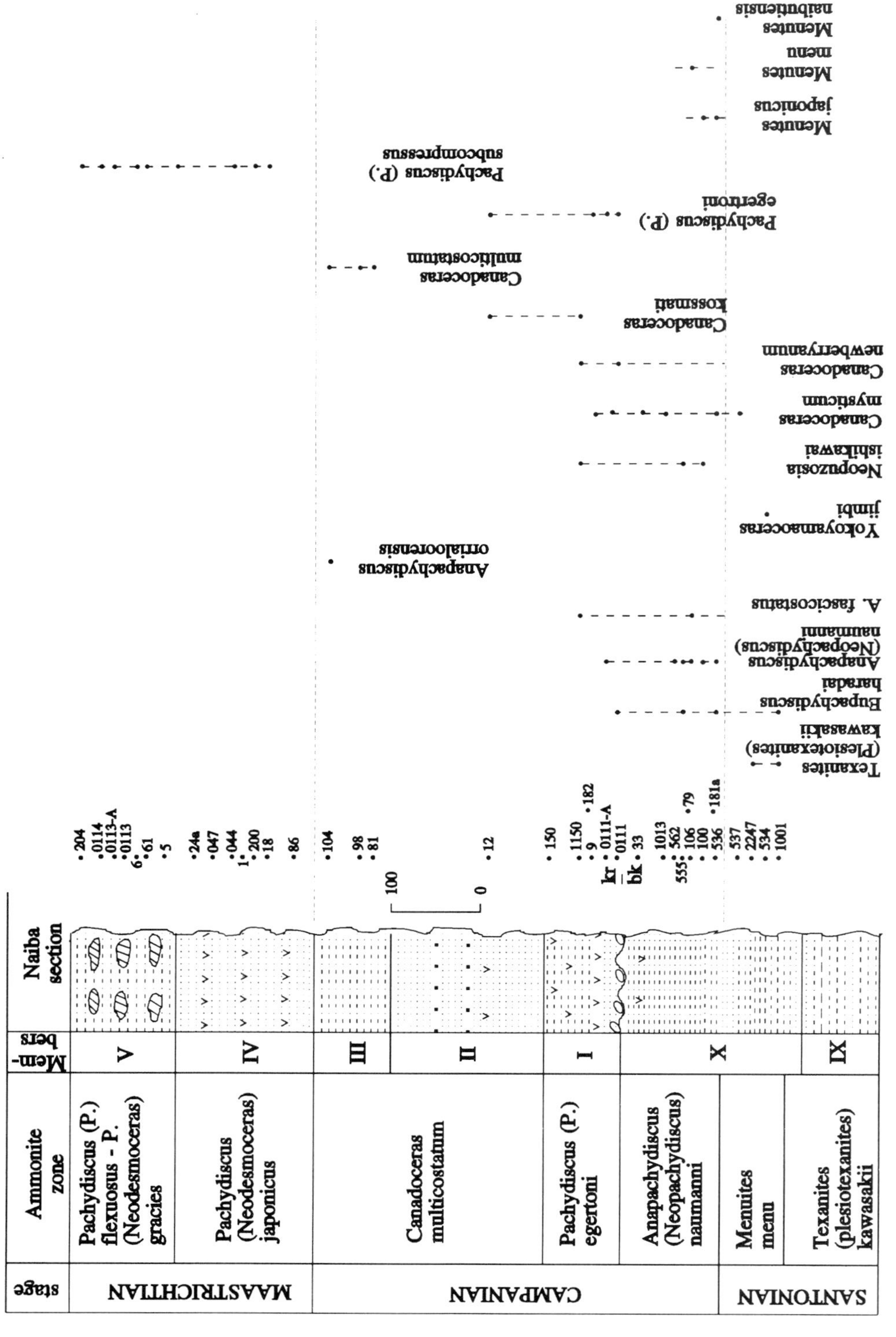

Fig. 1. The Santonian–Maastrichtian stratigraphic section with ammonoids, Bykovskaya and Krasnoyarkovskaya Formations.

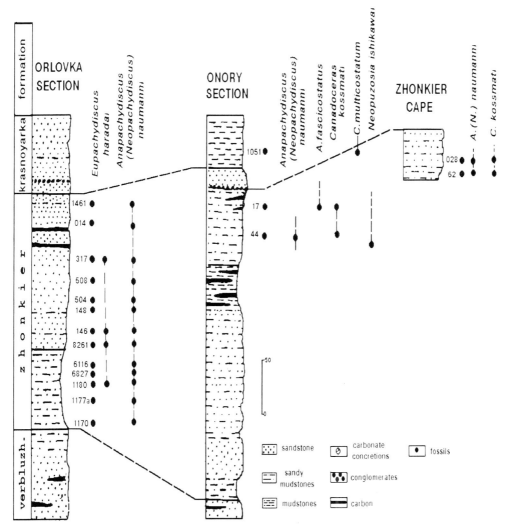

Fig. 2. Santonian–Campanian stratigraphic sections with ammonoids, zhonkierskaya formation.

chatka, Korjakia (northeastern Russia) and Shikotan Island (Kuril Islands) have also been used for comparative analysis.

Extinction

The Santonian ammonite fauna began to change from the beginning of the Late Santonian regression. In Sakhalin this event is recognized by an abrupt decrease of the taxonomical and morphological diversity of the Superfamily Acanthocerataceae (Order Ammonitida) in the section of the *Texanites (Plesiotexanites) kawasakii* Zone. The acanthoceratids which dominated from the Cenomanian to Coniacian interval by number of species and abundance in the sample localities is represented in the Santonian succession by only *Texanites (Plesiotexanites) kawasakii* (Kawada). Moreover, this species dis-appeared during the final extinctions at the Santonian–Campanian boundary. No taxa from the Superfamily Acanthocerataceae are known above this boundary in Sakhalin. However, rare representatives of this Superfamily have been found in the Lower Campanian succession of Japan (Matsumoto & Haraguchi 1978) and this suggests that their existence in far east Russia is still open to revision.

The other studied Superfamily is the Desmo-

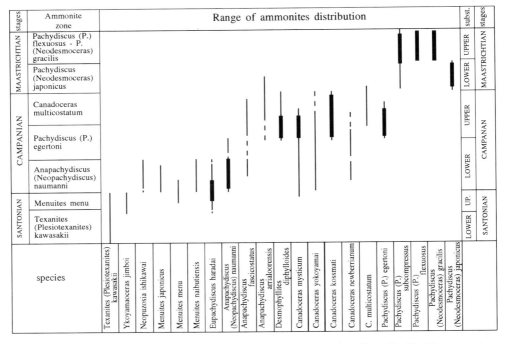

Fig. 3. Stratigraphic distribution of the Santonian–Maastrichtian ammonites (Superfamilies Desmocerataceae and Acanthocerataceae), Sakhalin Island.

cerataceae. Desmoceratids have a subordinate role in the Cenomanian to Early Santonian interval being represented only by *Yokoyamaoceras jimboi* Matsumoto and *Neopuzosia ishikawai* Jimbo. The former disappeared during the crisis event whereas the latter continued into the post-crisis interval and gave rise to a new lineage (Fig. 4).

New representatives of the Family Pachydiscidae (Superfamily Desmocerataceae) appeared during this event or immediately after it. They were *Eupachydiscus haradai* Jimbo and *Canadoceras mysticum* Matsumoto (Fig. 5).

In general, the pre-crisis community of the Coniacian to Santonian ammonites consisted of two Superfamilies (Acanthocerataceae and Desmocerataceae). They had large-size shells and distinct ornamentation with strong coarse riblets, 'plicae', 'spines' and high strong tubercles that were arranged in several rows on the shell surface. Some forms had a deep construction, high collars and pointed keel. Such coarse elements have not been found in the post-crisis ammonite community in Sakhalin. Similar features of the pre-crisis community are typical for northeast Russia that has been established by comparative analysis of the northeast Russia and Sakhalin material.

The crisis is expressed more clearly in sections described by Japanese geologists. According to their data there are two families (Collignoniceratidae and Muniericeratidae) of the Superfamily Acanthocerataceae in the Santonian *Texanites (Plesiotexanites) kawasakii* Zone (Matsumoto & Kanie 1979; Toshimitsu *et al.* 1991). The former consists of eight species (Matsumoto & Haraguchi 1978) and the latter contains approximately ten species (Matsumoto & Obata 1982). Some taxa which disappear in the early Campanian have also been found (Matsumoto & Haraguchi 1978). Such sharp differences in the number of described species between Japan and Sakhalin may be explained by the lack of detailed studies of the Sakhalin succession or by some differences in the environment. In either case, the last representatives of the Superfamily Acanthocerataceae contain only one species that disappears in the latest Santonian. The Superfamily Desmocerataceae survived to the post-crisis interval and reached its maximum abundance in the Campanian (Fig. 5). Two other Cretaceous ammonoid orders (Phylloceratida and Lytoceratida) formed long-lived lineages and were not significantly changed during this event (Fig. 4). Heteromorphic ammonites, which are widespread in Sakhalin are not discussed

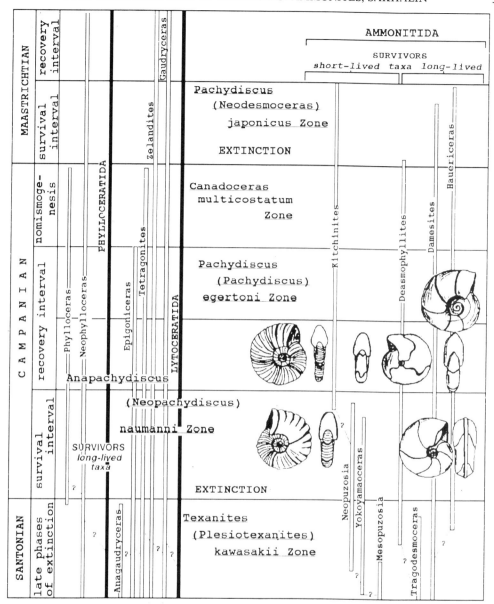

Fig. 4. Recovery of Campanian Ammonoidea in the Sakhalin sections.

here. The behaviour of this group across the event will be described in another paper.

Survival

This interval corresponds to the first half of the *Anapachydiscus (Neopachydiscus) naumanni* Zone (Figs 4, 5). During that time long-lived taxa of the Family Desmocerataceae (genera *Neopuzosia*, *Damesites*, *Desmophyllites* and *Hauericeras*) continued their existence and representatives of the genus *Kitchinites* appeared (Fig. 5). The genus disappeared during the Early Campanian. All these genera, except *Neopuzosia*, are conservative lines that continued up into Maastrichtian. The youngest sedimentary units containing *Neopuzosia* are of the Early Campanian age.

The first examples of the new Family Pachydiscidae (Superfamily Desmocerataceae) ap-

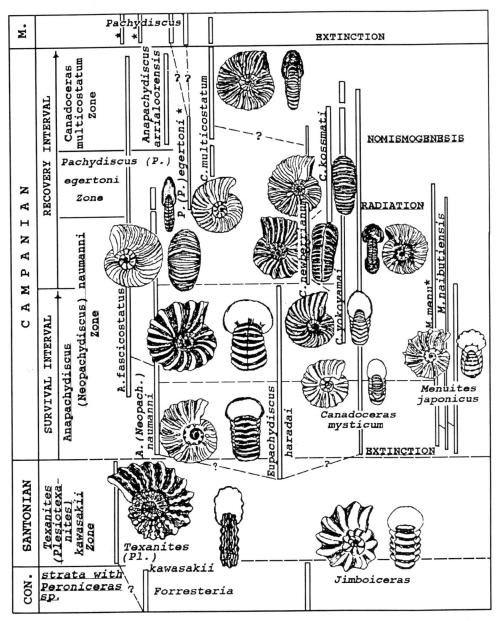

Fig. 5. Recovery of Campanian ammonoidea in the Sakhalin sections. —— Zone boundaries; –?– phyllogenetic links; · — · — interval boundaries; * immigrant species.

peared in the Late Santonian. They are represented by *Eupachydiscus haradai* Jimbo and a species which arose from it, *Canadoceras mysticum* Matsumoto. The boundary between the Santonian and Campanian stages is established by the first appearance of *Anapachydiscus (Neopachydiscus) naumanni* Yokoyama and *Inoceramus nagaoi* Matsumoto & Ueda. Both genera appear together in large numbers. There is some similarity between the Pacific Ocean species *A. (N.) naumanni* and *Eupachydiscus levyi* which is widespread in the Lower Campanian of Middle Asia. The same can be said of the Pacific Ocean species *Inoceramus nagaoi* and *I.*

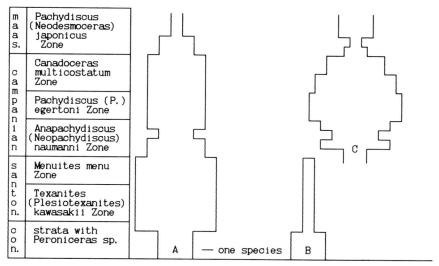

Fig. 6. Dynamics of the taxonomic diversity of some of the families of the Cretaceous Ammonoidea: A, Desmoceratidae; B, Collignoniceratidae; C, Pachydiscidae.

azerbaidjanensis. The latter has been found in the Lower Campanian of the southern USSR, Bulgaria and the USA.

Immediately after the extinction event there was the appearance of the short-lived taxa such as *Menuites menu* (Forbes), *M. naibutiensis* Matsumoto, *M. japonicus* Matsumoto. These continued the lineage of the genus *Anapachydiscus* (Kennedy & Klinger 1993) but without further development. The first species is widespread in Tethys whereas the other two are endemic. All of them disappeared in the Campanian (Fig. 5).

Recovery

This interval corresponds to the second half of the *Anapachydiscus (Neopachydiscus) naumanni* and *Pachydiscus (P.) egertoni* Zones. The beginning of this interval is characterized by the total extinction of *Eupachydiscus haradai*, gradual reduction of *Anapachydiscus (Neopachydiscus) naumanni* and by the appearance of new species of the genus *Canadoceras* in increasing numbers.

Morphological innovations are expressed by simplification of the shell sculpture ornamentation. It became finer as the coarse elements disappeared and the shell size decreased.

The genus *Canadoceras* is represented by five species during the time of its maximum abundance on Sakhalin and in northeastern Russia and by seven species in Japan. The Sakhalin assemblage consists of *Canadoceras mysticum* Matsumoto, *C. kossmati* Matsumoto, *C. yokoyamai* (Jimbo), *C. newberryanum* Meek, *C. multicostatum* Matsumoto. The latter is known from the Campanian of Madagascar and South Africa whereas the others are endemic.

Simultaneously, there was the development of a new species, *Anapachydiscus fasciostatus*, from *A. (Neopachydiscus) naumanni*. However, species immigrants *Anapachydiscus arrialoorensis* (Stoliczka) and *Pachydiscus (P.) egertoni* (Forbes) also appeared in the *Pachydiscus (P.) egertoni* Zone during the acme of the genus *Canadoceras*. *Desmophyllites diphylloides* (Forbes) appeared at the beginning of the Campanian associated with species of the Tethyan region as well as other species. These data indicate a significant level of ammonite migration.

Pachydiscus (P.) egertoni gave rise to many new species that were widespread in Maastrichtian of Sakhalin and northeast Russia, as well as in Japan.

Nomismogenesis

This interval corresponds to the *Canadoceras multicostatum* Zone. It is characterized by a relative stabilization of the ammonoid development. There are all the above-mentioned post-crisis species, together with some of the conservative, long-lived, lineages of phylloceratids, lytoceratids and desmoceratids (Figs 4, 5).

The number of specimens decreases abruptly directly before the Campanian–Maastrichtian

stage	substage	member	SAKHALIN Ammonite Zone — Desmocerataceae & Acanthocerataceae	member	JAPAN Ammonite Zone — Desmocerataceae	JAPAN — Selected associate
maastricht.	lower	4	Pachydiscus (Neodesmoceras) japonicus	K6b1	P. (P.) koboyashii	Gaudryceras izumiense
					P. (Neodesmoceras) japonicus	Nostoceras hetonaiense
campanian	upper	3 / 2	Canadoceras multicostatum	K6a4	P. (P.) awajiensis	Pravitocer. sigmoidale
					Patagios. laewis	Didimoceras awajiense
		1	Pachydiscus (P.) egertoni	K6a3	Anapachydiscus fascicostatus	M.subtilistriatum - Hoplitoplacent. monju
	lower	10	Anapachydiscus (Neopachydiscus) naumanni	K6a2	Canadoceras kossmati	Delawarella sp.
				K6a1	Anapachydiscus naumanni	Plesiotexanites shiloensis
			Eupachydiscus haradai	K5b2		Menabites mazenoti - Submortoniceras sf. condamyi
santonian	upper	9	Menuites menu		Eupachydiscus haradai	Plesiotexanites kawasakii - P. pacificus
	lower	8	Texanites (Plesiotexanites) kawasakii	K5b1	Anapachydiscus sutneri	Texanites collignoni
conjak.	upper	7	strata with Peroniceras sp.	K5a2	K.theobaldianum - E.keramasatoshii	Paratexanites orientalis

Fig. 7. Zonal ammonite divisions of the Santonian–Campanian in Sakhalin (by author) and in Japan (according to Toshimitsu *et al.* 1995).

boundary. This was another extinction event, probably related to the beginning of a Late Campanian regression. There was a reduction in the taxonomical diversity of the group and marked changes of the shell morphology occurred during that event. All but one species of the Family Pachydiscidae disappeared. A large specimen of *Canadoceras multicostatum* has been found in the Lower Maastrichtian sequences of Sakhalin.

The next taxonomical 'explosion' took place in the Late Maastrichtian, when the Family Pachydiscidae reached its acme. Moreover, at the same time there was an abundance of desmoceratids, phylloceratids and lytoceratids prior to the total extinction of ammonoids at the end of the Cretaceous.

Conclusions

The dynamics of taxonomic diversity of the Late Cretaceous ammonoids shows a definite sequence of events (Fig. 6). At first there was a mass extinction event followed by a survival interval with a single taxon which then gave rise to new morphotypes. These morphotypes indicate the beginning of a recovery phase that finishes with the acme of many groups. This is caused by an adaptive radiation based on local speciation as well as inward migration. After that the ammonoid development enters the nomismogenesis stage prior to a new extinction event. The author's zonal scheme of the Santonian to Campanian succession of Sakhalin clearly correlates with that of Japan (Fig. 7) and may be used in northeast Russia as well.

Research was supported by the Russian Fundamental Research Foundation (grant 94-05-17589, leader Professor T. N. Koren). I am very grateful to Dr T. D. Zonova (St Petersburg, VSEGEI) for helpful discussion and criticism.

References

HANCOCK, J. M. 1993. Transatlantic correlations in the Campanian–Maastrichtian stages by eustatic changes of sea-level. *In:* HAILWOOD, E. A. &

KIDD, R. B. (eds) *High Resolution Stratigraphy*. Geological Society, London, Special Publication, **70**, 241–256.

—— & KAUFFMAN, E. G. 1979. The great transgression of the Late Cretaceous. *Journal of the Geological Society, London*, **136**, 175–186.

KENNEDY, W. A. & KLINGER, H. C. 1993. On the affinities of Cobbanoscaphites Collignon, 1969 (Cretaceous Ammonoidea) *Annals of the South African Museum*, **102**, (7), 265–271.

MATSUMOTO, T. 1977. On so-called Cretaceous Transgressions. *Paleontological Society of Japan, Special Papers*, **21**, 75–84.

—— 1984. Some Ammonites from the Campanian (Upper Cretaceous) of the Northern Hokkaido. *Paleontological Society of Japan, Special Papers*, **27**, 1–34.

—— & HARAGUCHI, Y. 1978. A new Texanitinae ammonite from Hokkaido. *Transactions and Proceedings of the Paleontological Society of Japan, New Series*, **110**, 306–318.

—— & KANIE, J. 1979. Two ornate ammonites from the Urakawa Cretaceous area, Hokkaido. *Science Report of the Yokosuka City Museum*, **226**, 13–20.

—— & —— 1982. On three Cretaceous keeled ammonites from the Urakawa area, Hokkaido. *Science Report of the Yokosuka City Museum*, **229**, 9–22.

—— & OBATA, I. 1982. Some Interesting Ancanthoceras from Hokkaido. *Bulletin of the National Science Museum, Series C (Geology & Paleontology)*. **8(2)**, 67–92.

POYARKOVA, Z. N. (ed.) 1987. [The Reference section of Cretaceous deposits in Sakhalin (Naiba section).] Nauka, Leningrad [in Russian].

ZONOVA, T. D., KASINZOVA, L. I. & YAZYKOVA, E. A. 1993. *In:* ZONOVA, T. D. & ZHAMOIDA, A. I. (eds) *Atlas of the main groups of the Cretaceous fauna from Sakhalin*. Nedra, St Petersburg [in Russian].

Latest Cretaceous mollusc species 'fabric' of the US Atlantic and Gulf Coastal Plain: a baseline for measuring biotic recovery

CARL F. KOCH

Department of Geological Sciences, Old Dominion University, Norfolk, Virginia 23529, USA

Abstract: Studies of biosphere changes with time risk serious misinterpretation if they are based on a single section or closely spaced sections because lateral geographical changes in a fauna's distribution can be interpreted erroneously as profound temporal change. This problem can be ameliorated if the variation in fauna geographical distribution is well documented because such information would provide a reference to which temporal changes may be compared. That is, if temporal changes are greater than documented lateral changes, extinction or origination may have occurred. If temporal changes are within the observed lateral changes, then no compelling need to suggest evolutionary change exists.

Data for the *Haustator bilira* Assemblage Zone in the Atlantic and Gulf Coastal Plain are given in terms of nine areas distributed along a 2100 mile outcrop belt. The data result from examination of over 110 000 larger invertebrate fossil specimens, most of which are mollusc. Each mollusc species encountered was classified by feeding types, habitat and size. Because the *Haustator bilira* Assemblage Zone represents the youngest Cretaceous sediments of the Coastal Plain, the data assembled provide a reference plane just prior to the K/T boundary. It is hoped that these data will be used to evaluate the significance of biosphere changes in the vicinity of the K/T boundary.

Many past studies of biosphere changes with time use data from a single vertical section (e.g. Ward *et al.* 1986; Bryan & Jones 1989) or several sections confined to a restricted area (e.g. Elder 1987; Hansen *et al.* 1993). Use of such data to characterize biosphere change for large areas such as the Western Interior epicontinental sea, the Atlantic and Gulf Coastal Plain, or the world, may lead to incorrect conclusions because observed changes may merely reflect the migration of new environments into the studied local area.

Analyses of more widely dispersed sections result in limitations of the temporal precision because of correlation difficulties. Although change within each section can be measured centimetre by centimetre, the overall picture cannot be so precise. The choice seems to be between the temporal precision of closely spaced sections and the reliability of studies based on widely dispersed sections.

Another approach is to document thoroughly a 'time slice' over a large geographical area and to use information from the 'time slice' to make judgements about faunal patterns observed in a single vertical section. If the changes in the single vertical sequence exceed any changes in space documented for the 'time slice' then suggestions that the single vertical sequence represents evolutionary change is warranted. However, if the reverse is true, that is, that the observed temporal changes are similar to spatial changes of the 'time slice', then it is most likely that laterally moving environments caused the observed pattern.

This study will document the 'time slice' defined by the *Haustator bilira* Assemblage Zone

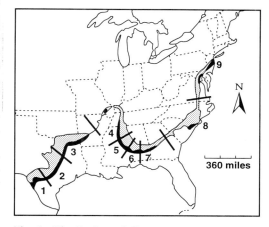

Fig. 1. Distribution of Cretaceous deposits on the emerged Atlantic and Gulf Coastal Plains of the United States (stippled). Darkened areas oceanward indicate occurrence of sediments of the *Haustator bilira* Assemblage Zone of Sohl (1977). This outcrop belt is divided into Areas 1 through 9.

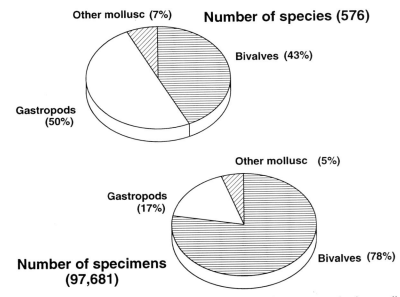

Fig. 2. Diagram illustrating the percentage distribution of species and specimens among the three mollusc groups (bivalves, gastropods and other molluscs) for this study's material.

for the Atlantic and Gulf Coastal Plain (Fig. 1). The presentation of data is organized for areas 1 through 9 to facilitate its use in comparisons with data from vertical sequences. Since the *Haustator bilira* Assemblage Zone represents the youngest Cretaceous sediments in the coastal plain it has particular usefulness for studies just before and just after the K/T boundary. Further details about the relationship of this assemblage zone and the K/T boundary are given in Koch (1991).

Material

Well over 110 000 fossil specimens were examined by the late Norman Sohl and the author between 1977 and 1987. The specimens were from collections made by US Geological Survey scientists over the last 100 years in the Atlantic and Gulf Coastal Plain uppermost Cretaceous sediments (Fig. 1). Most of the specimens were from bulk collections, thus observed relative abundance of species probably represents reasonably well the true species relative abundance. Taxonomic consistency is high since all specimens were identified by one individual. The results of these examinations are documented in three USGS Open File Reports (Sohl & Koch 1983, 1984, 1987).

These data or subsets of these data have been used for a variety of studies. Preservation effects in palaeoecological studies were measured and discussed in Koch & Sohl (1983). Sampling effects are examined in Koch (1987) and expected distribution of species ranges are discussed in Koch & Morgan (1988). Also given in Koch & Morgan (1988) is the distribution of sediment types for the nine areas of Fig. 1.

These data were compared to specimens from sediments of similar age and housed in European museums (Koch 1995). Hierarchical fabric similarities between these data and similar data for the Pleistocene (Valentine 1989) are given in Koch & Lundquist (1990).

These previous studies are mentioned here because they document characteristics of the dataset. Such information supplements this study and should help investigators understand the underlying factors which affect distribution of these species.

Not all of these 110 000 specimens were suitable for use in this study. Taxa and specimens which could not be identified confidently to the species level were eliminated. Non-mollusc specimens were also set aside. Quantities of the remaining species and specimens are given in Fig. 2. In all over 97 000 specimens were assigned to 576 mollusc species. Bivalve specimens comprise 78% of the specimens examined but gastropods were found to be more diverse.

Species relative abundance is known to fit the log-normal distribution, and occurrence frequency is known to fit the log-series distribution (Buzas *et al.* 1982; Koch 1987). This means that

LATEST CRETACEOUS MOLLUSCS, US COASTAL PLAIN

Area	South Texas	Central Texas	N.E. Texas, Arkansas	Missouri, Tennessee- N.E. Mississippi	E. Central Mississippi- N.E. Alabama	Central Alabama	E. Alabama- W. Georgia	N. and S. Carolina	Maryland, New Jersey	
Formations	Escondido	Kemp-Corsicanna	Kemp-Corsicanna Arkadelphia	Owl Creek, Chiwapa Member	Prairie Bluff	Prairie Bluff	Providence	Peedee (upper part)	Severn, Tinton-Red Bank	
General Lithologic Character										
Area Number	1	2	3	4	5	6	7	8	9	Total
No. of Collections	117	105	32	39	37	100	69	22	18	539
No. of Species	114	233	91	253	132	306	232	108	227	576
No. of Occurrences	696	1,128	271	1,121	753	2,390	1,659	426	968	9,412
No. of Specimens	13,135	8,843	1,754	8,572	4,407	19,493	22,025	5,807	13,645	97,681
No. of Bivalve Specimens	11,592	7,158	1,088	6,524	2,823	14,772	15,560	5,472	11,394	76,383
No. of Gastropod Specimens	1,198	1,234	577	1,502	1,172	3,515	5,854	316	1,579	16,947
Other Mollusc Specimens	345	451	89	546	412	1,206	611	19	672	4,351
Median Specimen Size	3	2	4	3	4	3	2	2	2	3
% Bivalve Specimens	88%	81%	62%	76%	64%	76%	71%	94%	84%	78%
% Aragonite Specimens	29%	67%	68%	64%	62%	63%	82%	37%	84%	73%

Fig. 3. Comparative quantities for the nine areas of Fig. 1 showing the formations sampled, general lithological character of the sediments and amount of data obtained from each area. Also given are data on proportion of various mollusc groups, specimen size and specimen mineralogy for each area.

most species are represented by few specimens and most species occur in few collections among the 539 collections examined.

Most of these 97 000 specimens are listed in Appendix 1. All species which are represented by more than 1% of the specimens in one of the nine areas of Fig. 1 are listed along with the number of specimens identified in each of the nine areas. These 84 species comprise more than 87% of the specimens and thus characterize the faunal content well.

Further insight into these data can be obtained from Fig. 3. This figure presents data quantities for each area in terms of numbers of collections, species, occurrences and specimens. The specimen counts are further subdivided into bivalves, gastropods and other molluscs.

Analysis and results

Changes in faunal patterns with time are often interpreted to indicate environmental changes with time. For example, if deposit feeders suddenly become the dominant fauna when suspension feeders had previously been domi-

nant, then stress is proposed to account for extinction. In other situations, changes in average specimen size have suggested stressed fauna. The question is: What variations in specimen numbers for feeding types, habitats or size over distance for a short period of time or a 'time slice' can be documented?

To answer the question, each of the 576 mollusc species was classified as to feeding type, habitat and size. Feeding type and habitat were assigned based on shell morphology and its relationship to the shell morphology of extant species or based on inferred functional morphology. Number of individuals for each feeding type, habitat and size were summed for the entire dataset and for each of the nine areas. For this analysis each bivalve specimen was considered to be one half of an individual.

Feeding types

Results of summation by feeding types are presented in Fig. 4A for the entire dataset, and in Fig. 4B for each of the 9 areas of Fig. 1. Suspension feeders dominate, comprising 81%

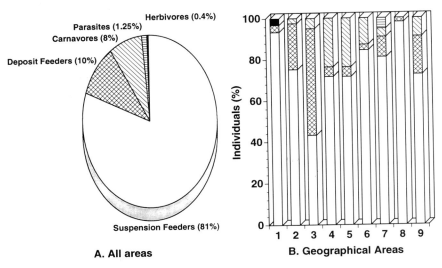

Fig. 4. Diagrams illustrating the proportion of individuals by feeding type for (**A**) the entire dataset and (**B**) each of the nine areas. For these data the number of bivalve specimens was divided by two before computing percentages.

of all mollusc individuals for the nine areas combined.

Among the individual areas variation is noted. Proportion of suspension feeding individuals varies from 98% in the Carolinas (area 8) to 43% in north Texas and Arkansas (area 3). Conversely, deposit feeders reach a maximum proportion of 52% in north Texas and Arkansas. Carnivores reach the maximum value of 24% in areas 4 and 5 (Missouri eastward to western Alabama). Herbivores are at the 3.2% level for south Texas (area 1) but do not reach significant numbers elsewhere.

Parasites comprised about 5% of the specimens from the Providence sands of eastern Alabama and western Georgia. This is attributed to the excellent preservation in the silty fine grained quartzose sands that contain sufficient clay to seal the shells from dissolution by ground water. Thus, more delicate aragonitic forms are preserved.

Habitats

Results of summations by habitat are presented in Fig. 5A for the entire dataset and in Fig. 5B for the various areas. More than half (55%) of the mollusc individuals utilize the shallow infaunal habitat. Very few (0.2%) are from the deep infaunal habitat. The remaining 45% are epifauna with the epifaunal groups nearly evenly represented. Cemented forms comprise 15%, attached forms 18% and free epifauna 12%.

The proportion of shallow infauna is maximum (68%) in eastern Alabama and western Georgia (area 7) and minimum (31%) in south Texas (area 1). Cemented epifauna reach their highest value (25%) in the marls of central Alabama (area 6), but are few (3%) in eastern Alabama to western Georgia (area 7). Maximum proportion of attached epifauna (49%) are noted in south Texas (area 1), and minimum proportions (5%) in north Texas and Arkansas (area 3). The north Texas and Arkansas area (area 3) has the largest proportion of free epifauna (27%) and south Texas (area 1) the smallest value (3%).

Although deep dwelling infauna represents only 0.2% of the individuals, these are concentrated in areas 3 (3.6%) and 4 (1.8%).

Size classes

Size classes are based on species bulk and not on any particular measurement. Table 1 gives examples of some well known species within each size class. For example, *Crenella serica* Conrad is subspherical in outline but *Cadulus*

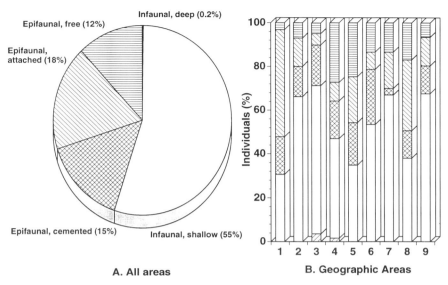

Fig. 5. Diagram illustrating the proportion of individuals by habitat for (A) the entire dataset and (B) each of the nine areas. For these data the number of bivalve specimens was divided by two before computing percentages.

obnustus (Conrad) is more elongate. Both are assigned to size class 1 because they represent about the same biomass.

Table 1. *Commonly occurring Late Cretaceous mollusc species which illustrate the size classes 1 through 6*

1.	*Crenella serica* Conrad
	Cadulus obnustus (Conrad)
2.	*Cymbophora berryi* Gardner
	Euspira rectilabrum Gardner
3.	*Nucula percrassa* Conrad
	Turritella (Sohlitella) bilira (Stephenson)
4.	*Leptosolen biplicata* Conrad
	Turritella (Sohlitella) trilira (Conrad)
5.	*Exogyra costata* Say
	Belemnitella americana (Morton)
6.	*Crassostrea cortex* (Conrad)
	Sphenodiscus lobatus (Tourney)

Results of summations by size class are given in Fig. 6A for all data and Fig. 6B for each of the nine areas. The median size for all data is about at the dividing line between size classes 2 and 3, that is, about half of all specimens are small (size classes 1 and 2) while the remainder are medium to large in size.

The largest proportions of small mollusc individuals are found in central Texas (area 2), eastern Alabama and western Georgia (area 7), and Maryland/New Jersey (area 9) – 36%, 19% and 27% respectively. These values owe much to the better preservation of aragonitic shelled forms in the muddy fine sands and marls.

South Texas has the most larger individual molluscs (26.5%) of all the areas. Three estuarine forms, *Crassostrea cortex*, *Pachymelania* n.sp. and *Turritella* cf. *T. forgemolli*, comprise most of these larger individuals.

Discussion

Data about variation in proportions of fossil individuals by feeding type, habitat and size class based on very large, high quality data from an outcrop belt that meanders over 2100 miles from just south of New York City to the Rio Grande River in Texas are provided in Figs 4, 5 and 6. The spatial variations documented provide a standard by which variation in time may be judged. If biosphere variations with time do not exceed the variations demonstrated in these figures then there is no reason to propose that evolutionary change has occurred.

The data of this study represent a variety of marine environments, mostly shelfal. The chalks and marls of east central Mississippi and

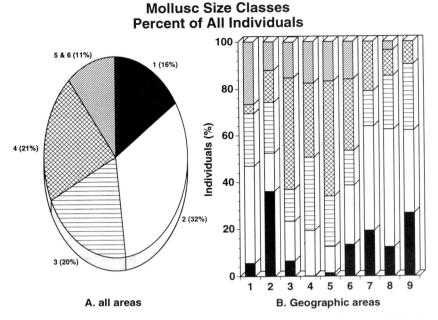

Fig. 6. Diagram illustrating the proportion of individuals by size for (A) the entire dataset and (B) each of the nine areas. For these data the number of bivalve specimens was divided by two before computing percentages.

Alabama undoubtedly represent environments further from the shoreline while an estuarine environment is suggested for the Escondido Formation of south Texas (see Sohl & Koch 1986; Koch & Morgan 1988 for more detail and discussion). Obviously, other marine environments are possible during this 'time slice' that are not represented by the dataset. Thus, if temporal variation in faunal content is encountered which exceeds the spatial variation documented, the temporal variation may still only represent shifting environments.

In addition to the raw data about numbers of collections, species, occurrences and specimens, Fig. 3 gives median specimen size, percent bivalve species and percent aragonite specimens for each area. These three quantities are related to preservational factors (Koch & Sohl 1983) as well as environments. For example, more small aragonitic specimens are obtained when preservation is excellent, areas 2, 7 and 9. Proportions of aragonitic specimens are lowest for south Texas and the Carolinas. In both areas the proportions of bivalve specimens are high reflecting the presence of the more often preserved calcitic bivalves.

The advantages of having data from a larger area are illustrated by recent work on the uppermost Cretaceous of northern Spain and western France. Ward et al. (1986) measured a single section near Zumaya, Spain, observed that ammonites disappeared at least 10 m below the K/T boundary, and suggested that this might represent 'a very different mode of extinction in a more tropical, Tethyan environment, as compared to our current information based largely on boreal chalk fauna and facies'. However, Ward (1988) presents a somewhat different picture based on the Zumaya section and four new sections between Bidart, France and Sopelana, Spain, a distance of about 110 km. He concluded that 'as many as seven species of ammonites were apparently present in this region during sedimentation of the final metre of Maastrichtian'.

Hansen et al. (1993) suggest that a prolonged period of stress followed the K/T boundary extinction and that the stressed period is characterized by molluscan faunas with low species richness, low abundance of individuals, high species turnover and a dominance of deposit feeders. These suggestions are based on sections and cores within a 10 km belt near the Brazos River in central Texas. Proportions of deposit feeding molluscs increased from about 12% just prior to the K/T boundary to an average of 50% just after the K/T boundary. Two post boundary levels are shown as having

LATEST CRETACEOUS MOLLUSCS, US COASTAL PLAIN

75% and 85% deposit feeders respectively.

However, the nearby area 3 (north Texas and Arkansas), measures 52% deposit feeders (Fig. 4B) and several individual collections have over 80% deposit feeding molluscs. For example, a collection from Palms Spring, Travis County, has 87% and one from West Zuehl, Bexar County measured 84% deposit feeders.

Thus, the migration of nearby environments into the Brazos River area just after the K/T boundary, might account for the observed temporal patterns and not some profound extinction of suspension feeding forms. There is a need for other data to distinguish between extinction patterns and shifting environments.

Unfortunately, only one other section in the Atlantic and Gulf Coastal Plain is thought to have a hiatus under 1 Ma at the K/T boundary. The section at Braggs, Alabama may be missing only about 0.8 Ma at the K/T boundary (Bryan & Jones 1989). Worldwide, very few 'contin-uous sections are known at the K/T boundary (MacLeod & Keller 1991).

Thus, uncertainty cannot always be resolved. The data provided in Figs 4, 5 & 6 can serve a null hypothesis against which observed patterns such as those of the Brazos River sections can be compared.

Summary

Data from a large, high quality and widely dispersed 'time slice' just prior to the K/T boundary for the Atlantic and Gulf Coastal Plain provide a reference plane for studies of biosphere changes in this region and for approximately this time. The intent is that this reference plane will serve as one tool to differentiate between changes in time versus changes in space.

I thank S. J. Culver for reading the manuscript and for many helpful suggestions.

Appendix 1. List of 84 mollusc species which represent at least 1% of specimens in at least one area. More than 87% of all specimens are included.

Mollusc species	Area								
	1	2	3	4	5	6	7	8	9
Bivalves									
Nucula percrassa Conrad	56	138	5	290	22	89	262	7	754
Nuculana whitfieldi (Gardner)	–	197	35	85	1	488	266	97	560
Striarca webbervilleensis (Stephenson)	692	142	–	–	–	–	–	–	–
Striarca cuneata Gabb	4194	–	–	2	–	172	2452	–	5
Striarca saffordi Gardner non Gabb	–	29	–	28	–	101	280	43	143
Cucullaea (Idonearca) capax Conrad	4	202	18	141	573	244	19	8	70
Glycymeris rotundata (Gabb)	–	–	–	142	8	118	314	314	14
Postligata wordeni Gardner	–	–	–	63	–	14	54	19	391
Lycettia tippana (Conrad)	–	–	–	23	5	1	11	18	175
Crenella serica Conrad	37	1161	60	126	75	493	104	200	1393
Tenuipteria argentea (Conrad)	–	–	1	135	–	1	1	–	2
Syncyclonema simplicius (Conrad)	19	198	12	263	28	830	628	637	496
Camptonectes (Camptonectes) bubonis Stephenson	18	343	9	113	17	100	282	238	24
Plicatula (Plicatula) mullicaensis Weller	96	9	–	–	45	1125	–	–	–
Plicatula (Plicatula) tetrica Conrad	–	3	18	10	15	330	–	–	–
Spondylus (Spondylus) munitas Stephenson	–	9	–	–	7	218	–	–	–
Atreta melleni (Stephenson)	–	–	–	–	–	877	–	–	–
Anomia (Anomia) argentaria Morton	121	182	32	242	143	672	430	1220	436
Anomia (Anomia) ornata Gabb	1079	3	–	30	–	22	331	6	7
Paranomia scabra (Morton)	–	–	–	–	101	37	–	–	–
Lima (Lima) reticulata Lyell & Forbes	–	15	5	62	69	128	23	133	52
Lima (Lima) pelagica (Morton)	46	241	–	4	9	15	2	300	1
Pycnodonte vesicularis Lamark	6	235	88	43	165	349	2	1	94
Pycnodonte belli Stephenson)	6	41	24	47	15	–	–	–	–
Exogyra costata Say	44	341	219	712	349	921	39	82	28
Gryphaeostrea vomer (Morton)	24	541	117	102	156	244	–	–	23
Agerostrea falcata (Gabb)	–	6	3	6	100	34	–	–	14
Cubitostrea tecticosta (Gabb)	80	185	5	296	114	427	310	565	1000

Ostrea mesenterica Morton	3	52	1	28	43	30	7	–	654
Crassostrea cortex Stephenson	24	1	5	39	54	27	1	26	10
Ostrea plumosa Morton	2130	1	–	–	–	1	25	–	–
Pterotrigonia (Scabrotrigonia) angulicostata Gabb	–	–	–	72	1	–	252	–	2
Pterotrigonia (Scabrotrigonia) eufaulensis (Gabb)	–	10	8	312	85	154	47	69	309
Nymphalucina linearia (Stephenson)	–	78	20	1	–	51	869	51	39
Vetericardiella webbervillensis (Stephenson)	–	281	4	–	–	3	–	–	–
Crassatella vadosa vadosa Morton	25	379	87	490	80	861	224	427	1084
Uddenia texana Stephenson	–	2	–	1	–	17	2261	8	12
Granocardium (Granocardium) lowei Stephenson	–	4	2	137	–	106	166	22	69
Granocardium (Granocardium) cf. *G. (G.) bowenae* Stephenson)	–	–	–	–	–	3	1	72	–
Trachycardium eufaulense (Conrad)	1712	58	77	248	240	221	765	36	294
Cymbophora berryi (Gardner)	212	90	14	38	–	257	655	24	496
Cymbophora inflata Stephenson	131	1	–	–	–	–	–	–	–
Aenoa eufaulensis (Conrad)	–	1	–	74	–	28	340	13	233
Veniella conradi (Morton)	65	26	22	148	47	297	49	64	91
Tenea parilis (Conrad)	–	5	5	134	8	23	49	–	122
Cyclorisma parva (Gardner)	–	2	–	56	–	86	138	73	35
Cyprimeria depressa Conrad	6	5	1	63	21	49	289	5	67
Cyprimeria major Gardner	–	–	–	–	–	–	102	48	–
Lequmen ellipticus Conrad	–	3	–	108	50	65	22	–	65
Corbula monmouthensis Gardner	102	19	74	15	–	222	719	29	49
Corbula n. sp (large)	159	–	–	–	–	–	–	–	–
Caesticorbula (Caesticorbula) crassiplica (Gabb)	1	1002	47	29	8	1675	199	362	609
Caesticorbula (Parmicorbula) terrimaria (Gardner)	–	4	1	10	–	10	238	7	13
Caesticorbula (Parmicorbula) percompressa (Gardner)	–	–	–	3	–	104	3	–	216
Eufistulana ripleyana (Stephenson)	–	3	24	60	10	13	1	–	–
Liopistha protexta (Conrad)	–	5	2	165	78	310	40	6	72

Gastropods

Laxispira monilifera Sohl	–	25	1	64	–	478	19	98	–
Turritella (Sohlitella) bilira Stephenson	77	138	3	18	104	212	1590	20	106
Turritella (Sohlitella) trilira Conrad	265	18	6	40	210	183	1820	36	101
Turritella vertebroidses Morton	6	44	14	109	196	423	23	13	32
Turritella tippana Conrad	–	–	–	188	21	30	19	2	2
Turritella cf. *T. forgemolli* Coquand	316	–	–	–	–	88	–	–	–
Turritella n. sp. (multilirate form)	–	18	–	–	–	4	459	–	–
Pachymelania n. sp. *(Escondido Fm.)*	196	–	–	–	–	–	–	–	–
Xenophora leprosa (Morton)	1	1	2	–	66	91	–	–	1
Graciliala cooki (Stephenson)	–	–	76	–	–	–	–	–	–
Arrhoges (Latiala) ciboloensis (Stephenson)	–	239	384	–	–	–	–	–	–
Anchura noackensis Stephenson	–	173	–	8	53	–	1	–	–
Gyrodes petrosus (Morton)	–	3	4	2	169	122	–	–	–
Gyrodes abyssinus Morton	–	–	18	2	137	150	–	–	–
Gyrodes spillmani Gabb	33	4	32	11	9	20	20	1	32
Euspira rectilabrum (Conrad)	4	69	–	92	12	43	423	26	180
Paladmete cancellaria (Conrad)	2	26	-	34	-	206	42	5	74
Ringicula (Ringicula) clarki Gardner	–	–	–	13	–	4	90	11	–
Ringicula (Ringicula) pulchella Shumard	–	98	–	7	–	4	–	–	330
Bullopsis cretacea (Conrad)	–	–	–	148	–	–	–	–	–
Creonella triplicata Wade	–	2	–	–	–	12	528	–	27

Other molluscs

Eutrephoceras perlata Morton	–	–	13	8	69	–	–	–	–
Baculites tippahensis Conrad	–	–	1	–	33	420	–	–	–
Baculites (Eubaculites) carinatus Morton	–	18	–	188	108	361	–	–	39
Discoscaphites iris Conrad	–	–	–	220	–	3	–	–	3
Discoscaphites conradi (Morton)	–	43	2	1	68	113	–	–	65
Cadulus obnustus (Conrad)	–	3	1	1	–	11	276	–	313
Dentalium level Stephenson	209	279	47	24	–	36	324	6	506

References

BRYAN, J. R. & JONES, D. S. 1989. Fabric of the Cretaceous–Tertiary marine macrofaunal transition at Braggs, Alabama. *Palaeogeography, Palaeoclimatology, Palaeoecology*, **69**, 279–301.

BUZAS, M. A., KOCH, C. F., CULVER, S. J. & SOHL, N. F. 1982. On the distribution of species occurrence. *Paleobiology*, **8**, 143–150.

ELDER, W. P. 1987. The paleoecology of the Cenomanian–Turonian (Cretaceous) stage boundary extinctions at Black Mesa, Arizona. *Palaios*, **2**, 24–40.

HANSEN, T. A., FARREL, B. R. & UPSHAW, B. 1993. The first 2 million years after the Cretaceous–Tertiary boundary in east Texas: rate and paleoecology of molluscan recovery. *Paleobiology*, **19**(2), 251–265.

KOCH, C. F. 1987. Prediction of sample size effects on the measured temporal and geographic distribution of species. *Paleobiology*, **13**(1), 100–107.

—— 1991. Species extinctions across the Cretaceous–Tertiary boundary: observed patterns versus predicted sampling effects, stepwise or otherwise? *Historical Biology*, **5**, 355–361.

—— 1995. Bivalve species' distribution in uppermost Cretaceous boreal marine beds, Europe and North America and the implied taxonomic problems. *In: Proceedings of the 4th International Cretaceous Symposium, Hamburg, Germany, September, 1992*. Special issue of 'Mitteilungen aus dem Geologisch-Paläontologischen Institut der Universität Hamburg', in press.

—— & LUNDQUIST, J. J. 1990. Hierarchical fabric similarities of a Late Cretaceous and a Pleistocene mollusc data set. *Annual Meeting of the Geological Society of America*, **22**(7), A266.

—— & MORGAN, J. P. 1988. On the distribution of species' ranges. *Paleobiology*, **14**(2), 126–138.

—— & SOHL, N. F. 1983. Preservational effects in paleoecological studies: Cretaceous mollusc examples. *Paleobiology*, **9**(1), 26–34.

MACLEOD, N. & KELLER, G. 1991. How complete are Cretaceous/Tertiary boundary sections? A chronostratigraphic estimate based on graphic correlation. *Geological Society of America Bulletin*, **103**, 1439–1457.

SOHL, N. F. 1977. Utiligy of gastropods in biostratigraphy. *In:* KAUFFMAN, E. G. & HAZEL, J. E. (eds) *Concepts and Methods of Biostratigraphy*. Dowden, Hutchinson and Ross Inc; Stroudsburg, PA.

—— & KOCH, C. F. 1983. *Upper Cretaceous (Maastrichtian) mollusca from the* Haustator bilira *Assemblage Zone in the East Gulf Coastal Plain.* United States Geological Survey Open File Report, **83**-451.

—— & —— 1984. *Upper Cretaceous (Maastrichtian) larger invertebrates from the* Haustator bilira *Assemblage Zone in the west gulf coastal plain.* United States Geological Survey Open File Report, **84**-687.

—— & —— 1986. Molluscan biostratigraphy and Biofacies of the *Haustator bilira* Assemblage Zone (Maastrichtian) of the East Gulf Coastal Plain, *In:* REINHARDT, J. (ed.) *Stratigraphy and Sedimentology of Continental Near Shore and Marine Cretaceous Sediments of the Eastern Gulf Coastal Plain, Field Trip No. 3.* SEPM and AAPG Annual Meeting, June 18–20, 45–56.

—— & —— 1987. *Upper Cretaceous (Maastrichtian) Larger invertebrates from the* Haustator bilira *Assemblage Zone in the Atlantic Coastal Plain with Further Data from the east Gulf.* United States Geological Survey Open File Report, **87**-194.

VALENTINE, J. W. 1989. How good was the fossil record? Clues from the California Pleistocene. *Paleobiology*, **15**(2), 83–94.

WARD, P. 1988. Maastrichtian ammonite and inoceramid ranges from Bay of Biscay Cretaceous–Tertiary boundary sections, *In:* LAMOLDA, M., KAUFFMAN, E. & WALLISER, O. (eds) *Paleontology and Evolution: Extinction events.* Revista Española de Paleontologia, n. estraordinario, 119–126.

——, WIEDMANN, J. & MOUNT, J. F. 1986. Maastrichtian molluscan biostratigraphy and extinction patterns in a Cretaceous/Tertiary boundary section exposed at Zumaya, Spain. *Geology* **14**, 899–903.

Phenotypic experiments into new pelagic niches in early Danian planktonic foraminifera: aftermath of the K/T boundary event

EDUARDO A. M. KOUTSOUKOS

Petrobrás-Cenpes, Cidade Universitária, Quadra 7, Ilha do Fundão,
21949-900 Rio de Janeiro, Brazil

Abstract: In the aftermath of the Cretaceous–Tertiary (K/T) boundary event it appears that unusual phytoplankton blooms were boosted by the abundant nutrient levels left over from a short-lived 'Strangelove Ocean' period, which possibly gave rise to the negative-to-positive $\delta^{13}C$ surface-to-bottom gradient in the Danian.

The highly plastic morphology of the planktonic foraminifera at the base of the Palaeocene appeared, most likely, as an adaptive response of functional morphology to the pelagic habitat of the species and its mode of life. The diversification episodes of Danian planktonic foraminifera appear to correspond to times of reduced oceanic mixing, stratified water masses, recovery of the surface water productivity and a progressive expansion of the trophic-resource continuum, with periods of maximum diversity corresponding to widespread oceanic surface-water oligotrophic conditions. These suggest a parallel adaptive trend of feeding strategies caused by the changing availability of nutrients in the water column, with probable increasing reliance in symbiosis under low nutrient conditions later in the Danian.

Guembelitria cretacea, one of the few Cretaceous survivors, has been recorded in northeastern Brazil up to the lower part of the P2 Zone. A triserially coiled planktonic morphotype, this species started to diversify immediately after the K/T transition. After only a few thousand years, in the earliest Danian, the first Tertiary taxa (the *Guembelitria–Woodringina* lineage) appeared; these are microperforate non-spinose forms that were probably restricted to shallow epipelagic waters. There is a trend towards uncoiling of the triserial pattern by progressively changing the positioning of the aperture, without any detectable ecophenotypic preference at first.

The first specimens of the *Eoglobigerina–Pseudosubbotina*, *Eoglobigerina–Subbotina* and *Praemurica* lineages, cancellate spinose and non-spinose forms, also evolved in the Danian in a nearly coeval and parallel evolutionary trend from a probably ancestral *Hedbergella* stock.

A revised phylogeny is suggested for the early Palaeocene planktonic foraminifera based on recorded stratigraphic ranges and morphological affinities (chamber arrangement and surface texture ornamentation).

Never before in the stratigraphic history of the planktonic foraminifera have such patterns of rapid speciation rates been seen as in the early Danian. Biochronostratigraphic zonation in the Cretaceous, based on concurrence of FADs and LADs of diagnostic taxa, allows a mean resolution of about 2 Ma, approaching 0.5 Ma at the best in specific intervals (Sliter 1989). However, just after the Cretaceous–Tertiary (K/T) extinction event, when nearly all planktonic taxa perished, the few survivors (possibly only three species) found in the early Danian a complete range of new vacated pelagic niches waiting to become populated. Rapid speciation followed, reflected by small morphological changes affecting one or two characters at a time, which permits biochronostratigraphic resolution to about 30 000–100 000 years (Berggren *et al.* 1995).

The early Danian planktonic ecosystems characteristically contain small species with highly variable morphologies and, consequently, species definition often becomes blurred. There are still many problems regarding the taxonomy and evolutionary lineages of Danian planktonic foraminifera. No detailed knowledge of the ontogeny of the species and their intra- and interspecific variations is at present available.

This work presents an attempt to unveil patterns of biotic recovery and possible driving-evolutionary mechanisms of the planktonic foraminifera after the K/T boundary event in low-latitude regions. The biostratigraphic, phylogenetic and environmental data drawn from specimens recovered from a K/T boundary section near Recife (the Poty section), in Pernambuco, northeastern Brazil (Fig. 1), is integrated with reported evidence from the

From Hart, M. B. (ed.), 1996, *Biotic Recovery from Mass Extinction Events,*
Geological Society Special Publication No. 102, pp. 319–335

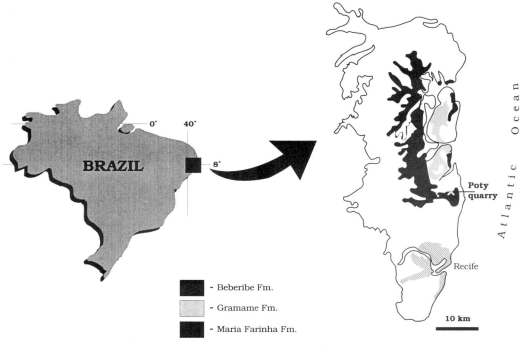

Fig. 1. Location map of the K/T boundary Poty section (UTM 9 152 000N/300 000 E), in northeastern Brazil. (Starting points for UTM coordinates: 10 Mm S of the Equator, 0.5 Mm W from the meridian 39°W of Greenwich, respectively.)

literature, thus providing a more reliable picture of the sequence of events.

The P-prefixed foraminiferal zonation used in this study follows Berggren et al. (1995). Figure 2 shows the main planktonic foraminiferal biostratigraphic datum events for the Danian along with the correlation of Berggren et al.'s zonation to the zonal schemes of Blow (1979), Smit (1982), Berggren & Miller (1988), Keller (1988), Canudo et al. (1991), Keller & Benjamin (1991) and Keller (1993). Specimens illustrated in this study will be deposited in the micropalaeontological collections of the Museu Nacional in Rio de Janeiro, Brazil.

Danian negative-to-positive $\delta^{13}C$ surface-to-bottom gradient

Carbon isotope analysis of foraminifera tests indicates a drastic reduction or an inversion of the $\delta^{13}C$ gradient patterns between surface and bottom waters in the Danian from the P0 zone through about the P1a/P1b zonal interval (Boersma et al. 1979; Hsü & McKenzie 1985, 1991; Zachos & Arthur 1986; Keller & Lindinger 1989; Barrera & Keller 1990; Schmitz et al. 1992; MacLeod 1993). This inverted gradient is marked by a greater enrichment of ^{12}C in surface waters, with benthic foraminifera being differently enriched in ^{13}C relative to planktonic species (Boersma et al. 1979; Boersma & Shackleton 1981; Hsü & McKenzie 1985). The inverted surface-to-deep $\delta^{13}C$ trend is also recorded throughout the planktonic assemblages, with shallow-dwelling species possessing lighter $\delta^{13}C$ values than intermediate-water taxa (Boersma et al. 1979; Boersma & Shackleton 1981). In addition, carbonates (whole-rock record) and marine organic matter deposited at K/T boundary sequences indicate synchronous shifts first in an isotopically lighter direction and then becoming isotopically heavier just a few centimetres above the K/T boundary (e.g. Keller & Lindinger 1989; Gilmour et al. 1990; Schmitz et al. 1992; Hollander et al. 1993).

A mass mortality of oceanic plankton at the K/T boundary with a subsequent drastic decrease in primary productivity and the accumulation of nutrients, including dissolved CO_2, in the photic zone, the 'Strangelove Ocean' effect, has been proposed as a mechanism to produce the negative $\delta^{13}C$ shift in tests of planktonic foraminifera and nannoplankton ooze and the

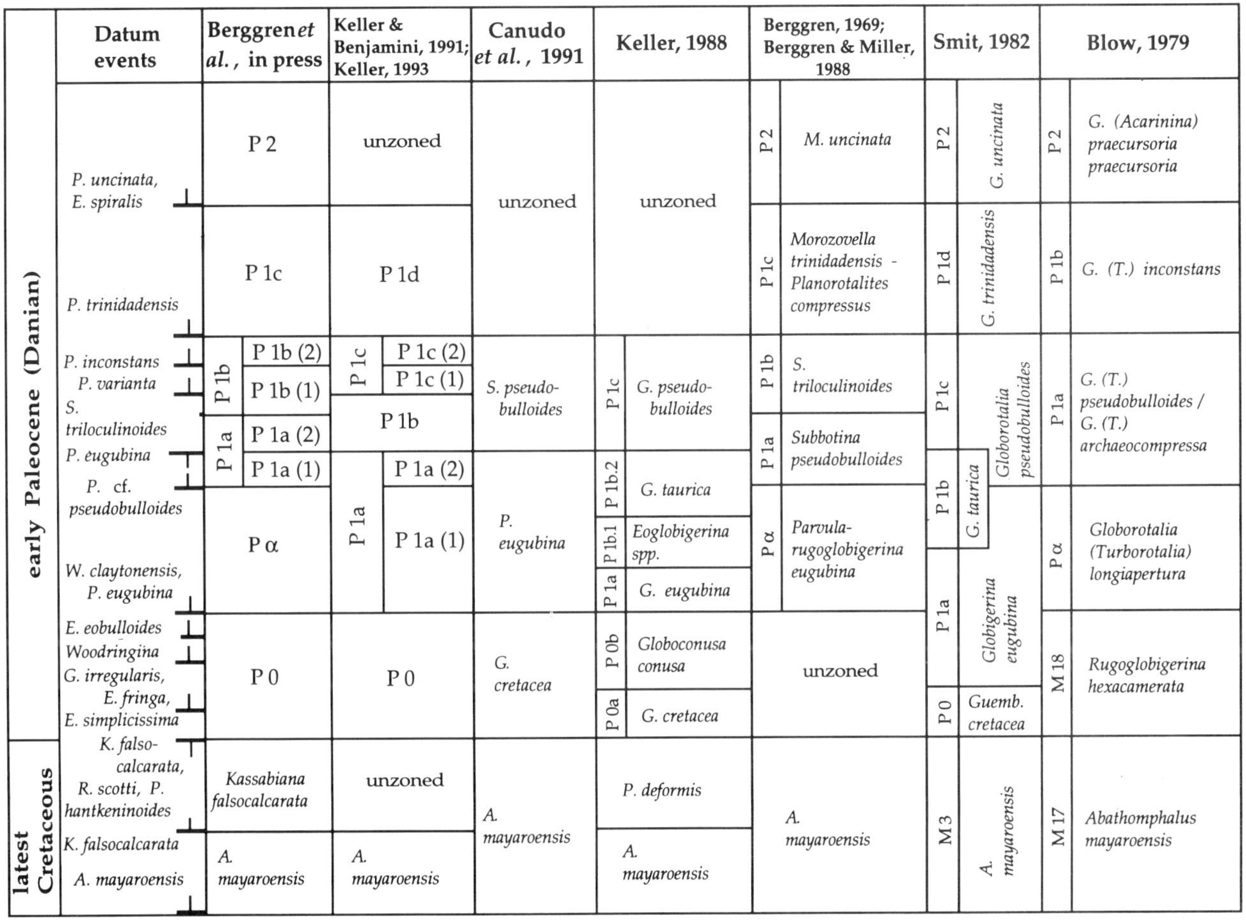

Fig. 2. Integrated foraminiferal zonal schemes. The datum level sequence in the left-hand column is inferred after MacLeod & Keller (1991), Liu & Olsson (1992), Olsson et al. (1992), Keller (1993), Berggren et al. (1995) and this study. Berggren et al. (1995) zonal scheme is complemented for the latest Cretaceous with data from Solakius et al. (1984).

Datum events	Berggren et al., in press	Keller & Benjamini, 1991; Keller, 1993	Canudo et al., 1991	Keller, 1988	Berggren, 1969; Berggren & Miller, 1988	Smit, 1982	Blow, 1979
early Paleocene (Danian)							
P. uncinata, E. spiralis	P2	unzoned	unzoned	unzoned	*M. uncinata* (P2)	*G. uncinata* (P2)	*G. (Acarinina) praecursoria praecursoria* (P2)
P. trinidadensis	P1c	P1d			*Morozovella trinidadensis - Planorotalites compressus* (P1c)	*G. trinidadensis* (P1d)	*G. (T.) inconstans* (P1b)
P. inconstans, P. varianta	P1b (2) / P1b (1)	P1c (2) / P1c (1)	*S. pseudobulloides*	*G. pseudobulloides* (P1c)	*S. triloculinoides* (P1b)	*Globorotalia pseudobulloides* (P1c)	*G. (T.) pseudobulloides / G. (T.) archaeocompressa* (P1a)
S. triloculinoides	P1a (2) / P1a (1)	P1b		*G. taurica* (P1b.2) / *Eoglobigerina spp.* (P1b.1)	*Subbotina pseudobulloides* (P1a)	*G. taurica* (P1b)	
P. eugubina	Pα	P1a (2) / P1a (1)	*P. eugubina*	*G. eugubina* (P1a)	*Parvula-rugoglobigerina eugubina* (Pα)	*Globigerina eugubina* (P1a)	*Globorotalia (Turborotalia) longiapertura* (Pα)
P. cf. *pseudobulloides*							
W. claytonensis, P. eugubina	P0	P0	*G. cretacea*	*Globoconusa conusa* (P0b)	unzoned	*Guemb. cretacea* (P0)	*Rugoglobigerina hexacamerata* (M18)
E. eobulloides, Woodringina, G. irregularis, E. fringa, E. simplicissima				*G. cretacea* (P0a)			
K. falso-calcarata, R. scotti, P. hantkeninoides	*Kassabiana falsocalcarata*	unzoned		*P. deformis*			
Latest Cretaceous							
K. falsocalcarata	*A. mayaroensis*	*A. mayaroensis*	*A. mayaroensis*	*A. mayaroensis*	*A. mayaroensis*	*A. mayaroensis* (M3)	*Abathomphalus mayaroensis* (M17)
A. mayaroensis							

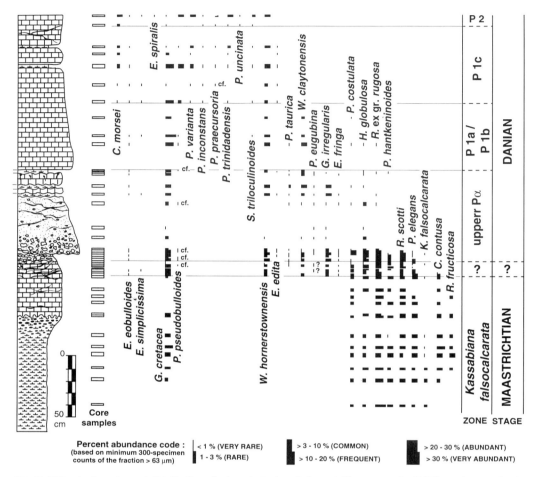

Fig. 3. Lithostratigraphy and distribution of selected planktonic foraminifera across the K/T boundary section in the Poty quarry (samples from core Poty #1).

breakdown of the normal surface-to-bottom-water carbon-isotope gradient (Broecker & Peng 1982; Hsü & McKenzie 1985; Hollander et al. 1993). A subsequent recovery period with reduced numbers of zooplankton grazers, the 'Respiring Ocean' (Hsü & McKenzie 1991; Hollander et al. 1993), is interpreted to reflect bacterial decomposition of slowly sinking particulate organic matter derived from massive phytoplankton blooms, releasing dissolved CO_2 enriched in ^{12}C into near surface waters.

Hence, it appears that unusual phytoplankton blooms were boosted by the abundant nutrient levels (eutrophic surface waters) left over from a 'Strangelove Ocean' period, in the aftermath of the K/T boundary event, possibly giving rise to the negative-to-positive $\delta^{13}C$ surface-to-bottom gradient.

The progressive recovery of the surface water productivity and the oceanic carbon cycle restored the normal positive carbon-isotope gradients, similar to those observed in the latest Cretaceous (Perch-Nielsen et al. 1982; Hsü & McKenzie 1985; Keller and Lindinger 1989), by the P1a/P1b zonal interval. That period also corresponds to widespread oceanic surface-water oligotrophic conditions (expanded Trophic Resource Continuum, TRC; Hallock 1987), and to a radiation maximum in the planktonic foraminifera (recovery interval) with development of several intermediate and deep-water dwellers (see below).

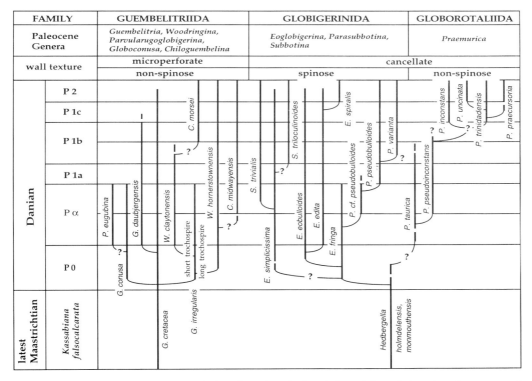

Fig. 4. Suggested phylogenetic lineages of selected Danian planktonic foraminifera. Inferred species ranges and lineages after Hemleben et al. (1991), Olsson et al. (1992), Liu & Olsson (1992), MacLeod (1993) and this study.

The Danian biostratigraphic record and phylogeny

The percentage abundance curve of the planktonic foraminifera (Fig. 3) shows that they are not evenly distributed throughout the Danian section in northeastern Brazil, mostly due to facies variations and episodic sedimentation (Koutsoukos 1996b). Apparently the entire P0 and most of the Pα Zone (lower part) are missing (Fig. 3). The missing section may have been eroded away and/or mixed within the lower Danian beds. Therefore, in order to obtain a complete picture of the biostratigraphic record and evolutionary lineages of Danian planktonic foraminifera (Fig. 4), complementary data are compiled and integrated from the available literature (mainly after the works of Blow 1979; MacLeod & Keller 1991; Liu & Olsson 1992; Olsson et al. 1992).

The phylogenetic succession of Danian planktonic foraminifera species is still a matter of discussion and controversy (e.g. Smit 1982; Keller 1988, 1989; D'Hondt 1991; MacLeod & Keller 1991; Liu & Olsson 1992; Olsson et al. 1992; MacLeod 1993). This confusion reflects taxonomic uncertainties and different concepts of generic and sometimes species assignments (e.g. Luterbacher & Premoli-Silva 1964; Blow 1979; Olsson et al. 1992), sampling biases (e.g. Signor & Lipps 1982), local facies variations and biogeographical patterns. Detailed knowledge of phylogenetic relationships throughout the planktonic foraminifera is needed before many questions on their evolutionary history can be sensibly addressed.

Previous studies (Hemleben 1975; Hemleben et al. 1989, 1991) have demonstrated the wall structure and texture of the planktonic foraminifera to reflect evolutionary functional adaptations for different feeding strategies, spinose species with cancellate wall textures first evolving in the early Palaeocene (Hemleben et al. 1991). Therefore, the wall texture classification recently proposed by Olsson et al. (1992) is adopted in this study as a means of following a more biological classification for generic assign-

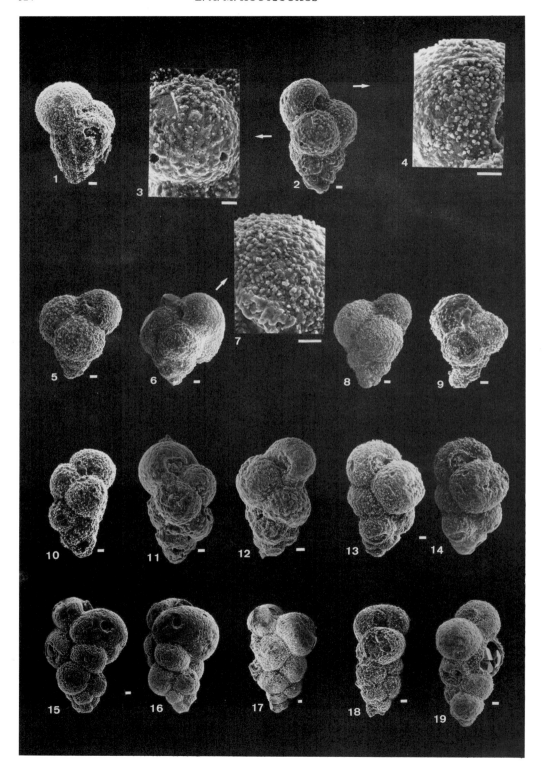

ments of Palaeocene foraminifera. The phylogenetic sequence proposed in Fig. 4 resembles in general the evolutionary steps postulated by Liu & Olsson (1992) and Olsson *et al.* (1992), revised, after this study, for phenotypic–phyletic relationships and the recorded biostratigraphic ranges of selected species.

The Campanian and Maastrichtian were times of long-term environmental stability, with well-oxygenated pelagic conditions, high sea-levels and newly established global, intermediate and deep oceanic circulation patterns, which promoted high specific and intra-specific diversification (e.g. Haq *et al.* 1987; Mello *et al.* 1989; Koutsoukos 1989, 1992). Nearly all species of planktonic Foraminiferida became extinct at the K/T boundary (about 12 genera and 27 species, with probably 5 surviving species or fewer) (e.g. Tappan & Loeblich 1988; Liu & Olsson 1992; Olsson & Liu 1993; Huber *in press*). *Guembelitria cretacea* Cushman, one of the few true Cretaceous survivors, has been recorded from the Coniacian to Campanian section near El Kef, Tunisia (Kroon & Nederbragt 1990), up to the lower part of the P2 Zone (upper Danian), in northeastern Brazil (Fig. 3). A triserially coiled planktonic morphotype (Fig. 5, parts 1–8), this species remained morphologically unchanged throughout the Coniacian to Maastrichtian, but started to diversify immediately after the K/T transition.

The planktonic foraminifera in the P0 zonal interval consists, at its base (survival interval), of a few, small sized (< 100 μm) Cretaceous survivor species such as the zonal marker, *G. cretacea*, and *H. monmouthensis* (Fig. 4), both species considered to have been surface-dwelling forms (Smith & Pessagno 1973; Kroon & Nederbragt 1990). The abundance of *G. cretacea* along with a significant decrease in other Late Cretaceous species has been used to define the K/T boundary and the lower P0 Zone of several

sections (e.g. Gredero, Spain and El Kef, Tunisia: Smit 1982; Agost and Caravaca, Spain: Canudo *et al.* 1991; Miller's Ferry, Alabama, USA: Liu & Olsson 1992). In addition, the stratigraphic distribution of *Guembelitria* appears to suggest a preference for dysoxic, shallow neritic environments (Leckie 1987; Koutsoukos 1989, 1994; Kroon & Nederbragt 1990). The dominance of survivor Cretaceous guembelitriids at this early interval, an opportunistic taxon which has 'the potential to persist during times of drastically changing marine conditions' (Kroon & Nederbragt 1990, p. 31), may be interpreted as a time when surface water eutrophic conditions prevailed in the aftermath of the K/T boundary event (the 'Strangelove Ocean' and 'Respiring Ocean' periods; see above).

It is important to note a significant increase in abundance of *G. cretacea* observed in the uppermost Maastrichtian, dark-grey marlstones (probably enriched in organic matter), between 85 cm and 30 cm (maximum at about 55 cm) below the K/T boundary (Fig. 3). A similar distribution pattern has been observed from other K/T boundary sequences examined elsewhere (G. Keller pers. comm., Angers, July 1994; Schmitz *et al.* 1992). In the Poty section this is recorded together with an abundance increase of deeper water dwellers such as *Racemiguembelina* and *Contusotruncana* (Fig. 3). This may suggest an increase in primary productivity perhaps associated with a sea-level highstand and oxygen-depleted bottom waters, just before the K/T transition.

After only a few thousand years, in the earliest Danian, the first Tertiary taxa (the *Guembelitria–Woodringina–Chiloguembelina* lineage) appeared (Olsson 1982; D'Hondt 1991; Liu & Olsson 1992; MacLeod 1993; Berggren *et al.* 1995) (Fig. 4). These are all microperforate non-spinose forms evolving progressively from

---

Fig. 5. All illustrations are scanning electron photomicrographs. Scale bars = 10 μm. **(1)** *Guembelitria cretacea*. Lateral view. Sample Poty core #1, depth 478.5–480 cm (upper Pα Zone). **(2–4)** *G.* cf. *cretacea*. Sample Poty core #1, depth 93–95 cm (P2 Zone). 2 Lateral view; 3. detail of pores at the edges of poreless pustules, first chamber of the last whorl; 4. detail of ultimate chamber with poreless pustules. **(5)** *G.* aff. *cretacea*. Lateral view. Sample Poty core #1, depth 93–95 cm (P2 Zone). **(6–7)** '*Guembelitria* sp. A'. Lateral view. Sample Poty core #1, depth 93–95 cm (P2 Zone). 6. Lateral view; 7. detail of ultimate chamber with poreless pustules. **(8)** '*Guembelitria* sp. A'. Lateral view. Sample Poty core #1, depth 93–95 cm (P2 Zone). **(9)** *Guembelitria trifolia*. Sample Poty core #1, depth 414.5–415.5 cm (upper Pα Zone). **(10)** *Guembelitria* cf. *danica*. Sample Poty core #1, depth 478.5–480 cm (upper Pα Zone). **(11–14)** Ex interc. *G. cretacea* and *G. irregularis*. Lateral views. 11–12. Both specimens from sample Poty core #1, depth 487–489 cm (upper Pα Zone). 13–14. Sample 1-ACP-11 (base of Bed D; upper Pα Zone). **(15–19)** *Guembelitria irregularis*. Lateral views. 15–16. Sample Poty core #1, depth 487–489 cm (upper Pα Zone). 17. Sample Poty core #1, depth 480–482 cm (upper Pα Zone). 18. Sample Poty core #1, depth 483.5–485.5 cm (upper Pα Zone). 19. Sample Poty core #1, depth 487–489 cm (upper Pα Zone).

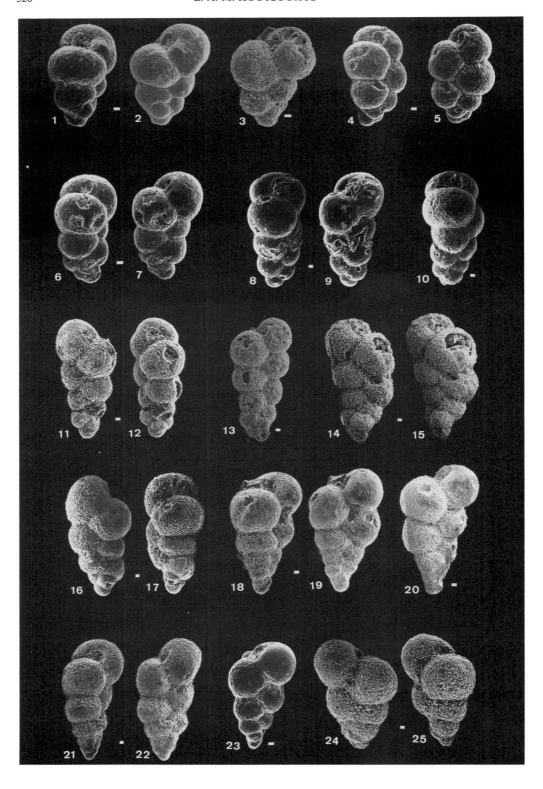

PELAGIC NICHES IN EARLY DANIAN

shallow neritic into open-ocean environments (Olsson 1970; Smit 1982; D'Hondt 1991; Liu & Olsson 1992), with several species of *Chiloguembelina* probably occupying intermediate water depths within the photic zone (Corfield & Cartlidge 1991; MacLeod 1993). There is a trend towards uncoiling of the triserial pattern by progressively changing the positioning of the aperture, without any detectable ecophenotypic preference at first (Fig. 5, parts 10–19; Fig. 6, parts 1–13). The $\delta^{13}C$ signature of *Woodringina hornerstownensis* Olsson and *Chiloguembelina morsei* (Kline) (Fig. 6, parts 24–25), slightly heavier than those on specimens of *G. cretacea* (cf. MacLeod 1993, figs 10–11), suggests that these species were preferentially dwelling in shallow epipelagic, distal-to-shore, oligotrophic water masses (Fig. 7). Intermediate forms (Fig. 6, parts 14–17) between *W. hornerstownensis* and *Chiloguembelina waiparaensis* (Jenkins) (Fig. 6, parts 18–23), the latter an intermediate-dwelling species (MacLeod 1993), demonstrate that a shift in preferred depth habitat occurred in the speciation event.

The first specimens of the *Eoglobigerina–Pseudosubbotina* (Fig. 8, parts 1–12, 16–20; Fig. 9, parts 1–7), *Eoglobigerina–Subbotina* (Fig. 8, parts 13–15) and *Praemurica* (Fig. 9, parts 8–12) lineages, cancellate spinose and non-spinose forms, also evolved in the Danian in a nearly coeval and parallel evolutionary trend from a probably ancestral *Hedbergella* stock (*H. holmdelensis* and *H. monmouthensis*) (Fig. 4). The early Danian Globigerinidae and Globorotaliidae species are characteristically represented by specimens with small test sizes and coarse cancellate surface ornamentation (e.g. Fig. 8, parts 1–5, 16–20; Fig. 9, parts 8–12). These seem to have had a preference for slightly more open (distal-to-shore), probably eutrophic to mesotrophic, shallow water masses, as suggested by their apparently conspicuous stratigraphic distribution to carbonate-rich neritic-bathyal successions (e.g. Fig. 3).

The return to 'normal' oligotrophic, surface-oceanic conditions in the Danian is suggested by the development of more diverse planktonic foraminiferal faunas near the P1b/P1c zonal transition (recovery interval), which included several intermediate and deep-water dwellers (e.g. *Chiloguembelina midwayensis*, *Praemurica inconstans*, *P. trinidadensis*, *P. uncinata*, *Parasubbotina pseudobulloides* and *S. triloculinoides*; Fig. 8, part 15; Fig. 9, parts 1–7, 13–21) (Fig. 7).

In the Poty section a few Cretaceous planktonic taxa (*Pseudoguembelina costulata*, *Heterohelix* ex gr. *globulosa*, *Rugoglobigerina reicheli* and *R.* ex gr. *rugosa*) show unconspicuous scattered occurrences, usually only one or two specimens per sample, up to near the top of the P1a/P1b Zones (Fig. 3). Similar microfaunal patterns have also been observed in other lowermost Palaeocene sections (e.g. Keller 1988, 1989, 1993; Barrera & Keller 1990; Canudo *et al.* 1991; Liu & Olsson 1992). These may represent either reworked specimens or Cretaceous survivor species, although neither could be determined to be the case in this study. However, the presence of only these species at such high levels gives further strength to the second alternative. If these species are true Cretaceous survivors they clearly did not repopulate the Danian pelagic realm, nor did they give rise to any obvious immediate descendants, such as in the *Guembelitria cretacea–G. irregularis–Woodringina–Chiloguembelina* lineage. It is more likely that these taxa remained out-of-place, under unfavourable environmental conditions, and were destined to rapid extinction from the Palaeocene scenario.

Evolutionary patterns

The very low diversity, early Palaeocene planktonic foraminiferal fauna contains a very high percentage of new species (Hart 1990; Fig. 4). The earliest Palaeocene diversification of planktonic foraminifera has been often described as

---

Fig. 6. All illustrations are scanning electron photomicrographs. Scale bars = $10\,\mu m$. **(1–3)** *Woodringina claytonensis*. Lateral views. 1–2. Sample Poty core #1, depth 483.5–484 cm (upper Pα Zone). 3. Sample Poty core #1, depth 487–489 cm (upper Pα Zone). **(4–5)** Ex interc. *W. hornerstownensis* and *W. claytonensis*. Lateral views. Sample Poty core #1, depth 483.5–485.5 cm (upper Pα Zone). **(6–13)** *W. hornerstownensis*. Lateral views. 6–7. Sample Poty core #1, depth 483.5–485.5 cm (upper Pα Zone). 8–9. Sample Poty core #1, depth 483–483.5 cm (upper Pα Zone). 10. Sample Poty core #1, depth 468.5–469 cm (upper Pα Zone). 11–12. Sample Poty core #1, depth 480–482 cm (upper Pα Zone). 13. Sample Poty core #1, depth 487–489 cm (upper Pα Zone). **(14–17)** Ex interc. *W. hornerstownensis* and *Chiloguembelina waiparaensis*. Lateral views. 14–15. Sample Poty core #1, depth 432.5–433.5 cm (upper Pα Zone). 16–17. Sample Poty core #1, depth 480–482 cm (upper Pα Zone). **(18–23)** *C. waiparaensis*. Lateral views. 18–19. Sample Poty core #1, depth 487–489 cm (upper Pα Zone). 20. Sample Poty core #1, depth 480–482 cm (upper Pα Zone). 21–22. Sample Poty core #1, depth 487–489 cm (upper Pα Zone). 23. Sample Poty core #1, depth 483.5–485.5 cm (upper Pα Zone). **(24–25)** *C. morsei*. Lateral views. Sample Poty core #1, depth 93–95 cm (upper P2 Zone).

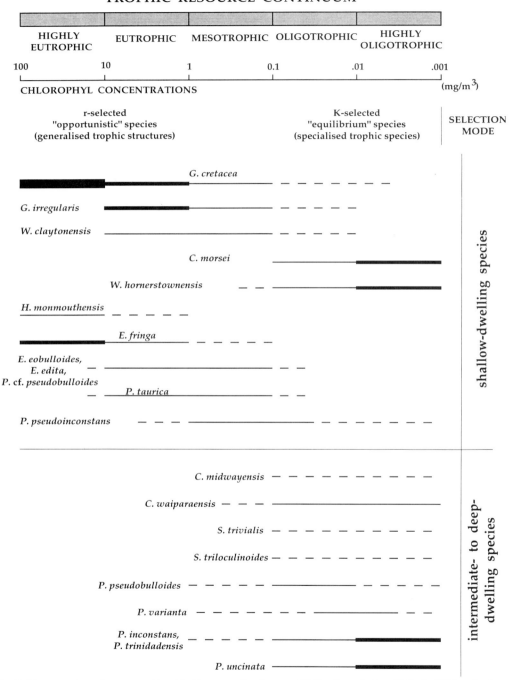

Fig. 7. The oceanic surface-water Trophic-Resource Continuum (TRC; adapted after Hallock 1987), selection mode (after Pianka 1970), schematic distribution and relative abundance of selected planktonic foraminifera along the TRC and water-depth habitats. Tentatively inferred after stable isotope data from Barrera & Keller (1990), Corfield & Cartlidge (1991), Schmitz et al. (1992), MacLeod (1993) and Huber & Barrera (1995), and empirically complemented from the distribution patterns (through space and time) after this study.

an 'evolutionary adaptive radiation' through 'recolonization events' (e.g. Hart 1990; Liu & Olsson 1992) of vacated pelagic niches within the same water depth zone or downward (Caron & Homewood 1983). This interpretation has been disputed more recently by MacLeod (1993), based on the absence of positive stable isotope evidence for patterns of morphological phylogenetic change and habitat diversification (= changes in water depth habitats) for the majority of Palaeocene speciation events of triserial and biserial morphotypes. The 'taxon pulse' model (Erwin 1985; MacLeod 1993), where speciation events took place at a much higher rate than changes in preferred habitat, may be a suitable explanation and suggest a complex set of driving evolutionary mechanisms. Moreover, the evolutionary strategy of the Mesozoic planktonic foraminifera has been considered as 'an adaptation to trophic resource variations in the oceanic environment' (Caron & Homewood 1983, p. 461). An alternative, or perhaps concurrent, pathway for speciation may have occurred, initially, within the same water depth zone. As the surface-water nutrient spectrum (the TRC) gradually expanded, it appears to have induced the progressive migration of the evolving planktonic species from proximal-to-shore, eutrophic–mesotrophic water masses in the early Danian, to distal-to-shore, oligotrophic water masses later in the Danian (see also discussion above) (Figs 4, 7 & 10). Intermediate and deep-water dwellers slowly evolved from shallower ancestors.

Eutrophication of surface waters in the earliest Danian may have inhibited the appearance of endosymbiont-bearing planktonic taxa in the photic zone, favouring at first omnivorous and/or carnivorous feeding strategies (Hallock 1982; Tappan & Loeblich 1988). Such is probably the case for the development of spines in the Globigerinidae (Fig. 4) in order to support a more carnivorous feeding strategy (Hemleben *et al.* 1991).

A marked diversity increase of the planktonic foraminifera occurs within the upper part of the P0 Zone–lower Pα Zone interval (Fig. 4). The general trend is abruptly interrupted at about the Pα/P1a zonal boundary. This corresponds to approximately the biochronostratigraphic positioning of the bolide-impact-triggered tsunami beds in northeastern Brazil (Albertão *et al.* 1994; Koutsoukos 1996*a*, *b*), which may be significant.

A definite peak in speciation occurred by the P1a/P1b zonal transition (Fig. 4), at the same time when the planktonic foraminifera increased remarkably in abundance and mean test size (Fig. 3; e.g. Blow 1979; Olsson & Liu 1993). This radiation maximum in the planktonic communities coincides, apparently, with a nearly coeval change in sedimentary regime (climatically induced?) in low-latitude South Atlantic regions. Marly and siliciclastic dominated deep-water sequences in the latest Maastrichtian are replaced by more carbonate-rich neritic–bathyal successions in the Danian (e.g. Albertão 1992; Albertão *et al.* 1994), the latter strata being usually very condensed or even absent in offshore sections (Beurlen 1992). Moreover, such an event appears to correspond to a period of widespread oceanic surface-water oligotrophic conditions, stratified water masses and recovery of the normal positive carbon-isotope gradients, similar to those observed in the latest Cretaceous (e.g. Hsü & McKenzie 1985). These suggest an increasing reliance, late in the Danian, on symbiosis under conditions of nutrient-poor waters in the photic zone (e.g. Brasier 1986; Tappan & Loeblich 1988).

Conclusion

The diversification episodes of Danian planktonic foraminifera appear to correspond to times of reduced oceanic mixing, stratified water masses, recovery of surface water productivity and a progressive expansion of the trophic-resource spectrum (the TRC), with periods of maximum diversity, such as the P1a/P1b zonal transition, relating to widespread oceanic surface-water oligotrophic conditions (Figs 4, 7 & 10). As the TRC gradually expanded it appears initially to have propiciated the progressive migration of the evolving planktonic species within the same water depth zone, from proximal-to-shore, eutrophic–mesotrophic, surface water masses, to distal-to-shore, oligotrophic, surface oceanic water masses. Intermediate and deep-water dwellers slowly evolved from shallower ancestors.

Furthermore, these suggest a parallel adaptive trend of feeding strategies caused by the changing availability of nutrients in the water column, from largely dominant planktonic assemblages of herbivorous or passive grazers (probably most of the non-spinose, microperforate, pustulose Guembelitriidae; e.g. Anderson *et al.* 1979) and subordiante carnivorous (spinose, cancellate-walled Globigerinidae) in the early Danian, with probable increasing reliance in symbiosis (likely the non-spinose, cancellate-walled, globular-chambered Globorotaliidae) under low nutrient conditions in the late Danian.

The early Danian impact event strongly disturbed the geological record at the K/T boundary in low latitude Atlantic regions,

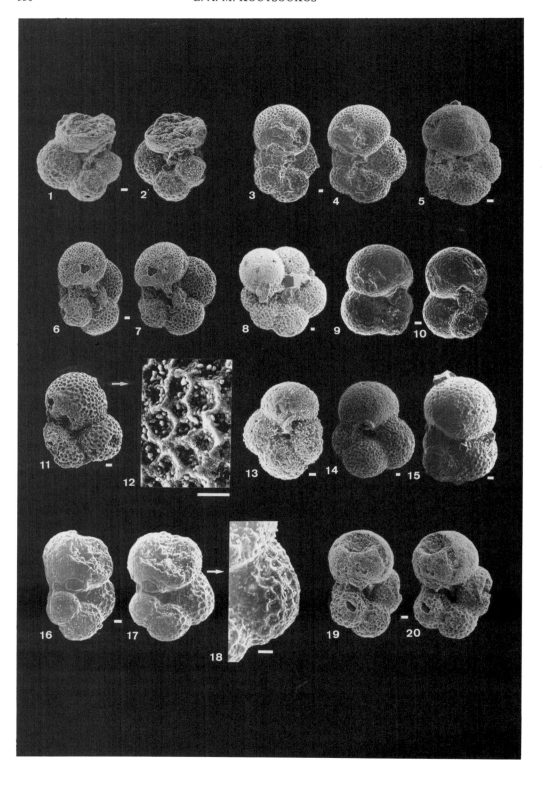

PELAGIC NICHES IN EARLY DANIAN

causing extensive mixing of microfossils of different ages and environments. Such mixing of microfossils can easily result in misleading interpretations of species survivorship and phylogenetic trends, which may account for many of the conflicting arguments in the literature.

One of the questions that remains unanswered concerns the possible extent and biotic effects of the early Danian impact event. The biochronostratigraphic positioning of the event, near or at the boundary between the Palaeocene $P\alpha$ and P1a foraminiferal zones, is significant. At this level the last appearances of a few Danian taxa and most 'reworked' Cretaceous species are recorded, just preceding the major radiation episode of planktonic foraminifera at the P1a/P1b interval. This event may have caused the elimination of most of the Cretaceous survivors from the Cenozoic scenario, as well as some 'unsuitable' (or 'unlucky') Danian planktonic foraminifera, which apparently did not give rise to any known descendants (dead-end branches on the phylogenetic tree) such as *P. eugubina*.

A sudden event, for example a bolide impact, can only cause extinctions of all individuals of a group of species if the environmental effects (such as climate changes, overturn of deep waters) are widespread, at least affecting the entire species domain, and last long enough to disturb the life cycle and/or the trophic structure of those taxa. The well-adapted and specialized species (k-selected) have greatly restricted geographical distributions, latitudinally as well as bathymetrically, with specific requirements for a more complex and/or sensitive life cycle and trophic structure (highly interdependent and specialized food chains). These taxa would be primary extinction targets during catastrophic environmental turnovers. The more cosmopolitan (r-selected) a species is, the greater its chance of survival during such major events, a kind of natural selection in reverse, selecting the 'overly' adapted and specialized taxa for elimination. The impact event in the earliest Danian hap-

pened after the near complete demise of the highly specialized Cretaceous planktonic foraminifera. It apparently did not have a great effect on the Danian assemblages, which had just begun to speciate and specialize and would be fairly cosmopolitan in character during that early phase.

This work is part of a study undertaken at the Geologisch–Paläontologisches Institut of the University of Heidelberg and supported by a research fellowship from the Alexander von Humboldt Foundation, which is gratefully acknowledged. I would like to thank Petroleo Brasileiro S.A. (Petrobras), Rio de Janeiro, for permission to publish the paper. I thank P. Bengtson (University of Heidelberg) and M. B. Hart (University of Plymouth) for the critical reading of parts of the manuscript and several constructive suggestions. I am also deeply grateful to N. Paweletz, K. Hartung and D. Schroeter (Deutsches Krebsforschungszentrum, Heidelberg) for assistance with the SEM microphotography. This paper is a contribution to IGCP Project 335 'Biotic recovery from mass extinction events' and to IGCP Project 381 'South Atlantic Mesozoic Correlations'.

References

ALBERTÃO, G. A. 1992. *Abordagem interdicipliner e epistemológica sobre as evidências do limite Cretáceo–Terciário, com base em leituras efetuadas no registro sedimentar das bacias da costa leste brasileira*. MSc dissertation, University of Ouro Preto.

———, KOUTSOUKOS, E. A. M., REGALI, M. P. S., ATTREP JR, M. & MARTINS JR, P. P. 1994. The Cretaceous–Tertiary boundary in southern low-latitude regions: preliminary study in Pernambuco, northeastern Brazil. *Terra Nova*, **6**, 366–375.

ANDERSON, O. R., SPINDLER, M., BÉ, A. W. H. & HEMLEBEN, C. 1979. Trophic activity of planktonic foraminifera. *Journal of Marine Biological Association*, **59**, 791–799.

BARRERA, E. & KELLER, G. 1990. Stable isotope evidence for gradual environmental changes and species survivorship across the Cretaceous/Tertiary boundary. *Paleoceanography*, **5**, 867–890.

BERGGREN, W. A. & MILLER, K. G. 1988. Paleogene

Fig. 8. All illustrations are scanning electron photomicrographs. Scale bars = 10 μm. **(1–2)** *Eoglobigerina fringa*. Umbilical and lateral views. Sample Poty core #1, depth 487–489 cm (upper $P\alpha$ Zone). **(3–5)** *E. eobulloides*. 3–4. Lateral and umbilical views. Sample Poty core #1, depth 480–482 cm ($P\alpha$ Zone). 5. Umbilical view. Sample Poty core #1, depth 487–489 cm ($P\alpha$ Zone). **(6–7)** *E. edita*. Lateral and umbilical views. Sample Poty core #1, depth 487–489 cm ($P\alpha$ Zone). **(8)** *E. spiralis*. Umbilical view. Sample 3-CPMF-10 (P2 Zone). **(9–12)** *E. simplicissima*. 9–10. Lateral and umbilical views. Sample Poty core #1, depth 496–498 cm ($P\alpha$ Zone). 11–12. Umbilical view and detail of ultimate chamber; note honeycomb cancellate wall texture with elevated ridges around pores. Sample Poty core #1, depth 93–95 cm (P2 Zone). **(13–14)** *Subbotina trivialis*. Umbilical views. Both specimens from sample Poty core #1, depth 93–95 cm (P2 Zone). **(15)** *S. triloculinoides*. Umbilical view. Sample 3-CPMF-9 (P2 Zone). **(16–18)** Ex interc. *E. fringa* and *Parasubbotina* cf. *pseudobulloides*. 16–17. Lateral and umbilical views. 18. Detail of cancellate wall texture. Sample Poty core #1, depth 483.5–485.5 cm ($P\alpha$ Zone). **(19–20)** *P.* cf. *pseudobulloides*. Lateral and umbilical views. Sample Poty core #1, depth 487–489 cm ($P\alpha$ Zone).

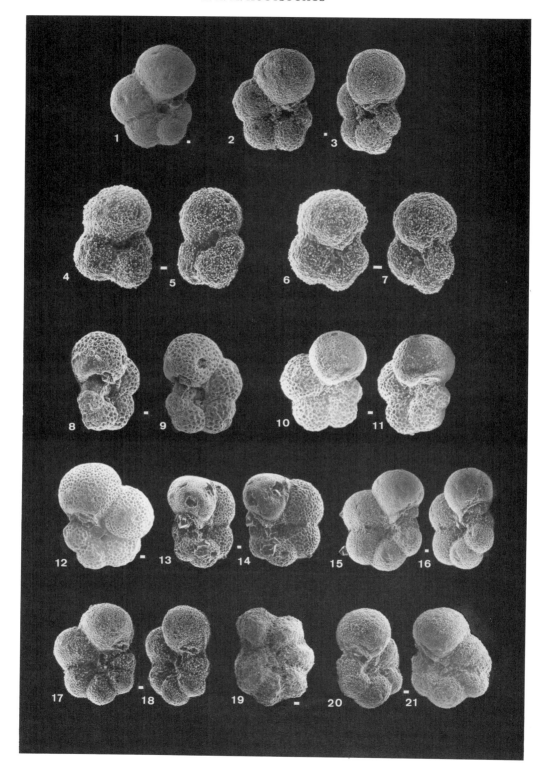

tropical planktonic foraminiferal biostratigraphy and magnetobiochronology. *Micropaleontology*, **34**, 362–380.

———, KENT, D. V., SWISHER, C. III & MILLER, K. G. 1996. A revised geochronology and chronostratigraphy. *In:* BERGGREN, W. A., KENT, D. V. & HARDENBOL, J. (eds) *Geochronology, Time Scales and Stratigraphic Correlation Framework for an Historical Geology*. Society of Economic Geologists and Paleontologists, Special Publication, in press.

BEURLEN, G. 1992. *Bioestratigrafia das Bacias Mesozóicas–Cenozóicas Brasileiras*. Texto Explicativo das Cartas Bioestratigráficas, PETROBRÁS/CENPES/DIVEX/SEBIPE, Rio de Janeiro, unpublished company report.

BLOW, W. H. 1979. *The Cainozoic Globigerinida*. Leiden.

BOERSMA, A. & SHACKLETON, N. 1981. Oxygen- and carbon-isotope variations and planktonic foraminifer depth habitats, Late Cretaceous to Paleocene, central Pacific, Deep Sea Drilling Project Sites 463 and 465. *Initial Reports of the Deep Sea Drilling Project*, **62**, 513–526.

———, ———, HALL, M. & GIVEN, Q. 1979. Carbon and oxygen isotope records at DSDP Site 384 (North Atlantic) and some Paleocene paleotemperatures and carbon isotope variations in the Atlantic Ocean. *Initial Reports of the Deep Sea Drilling Project*, **43**, 695–717.

BROECKER, W. S. & PENG, T. H. 1982. *Tracers in the Sea*. Eldigio Press, Columbia University, New York.

BRASIER, M. D. 1986. Form, function, and evolution in benthic and planktic foraminiferal test architecture. *In:* LEADBEATER, B. S. C. & RIDING, R. (eds) *Biomineralization in Lower Plants and Animals*. The Systematics Association, Special Volume, **30**, Clarendon Press, Oxford, 251–268.

CANUDO, J. I., KELLER, G. & MOLINA, E. 1991. Cretaceous–Tertiary boundary extinction pattern and faunal turnover at Agost and Caravaca, S.E. Spain. *Marine Micropaleontology*, **17**, 319–341.

CARON, M. & HOMEWOOD, P. 1983. Evolution of early planktic foraminifers. *Marine Micropaleontology*, **7**, 453–462.

CORFIELD, R. M. & CARTLIDGE, J. E. 1991. Isotopic evidence for the depth stratification of fossil and recent Gobigerinina: A Review. *Historical Biology*, **5**, 37–63.

D'HONDT, S. L. 1991. Phylogenetic and stratigraphic analysis of earliest Paleocene biserial and triserial planktonic foraminifera. *Journal of Foraminiferal Research*, **21**, 168–181.

ERWIN, T. 1985. The taxon-pulse: a general pattern of lineage radiation and extinction among carabid beetles. *In:* BALL, G. E. (ed.) *Taxonomy, Phylogeny, and Zoogeography of Beetles and Ants*. Junk, The Hague, 437–472.

GILMOUR, I., WOLBACH, W. S. & ANDERS, E. 1990. Early environmental effects of the terminal Cretaceous impact. *In:* SHARPTON, V. L. & WARD, P. D. (eds) *Global Catastrophes in Earth History; An Interdisciplinary Conference on Impact, Volcanism, and Mass Mortality*. Geological Society of America, Special Paper, **247**, 383–390.

HALLOCK, P. 1982. Evolution and extinction in larger foraminifera. *In: Proceedings of the Third North American Paleontological Convention*, **1**, 221–225.

——— 1987. Fluctuation in the trophic resource continuum: A factor in global diversity cycles? *Paleoceanography*, **2**, 457–471.

HAQ, B. U., HARDENBOL, J. & VAIL, P. R. 1987. Chronology of fluctuating sea levels since the Triassic. *Science*, **225**, 1156–1167.

HART, M. B. 1990. Major evolutionary radiations of planktonic foraminifera. *In:* TAYLOR, P. D. & LARWOOD, G. P. (eds) *Major Evolutionary Radiations*. Systematics Association, Special Volume, **42**, Clarendon, Oxford, 59–72.

HEMLEBEN, C. 1975. Spine and pustule relationships in some recent planktonic foraminifera. *Micropaleontology*, **21**, 334–341.

———, SPINDLER, M. & ANDERSON, O. R. 1989. *Modern planktonic foraminifera*. Springer, New York.

———, MÜHLEN, D., OLSSON, R. K. & BERGGREN, W. A. 1991. Surface texture and the first occurrence of spines in planktonic foraminifera from the early Tertiary. *Geologisches Jahrbuch*, **A128**, 117–146.

HOLLANDER, D. J., McKENZIE, J. A. & HSÜ, K. J. 1993. Carbon isotope evidence for unusual plankton blooms and fluctuations of surface water CO_2 in 'Strangelove Ocean' after terminal Cretaceous event. *Palaeogeography, Palaeoclimatology, Palaeoecology*, **104**, 229–237.

HSÜ, K. J. & McKENZIE, J. A. 1985. A 'Strangelove' ocean in the earliest Tertiary. *In:* SUNDQUIST, E. T. & BROECKER, W. S. (eds) *The Carbon Cycle and Atmospheric CO_2: Natural Variations Archean to Present*. American Geophysical Monographs,

---

Fig. 9. All illustrations are scanning electron photomicrographs. Scale bars = 10 µm. **(1–3)** *Parasubbotina pseudobulloides*. 1. Umbilical view. 2–3. Umbilical and lateral views. Both specimens from sample Poty core #1, depth 93–95 cm (P2 Zone). **(4–7)** *P. variata*. 4–5. Umbilical and lateral views. 6–7. Umbilical and lateral views. Both specimens from sample Poty core #1, depth 93–95 cm (P2 Zone). **(8–9)** *Praemurica taurica*. Lateral and umbilical views. Sample Poty core #1, depth 487–489 cm (Pα Zone). **(10–11)** Ex interc. *P. taurica* and *P. pseudoinconstans*. Umbilical and lateral views. Sample Poty core #1, depth 480–482 cm (Pα Zone). **(12)** *P. pseudoinconstans*. Umbilical view. Sample Poty core #1, depth 480–482 cm (Pα Zone). **(13–14)** *P. inconstans*. Lateral and umbilical views. Sample 3-CPMF-10 (P2 Zone). **(15–18)** *P. trinidadensis*. 15–16. Umbilical and lateral views. 17–18. Umbilical and lateral views. Both specimens from sample Poty core #1, depth 93–95 cm (P2 Zone). **(19)** *P. praecursoria*. Umbilical view. Sample Poty core #1, depth 355–357 cm (P1c Zone). **(20–21)** *P. uncinata*. Lateral and umbilical views. Sample Poty core #1, depth 93–95 cm (P2 Zone).

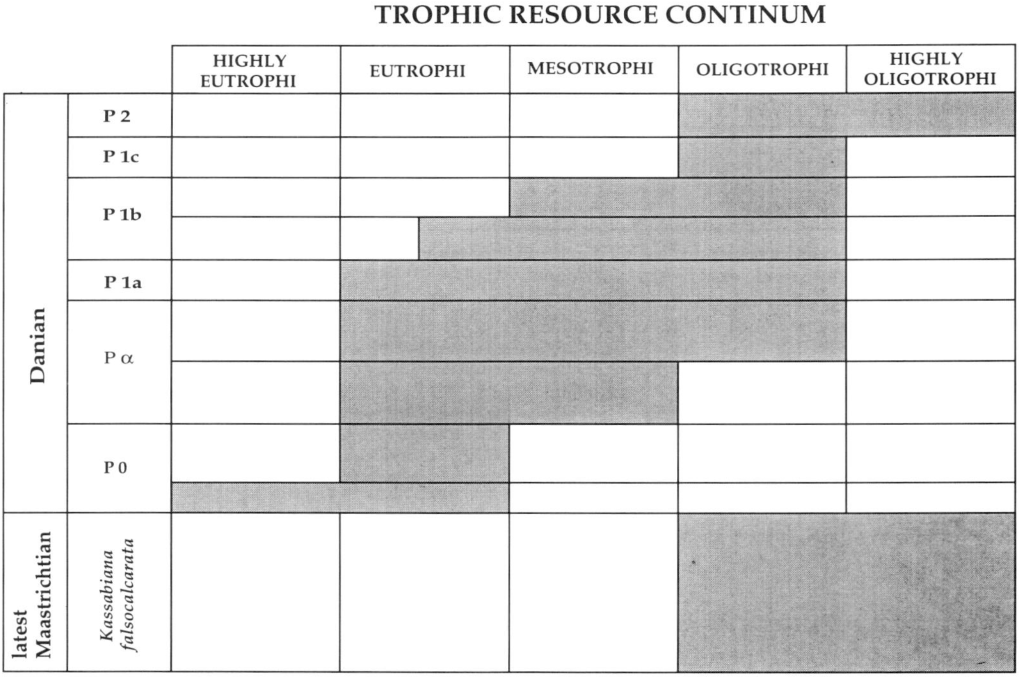

TROPHIC RESOURCE CONTINUM

Fig. 10. Suggested fluctuations of the TRC spectrum in the Danian oceanic surface-water masses at low latitude regions. These are interpreted as possible driving evolutionary mechanisms for adaptive radiation episodes of the planktonic foraminifera in the aftermath of the K/T boundary mass extinction event. The diversification episodes correspond to times of reduced oceanic mixing and expanded TRC, with widespread oceanic surface-water oligotrophic conditions.

Washington, DC, **32**, 487–492.

—— & —— 1991. Carbon-isotope anomalies at era boundaries. Global catastrophes and their ultimate cause. In: SHARPTON, V. L. & WARD, P. D. (eds) Global Catastrophes in Earth History; An Interdisciplinary Conference on Impact, Volcanism, and Mass Mortality. Geological Society of America, Special Paper, **247**, 61–70.

HUBER, B. T. & BARRERA, E. 1997. Evidence for planktonic foraminifera reworking vs. survivorship across the Cretaceous–Tertiary boundary at high latitudes. In: RYDER, G., FASTOVSKY, D. & GARTNER, S. (eds) The Cretaceous–Tertiary Event and Other Catastrophes in Earth History. Geological Society of America, Special Paper **307**.

KELLER, G. 1988. Extinction, survivorship and evolution of planktic foraminifers across the Cretaceous/Tertiary boundary at El Kef, Tunisia. Marine Micropaleontology, **13**, 239–263.

—— 1989. Extended Cretaceous/Tertiary boundary extinctions and delayed population change in planktonic foraminifera from Brazos River, Texas. Palaeoceanography, **4**, 287–332.

—— 1993. The Cretaceous–Tertiary boundary transition in the Antarctic Ocean and its global implications. Marine Micropaleontology, **21**, 1–45.

—— & BENJAMINI, C. 1991. Paleoenvironment of the eastern Tethys in the Early Paleocene. Palaios,

6, 439–464.

—— & LINDINGER, M. 1989. Stable isotope, TOC and CaCO$_3$ record across the Cretaceous/Tertiary boundary at El Kef, Tunisia. Palaeogeography, Palaeoclimatology, Palaeoecology, **73**, 243–265.

KOUTSOUKOS, E. A. M. 1989. Mid- to late-Cretaceous microbiostratigraphy, palaeo-ecology and palaeogeography of the Sergipe Basin, northeastern Brazil. PhD Thesis, 2 vols, Council for National Academic Awards/Polytechnic South West, Plymouth, England.

—— 1992. Late Aptian to Maastrichtian foraminiferal biogeography and palaeoceanography of the Sergipe Basin, Brazil. In: MALMGREN, B. A. & BENGTSON, P. (eds) Biogeographic Patterns in the Cretaceous Ocean. Palaeogeography, Palaeoclimatology, Palaeoecology, Special Issue, **92**, 295–324.

—— 1994. Early stratigraphic record and phylogeny of the planktonic genus Guembelitria Cushman, (1933). Journal of Foraminiferal Research, **24**, 288–295.

—— 1996a. A cometary impact in the early Danian: A secondary K/T boundary event? Geology, in press.

—— 1996b. The Cretaceous–Tertiary boundary at Poty, NE Brazil: event stratigraphy and palaeoenvironments. In: Compte-rendu du 12ᵉ Colloque de Stratigraphie et de Paleogeographie de l'Atlantique Sud (Angers, 16–20 July 1994), Recueil des Communications. Bulletin des Centres de Recherches Exploration–Production Elf Aquitaine, in press.

KROON, D. & NEDERBRAGT, A. J. 1990. Ecology and paleoecology of triserial planktic foraminifera. *Marine Micropaleontology*, **16**, 25–38.

LECKIE, M. R. 1987. Paleoecology of mid-Cretaceous planktonic foraminifera: A comparison of open ocean and epicontinental sea assemblages. *Micropaleontology*, **33**, 164–176.

LIU, C. & OLSSON, R. K. 1992. Evolutionary radiation of microperforate planktonic foraminifera following the K/T mass extinction event. *Journal of Foraminiferal Research*, **22**, 328–346.

LUTERBACHER, A. R. & PREMOLI-SILVA, I. 1964. Biostratigrafia del limite Cretaceo–Terziario nell' Appennino centrale. *Rivista Italiana di Paleontologia e Stratigrafia*, **70**, 67–128.

MACLEOD, N. 1993. The Maastrichtian–Danian radiation of triserial and biserial planktic foraminifera: testing phylogenetic and adaptational hypotheses in the (micro)fossil record. *Marine Micropaleontology*, **21**, 47–100.

—— & KELLER, G. 1991. How complete are Cretaceous/Tertiary boundary sections? A chronostratigraphic estimate based on graphic correlation. *Geological Society of America Bulletin*, **103**, 1439–1457.

MELLO, M. R., KOUTSOUKOS, E. A. M., HART, M. B., BRASSELL, S. C. & MAXWELL, J. R. 1989. Late Cretaceous anoxic events in the Brazilian continental margin. *Organic Geochemistry*, **14**, 529–542.

OLSSON, R. K. 1970. Planktonic foraminifera from the base of Tertiary, Miller's Ferry, Alabama. *Journal of Paleontology*, **44**, 598–604.

—— 1982. *Cenozoic planktonic foraminifera: A paleobiogeographic summary. Notes for a Short Course.* Organized by BUZAS, M. A. & SEN GUPTA, B. K., University of Tennessee, 1–26.

—— & LIU, C. 1993. Controversies on the placement of Cretaceous–Paleogene boundary and the K/P mass extinction of planktonic foraminifera. *Palaios*, **8**, 127–139.

——, HEMLEBEN, C., BERGGREN, W. A. & LIU, C. 1992. Wall texture classification of planktonic foraminifera genera in the lower Danian. *Journal of Foraminiferal Research*, **22**, 195–213.

PERCH-NIELSEN, K., MCKENZIE, J. A & HE, Q. 1982. Bio- and isotope-stratigraphy and the 'catastrophic' extinction of calcareous nannoplankton at the Cretaceous/Tertiary boundary. *In:* SILVER, L. T. & SCHULTZ, P. H. (eds) *Geological implications of impacts of large asteroids and comets on the Earth.* Geological Society of America, Special Paper, **190**, 353–371.

PIANKA, E. R. 1970. On r- and K-selection. *American Naturalist*, **104**, 592–597.

SCHMITZ, B., KELLER, G. & STENVALL, O. 1992. Stable isotope and foraminiferal changes across the Cretaceous–Tertiary boundary at Stevn Klint, Denmark: Arguments for long-term oceanic instability before and after bolide-impact event. *Palaeogeography, Palaeoclimatology, Palaeoecology*, **96**, 233–260.

SIGNOR, P. W. & LIPPS, J. H. 1982. Sampling bias, gradual extinction patterns and catastrophes in the fossil record. *In:* SILVER, L. T. & SCHULTZ, P. H. (eds) *Geological implications of impacts of large asteroids and comets on the Earth.* Geological Society of America, Special Paper, **190**, 291–296.

SLITER, W. V. 1989. Biostratigraphic zonation for Cretaceous planktonic foraminifers examined in thin section. *Journal of Foraminiferal Research*, **19**, 1–19.

SMIT, J. 1982. Extinction and evolution of planktonic foraminifera after a major impact at the Cretaceous/Tertiary boundary. *In:* SILVER, L. T. & SCHULTZ, P. H. (eds) *Geological implications of impacts of large asteroids and comets on the Earth.* Geological Society of America, Special Paper, **190**, 329–352.

SMITH, C. C. & PESSAGNO, E. A. JR. 1973. Planktonic foraminifera and stratigraphy of the Corsicana Formation (Maestrichtian), North-Central Texas. Cushman Laboratory for Foraminiferal Research, Special Publication, **12**, 5–68.

SOLAKIUS, N., MAAMOUM, A. L. & BENSALEM, H. Planktic foraminiferal biostratigraphy of the Maastrichtian sedimentary beds at Ain Mdeker, northeastern Tunisia. *Geobios*, **17**, 583–591.

TAPPAN, H. & LOEBLICH, A. R. JR. 1988. Foraminiferal evolution, diversification, and extinction. *Journal of Paleontology*, **62**, 695–714.

ZACHOS, J. C. & ARTHUR, M. A. 1986. Paleoceanography of the Cretaceous/Tertiary boundary event: Inferences from stable isotopic and other data. *Paleoceanography*, **1**, 5–26.

Recovery of North Caucasus foraminiferal assemblages after the pre-Danian extinctions

E. M. BUGROVA

All-Russian Geological Research Institute (VSEGEI), Srednyi prospect 74, St Petersburg 199026, Russia

Abstract: The planktonic changes associated with the Maastrichtian–Danian boundary are described from successions in the North Caucasus, Russia. The extinctions recorded in the benthonic fauna are less sudden than those recorded in the plankton, with many pre-dating the end of the Maastrichtian. The Danian benthonic fauna contains a number of survivor taxa with few new species. The recovery interval for the benthonic fauna was much shorter and the increase in diversity occurred earlier than that of the planktonic fauna. Within the *Globorotalia angulata* Zone the recovery of the foraminiferal assemblage was almost complete.

The extinction and subsequent recovery of the planktonic and benthonic foraminifera across the Maastrichtian/Danian boundary have been studied in the well documented successions exposed in the basin of the Kuban and Heu Rivers (North Caucasus). This is the area near the towns of Cherkessk and Nalchick (Fig. 1). In these sections the Maastrichtian limestones and marls (thickness about 30–40 m) with the remains on *Inoceramus tegulatus* Hagenow and echinoids of Cretaceous aspect (see Alimarina 1963) are overlain, without visible break, by the Palaeocene marls of the Elburgan Formation. These contain a fauna which includes *Echinocorys edhemi* Boehm, *E. renngarteni* Moskvin and *Hercoglossa danica* (Schlotheim). Despite the fact that a break is not apparent in the field, it would appear that there is a small boundary hiatus and that sediments of *Globigerina eugubina* Zone age have either been removed by erosion or were never deposited in the area.

The Palaeocene zonation used in this paper (Fig. 2) is based on the planktonic foraminifera and has been adopted by all those working in the area (Bugrova *et al.* 1991). Until recently the benthonic foraminifera from the Palaeocene have not been studied in sufficient detail and no zonation using them has been produced. The

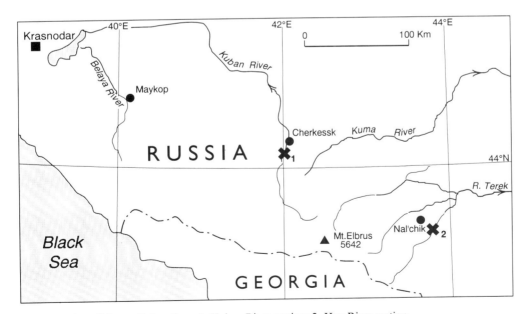

Fig. 1. Location of the studied sections: 1, Kuban River section; 2, Heu River section.

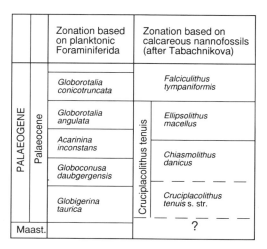

Fig. 2. Regional zonation using planktonic foraminifera and calcareous nannofossils (based on the work of Bugrova and Tabachnikova).

Elburgan Formation does, however, contain two quite distinctive benthonic assemblages and these have been regarded as zones, albeit provisionally (Bugrova et al. 1988; Bugrova 1992). The lowest, *Anomalina danica* Zone, corresponds to the three planktonic zones of the Danian Stage. It seems probable that this zone can be further subdivided. The uppermost, *Pyramidina crassa* Zone, corresponds to the *Globorotalia angulata* and *Globorotalia conicotruncata* Zones. The sub-division was identified as the 'beds with *Brotzenella* aff. *monterelensis*' by Bugrova et al. (1991).

The Palaeogene foraminifera from this region have been studied by various authors, including Subbotina, Morozova, Alimarina and Schutzkaya. Subbotina (1953) has, in particular, described the evolutionary trends in the development of the planktonic foraminifera immediately after the K/T extinction level. Morozova (1960) has shown that the earliest stage, with minute smooth *Eoglobigerina*, is followed by *Globigerina* and more specialized *Globoconusa* and *Acarinina*. Schutzkaya (1970) has noted the same faunal succession in Central Asia. The extinction and recovery of the benthonic fauna were not studied by these or any other micropalaeontologists. The preliminary results of such a study are presented in this paper.

Benthonic and planktonic foraminifera

A literature survey has shown that Late Maastrichtian foraminiferal assemblages from the North Caucasus contain about 100 species, the majority of which appear to be valid taxa. A substantial part of this fauna disappeared at the Maastrichtian/Danian boundary. The data collected for this study indicate that this extinction event was gradual. In the Maastrichtian sediments globotruncanids are diverse, but immediately below the boundary no planktonic species are observed. At this level there were only 20 benthonic species belonging to 14 genera, a fauna which includes *Verneuilina kelleri*, *Heterostomella foveolata*, *Marssonella oxycona*, *Ataxophragmium variabilis*, *Arenobulimina preslii*, *Plectina convergens*, *Palmula reticulata*, *Globorotalites michelinianus*, *Oridosalis frankei*, *Gavelinella umbilicata*, *Stensioeina caucasica*, *Stensioeina whitei*, *Bulimina ventricosa*, *Bolivinoides incrassatus crassus* and *Bolivinoides delicatulus* (Fig. 3).

In the Heu River section the author has observed a decrease in the species richness values and relative abundances of the planktonic fauna. Foraminiferal tests constitute only 15–20% of the prepared sample volume. The poorness and low diversity of the foraminiferal assemblages is quite marked with only very rare *Globotruncana* spp. and rare heterohelicids only. In the final phase of the extinction event these individuals also disappeared. This final extinction took place immediately below the local Cretaceous/Tertiary boundary.

Alimarina (1963), Leonov & Alimarina (1963) and Schutzkaya (1970) have all noted the presence of some small specimens of *Globotruncana* spp. immediately above the boundary. Very rare tests of *Globotruncana* sp. indet. have been found in the lower and middle parts of the Danian in the Heu and Kuban sections, together with small, fragile tests of chiloguembelinids. These individuals are not thought to be *in situ*, and the globotruncanids are identified as being reworked by their poor preservation. Only a few species with a primitive appearance (*Chiloguembelina* spp. and *Pseudotextularia elegans* Glaessner) appear to have survived into the Early Palaeocene. These species are found in very small numbers and never became abundant.

A non-planktonic zone and the *Globigerina eugubina* Zone are not identified at the base of the Danian in the studied sections. It would appear that the sediments of *G. eugubina* Zone age were removed by erosion. A thin bed of limestone, with the rare remains of minute unidentified planktonic foraminifera, has been reported from the Kuban River section just above the boundary at the base of the *Globigerina taurica* Zone. This zone is defined as the interval from the first appearance of *G. taurica*

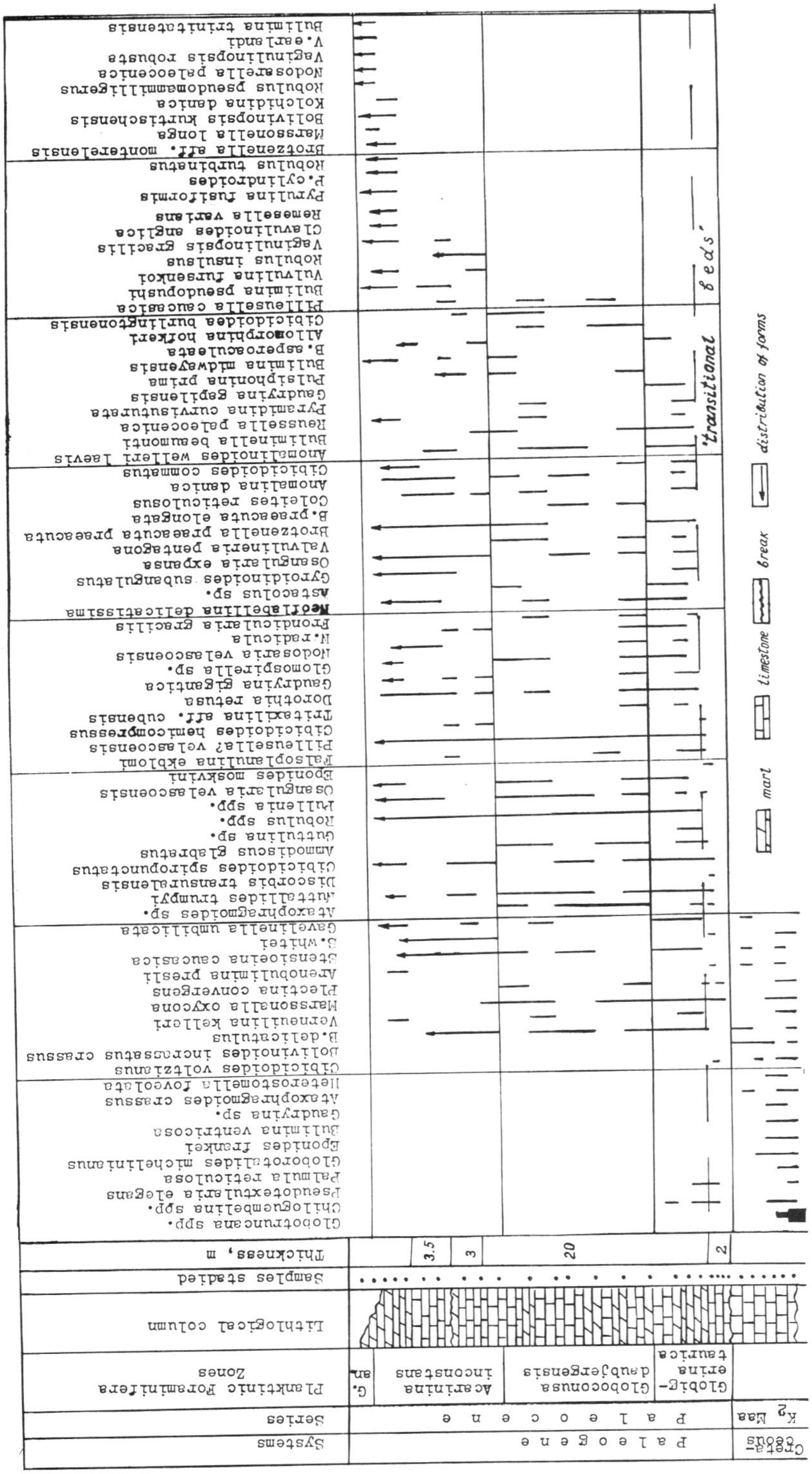

Fig. 3. The distribution of some Early Paleocene foraminifera in the Heu River section.

340   E. M. BUGROVA

up to the first appearance of *Globoconusa daubjergensis*.

After the extinction of the Cretaceous planktonic fauna the new species which appeared make up more than 90% of the Danian assemblage. The first phase in the recovery of the planktonic taxa began with the appearance of minute forms of *Eoglobigerina* within the *Globigerina (Eoglobigerina) taurica* Zone. In the Heu River section, near the base, are found very rare tests of *E. taurica, E. fringa* and *E. eobulloides*. Later, in the middle of this zone, *Globorotalia pseudobulloides, Eoglobigerina microcellulosa, Globigerina varianta, G. triangularis, Subbotina triloculinoides* and *Planorotalites planocompressus* are also recorded. Such forms as small *Acarinina* sp., minute *Globoconusa daubjergensis* and *Globorotalia compressa* appear near the top of the zone.

Within the *G. taurica* Zone the first appearance of 7 genera and 14 species is reported. Throughout the interval there is an increase in the size of the tests and a gradual improvement in their state of preservation. Aside from the survival taxa (*Chiloguembelina* spp. and *Pseudotextularia elegans*) two new species (*Guembelina taurica* and *Chiloguembelina* sp. nov.) appeared. However, throughout, heterohelicids are neither abundant nor diverse. It is suggested that at the base of the *G. taurica* Zone there is a brief survival interval (Fig. 3 shows this as the 'transitional beds') during which a few rare benthonic survivors (*Verneuilina kelleri, Plectina convergens, Arenobulimina preslii* and *Marssonella oxycona*) persisted. Although some immigrant taxa appeared and became more abundant, the generic and species diversity of the fauna remained low.

A much longer recovery phase began from the end of the *G. taurica* Zone and continued through the next two zones. These are (in ascending order) the *Globoconusa daubjergensis* Zone (the interval from the flood appearance of the index taxon up to the appearance of *Acarinina inconstans*) and the *Acarinina inconstans* Zone (from the first appearance of the index talon up to the appearance of *Globorotalia angulata*). Throughout this interval the planktonic forms prevailed, and a gradual growth of specific and systematic diversity can be observed. In this interval all the eoglobigerinids, some species of *Globigerina* and *Chiloguembelina*, and *Guembelina taurica* disappeared. The large size of the tests of *Globigerina* and *Globoconusa* also characterize this interval. Certain species which appeared during the first phase of recovery became very abundant in the *G. daubjergensis* Zone (e.g. *Globigerina* spp.,

Globoconusa spp., *Subbotina triloculinoides* and *Globorotalia pseudobulloides*). In the overlying *A. inconstans* Zone they gave way to the new taxa. This zone was the interval of diversification for species of *Acarinina* (e.g. *A. inconstans, A. praecursoria*) with large tests. The smaller *A. trinidadensis* also becomes more common as well as some transitional forms. *A. inconstans* tends to be the dominant species in this interval.

With the evolution of new groups, diversity increases and the geographical distribution of taxa tends to characterize this phase of the recovery. These foraminiferal assemblages contain approximately 20 species belonging to 5 genera. Just below the upper boundary of the *A. inconstans* Zone the new species, *Globorotalia angulata* (nominate species of the overlying zone), appears for the first time. This species represents a new group of angular–conical globorotalids which developed in the Middle Palaeocene. By this time the planktonic diversity was nearly recovered. However, during the Late Palaeocene, the environment in this basin degenerated and became unfavourable for the further development of the planktonic foraminifera.

Changes in the benthonic foraminiferal assemblages also took place during this interval. The K/T extinction resulted in the disappearance of many benthonic taxa, especially the highly specialized forms. Several of these species left no descendants while others were replaced by new, initially more primitive, Palaeogene groups. One such specialized group is the Orbitoididae. These were present in the Transcaucasus Region and were replaced by the Discocyclinidae which may have evolved from some surviving lineage. Very rare, small, specimens of primitive *Discocyclina* sp. have been found (Bugrova 1984) in the lowest *G. taurica* Zone in the Kuban River section and in the overlying *G. daubjergensis* Zone in the Heu River section. These are the earliest forms recorded up to the present time.

The Late Maastrichtian benthonic assemblages were rich and varied. Several of these species disappeared during the uppermost Cretaceous. Towards the end of the Maastrichtian the benthonic assemblage was much reduced. It contains only 25 species belonging to 12 genera. Large *Cibicidoides* spp., *Osangularia* spp. and *Stensioeina* spp. are dominant. There are also agglutinated species belonging to the genera *Verneuilina, Marssonella, Ataxophragmoides, Hagenowina, Plectina* and *Orbignyna*. Species surviving from the Late Maastrichtian are quite numerous in the Lower Danian strata. *Verneuilina kelleri, Ataxophragmoides variabilis, Plec-*

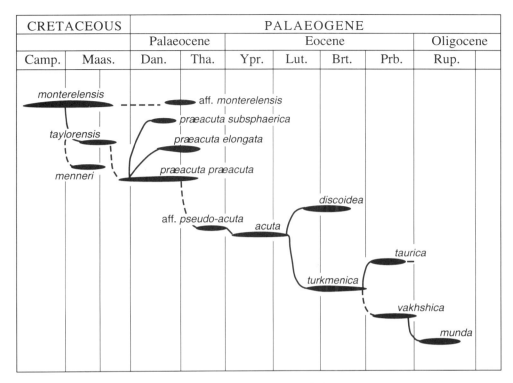

Fig. 4. Evolutionary lineages of the genus *Brotzenella* Vassilenko 1958.

tina convergens, *Hagenowina* sp., *Bolivinoides delicatulus*, *Osangularia lens*, *Cibicidoides voltzianus* and *Stensioeina caucasica* survived into the Early Danian. The benthonic assemblage just above the base of the Danian may be described as a reduced version of the Maastrichtian fauna. Only one genus of benthonic foraminifera (*Pulsiphonina*) appeared in the lowermost Danian. At the start of the survival interval benthonic survivors are a significant component of the foraminiferal assemblage and are widely distributed. During this period the fauna was gradually decreasing and the evolutionary lineages are rather fragmented.

In the middle of the *G. taurica* Zone the first new species of *Gaudryina*, *Pullenia*, *Nodosaria*, *Cibicidoides*, *Pilleussella*, *Brotzenella*, and *Pyramidina* appeared. The majority of these are immigrant forms and had ancestors outside the Cretaceous fauna of the Caucasus region. The appearance of several new taxa and two new genera (*Clavulinoides* and *Valvulineria*) may indicate the initial stage of the recovery. This process was more rapid than the recovery of the planktonic foraminifera. At the end of the *G. taurica* Zone the benthonic component of the assemblages was 75–80% of the processed sample volume. These assemblages contain about 60 species and represent some 30 genera. The populations of some species appear to be polymorphic and there are a number of non-typical forms which are considered valid species or subspecies (e.g. three subspecies of *Brotzenella praeacuta*).

Some progenitors give rise to a number of new taxa (Fig. 4). *Brotzenella praeacuta* is regarded as the ancestor of a lineage that underwent a marked diversification at the beginning of the Middle Eocene (Bugrova 1983). *Falsoplanulina ekblomi* also gave rise to several new lineages.

The diversity of benthonic assemblages increased in the Late Danian (*Acarinina inconstans* Zone and in the beginning of the *Globorotalia angulata* Zone). An examination of this assemblage reveals many elements in common with the 'Midway fauna' of the Gulf Coastal Plain and Northern Europe (e.g. *Citharina plumoides*, *Vaginulina robusta*, *V. longiforma*, *Pulsiphonia prima, Falsoplanulina ekblomi, Anomalina danica, Coleites reticulosus, Bulimina midwayensis, Loxostomoides applinae, Tappanina selmensis, Pyramidina crassa* and *P. curvisuturata*). In the rather shallow-water deposits of the Kuban River section many of these taxa are more abundant. Some endemic species are also present.

The high taxonomic and morphological diversity, increasing rates of evolutionary radiation and the wide geographic distribution (across the North Caucasus, the southern part of the Russian Platform, North Kasakhstan and the West Siberian Plain) of this rich association may indicate that the composition of the benthonic fauna after the K/T extinction event was at its acme. Higher in the succession the depositional environment was not favourable for these benthonic groups and many primitive agglutinated species appeared.

Summary

The changes to the planktonic and benthonic foraminiferal assemblages across the Maastrichtian/Danian boundary are documented. The extinction of the planktonic fauna was more marked than that of the benthonic assemblage. The majority of forms disappeared, except the species that possessed a more primitive morphology. The survival interval was brief (the beginning of the *G. taurica* Zone) and in the recovery interval about 20 species belonging to 5 genera appeared.

The extinction of the planktonic group was more gradual. Some quite specialized genera crossed over the boundary. The earliest Danian fauna was very similar to a reduced version of the Maastrichtian assemblage. The recovery interval began sooner and the increase in diversity was more rapid. At the end of the *Acarinina inconstans* Zone, and in the *Globorotalia angulata* Zone, many immigrants joined some of the survivors, together with many new taxa. By this time the recovery of the foraminiferal assemblage was almost complete.

I express my thanks to Drs L. Panova and I. Nikolaeva who collected the material for the investigation. The study was completed with the support of RFFI grant 17589. I would also like to thank the unknown referee and Prof. Malcolm Hart for the careful work that has assisted the publication of this paper.

References

ALIMARINA, V. P. 1963. Some peculiarities of the development of planktonic foraminifera in connection with the zonal differentiation of the Lower Palaeogene of the North Caucasus. *Questions of Micropaleontology*, **7**, 158–195.

BUGROVA, E. M. 1983. Variation and formation of the Paleogene foraminifers. *Questions of Micropaleontology*, **26**, 55–62.

—— 1984. Nummulites in the Paleogene of the Kuban river (North Caucasus). *Doklady Akademii Nauk SSSR*, **274**, 376–378.

—— 1992. Paleocene and Eocene Benthic Smaller Foraminifers and Biostratigraphy of the South USSR, *In: Studies in Benthic Foraminifera*. Proceedings of the Fourth International Symposium on Benthic Foraminifera, Sendai, Japan, 309–312.

——, NIKOLAEVA, I. A., PANOVA, L. A. & TABACHNIKOVA, I. P. 1988. On zonal differentiation of the Paleogene in the Southern regions of the USSR. *Soviet Geology*, **4**, 96–107.

——, TABACHNIKOVA, I. P., NIKOLAEVA, I. A. *ET AL.* 1991. Paleogene System. *In:* KOREN, T. N. (ed.) *Zonal Stratigraphy of the Phanerozoic of the USSR*. 'Nedra', Moscow.

LEONOV, G. P. & ALIMARINA, V. P. 1963. Stratigraphy and planktonic foraminifera of Cretaceous–Paleogene 'Transitional' beds of the Central part of the North Caucasus. *Transactions of the Geological Department of the Moscow University to XXI session of the International Geological Congress*, 29–50.

MOROZOVA, V. G. 1960. Stratigraphical zonation of Danian–Montian deposits in the USSR and the Cretaceous–Paleogene boundary. *International Geological Congress, XXI Session, Contributions to Soviet Geologists*, 83–100.

SCHUTZKAYA, E. K. 1970. *Stratigraphy, foraminifers and paleogeography of the Lower Paleogene of the Crimea, North Caucasus and the Western part of Central Asia*. 'Nedra', Moscow.

SUBBOTINA, N. N. 1953. *Globigerinidae, Hantkeninidae and Globorotalitidae*. Gostoptechisdat, Leningrad–Moscow.

Extinction and survivorship of southern Tethyan benthic foraminifera across the Cretaceous/Palaeogene boundary

R. P. SPEIJER[1,2] & G. J. VAN DER ZWAAN[1]

[1]*Department of Geology, Institute of Earth Sciences, Utrecht University, PO Box 80.021, 3508 TA Utrecht, The Netherlands*
[2]*Present address: Department of Marine Geology, Earth Sciences Centre, Göteborg University, S-413 81 Göteborg, Sweden*

Abstract: The benthic foraminiferal record from the Cretaceous/Palaeogene boundary stratotype of El Kef, Tunisia, shows a succession of three distinct assemblages. The late Maastrichtian upper bathyal assemblage is highly diversified and shows no prominent signs of gradual change towards the boundary. The earliest Palaeocene is marked by the disappearance of more than 50% of the taxa, resulting in a strongly impoverished fauna, tolerant to low oxygen conditions and with a shallower water affinity. Sequential (re-) appearance of many taxa in the early Palaeocene signifies the restoration towards normal Palaeocene upper bathyal faunas. The faunal changes reflect major perturbations in redox and trophic conditions at the sea-floor.

At least locally, and perhaps even on a regional Tethyan scale, the extinctions can be related to a sharp decrease in oxygen supply, in combination with strongly reduced nutrient resources. It is suggested that a prolonged reduction in surface fertility and food flux to the sea-floor invoked worldwide (but diachronous) benthic extinctions. In particular endo-benthic deposit feeders and other taxa adapted to high and perhaps heterogeneous nutrient resources suffered extinction.

Since the asteroid impact hypothesis was proposed to explain the mass-extinction at the end of the Cretaceous (Alvarez *et al.* 1980), the Cretaceous/Palaeogene (K/Pg) boundary has become one of the most controversial themes in Earth Sciences. Huge amounts of data support this extra-terrestrial event, a more (gradual) earthly cause, or a combination of both (see e.g. overviews in Sharpton & Ward 1990; Sutherland 1994). In recent years, the focus has shifted somewhat in the direction of gaining a better understanding of extinction selectivity and of ecosystem recovery in the aftermath of the event (e.g. Gerstel *et al.* 1987; Archibald & Bryant 1990; Gallagher 1991; Rhodes & Thayer 1991; Sheehan & Fastovsky 1992; Jäger 1993; Hansen *et al.* 1993; Raup & Jablonski 1993).

In particular the well-documented planktonic foraminiferal extinction at the K/Pg boundary has been the subject of intensive research (e.g. Luterbacher & Premoli-Silva 1964; Smit 1982; Brinkhuis & Zachariasse 1988; Keller 1988*a*, 1989*a, b*, 1993; D'Hondt & Keller 1991; Huber 1991). In combination with a major turnover in calcareous nannofossils (e.g. Bramlette & Martini 1964; Romein 1979*b*; Perch-Nielsen 1981; Alcala-Herrera *et al.* 1992; Pospichal 1994), a

negative excursion in the $\delta^{13}C$ record (e.g. Stott & Kennett 1989; Zachos *et al.* 1989; Magaritz *et al.* 1992; Keller *et al.* 1993), and a drop in $CaCO_3$ accumulation (e.g. Arthur *et al.* 1987; Keller & Lindinger 1989), the extinction has been related to a major productivity crisis of the pelagic ecosystem (Perch-Nielsen *et al.* 1982; Hsü & McKenzie 1985; Arthur *et al.* 1987; Meyers & Simoneit 1989; Zachos *et al.* 1989). The productivity crisis appears to have been most profound at middle and low latitudes, but less dramatic at some high latitude sites (Hollis 1993; Keller 1993; Barrera & Keller 1994). Since benthic foraminiferal communities largely depend on the vertical food supply from the overlying surface waters (see e.g. Van der Zwaan *et al.* 1992 and references therein) it seems plausible to expect that, in particular in tropical and subtropical regions, a drop in primary production had a profound effect on benthic communities.

Several studies discussed benthic foraminiferal changes across the K/Pg boundary (Dailey 1983; Keller 1988*b*, 1992; Thomas 1990*b*; Nomura 1991; Kaiho 1992; Schmitz *et al.* 1992; Widmark & Malmgren 1992*a*; Kuhnt & Kaminski 1993; Coccioni & Galeotti 1994). There is a threefold general outcome of these studies: the first is that

From Hart, M. B. (ed.), 1996, *Biotic Recovery from Mass Extinction Events*, Geological Society Special Publication No. 102, pp. 343–371

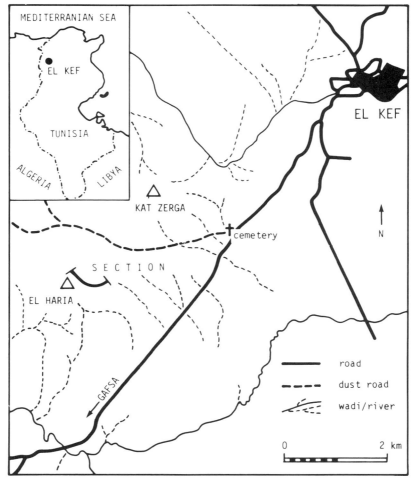

Fig. 1. Location map of the El Kef section (from Brinkhuis & Zachariasse 1988).

in particular biserial and triserial species became extinct in the earliest Palaeocene (e.g. species of *Praebulimina*, *Sitella*, *Eouvigerina*, *Orthokarstenia* and *Bolivinoides*), next to some trochospiral species (e.g. species of *Gavelinella* and *Stensioeina*) (cf. Loeblich & Tappan 1988), suggesting ecological selectivity. The second outcome is that the extinctions appear to have occurred over a prolonged time-span (Dailey 1983; Keller 1988b, 1992; Thomas 1990b; Nomura 1991; Schmitz et al. 1992; Widmark & Malmgren 1992a). However, in some cases reworking of Cretaceous material into the Palaeocene may have contributed to this gradual pattern (Dailey 1983; Keller 1992). The third general result of previous studies is that deep-sea assemblages were less severely affected than continental margin assemblages (Thomas 1990b; Kaiho 1992). Although some of the species previously mentioned are restricted to neritic deposits many others are only much less common in deep-sea deposits. These lower frequencies contribute to a less conspicuous turnover in quantitative studies. Thomas (1990a) suggested that the differential effects on the deep and shallower benthic communities might be related to the pelagic productivity breakdown, arguing that relatively oligotrophic deep-sea ecosystems would hardly be affected by a decrease in food supply, in contrast to shallower more eutrophic ecosystems. We discuss this hypothesis on the basis of a quantitative benthic foraminiferal analysis of the El Kef section (Tunisia) and semi-quantitative data from four K/Pg boundary profiles in the Middle East.

The El Kef section provides an expanded and probably continuous record across the K/Pg boundary (Perch-Nielsen 1981; Romein & Smit 1981; MacLeod & Keller 1991a, b; Olsson & Liu 1993), enabling a detailed faunal analysis. A similar study on this section has been performed earlier by Keller (1988b). In that paper a major fall in sea-level and oxygen content was postulated to explain the benthic foraminiferal sequence across the K/Pg boundary. Apart from using rather different taxonomic concepts, our interpretation of the data is substantially different. We also expanded the research further down into the Maastrichtian sequence to discern pre-boundary background variation and possible long-term trends (e.g. sea-level change). In order to improve our understanding of the faunal sequence of El Kef we studied a variety of different upper Maastrichtian and lower Palaeocene assemblages from Egypt and Israel. At the same time, these enable us to evaluate overall changes in southern Tethyan palaeocommunities in a broader context. (A list of species discussed can be found in Appendix 1.)

Material and methods

El Kef profile

Samples were obtained from the Maastrichtian–Palaeocene El Haria Formation (Salaj 1980), exposed in a section near the town of El Kef, Tunisia (Fig. 1). This paper focuses on the uppermost 30 m of the Maastrichtian marls and the lowermost 10 m of the Palaeocene shales and marls (Fig. 2). Basically, the widely distributed and studied AFN-coded samples were used. These samples were collected in 1982 by Drs A. J. T. Romein and J. Smit as representatives of the Cretaceous/Palaeocene Working Group of the International Committee on Stratigraphy. Drs A. J. Nederbragt and J. Smit kindly provided additional SN-coded samples, collected in 1992.

Apart from a 3 m thick basal Palaeocene unit, the sequence mainly consists of homogeneous grey to greenish-brown marls, without distinct sedimentological features. The K/Pg boundary is marked by a thin (2 mm) reddish ferruginous layer at the base of a 50 cm thick black shale bed (boundary clay). The boundary clay and the overlying 2.5 m dark grey to grey shales show preservation of sedimentary lamination and few pyritized and hematitic burrow molds. Abundant pyritized *Chondrites*-type burrow molds (diameter up to 1 mm) are present in the overlying 6 m of marls, whereas these are absent in the upper metre of the profile studied.

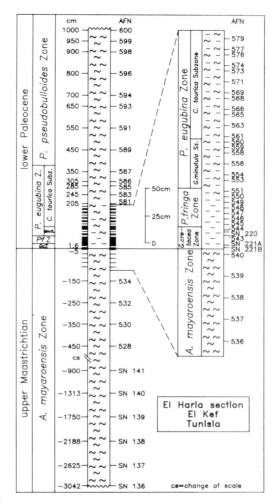

Fig. 2. Lithology and biostratigraphy of the El Kef section (modified from Brinkhuis & Zachariasse 1988). Note the change in scale in the lower part of the section; figures on the left side of the column indicate distance (cm) relative to the K/Pg boundary.

The samples were washed and sieved into four size fractions. A split of the 125–595 μm fraction, containing 200–400 specimens, was used for benthic foraminifera counts. All specimens were picked and stored in Chapman-slides. The benthic foraminifera were, whenever possible, identified at species level. Nodosariacea and poorly preserved agglutinants were lumped in a higher taxonomic level. For generic classification we largely followed Loeblich & Tappan (1988) and in general, species concepts of Cushman (1946), Aubert & Berggren (1976), and Salaj *et al.* (1976) were adopted.

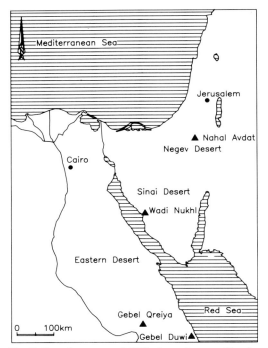

Fig. 3. Location map of K/Pg boundary profiles in Egypt and Israel

After finishing census counts the initial dataset (containing 132 taxa) was condensed by lumping all highly infrequent taxa into higher order groupings. Fisher-α diversity (Murray 1991), however, is based on the original census data. Foraminiferal numbers (specimens/gr dry sediment) are calculated from separately dried and weighed samples, which were washed over a 63 μm sieve. The >63 μm fraction was used in order to be able to establish planktonic foraminiferal numbers for the lowermost Palaeocene, because up to 3 m above the K/Pg boundary planktonic foraminifera are <125 μm in diameter (Brinkhuis & Zachariasse 1988).

Israeli and Egyptian profiles

Samples from other marly sequences covering the K/Pg boundary interval were obtained from a profile in Israel (Nahal Avdat) and from three sections in Egypt (Gebel Duwi, Gebel Qreiya and Wadi Nukhl; Fig. 3). Calcareous nannoplankton studies of the Nahal Avdat profile indicate the K/Pg boundary at about 1.5 m above the top of the Ghareb chalk (Romein 1979a, b). The K/Pg boundary in Wadi Nukhl is situated about 1 m above a ferruginous hardground, in the basal part of a Maastrichtian– upper Palaeocene marl-shale unit (cf. Shahin 1992). At Gebel Duwi the K/Pg boundary is situated within the marls just below a distinct 1.2 m thick, black, decalcified shale (cf. Faris 1982). Samples from Nahal Avdat, Wadi Nukhl, and Gebel Duwi were processed in the same way as the ones from El Kef. At Gebel Qreiya the K/Pg boundary lies approximately 2 m above a limestone bed in the Dakhla Formation (Luger 1988). Foraminiferal associations of this section were examined during a visit of the first author to Dr Luger in Bremen in 1993.

From all sections several upper Maastrichtian as well as lower Palaeocene samples were investigated. Faunal abundances of these samples are counted from pickings or estimated from strewings; all are treated semi-quantitatively and averaged per time interval. Since these sequences are either stratigraphically incomplete across the boundary, or show mixed Cretaceous and lowermost Palaeocene assemblages (see next section), it was not considered useful to study extinction and survivorship patterns in greater detail. Samples that contained mixed faunas were omitted in this study.

Biostratigraphy

El Kef profile

Since the El Kef section provides one of the most expanded and complete K/Pg boundary profiles (Perch-Nielsen 1981; Romein & Smit 1981; MacLeod & Keller 1991a, b; Olsson & Liu 1993), it was chosen to serve as a Global Stratotype Section and Point (GSSP) for the K/Pg boundary (Smit 1990). The position of the 'golden spike' is at the base of the boundary clay (i.e in the thin red ferruginous layer; Smit 1990). Our uppermost Maastrichtian sample spans the upper 10 cm of marls below the boundary clay, while the lowermost Palaeocene sample spans the lower 2.5 cm of the boundary clay. Therefore, the K/Pg boundary as delineated by our samples roughly corresponds to the officially defined boundary.

Many biostratigraphers studied this section, or a nearby parallel section (e.g. Verbeek 1977; Salaj 1978, 1980; Wonders 1980; Perch-Nielsen 1981; Brinkhuis & Zachariasse 1988; Keller 1988a; Pospichal 1994). We adopted the planktonic foraminiferal biozonation (Fig. 2) of Brinkhuis & Zachariasse (1988). For an elaborate account of biozonal definitions we refer to Brinkhuis & Zachariasse (1988); we confine ourselves to some brief remarks.

The lower 30.4 m of the section belongs to the uppermost Maastrichtian *Abathomphalus*

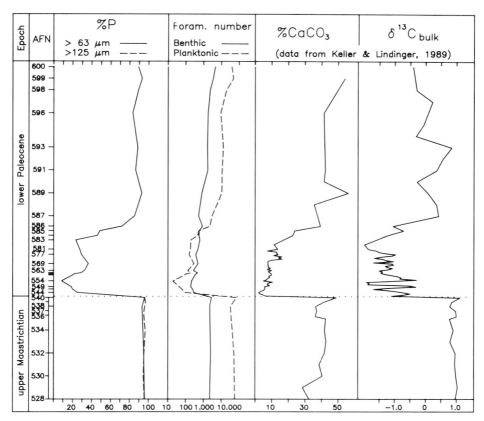

Fig. 4. The proportion of planktonic foraminifera (%P) and palaeoproductivity proxies for the K/Pg boundary sequence of El Kef. Maastrichtian samples for which %P in the >125 μm fraction has been calculated are indicated by sample ticks on the left side of the vertical line; samples for which the other calculations have been performed are marked by the sample ticks on the right side of this line.

mayaroensis Zone (Brinkhuis & Zachariasse 1988). Since the zonal marker is very rare in El Kef, the local *Racemiguembelina fructicosa* Zone was proposed, and considered to overlap with the *A. mayaroensis* Zone (Nederbragt 1992). The base of the *R. fructicosa* Zone, is at approximately 100 m below the K/Pg boundary; a single specimen of *A. mayaroensis* was found eight metres higher up in the section (A. Nederbragt pers. comm. 1992). The presence of the calcareous nannoplankton species *Micula prinsii* in the upper 20 m of the Maastrichtian suggests that the youngest part of the Maastrichtian is present in the El Kef section (Perch-Nielsen et al. 1982; Pospichal 1994). Slightly deviating from Brinkhuis & Zachariasse (1988), we define the base of the *Guembelitria cretacea* Zone by the first common occurrence of the nominate species in the >63 μm size fraction in the studied interval.

This level corresponds to the K/Pg boundary. The subsequent Palaeocene zones as well as the *Chiloguembelina taurica* Subzone are marked by the entry of their respective nominate (sub)zonal markers at the base of each (sub)zone; the base of the *Globoconusa minutula* Subzone is defined by the entry of *Parvularugoglobigerina eugubina*. In this paper the lowermost two biozones and the lower to middle part of the *P. eugubina* Zone are generally taken together and referred to as lowermost (or earliest) Palaeocene. The upper part of the studied interval of the El Kef section comprises the basal 7 m of the 40 m thick *Parasubbotina pseudobulloides* Zone (Brinkhuis et al. 1994). For convenience, the upper part of the *P. eugubina* Zone (samples AFN 585–587), together with the *P. pseudobulloides* Zone will generally be referred to as lower (or early) Palaeocene.

Israeli and Egyptian profiles

Abathomphalus mayaroensis has been recovered neither from the Israeli nor from the Egyptian Cretaceous samples. Environmental circumstances (e.g. too shallow conditions) have been proposed to explain the general absence or rarity of *A. mayaroensis* in upper Maastrichtian deposits from the southern Tethys and the Gulf Coastal plain (Keller 1989*a*, *b*). Therefore the uppermost Maastrichtian has to be differentiated in another way. *Plummerita reicheli* has been considered as an alternative index species for the uppermost Maastrichtian in many low latitude sites (Luger 1988; Rosenfeld *et al.* 1989; Masters 1993). According to Masters (1993) the first appearance of *P. reicheli* is well after the first appearance of *A. mayaroensis*. The former species has been observed in all Cretaceous samples studied, indicating that the uppermost Maastrichtian is present indeed in the Israeli and Egyptian sections. In Gebel Qreiya an uppermost Maastrichtian assignment is confirmed independently by the presence of *M. prinsii* (Luger 1988). The other sections yield other upper Maastrichtian calcareous nannoplankton markers such as *Micula murus, Lithraphidites quadratus*, or *Nephrolithus frequens* (Romein 1979*b*; Faris 1982; A. Henriksson pers. comm. 1993).

In Wadi Nukhl and Nahal Avdat we did not encounter the lowermost Palaeocene biozones. Although this may be related to a relatively wide sample spacing, it indicates that in the best case the lowermost Palaeocene is very condensed compared to El Kef. At Gebel Qreiya Palaeocene planktonic foraminifera of the *P. eugubina* Zone are mixed with up to 95% of Maastrichtian foraminifera (Luger 1988). Although some of the Cretaceous species may be true survivors (e.g. *G. cretacea*; cf. Smit 1982; Keller 1989*a, b*; Olsson & Liu 1993), this most likely does not apply to large specimens of *Globotruncana* and *Rugoglobigerina* (Luger 1988; and cf. MacLeod & Keller 1991*b*). More or less the same applies to Gebel Duwi, where we encountered extremely rare (< 1%) specimens of *Parvularugoglobigerina fringa* within an otherwise Maastrichtian fauna. Severe reworking of Maastrichtian foraminifera in lower Palaeocene deposits is a common feature in the Middle East and has been explained by sea-level related erosional events (Luger 1988; Keller *et al.* 1990). In all sections the *P. pseudobulloides* Zone was present and yielded samples with few, if any, reworked Cretaceous foraminifera.

Palaeoproductivity

Relative abundances of planktonic foraminifera (P/B ratios expressed as %P) and both planktonic as well as benthic foraminiferal numbers (named PFN and BFN, respectively) for a selected number of samples are shown in Fig. 4. Maastrichtian P/B ratios are high and stable, varying between 93%P and 97%P. Also BFN and PFN values are high: BFN values vary between 2300 and 2900 specimens/gr, whereas PFN ranges between 33 000 and 73 000 specimens/gr. The K/Pg boundary is marked by a sharp fall in all three variables. The proportion of planktonic foraminifera drops to 20%; BFN values are approximately 200–300 specimens/gr, while PFN values go down to 20–100 specimens/gr, i.e. a reduction by one and three orders of magnitude, respectively. The proportion of planktonic foraminifera remains low up to 2.5 m above the K/Pg boundary, from where the P/B ratio gradually increases up to around 90%P in the top of the profile. The restoration towards approximately pre-boundary levels in BFN (up to 4700 specimens/gr) and PFN (up to 49 000 specimens/gr) is similar.

Fine fraction $\delta^{13}C$ and $CaCO_3$ records of El Kef (Fig. 4) indicate relatively high surface water productivity during the Maastrichtian followed by strongly reduced productivity from the K/Pg boundary onwards into the earliest Palaeocene (Keller & Lindinger 1989). Planktonic foraminiferal accumulation rates (PFARs) have also been used to estimate surface productivity levels (Berger & Diester-Haass 1988), but in order to obtain reliable PFAR values a good time control for calculating sedimentation rates is crucial. Unfortunately, there is considerable uncertainty with respect to the exact amount of time involved within the El Kef sequence (e.g. MacLeod & Keller 1991*b*; Olsson & Liu 1993; Berggren *et al.* 1995), and estimated average sedimentation rates vary accordingly. The estimated average sedimentation rate for the Upper Maastrichtian is about 4 cm/ka (Brinkhuis & Zachariasse 1988). For the boundary clay estimates vary between 0.7 and 1.7 cm/ka, whereas for the *P. eugubina* Zone they vary between 1.9 and 4 cm/ka (Brinkhuis & Zachariasse 1988; MacLeod & Keller 1991*b*; Olsson & Liu 1993). In the lower part of the *P. pseudobulloides* Zone average sedimentation rate may have decreased to 1.1 cm/ka (MacLeod & Keller 1991*b*). Due to these uncertainties we refrain from calculating PFARs and consider planktonic foraminiferal numbers as an alternative, though very crude, way of assessing productivity changes in a relative sense. We calculated an almost three orders of magnitude decrease in PFN at the K/Pg boundary, at a level where sedimentation rates are generally considered to have reduced. Although some taphonomical loss by dissolution probably contributed to this decrease, it concurs with the geochemical interpretation of a sharp pelagic production decline during the earliest Palaeocene. At about 2.5 m above the K/Pg boundary (i.e. in the upper part of the *P. eugubina* Zone), PFN values gradually increase, together with an increase in the $CaCO_3$ and $\delta^{13}C$ records. This could indicate a gradual restoration towards former (Maastrichtian) surface fertility levels as proposed by Keller & Lindinger (1989). However, both PFN and $\delta^{13}C$ values remained slightly depressed relative to the Maastrichtian, suggesting that surface productivity probably did not fully recover within the studied interval.

Benthic foraminiferal numbers decrease in a similar way as PFN, though with lesser magnitude, across the K/Pg boundary. This coincidence is not surprising since benthic foraminifera that live below the euphotic zone primarily depend on the vertical organic carbon flux for their food supply (Van der Zwaan *et al.* 1990; Herguera & Berger 1991). In addition, oxygen deficiency may have contributed to reduced benthic productivity as well (Van der Zwaan *et al.* 1990). With increasing surface productivity, BFN values also increase again.

Benthic foraminiferal assemblages from El Kef

Stratigraphic distribution

The frequency data (Table 1) of the sixty most common taxa in the K/Pg boundary interval are displayed graphically in Fig. 5. The taxa are arranged and grouped according to their presence within three biostratigraphic intervals. These intervals are approximately the *A. mayaroensis* Zone, the *G. cretacea* Zone to the upper part of the *P. eugubina* Zone (=lower Palaeocene). In this way, we discriminate six stratigraphic assemblages (SA 1–6), each having its specific range, but note that this does not mean that the ranges of individual taxa span the entire range of the stratigraphic assemblages to which they belong.

Stratigraphic assemblage 1 (SA 1) consists of taxa that are virtually restricted to the *A. mayaroensis* Zone. *Gavelinella martini* occurs persistently in all Maastrichtian samples; all other taxa have a discontinuous distribution within the data set. Only one species (*Stensioeina*

pommerana) disappeared permanently well below the K/Pg boundary. Six taxa (*Bolivinopsis clotho, Heterostomella* spp., *Eouvigerina subsculptura, Praebulimina reussi, Coryphostoma plaitum* and *Anomalinoides* sp. 1) disappeared at the K/Pg boundary. Scattered occurrences (only one or two specimens per sample) of six other species (*Bolivinoides draco draco, Sitella colonensis, S. cushmani, Cibicides beaumontianus, Gyroidinoides tellburmaensis* and *G. martini*) were encountered up to 3.6 cm above the boundary; only *Sitella fabilis* was found in considerable numbers (2.5%) at 1.2 cm above the boundary. One poorly preserved specimen of *Coryphostoma incrassata gigantea* was encountered at 12.5 cm above the boundary.

Stratigraphic assemblage 2 (SA 2) contains taxa that are virtually absent in the lowermost Palaeocene, whereas they are more or less common in the *A. mayaroensis* Zone as well as in the lower Palaeocene. Four species (*Loxostomoides applinae, Cibicidoides* sp. 1, *Oridorsalis plummerae* and *Anomalinoides* sp. 2) of this assemblage disappeared temporarily well below the K/Pg boundary. Most taxa, however, disappeared near the K/Pg boundary: three taxa at the boundary (*Cibicidoides* cf. *hyphalus* – biconvex morphology –, *Gyroidinoides* spp., and 'calcareous agglutinants rest'), four taxa between 1.2 and 3.6 cm above the boundary (*Pseudouvigerina plummerae, Cibicidoides suzakensis, Pullenia* spp. and *Anomalinoides affinis*), and seven taxa (*Gaudryina pyramidata, Cibicidoides abudurbensis, C.* cf. *hyphalus* – planoconvex morphology –, *Valvalabamina depressa, Gyroidinoides octocameratus*, miliolids and 'trochospiral rest') between 3.6 and 12.5 cm above the boundary. Only very low frequencies of most of these taxa (one to three specimens per sample) are recorded in the lowermost Palaeocene. The uppermost part of the *P. eugubina* Zone (from AFN 583 onwards) and the *P. pseudobulloides* Zone show a gradual reappearance of all (socalled Lazarus) taxa of this assemblage.

The taxa in stratigraphic assemblage 3 (SA 3) show a fairly continuous distribution pattern along the entire sequence and most of them have relatively constant frequencies as well. Note that most taxa in SA 3, except for *Bolivinoides decoratus* and *Anomalinoides simplex,* are lumped categories. Taxa of stratigraphic assemblage 4 (SA 4) have a similar stratigraphic distribution to SA 3, but all show either a highest abundance or a frequency increase in the lowermost Palaeocene. Most taxa of SA 4 decrease in relative abundance in the lower Palaeocene. Stratigraphic assemblage 5 (SA 5) consists of taxa that first appear in the basal part

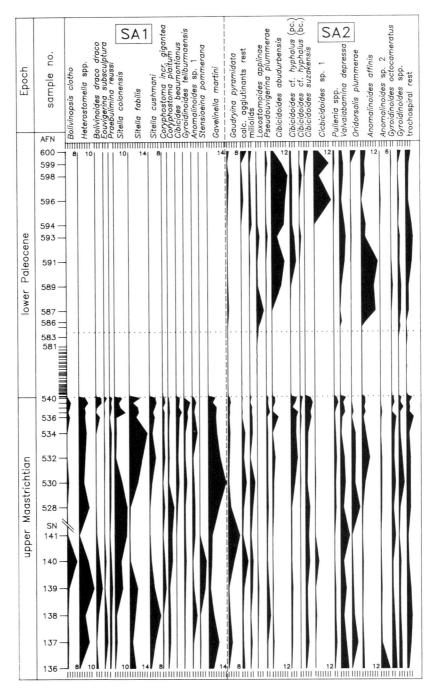

Fig. 5. Frequency range chart of the 60 most common taxa in the El Kef K/Pg boundary profile. SA 1–6 indicate groupings of taxa with corresponding distributional patterns. Note the change of scale in the upper Maastrichtian as in Fig. 2.

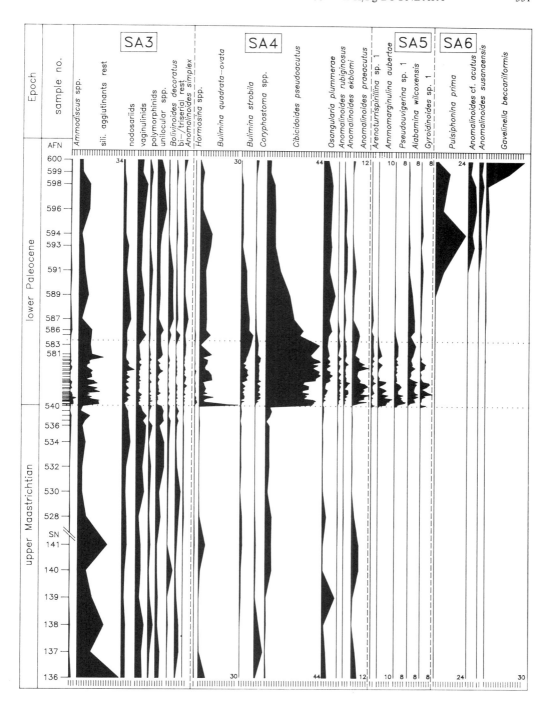

Table 1. *Frequency data of the 60 most common taxa in the El Kef K/Pg boundary profile*

Sample AFN (or SN in bold)	136	137	138	139	140	141	528	530	532	534	536	537	538	539	540	**221b**	**221a**	543	220	544	545	546	547	548	549	550	551	553	554
cm rel. to K/Pg boundary	-3042	-2625	-2188	-1750	-1313	-900	-450	-350	-250	-150	-85	-65	-45	-25	-5	1.2	3.6	12.5	15.6	17.5	22.5	27.5	32.5	37.5	42.5	47.5	52.5	62.5	67.5
Bolivinopsis clotho	0.4	0.8	0.4	1.2	8	0.5	0	1.3	0	0	2.2	0.9	2.2	3	0.9	0	0	0	0	0	0	0	0	0	0	0	0	0	0
Heterostomella spp.	0	7.9	4.5	11.9	7.3	0	8.2	3.9	4.9	6.1	4.8	5.2	4.5	7.4	5.1	0	0	0	0	0	0	0	0	0	0	0	0	0	0
Bolivinoides draco draco	1.3	0	1.5	4.8	2.5	0.5	0	0	0	0	3.1	0	0.7	4.4	4.2	0	0.3	0	0	0	0	0	0	0	0	0	0	0	0
Eouvigerina subsculptura	0	2.4	1.9	0	1.1	0	0	0	0.4	0	0	0.4	0	0	1.9	0	0	0	0	0	0	0	0	0	0	0	0	0	0
Praebulimina reussi	0.4	1.2	1.1	2	1.1	0	0.8	0.9	0.9	1.5	0	0.9	1.5	1	1.9	0	0	0	0	0	0	0	0	0	0	0	0	0	0
Sitella colonensis	0	1.6	1.1	6.3	5.1	9.7	10.2	6.1	0.9	1.5	1.3	8.7	4.5	3.4	6	0.4	0	0	0	0	0	0	0	0	0	0	0	0	0
Sitella fabilis	4.6	2	4.2	6.3	0.4	0	0.4	6.5	10.6	14.6	5.3	7.4	6.7	2.5	8.3	2.5	0.3	0	0	0	0	0	0	0	0	0	0	0	0
Sitella cushmani	0.8	4	8.7	3.6	1.1	0	1.7	4.9	2	4	2.6	4.1	5.4	4.6	0.7	0.3	0	0	0	0	0	0	0	0	0	0	0	0	0
Coryphostoma incrassata gigantea	0	0.4	0	2.4	1.5	0	1.6	0	1.8	1	1.3	1.3	0	2	1.4	0.5	0	0	0	0	0	0	0	0	0	0	0	0	0
Coryphostoma plaitum	0	1.6	1.5	0.8	3.3	2.9	4.7	0.4	0.9	0.5	0.9	1.3	1.1	1	0.5	0	0	0	0	0	0	0	0	0	0	0	0	0	0
Cibicides beaumontianus	0	0	0	0	0	0	0.8	2.2	1.3	2	2.6	4.3	1.9	2.5	2.8	0	0.3	0	0	0	0	0	0	0	0	0	0	0	0
Gyroidinoides tellburmaensis	0	0.4	0.4	0.8	1.5	1	0.8	2.2	0.9	1	3.1	2.6	3.7	4.9	0	0.4	0	0	0	0	0	0	0	0	0	0	0	0	0
Anomalinoides sp. 1	0	0.4	3.4	0.8	0	1.9	2	1.3	0.4	4	2.2	1.3	1.9	3	1.4	0	0	0	0	0	0	0	0	0	0	0	0	0	0
Stensioeina pommerana	0	0.4	0	4	5.1	0	0	0	0	0	0	0	0	0	0	0	0	0	0	0	0	0	0	0	0	0	0	0	0
Gavelinella martini	3.8	8.3	4.2	1.2	2.5	2.9	8.6	14.3	9.7	7.1	6.2	6.5	6.7	3.4	5.6	0.4	0.3	0	0	0	0	0	0	0	0	0	0	0	0
Gaudryina pyramidata	0	2.8	1.1	0	4.4	9.2	2	0.4	0.9	1.5	0.9	1.3	1.1	2	1.4	0	0.3	0	0	0	0	0	0	0	0	0	0	0	0
calcareous agglutinants rest	1.7	0.8	0.4	1.2	4.7	0.5	1.2	3.5	1.3	2.5	0	0.4	0	0	0.5	0	0.3	0	0	0	0	0	0	0	0	0	0.3	0	0
miliolids	0	1.2	0	1.6	2.2	0.5	2	3.9	0	2	1.8	1.3	1.5	0	0.9	0	0.3	0	0	0	0	0	0	0	0	0	0	0	0
Loxostomoides applinae	0	0	0.4	0.8	0	0	0	0	0	0	0	0	0	0	0	0	0	0	0	0	0	0	0	0	0	0	0	0	0
Pseudouvigerina plummerae	0.4	0.4	2.7	1.6	0.4	0.5	1.2	0.9	0.9	1	0.4	2.6	1.5	1	1.4	0.4	0	0	0	0	0	0	0	0	0	0	0	0	0
Cibicidoides abudurbensis	0	0.8	0	0.4	0	0	0.8	0.9	2.2	0	4.8	0	0.4	2	1.9	0	0.3	0	0	0	0	0	0	0	0	0	0	0	0
Cibicidoides cf. hyphalus (planoc.)	0	0.8	0.8	1.2	0.4	0.5	0.4	3	5.3	4	4.4	5.6	1.1	3.4	4.6	0	0.6	0	0	0	0	0	0	0	0	0	0	0	0
Cibicidoides cf. hyphalus (biconv.)	0	0	0	0	0	0	0.4	0	0	0	0.9	0.4	0.7	1	1.9	0	0	0	0	0	0	0	0	0	0	0	0	0	0
Cibicidoides suzakensis	0.4	2.4	1.5	1.2	3.3	2.4	3.5	1.3	1.8	2	4.8	3.9	4.9	5.9	4.2	1.1	0	0	0	0	0	0	0	0	0	0	0	0	0
Cicbicidoides sp. 1	0	0	0.8	0	2.9	0	0.4	0	0	0	0	0	0	0	0	0	0	0	0	0	0	0	0	0	0	0	0	0	0
Pullenia spp.	0	1.6	0.8	0.8	0	0	1.2	0.9	1.8	1	0.4	0.4	1.5	3	1.9	0.4	0	0	0	0	0	0	0	0	0	0	0	0	0
Valvalabamina depressa	6.3	4.3	2.3	4.4	2.2	7.2	2.3	3.9	3.1	0.5	1.3	3.5	3	2	4.2	0.4	0.3	0	0	0	0	0	0	0	0	0	0	0	0
Oridorsalis plummerae	0	4.7	1.9	1.2	0.7	0	6.6	2.6	3.1	3	0	1.3	0.7	0	0	0	0	0	0	0	0	0	0	0	0	0	0	0	0
Anomalinoides affinis	1.3	0.8	0	0.4	0.7	0	0.4	2.2	6.2	3	0.9	2.6	1.5	2.5	2.3	1.1	0	0	0	0	0	0	0	0	0	0	0	0	0
Anomalinoides sp. 2	7.1	0.8	0.4	0.8	0.4	0	0.8	0.9	0.9	0	0.4	0.9	0.4	0	0	0	0	0	0	0	0	0	0	0	0	0	0	0	0
Gyroidinoides octocameratus	3.8	2.8	1.9	0.8	0.7	2.9	1.2	3	1.8	2	3.5	0	0.4	2	1.4	0.4	0.3	0	0	0	0	0	0	0	0	0	0	0	0
Gyroidinoides spp.	2.1	2	1.1	0.8	1.5	0	0.8	4.3	2.7	1	0.9	0.9	3.4	1	1.9	0	0	0	0	0	0	0	0	0	0	0	0	0	0
trochospiral rest	1.3	3.2	4.2	3.2	1.5	1	3.1	3	3.1	2	5.7	2.2	2.6	2	4.2	1.1	0.9	0	0	0	0	0	0	0	0	0	0	0	0
Ammodiscus spp.	1.7	0.4	0.8	0	0.4	0	0.4	0	0	0.5	0.9	0.9	1.1	0	0.5	0.4	2.6	2.4	2.2	2.2	1.7	4.7	0.9	1.9	1.4	1.2	0	0	1.3
silicious agglutinants rest	34.2	10.7	21.6	11.9	6.9	24.2	7	3.5	3.5	6.1	3.5	5.6	9.7	2	3.7	12.3	17	16.5	14.4	17.2	11.4	7.9	11	5.4	9.2	5	17.5	11.5	15.4
nodosariids	2.9	2.8	3	1.6	2.2	3.9	3.5	2.6	3.1	6.6	5.3	5.2	4.5	4.4	2.3	2.1	0.9	2.4	4	3.3	4.7	3.7	3	6.2	5.1	3.3	2.6	3.4	4.9
vaginulinids	5.4	4.7	6.8	2.8	4	3.9	2	6.5	4.9	7.1	7.9	5.2	8.2	4.9	3.7	1.1	1.8	1.4	2.9	2.8	1.7	3.3	1.8	3.5	4.1	2.5	1.4	1.5	2.3
polymorphinids	0.8	1.6	4.2	1.6	0.7	1.9	0.8	1.3	2.2	0	3.1	0	0.4	0.5	2.3	0	1.2	0.7	2.9	0.6	1.3	1.4	0.6	1.9	0	0.4	1.4	0.8	0.8
unilocular spp.	2.9	3.6	1.5	2	1.1	1	3.9	2.6	6.6	6.6	2.6	6.1	5.9	4.2	0.7	0.6	0.3	0.3	1.3	0.8	0	0.4	0.9	1.2	0	1.8	3.3	1.4	1.9
Bolivinoides decoratus	0	0	0.8	0	4.4	0	0.4	0.4	0.9	1	1.8	2.2	0.4	1	1.4	0	0.4	0.3	0.3	0	0.8	0	0.3	0	0.9	1.2	0.3	1.1	0.5
bi-/triserial rest	0.4	3.2	3	1.2	0.4	1.4	2.7	4.3	1.3	0.5	0.9	2.2	0.7	2.5	0.5	1.1	0.3	0.3	0.7	0.8	0	0.9	0	1.4	0.4	0.3	1.1	0.8	0.5
Anomalinoides simplex	0	0	0	0	0	0	0.8	0.4	0.4	0.5	0.4	0	0.5	0	0.7	0	0.3	0.3	0.7	0.8	0	0.9	0	0	1.4	0.4	0.3	0.8	0.5
Hormosina spp.	0.4	0	0	0	0	0.5	0	0	0	0	0	0	0	0	0	0	0.3	1.7	0.7	1.4	0.4	0.9	0.9	0	0.6	0.4	0	0.4	0.5
Bulimina quadrata-ovata	5.8	0.4	0	0	1.1	5.3	0.4	0	0	0	0.4	0.4	0	0	0	31.7	21.7	4.5	3.6	3.3	3.8	6	11.3	4.6	0.9	6.2	3.7	4.6	5.1
Bulimina strobila	0	0	0	0.8	0.7	1.9	1.6	0	0.9	0.5	0.9	0	0.7	1	0.5	0.4	0.6	1	2.9	0.8	0	2.8	3	3.1	2.3	2.5	1.1	3.1	1.5
Coryphostoma spp.	0.8	6.7	3.4	0.4	0.7	1	0.4	0	0	0	0	0	0	0	0	0	0.3	1	2.5	0.3	1.7	3.7	4.2	1.5	4.1	3.7	1.7	3.4	2.8
Cibicidoides pseudoacutus	0.8	0	0.8	0.4	5.1	4.8	5.1	1.7	3.1	3	3.1	0.4	3.4	4.9	3.2	29.2	37.8	39.5	26.6	40.8	38.6	28.8	31.6	33.5	29	35.1	29.6	30.3	34.9
Osangularia plummerae	3.8	2.8	1.1	10.3	0.4	0.5	3.1	0.4	0	0.5	0	0.4	0.7	1	0.9	0	0.3	0.3	0.7	0	2.1	0	0.9	1.9	0.9	2.1	2.9	3.8	1.8
Anomalinoides danicus	0	0	0	0	0	0	0.4	0	0	0	0	0.4	0	0	0	0	0	1	0	0.6	0.8	1.9	0.9	1.5	1.8	0.8	2	0.8	1.3
Anomalinoides ekblomi	0	0	0	0	0	0	0.4	0	0	0	0.4	0	0	0	0	0	0	0.7	0.3	1.7	0.9	0.9	2.7	0.9	3.3	3.7	3.4	0	0.3
Anomalinoides praeacutus	4.6	2.8	0	0.8	5.8	5.8	2	0	0.4	0	0.4	0.4	0.4	0	0	0	0.6	7.2	11.9	3.6	5.5	9.8	9.6	10	13.8	4.1	8.3	10	8.5
Arenoturrispirillina sp. 1	0	0	0	0	0	0	0	0	0	0	0	0	0	0	0	2.5	0.9	0.3	0	0	0.4	0	1.2	0	1.2	2.3	1.1	1	0
Ammomarginulina aubertae	0	0	0	0	0	0	0	0	0	0	0	0	0	0	0	1.4	6.2	7.2	6.8	9.2	9.3	11.6	4.2	5.8	8.3	2.5	3.4	6.9	5.1
Pseudouvigerina sp. 1	0	0	0	0	0	0	0	0	0	0	0	0	0	0	0	2.1	1.8	8.2	9	6.7	6.4	4.7	5.7	5.3	6.9	2.9	2.6	2.7	3.8
Alabamina wilcoxensis	0	0	0	0	0	0	0	0	0	0	0	0	0	0	0	0	0	2.1	5.8	4.7	5.5	3.7	5.7	8.1	4.1	7.9	5.5	4.6	4.9
Gyroidinoides sp. 1	0	0	0	0	0	0	0	0	0	0	0	0	0	0	0	4.6	0.6	0	0	0.6	0.8	0.9	1.2	3.8	1.8	9.9	6.9	3.1	0.5
Pulsiphonina prima	0	0	0	0	0	0	0	0	0	0	0	0	0	0	0	0	0	0	0	0	0	0	0	0	0	0	0	0	0
Anomalinoides cf. acutus	0	0	0	0	0	0	0	0	0	0	0	0	0	0	0	0	0	0	0	0	0	0	0	0	0	0	0	0	0
Anomalinoides susanaensis	0	0	0	0	0	0	0	0	0	0	0	0	0	0	0	0	0	0	0	0	0	0	0	0	0	0	0	0	0
Gavelinella beccariiformis	0	0	0	0	0	0	0	0	0	0	0	0	0	0	0	0	0	0	0	0	0	0	0	0	0	0	0	0	0
No. specimens	240	253	264	252	275	207	256	231	226	198	227	231	267	203	216	284	341	291	278	360	236	215	335	260	217	242	348	261	390
No. species	38	58	49	48	54	35	57	51	52	46	56	48	53	49	49	33	36	23	22	22	23	25	27	23	25	26	27	26	28

Sample AFN	556	558	559	560	561	563	565	566	568	569	571	573	574	576	577	579	581	583	585	586	587	589	591	593	594	596	598	599	600
cm rel. to K/Pg boundary	77.5	87.5	92.5	97.5	103	113	123	128	138	143	153	163	168	178	183	193	205	245	285	305	350	450	550	650	700	800	900	950	1000
Bolivinopsis clotho	0	0	0	0	0	0	0	0	0	0	0	0	0	0	0	0	0	0	0	0	0	0	0	0	0	0	0	0	0
Heterostomella spp.	0	0	0	0	0	0	0	0	0	0	0	0	0	0	0	0	0	0	0	0	0	0	0	0	0	0	0	0	0
Bolivinoides draco draco	0	0	0	0	0	0	0	0	0	0	0	0	0	0	0	0	0	0	0	0	0	0	0	0	0	0	0	0	0
Eouvigerina subsculptura	0	0	0	0	0	0	0	0	0	0	0	0	0	0	0	0	0	0	0	0	0	0	0	0	0	0	0	0	0
Praebulimina reussi	0	0	0	0	0	0	0	0	0	0	0	0	0	0	0	0	0	0	0	0	0	0	0	0	0	0	0	0	0
Stella colonensis	0	0	0	0	0	0	0	0	0	0	0	0	0	0	0	0	0	0	0	0	0	0	0	0	0	0	0	0	0
Stella fabilis	0	0	0	0	0	0	0	0	0	0	0	0	0	0	0	0	0	0	0	0	0	0	0	0	0	0	0	0	0
Stella cushmani	0	0	0	0	0	0	0	0	0	0	0	0	0	0	0	0	0	0	0	0	0	0	0	0	0	0	0	0	0
Coryphostoma incrassata gigantea	0	0	0	0	0	0	0	0	0	0	0	0	0	0	0	0	0	0	0	0	0	0	0	0	0	0	0	0	0
Coryphostoma platium	0	0	0	0	0	0	0	0	0	0	0	0	0	0	0	0	0	0	0	0	0	0	0	0	0	0	0	0	0
Cibicides beaumontianus	0	0	0	0	0	0	0	0	0	0	0	0	0	0	0	0	0	0	0	0	0	0	0	0	0	0	0	0	0
Gyroidinoides telliburmaensis	0	0	0	0	0	0	0	0	0	0	0	0	0	0	0	0	0	0	0	0	0	0	0	0	0	0	0	0	0
Anomalinoides sp. 1	0	0	0	0	0	0	0	0	0	0	0	0	0	0	0	0	0	0	0	0.4	1	0	0	0	0	0	0	0	0
Stensioeina pommerana	0	0	0	0	0	0	0	0	0	0	0	0	0	0	0	0	0	0	0	0	0	0	0	0	0	0	0	0	0
Gavelinella martini	0	0	0	0	0	0	0	0	0	0	0	0	0	0	0	0	0	0	0	0	0	0	0	0	0	0	0	0	0
Gaudryina pyramidata	0	0	0	0	0	0	0	0	0	0	0	0.4	0	0	0	0	0	0	0	0.4	0	0	0	0	0	0.6	1.3	0	0
calcareous agglutinants rest	0.4	0	0	0	0	0	0	0	0	0	0	0	0.3	0	0	0	0	0	0	0.4	0	0.4	0	0	0	0.6	0.9	0	1.5
miliolids	0	0	0	0	0	0	0	0.3	0	0	0	0	0	0	0	0	0	0	0	0	0	0.8	0.5	2.4	1.9	1.9	0.9	0.5	4.5
Loxostomoides applinae	0	0	0	0	0	0	0	0	0	0	0	0	0	0	0	0	0	0	1.4	3	5.6	0.4	0.5	1.4	0.4	1.3	0.9	1.5	1.9
Pseudouvigerina plummerae	0	0	0	0	0	0	0	0	0	0	0	0	0.3	0.8	0.4	0.5	0.4	0.7	0.5	1.3	1	0.4	0.5	1.4	0.4	0.6	0.4	0.5	0.4
Cibicidoides abudurbensis	0	0	0	0	0	0	0	0	11.6	9.9	6.6	7.2	9.2	8.7	10.2	20.3	11.1	9.8	10.6	11	2.6	4.9	2.7	3.8	10.1	11.3	8.2	13.4	1.5
Cibicidoides cf. hyphalus (planoc.)	0	0	0	0	2.7	5.9	10.8	9.7	1.7	9.4	5.7	3.8	4.2	4.2	3.1	2.8	2.7	3.6	3.2	3.4	6.6	5.7	3.2	4.3	3.3	1.6	1.3	8.5	0.8
Cibicidoides cf. hyphalus (bicov.)	0	0	0	0	1.9	2.8	2.5	2.1	0	2.7	2.4	3.8	5.2	1.9	3.5	2.8	2.7	1.8	7.3	4.6	5.1	5.7	2.7	3.8	0.8	1.6	2.2	1	0.4
Cibicidoides suzakensis	1.5	0.8	0	1.6	0.4	0.4	0.4	1.9	5.2	1.3	0.5	0.9	0.9	1.1	1.3	0.5	1.3	0.9	0.9	0.8	5.1	1.5	2.7	0.5	0.6	0.6	1.3	0	0.4
Cibicidoides sp. 1	0	0	0.3	0.5	0.8	0.8	1.7	2.4	0	3.1	2.4	2.9	2.9	0	0	3.3	4.9	3.3	3.2	4.6	1.5	2.7	2.3	2.4	3.5	7.5	5.6	6	3.4
Pullenia spp.	1.1	0	0	0	0	0	0.8	0.5	0.9	0.4	0.5	0.9	0.3	0.8	1.2	0.7	0	0	2.3	0.8	4.1	3.4	3.7	1	0.8	0.6	0.4	0.6	0.8
Valvalabamina depressa	2.6	0	0.7	1.1	0	0.2	0	0.8	0.4	1.3	0.9	0.4	0.6	0.4	0.8	0.5	1.3	1.8	3.2	1.7	3.6	0.8	0.5	0.8	2.3	0	0.4	0.5	0
Oridorsalis plummerae	0	0.4	0.3	1.1	1.5	0.2	0	0.3	0	0	0	0	0.3	0	0	0.5	0	0	0.9	0	3.6	0.8	0	2.4	0.8	1.3	0.4	0	2.3
Anomalinoides affinis	3.7	3.5	0	5.5	10.3	9.9	9.6	8.8	7.3	4.9	4.2	5.1	6.1	6.5	6.6	2.1	9.8	8.3	8.7	5.1	5.6	5.3	4.6	4.8	9.7	1.3	0.4	0	1.9
Anomalinoides sp. 2	0	0	0	3.3	4.3	6.3	6.3	5.4	6.5	2.8	2.8	1.7	3.8	3.8	6.6	4.5	9.8	8.1	10.1	5.1	7.1	4.9	1.7	0.5	2.3	1.9	1.7	0	4.2
Gyroidinoides octocameratus	1.2	0	0	2.4	2.7	2.8	2.9	4.2	7.3	3.1	3.8	6.1	4.3	3.8	3.9	1.8	5.8	8.3	5.8	3	5.6	5.3	13.2	2.9	3.5	0	1.3	2.5	3.8
Gyroidinoides spp.	0	0	0	0.5	0	0.4	2.6	2.6	41.3	37.7	41.3	42.6	35.4	44.5	4.3	38.5	1.8	2.5	1.4	30.4	23	19.3	11.4	0	0	0	0	0	0.5
trochospiral rest	0	0	0	0	0	0	0	0	0	0	0	0	0	0	0	0	0	0	1.4	0.8	0	1.1	0.9	1.4	4.7	0.9	2.2	2.5	2.3
Ammodiscus spp.	0.4	1.2	0	0	0	0	0.4	0	0	0.9	1.4	0	0.3	0.8	0.4	0.5	0.4	0.7	0.5	1.3	2	2.3	4.6	5.7	0	1.9	0.3	0.3	0.4
silicious agglutinants rest	5.5	15.4	13.4	12.6	5	11.1	10.8	9.7	0	9.9	6.6	7.2	9.2	8.7	10.2	20.3	11.1	9.8	10.6	11	1	1.9	0.9	3.8	2.3	8.5	9.1	2.5	4.2
nodosariids	2.9	2.3	0.4	3.3	2.7	5.9	2.5	2.1	1.7	2.7	5.7	3.8	4.2	4.2	3.1	2.8	2.7	3.6	3.2	3.4	2.6	8	2.7	4.3	1.9	8.5	2.2	0.3	2.7
vaginulinds	2.2	2.3	3.3	3.3	1.9	2.8	1.3	2.1	0	2.7	2.4	3.8	5.2	1.9	3.5	2.8	2.7	1.8	7.3	4.6	6.6	5.7	3.2	3.8	1.6	1.9	2.2	1	0.8
polymorphinids	0.7	0.8	1.1	1.1	0.4	0.4	0.4	0.4	0.4	0.9	0.5	1.7	0.9	0.8	0.8	1.2	2.2	1.8	0.9	4.6	5.1	9.8	2.7	0.5	5.8	0.6	1.3	4	4.9
unilocular spp.	1.5	0.8	1.6	1.6	2.3	0.8	1.7	2.4	5.2	3.1	1.4	1.3	2.9	1.1	1	0.5	1.3	1.8	2.3	4.6	1	1.5	3.2	2.4	0.8	1.3	5.6	2.5	0.4
Bolivinoides decoratus	1.1	0	0.3	0.5	0.4	0.4	0.8	0.5	0.9	0.4	13.2	0.9	0.4	6.1	1.2	3.3	4.9	4.3	3.2	5.5	3	2.7	3.7	1	3.5	7.5	0.4	0	3.4
bi-triserial rest	2.6	0	0	0	0	0.2	0	0.8	0.4	1.3	0.9	0.4	0.6	0.4	0.8	0.7	0	0	3.2	0.8	4.1	3.4	3.7	1	0.8	0.6	0.4	0.6	0
Anomalinoides simplex	0	0.4	0	1.1	1.5	2	0	0.5	0	0.4	0.5	0.9	0.3	0.8	0.8	0	1.3	1.8	3.2	1.7	3.6	0.8	0.5	2.4	2.3	0.6	0	0.5	0
Hormosina spp.	3.7	3.5	0.7	6.6	10.3	9.9	9.6	8.8	7.3	4.9	4.2	5.1	6.1	6.5	6.6	2.1	9.8	8.3	5.1	5.1	5.6	5.3	4.6	8.6	9.7	1.3	0.4	0	2.3
Bulimina quadrata-ovata	4	3.5	0	6.6	9.9	4.3	6.3	5.4	5.4	2.8	2.8	1.7	3.8	3.8	3.8	4.5	9.8	8.1	10.1	5.1	7.1	4.9	4.6	0.5	0.7	1.9	1.7	3	0
Bulimina strobila	1.2	0.7	0.5	0.5	1.9	2.8	2.9	2.2	4.7	2.8	1.7	5.1	6.1	3.8	3.9	4.3	1.8	2.5	8.7	7.1	5.6	5.3	1.7	7.1	0.7	1.9	1.3	0.5	0
Coryphostoma spp.	35.2	29.3	31.2	31.3	41.1	41.1	42.6	42.7	43.8	41.3	37.7	42.6	35.4	44.5	5.5	38.5	36.4	27.5	10.1	30.4	23	19.3	11.4	0.5	5.8	6.5	6.5	0.5	1.9
Cibicidoides pseudoacutus	1.8	2.3	2.7	4.4	6.1	5.1	7.9	5.6	0.6	3.6	2.8	0.6	3.5	5.7	5.5	3.5	5.3	1.4	1.4	7.2	5.1	9.8	5	7.1	6.6	7.2	7.4	6	3.4
Osangularia plummerae	2.2	2.7	0.5	1.1	1.6	0.4	0.6	0.4	0.4	1.4	1.4	1.7	0.9	0.8	0.8	2.8	4.6	0.5	4.6	7.2	5.1	9.8	6.7	6.7	5.8	6.6	6.6	6	3.8
Anomalinoides danicus	2.2	2.5	1.6	1.6	1.9	0.8	3.8	1.3	1.3	0.9	1.4	1.7	3.9	1.5	5.5	1.2	0.5	1.8	0.5	5.1	1.5	1.1	2.3	2.4	1.2	1.9	0.9	4	0
Anomalinoides ekblomi	12.8	3.5	6.4	0.6	1.9	3.8	4.3	4.3	4.7	5.4	13.2	6.4	10.1	6.1	5.7	5.7	4.3	4.3	0.6	5.1	1.5	0.4	2.7	0.5	0	7.5	5.6	0.5	1.5
Anomalinoides praeacutus	0.7	1.5	3.4	1.6	1.1	1.3	1.3	0.3	0.4	0.9	0.5	1.7	0.6	1.1	2.3	0.7	1.4	1.8	1.8	1.7	0	1.1	0	0	0	0	0.4	0.6	0
Arenoturispirillina sp. 1	0	4.2	0	1.6	1.2	1.2	0.5	0.5	0	0.9	0.5	1.7	0.2	0.4	0.2	0.2	2.2	1.4	0.9	0	0	0.4	0.4	0.4	0.8	0	0.4	0	0
Ammomarginulina aubertae	5.5	3.9	1.7	0.5	1.6	1.6	1.3	2.1	2.1	1.3	0.9	2.6	2.3	2.3	2.3	0.7	2.2	0.7	0.9	0.4	0	3.4	0.5	1	1.6	0	0.4	0	0
Pseudouvigerina sp. 1	5.5	6.9	4.7	7.7	0.8	0.8	2.4	2.4	0	2.7	1.9	0.9	3.5	1.5	2.3	0.6	2.2	2.9	1.8	5.1	5.1	0.8	1.8	18.6	24.5	6.9	3.9	11.5	2
Alabamina wilcoxensis	5.5	7.7	0	2.2	1.5	0.4	0.4	0.8	0	0	5.7	3.4	4.3	1.5	0	1.2	4	6.2	0.9	5.1	0.5	0.4	6.8	5.2	1.6	6.1	3.9	3.4	3.4
Gyroidinoides sp. 1	0	0	0	0	0	0	0	0	0	0	3.3	0	0	1.5	0.4	2.6	0	0	0	0.4	1.5	1.1	6.8	1.3	1.6	2.5	6.1	19	31.1
Pulsiphonina prima	0	0	0	0	0	0	0	0	0	0	0	0	0	0	0	0	0	0	0	0	0	0.4	0	0	0	0	0	0	0
Anomalinoides cf. *acutus*	0	0	0	0	0	0	0	0	0	0	0	0	0	0	0	0	0	0	0	5.1	0	0	0.5	1	0.4	0.6	9.5	11.5	1.9
Anomalinoides suzanensis	0	0	0	0	0	0	0	0	0	0	0	0	0	0	0	0	0	0	1.8	5.1	0	0.4	1.8	5.2	1.6	6.9	6.1	4	2.4
Gavelinella beccariiformis	0	0	0	0	0	0	0	0	0	0	0	0	0	1.5	0.4	1.2	0	0	0.9	0.4	0.5	0.4	6.8	2.4	1.6	2.5	1.3	19	3.4
No. specimens	273	259	298	182	261	253	240	373	233	223	212	235	347	263	256	423	225	276	218	237	196	264	219	210	257	319	231	200	264
No. species	22	22	27	25	21	22	21	30	21	24	25	24	26	26	23	27	22	20	28	26	30	37	37	34	33	39	36	34	40

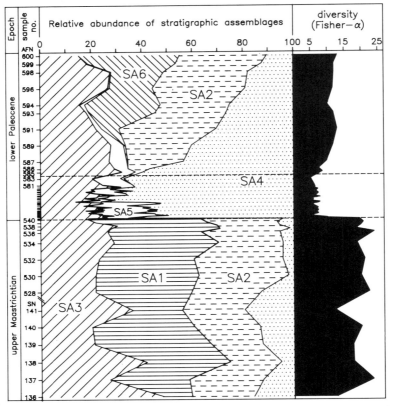

Fig. 6. Cumulative relative abundance of the stratigraphic assemblages (SA 1–6) and Fisher-α diversity visualize the abrupt community impoverishment at the K/Pg boundary and the more gradual community diversification at El Kef. Note the change of scale in the upper Maastrichtian as in Fig. 2.

of the boundary clay. Four species (*Arenoturrispirillina* sp. 1, *Ammomarginulina aubertae*, *Pseudouvigerina* sp. 1 and *Gyroidinoides* sp. 1) appear in the basal sample overlying the K/Pg boundary. *Alabamina wilcoxensis* appears 10 cm higher up. All taxa of SA 5 either disappear, or at least decline in relative numbers in the lower Palaeocene. Finally, stratigraphic assemblage 6 (SA 6) is composed of taxa that have their first appearance in the top of the *P. eugubina* Zone (*Gavelinella beccariiformis*) or in the *P. pseudobulloides* Zone (*Pulsiphonina prima*, *Anomalinoides susanaensis* and *A.* cf. *acutus*).

Faunal turnover

For each stratigraphic assemblage the cumulative frequencies are displayed in Fig. 6; in addition Fisher-α diversity values, based on the primary data set, are shown. The graph shows three main stages in benthic foraminiferal palaeocommunity development: the first stage comprises the late Maastrichtian interval, the second the earliest Palaeocene, and the third the early Palaeocene. The first and second stages display a relatively stable taxonomical composition (Fig. 5), whereas the third stage shows a gradual taxonomic change from bottom to top.

Stage one: the upper Maastrichtian assemblage (SN 136 to AFN 540) is highly diversified (average $\alpha = 19.2$). Apart from the abundant poorly preserved, and unspecified, silicious agglutinants in the lower part of the profile, the Maastrichtian assemblage is dominated by taxa of SA 1 (all together up to 46%): in particular, *Heterostomella* spp., *S. colonensis*, *S. fabilis* and *G. martini* reach peak abundances of 10–15%. Although frequencies of individual taxa vary strongly, the assemblage as a whole is relatively stable up to the K/Pg boundary (Fig. 6). Within less than 12.5 cm above the K/Pg boundary, 67% of the taxa that are present in the uppermost metre of the Maastrichtian disappear (partly temporarily) from the record

EXTINCTION AND SURVIVORSHIP ACROSS THE K/Pg BOUNDARY

(disregarding lumped taxa at supra-generic level). Whereas these taxa comprise more than 70% of the total assemblage in the uppermost Maastrichtian sample, they constitute only 8% of the sample 1.2 cm above the K/Pg boundary and only 3% at 3.6 cm. This rapid decrease marks a sharp faunal transition at the base of the boundary clay. The original transition, however, may well have been even sharper, since it is not unlikely that rare occurrences of SA 1 and SA 2 taxa in the boundary clay are reworked. Distinguishing reworked from *in situ* benthic foraminifera from this record by means of geochemical studies is very difficult, if not impossible, because of the state of preservation of the foraminifera tests. But, when we consider benthic foraminiferal numbers below and above the boundary (Fig. 4), it shows that merely a one percent mixing of Maastrichtian sediment into the basal boundary clay may account for the total number of SA1 and SA2 specimens found.

Stage two: the lowermost Palaeocene assemblage (SN 221b to AFN 583) is poorly diverse (average $\alpha = 6.6$). All samples are dominated by SA 4 taxa (together up to 77%), and for all but the lowermost sample the dominant species is *Cibicidoides pseudoacutus* (29–45%). The lowermost sample is dominated by *Bulimina quadrata–ovata*, showing a distinct spike in its distribution pattern at the base of the boundary clay. Other species reaching relative abundances greater than 10% are *Anomalinoides praeacutus* and *A. aubertae*. Most common taxa in this assemblage, except for the supra-generic groups, were either absent or infrequent in the Maastrichtian, in particular in the upper metre of this interval. Faunal composition is fairly constant within the lowermost Palaeocene. There is, however, a marked increase in relative abundance of *C. pseudoacutus*, coinciding with a decrease of SA 5 taxa (Figs 5, 6) at the base of the *C. taurica* Subzone.

Stage three: the faunal transition from the lowermost Palaeocene to the lower Palaeocene is very gradual compared to the K/Pg boundary transition. The boundary between the second and third stages is drawn at the first distinct decrease of *C. pseudoacutus* and the concomitant increase in diversity (at 2.65 m above the K/Pg boundary, in between AFN 583 and AFN 585). The lower Palaeocene assemblage is moderately diversified (average $\alpha = 10.9$), but gradually increasing upwards. Initially, SA 4 species are still dominant (in particular *C. pseudoacutus*), but the proportion of reappearing Lazarus taxa (SA 2) steadily increases up to the level where new taxa of SA 6 appear at the expense of the proportion of taxa of SA 3–5. *Cibicidoides* sp. 1,

C. abudurbensis, A. affinis, Bulimina strobila, C. pseudoacutus, P. prima and *G. beccariiformis* reach peak abundances of 10–31%. The lower Palaeocene assemblage clearly evolves within the *P. pseudobulloides* Zone and it does not appear to have stabilized at the top of the studied interval, judging from the increase of *G. beccariiformis* in the uppermost samples. Further developments of benthic foraminiferal assemblages in the Palaeocene are beyond the scope of this study and will be treated in detail in a separate paper.

Benthic foraminiferal assemblages from the Middle East

Semi-quantitative benthic foraminiferal data of the K/Pg boundary profiles of Nahal Avdat, Wadi Nukhul, Gebl Qreiya and Gebel Duwi are listed in Table 2. The table lists the taxa of El Kef, and in addition the more common taxa (average > 2%) occurring in any of the other localities. As explained earlier, we can only address upper Maastrichtian and lower Palaeocene distribution patterns and discuss the overall faunal change across the K/Pg boundary.

Upper Maastrichtian

Upper Maastrichtian assemblages from Wadi Nukhl and Nahal Avdat bear great resemblance to those from El Kef. Although relative abundances of many taxa may differ significantly between the localities, the overall taxonomical composition is very similar. In Wadi Nukhl the most common species (> 5%) are *P. reussi, C. incrassata gigantea, C. abudurbensis, C. pseudoacutus* and *Osangularia plummerae*. In Nahal Avdat the most common species are *B. draco draco, Eouvigerina subsculptura, S. cushmani* and *C. abudurbensis*. The proportion of planktonic foraminifera (80–95%P) in Wadi Nukhl and Nahal Avdat is similar to, though more variable than, those of El Kef.

The upper Maastrichtian assemblage from Gebel Qreiya differs significantly from the former assemblages; diversity is much lower and in particular most taxa of SA 1 are absent. Instead we find *Orthokarstenia oveyi, Pyramidina aegyptiaca, Elhasaella* cf. *allanwoodi, Neobulimina canadensis* and *Anomalinoides umboniferus*; none of these species was ever encountered in the assemblages from El Kef, Wadi Nukhl or Nahal Avdat. The most common species are *B. strobila, C. pseudoacutus,* and

R. P. SPEIJER & G. J. VAN DER ZWAAN

Table 2. *Semi-quantitative data of K/Pg profiles in Israel and Egypt*

	Stratigraphic interval:	Upper Maastrichtian				Lower Palaeocene			
SA	Locality: P/B ratio (%P):	Nukhl 80–95	Avdat 80–95	Qreiya 80	Duwi 50–80	Nukhl >90	Avdat >90	Qreiya 40–80	Duwi 50–90
1	*Bolivinopsis clotho*	*	*						
	Heterostomella spp.	*	*						
	Bolivinoides draco draco	*	C						
	Eouvigerina subsculptura	*	C						
	Praebulimina reussi	C	*						
	Sitella colonensis								
	Sitella fabilis	*							
	Sitella cushmani	*	C	*					
	Coryphostoma incrassata gigantea	C	*			C			
	Coryphostoma plaitum	*	*						
	Cibicides beaumontianus	*	*			*			
	Gyroidinoides tellburmaensis	*	*						
	Anomalinoides sp. 1	*	*						
	Stensioeina pommerana								
	Gavelinella martini	*	*						
2	*Gaudryina pyramidata*	*	*			C	*		
	calcareous agglutinants rest	*	*	*		C	*	*	*
	miliolids					*	*	*	*
	Loxostomoides applinae	*	*			*	*	*	
	Pseudouvigerina plummerae	*	*			*			
	Cibicidoides abudurbensis	C	C	*		C	*	*	
	Cibicidoides cf. *hyphalus* (planoc.)	*				*	*		
	Cibicidoides cf. *hyphalus* (biconvex)	*	*			*	*		
	Cibicidoides suzakensis	*	*			*	*		
	Cibicidoides sp. 1								
	Pullenia spp.	*	*			*	*		
	Valvalabamina depressa	*	*	*		*	C	*	*
	Oridorsalis plummerae	*	*			*	*		*
	Anomalinoides affinis	*	*			*		*	
	Anomalinoides sp. 2								
	Gyroidinoides octocameratus	*	*	*		*			
	Gyroidinoides spp.	*	*	*		*	*	*	
	trochospiral rest	*	*	*	*	*	*		*
3	*Ammodiscus* spp.								*
	silicious agglutinants rest	*	*	*	*	*	*	*	*
	nodosariids	*	*	*	*	*	*	*	*
	vaginulinids	*	*	*	*	C	*	*	*
	polymorphinids	*	*	*		*	*	*	*
	unilocular spp.	*	*	*		*	*	*	*
	Bolivinoides decoratus	*	*			*	*	*	*
	bi-/triserial rest	*	*	*	*	*	*	*	*
	Anomalinoides simplex		*						
4	*Hormosina* spp.	*	*	*		*	*	*	
	Bulimina quadrata-ovata	*		*		*	*	*	
	Bulimina strobila	*		C		*	*	*	
	Coryphostoma spp.	*				*	*	*	
	Cibicidoides pseudoacutus	C	*	C		C	C	C	*
	Osangularia plummerae	C	*	*		C	C		*
	Anomalinoides danicus					*	*		
	Anomalinoides ekblomi								
	Anomalinoides praeacutus	*				*	*	*	*
5	*Arenoturrispirillina* sp. 1								
	Ammomarginulina aubertae								
	Pseudouvigerina sp. 1								
	Alabamina wilcoxensis	*		*		*			
	Gyroidinoides sp. 1							*	
6	*Pulsiphonina prima*								
	Anomalinoides cf. *acutus*					*	*		
	Anomalinoides susanaensis					*		*	
	Gavelinella beccariiformis					C	C		
	Gyroidinoides cf. *cibaoensis*	*	*				*	C	*
	Gyroidinoides girardanus								
	Orthokarstenia oveyi			*	C				
	Pyramidina aegyptiaca			*	C				
	Elhasaella cf. *allanwoodi*			*	C				
	Anomalinoides umboniferus			C	C			*	C
	Neobulimina canadensis			*	C				*
	Cibicidoides sp. 2			*					C
	Alabamina midwayensis								*
	Pseudoeponides cf. *elevatus*							*	*
	Siphogenerinoides sp. 1								*

Data of the lowermost Palaeocene are not considered because of suspected reworking and/or hiatuses (see text).
* marks average abundance < 5%; C marks average abundance > 5%.

EXTINCTION AND SURVIVORSHIP ACROSS THE K/Pg BOUNDARY

A. umboniferus. P/B ratio's (80%P) in Gebel Qreiya are lower than in El Kef, Wadi Nukhl and Nahal Avdat.

The upper Maastrichtian assemblage of Gebel Duwi has even fewer taxa in common with El Kef, Wadi Nukhl and Nahal Avdat (actually only in the lump categories), whereas *O. oveyi, P. aegyptiaca, E. allanwoodi, N. canadensis* and *A. umboniferus* are highly abundant. The proportion of planktonic foraminifera (50–80%P) in Gebel Duwi is somewhat lower than in Gebel Qreiya.

Lower Palaeocene

Similar to the upper Maastrichtian situation, the lower Palaeocene faunas of Wadi Nukhl and Nahal Avdat bear great resemblance to the one from El Kef; most taxa of SA 2, 3, 4 and 6 are present, some of the more common ones also in similar relative frequencies: e.g. *C. abudurbensis, C. pseudoacutus, Osangularia plummerae* and *G. beccariiformis*. Next to these we find very common *C. incrassata gigantea, G. pyramidata*, other large calcareous agglutinants and vaginulinids in Wadi Nukhl, while *V. depressa* is highly frequent in Nahal Avdat. Two SA 1 species (*C. incrassata gigantea* and *C. beaumontianus*) range into the lower Palaeocene of Wadi Nukhl (in these cases reworking can be excluded), indicating that their extinction in El Kef is just a local feature. The same applies to *Bolivinopsis clotho* and *P. reussi*, which have occasionally been observed in lower and upper Palaeocene assemblages from Wadi Nukhl and Nahal Avdat (unpubl. data). P/B ratios are >90%P.

The lower Palaeocene assemblages of Gebel Qreiya are less diverse than the ones of El Kef, Wadi Nukhl and Nahal Avdat, although they are more similar than the Maastrichtian assemblages. *Cibicidoides pseudoacutus* is the dominant species, while *Gyroidinoides girardanus* is also a common species. The proportion of planktonic foraminifera (40–80%P) is considerably lower than in El Kef, Wadi Nukhl and Nahal Avdat.

The lower Palaeocene fauna of Gebel Duwi is again the one that is most different from El Kef, but as for Gebel Qreiya the similarity is greater and diversity is higher than in the Maastrichtian. *Cibicidoides* sp. 2 and *A. umboniferus* are the most abundant taxa. Relative numbers of planktonic foraminifera (50–90%P) are somewhat higher than in Gebel Qreiya.

Stages in southern Tethyan palaeoenvironment and benthic palaeocommunity development

Late Maastrichtian

As indicated in the previous section, the upper Maastrichtian benthic foraminiferal faunas from Wadi Nukhl and Nahal Avdat show a great resemblance to the ones from El Kef. Very similar assemblages were documented from other localities in southern Israel as well (Benjamini *et al.* 1992; Keller 1992), while the ones reported from northern Egypt (LeRoy 1953; Said & Kenawy 1956) also appear comparable.

Little is known about Maastrichtian palaeobathymetrical distribution of benthic foraminifera in North Africa. For a palaeobathymetrical estimate, we compared our assemblages with the ones documented in Campanian–Maastrichtian palaeoslope models of North America (Silter & Baker 1972; Olsson 1977; Nyong & Olsson 1984; Olsson & Nyong 1984). Our assemblages contain a few species that are most common in North American shelf environments, e.g. *C. incrassata gigantea, C. plaitum* and *V. depressa.* Other taxa are common in outer neritic and upper bathyal environments, e.g. *P. plummerae, B. draco draco, Anomalinoides* (mentioned as *Gavelinella*), *Gaudryina, Pullenia* and nodosariids. Bathyal taxa are, for example, *Stensioeina, Gyroidinoides, S. cushmani* and *G. martini.* From this comparison it follows that there is greatest similarity (though still limited) with upper bathyal assemblages from North America. Therefore, we suggest that the highly diverse Maastrichtian assemblages of El Kef, northern Egypt, and southern Israel are indicative for a similar upper bathyal (300–500 m) setting. High proportions of planktonic foraminifera (>90%P) concur with such an assignment. In a similar way Keller (1988b, 1992) inferred upper bathyal–outer neritic depths for assemblages from El Kef and southern Israel. According to Donze *et al.* (1982) the ostracoda microfauna of El Kef points to a palaeodepth of 400–500 m, whereas Peypouquet *et al.* (1986) suggested shallower depths (150–300 m).

Maastrichtian assemblages from Gebel Qreiya and Gebel Duwi, respectively, are increasingly different from the upper bathyal faunas. The combination of a decreasing diversity and proportion of planktonic foraminifera (80%P and 50%P, respectively) suggests that these localities were situated in shallower parts of the epicontinental basin that covered most of Egypt.

This view is substantiated by basinwide facies analysis (Hendriks *et al.* 1987; Luger 1988) and palaeogeographic reconstructions (Said 1990*a, b*). Gebel Qreiya was situated in an outer shelf environment, whereas Gebel Duwi was part of a shallower inner to middle neritic part of the basin (Hendriks *et al.* 1987: Luger 1988). In conclusion, the Gebel Qreiya assemblage may indicate 100–150 m palaeodepth, whereas the Gebel Duwi assemblage is indicative for an even shallower palaeodepth (50–100 m).

Although various studies have been performed on the distribution of Palaeocene benthic foraminifera in Tunisia (Aubert & Berggren 1976; Salaj *et al.* 1976; Saint-Marc & Berggren 1988; Saint-Marc 1993), very little has been documented on Maastrichtian distributions. To our knowledge, only one paper briefly describes Maastrichtian benthic foraminifera from an area 50 km southeast of El Kef (Said 1978); taxonomically the fauna shows considerable similarity to the one from El Kef. Without any more data from Tunisia, we cannot assess whether a palaeobathymetrically biofacial arrangement, similar to the one in Egypt, is present in Tunisia. However, since most of the shallow water taxa of Gebel Qreiya and Gebel Duwi (i.e. *Orthokarstenia, A. umboniferus, P. aegyptiaca* and *N. canadensis*) are also known from the Maastrichtian of West Africa (Petters 1982), it seems plausible that similar neritic assemblages are present in Tunisia. At any rate, the Maastrichtian assemblage of El Kef represents a relatively deep off-shore fauna, that is very similar to the deepest biofacies as seen in Egypt and Israel and has a wide distribution along the southern Tethyan margin.

Upper Maastrichtian planktonic foraminiferal assemblages from El Kef are characterized by a few dominant heterohelicids, whereas globotruncanids are relatively rare compared to open ocean sites (Brinkhuis & Zachariasse 1988; Keller 1988*a*; Nederbragt 1991). This planktonic assemblage, together with the $\delta^{13}C$ record and oceanic circulation models, has been interpreted to reflect high fertility in surface waters induced by (possibly intermittent) coastal upwelling (Parish & Curtis 1982; Reiss 1988; Keller & Lindinger 1989; Kroon & Nederbragt 1990; Nederbragt 1992). The resulting high organic carbon flux to the sea-floor may have favoured the most common taxa, such as *Heterostomella* spp., *Sitella* spp. and *G. martini.* A preference for a high nutrient supply of these taxa could explain why they are usually infrequent or even absent in more oligotrophic deep-sea assemblages. The presence of *E. subsculptura* in outer neritic to middle bathyal assemblages from mid

latitude ODP and DSDP sites has been interpreted to indicate more eutrophic deep water regimes (Widmark 1995). The high relative abundance (>20%) of this species in Nahal Avdat may indicate similar trophic conditions there as well. High nutrient availability may also explain why many other common bathyal taxa, such as *G. beccariiformis, C. hyphalus, Nuttallides truempyi* and *Gyroidinoides globosus*, that are associated with more oligotrophic deep-sea environments (Van Morkhoven *et al.* 1986; Widmark & Malmgren 1992*b*), were absent at southern Tethyan upper bathyal depths during the late Maastrichtian.

Judging from the high diversity of the benthic assemblage, a severe oxygen minimum zone (OMZ) did not develop at this depth, despite the inferred high organic carbon flux; concomitant oxygen advection was apparently sufficient to prevent the ecosystem from high stress levels. Our interpretation of trophic conditions and palaeoxygenation at the sea-floor during the late Maastrichtian concurs well with inferences from ostracoda assemblages. The ostracod microfauna points to slightly depressed oxygen saturation (4.5 ml/1 O_2) in a weakly developed OMZ, that probably resulted from high, upwelling related, surface productivity (Donze *et al.* 1982; Peypouquet *et al.* 1986).

Earliest Palaeocene

The rather peculiar lowermost Palaeocene assemblage of El Kef has, to our knowledge, not been identified from any other area. An exception might be one of the assemblages encountered by Said (1978); this author noted the presence of abundant *B. quadrata–ovata* and *Cibicidoides alleni* (which might correspond to our *C. pseudoacutus*) next to several Nodosariacea in lower(most) Palaeocene faunas. Without more detailed information we can, however, only speculate whether this assemblage is similar to the one from El Kef. Aubert & Berggren (1976) studied 160 samples from nine Palaeocene sequences, located in central Tunisia, but apparently none of them yielded an assemblage similar to ours from El Kef, although there are several species in common. The following common species in the lowermost Palaeocene of El Kef were not encountered in other localities in Tunisia: *Anomalinoides ekblomi, A. aubertae, Pseudouvigerina* sp. 1, *A. wilcoxensis* and *Gyroidinoides* sp. 1 (Aubert & Berggren 1976). *Ammomarginulina aubertae* and *A. ekblomi* have been recorded in boreal areas (Brotzen 1948; Gradstein & Kaminski 1989; Kuhn 1992), while we observed *Pseudouvigerina* sp. 1 in the lowermost Palaeocene of a K/Pg boundary profile in the Maastrichtian type-area (unpubl. data).

Of all assemblages studied, the Maastrichtian and the lower Palaeocene faunas of Gebel Qreiya show the greatest similarity with the lowermost Palaeocene fauna of El Kef. The presence and proportion of several SA 4 taxa, such as *C. pseudoacutus* and *B. strobila*, next to *B. quadrata–ovata* and *Osangularia plummerae*, is striking. On the other hand, they differ in the presence of SA 2 taxa, such as *C. abudurbensis*, *V. depressa*, *G. octocameratus* and *Oridorsalis plummerae* next to several shallow water taxa, that are more common in Gebel Duwi. Moreover, all SA 5 taxa, except *A. midwayensis*, are absent in Gebel Qreiya. Despite these differences, these shallow outer neritic (100–150 m) Egyptian assemblages show the best correspondence, and it could be argued that the lowermost Palaeocene assemblage of El Kef also indicates such palaeodepths (cf. Keller 1988b, 1992). Unfortunately, it is not possible to infer an independent palaeodepth estimate by means of ostracoda; during the earliest Palaeocene sea-floor redox conditions entirely control the composition of the assemblages (Donze *et al.* 1982; Peypouquet *et al.* 1986).

Low diversity, high dominance (*C. pseudoacutus*) and reduced numbers of burrowing organisms, judging from the preserved laminations, strongly suggest considerable oxygen deficiency at the sea-floor during the earliest Palaeocene. The abundance of *Tappanina selmensis* (up to 30%) in the 63–595 μm size fraction also suggested depressed oxygenation during the lowermost Palaeocene (Speijer 1992). Elsewhere, neritic and bathyal low diversity assemblages dominated by *T. selmensis*, have been considered to indicate oxygen deficiency as well (Olsson & Wise 1987; Thomas 1990b; Gibson *et al.* 1993; Gibson & Bybell 1995). The two samples from the base of the boundary clay (K/Pg + 1.2 and 3.6 cm) show a unique spike of *B. quadrata–ovata*. In morphological features this taxon is very similar to the modern *Globobulimina*. The latter is able to live deep within the sediment where it feeds on degraded organic matter and/or associated bacteria under nearly anoxic conditions (Corliss 1985; Buzas *et al.* 1993; Jorissen *et al.* 1994). *Bulimina quadrata–ovata* may have adopted a similar strategy; peak values of this species coinciding with a peak in total organic carbon (TOC) content (3–5%; Keller & Lindinger 1989), suggest that this taxon was relatively successful in coping with oxygen deficient conditions. The subsequent decrease in *B. quadrata–ovata* and TOC sig-

nalled slightly better ventilation, which favoured the proliferation of the epibenthic *C. pseudoacutus*. These conditions appear to have been fairly stable until the early Palaeocene, although a sharp decrease in SA 5 taxa 1 m above the K/Pg boundary (at the base of the *C. taurica* Subzone) marks a second order faunal change. This purely quantitative change in the benthic assemblage coincides with a minor rise in planktonic foraminiferal numbers and %P, and with a compositional turnover in the pelagic ecosystem, interpreted to reflect elevated nutrient levels in surface waters (Brinkhuis & Zachariasse 1988). Provided that these elevated nutrient levels were converted into a higher organic carbon flux to the sea-floor, this may indicate that SA 5 taxa were particularly well adapted to minimal food supply under oxygen deficient conditions. More distribution data for these taxa are required to test this hypothesis.

The ostracod record across the K/Pg boundary gives support to our interpretation. According to Donze *et al.* (1982) and Peypouquet *et al.* (1986), oxygen saturation dropped below 2 ml/l at the K/Pg boundary and it stepped up to 3 ml/l within the *P. eugubina* Zone. Furthermore, nutrient levels dropped at the K/Pg boundary and remained relatively low within the *P. eugubina* Zone.

Early Palaeocene

Faunas similar to the ones in the lower Palaeocene of El Kef, Wadi Nukhl and Nahal Avdat have also been recorded in other localities of northern Egypt, southern Israel and Tunisia (Said & Kenawy 1956; Aubert & Berggren 1976; Benjamini *et al.* 1992; Keller 1992). Many of the species (e.g. *Anomalinoides* spp., *Cibicidoides* spp., *Gyroidinoides* spp., *Osangularia plummerae* and *P. prima*) belong to the neritic Midway Fauna (Berggren & Aubert 1975). In contrast, *G. beccariiformis* is a typical Late Cretaceous–Palaeocene deep-sea species (Tjalsma & Lohmann 1983; Van Morkhoven *et al.* 1986), but it has also been observed in deposits from locally subsided basins in Tunisia (Aubert & Berggren 1976; Saint-Marc & Berggren 1988). The common presence of *G. beccariiformis*, next to low frequencies of other deep-sea species present in the lower Palaeocene (e.g. *Angulogavelinella avnimelechi* and *Nuttallides* sp., that are lumped in 'trochospiral rest') indicate a palaeodepth of 200–400 m, which is in agreement with Tunisian palaeobathymetrical models (Saint-Marc & Berggren 1988; Saint-Marc 1993), and with a palaeobathymetric estimate based on ostracoda (Donze *et al.* 1982). Based on the same ostracod dataset, Peypouquet *et al.* (1986) inferred a 75–150 m palaeodepth for this interval. Keller (1988*b*) interpreted even shallower depths (50–100 m), but abandoned this view in a later paper (Keller 1992). Palaeodepth estimates shallower than 200 m are in conflict with the well established Palaeocene upper depth limit of *G. beccariiformis* (Van Morkhoven *et al.* 1986). Already during the Maastrichtian *G. beccariiformis* was a ubiquitous cosmopolitan deep-sea species (e.g. Dailey 1983; Thomas 1990*b*; Nomura 1991; Widmark & Malmgren 1992*b*), and it has only been recorded in considerable numbers in deeper bathyal deposits of the Tethys (e.g. in Spain: Keller 1992; in Austria and Romania: unpubl. data). These distribution patterns suggest that *G. beccariiformis* acquired a shallower upper depth limit in the early Palaeocene.

The lower Palaeocene assemblages of Gebel Qreiya and Gebel Duwi show a strong resemblance with neritic assemblages from central Egypt and Tunisia (Aubert & Berggren 1976; Luger 1985; Saint-Marc & Berggren, 1988; Saint-Marc 1993). At Gebel Qreiya the palaeodepth was probably between 100 and 150 m, whereas Gebel Duwi was still somewhat shallower 75–100 m, although P/B ratios in these localities suggest more similar palaeodepths.

The sequential (re-)appearance of many taxa and the resulting diversification in El Kef clearly reflects progressive amelioration of the bottom environment. Increasing oxygen levels are also suggested by the more intensively bioturbated sediments. The return of only the less common Maastrichtian taxa and the appearance of the deep-sea marker *G. beccariiformis*, inferred to be an indicator of more oligotrophic conditions, may point at a slightly reduced food supply to the sea-floor compared to the late Maastrichtian. The $\delta^{13}C$ record as well as PFN values support this view. Assemblages of planktonic foraminifera and dinoflagellates also suggest a return to more stable conditions and higher nutrient availability (Brinkhuis & Zachariasse 1988). Based on the appearance of *Parakrithe*, Donze *et al.* (1982) and Peypouquet *et al.* (1986) speculated on high nutrient levels due to intermittent upwelling within the *P. pseudobulloides* Zone. Their data show, however, that this change did not occur at the base of this zone but higher up. Since we only studied the basal 7 m (of in total 40 m) of the *P. pseudobulloides* Zone, our interpretation of nutrient levels matches with the ostracoda record. Ostracod-based palaeoxygenation, however, is incompatible with our interpretation, since oxygen levels in the lower part of the *P. pseudobulloides* Zone

were considered to be as low as within the *P. eugubina* Zone.

Discussion

Sea-level change across the K/Pg boundary

Successive cycle charts and eustatic sea-level curves by the EXXON group (Vail *et al.* 1977; Haq *et al.* 1987, 1988; Donovan *et al.* 1988) as well as the sea-level curve established for El Kef (Brinkhuis & Zachariasse 1988) have played a central role in the discussion on sea-level related K/Pg boundary extinctions (e.g. Brinkhuis & Zachariasse 1988; Keller 1988*a*, *b*, 1989*a*, *b*, 1992; Schmitz 1988; Keller *et al.* 1990; MacLeod & Keller 1991*a*, *b*; Rohling *et al.* 1991; Habib *et al.* 1992; Schmitz *et al.* 1992). In the cycle chart by Vail *et al.* (1977) a major eustatic sea-level fall, in the order of 150–200 m, marks the K/Pg boundary, which is followed by a gradual rise in the early Palaeocene. A large number of papers is in support of a longer term sea-level fall or regression at the close of the Cretaceous (e.g. Hancock & Kauffman 1979; Matsumoto 1980; Ekdale & Bromley 1984; Hultberg & Malmgren 1986; Peypouquet *et al.* 1986). Other studies support a sharp sea-level fall close to the boundary, followed by sea-level rise during the earliest Palaeocene (e.g. Jones *et al.* 1987; Brinkhuis & Zachariasse 1988; Habib *et al.* 1992; Moshkovitz & Habib 1993).

In the revised cycle chart (Haq *et al.* 1987, 1988) a prominent 100 m sea-level fall is situated well below the K/Pg boundary (*G. contusa* Zone; sequence boundary age 68 Ma), and is followed by a rapid 75 m eustatic rise. A relatively minor fall (10–20 m) marks the uppermost Maastrichtian (*A. mayaroensis* Zone; sequence boundary at 67 Ma) and is followed by a 25 m sea-level rise from the K/Pg boundary onwards. Based on a study in Braggs, Alabama, Donovan *et al.* (1988) modified this part of the cycle chart of Haq *et al.* (1987, 1988) in such a way that the relative magnitudes of these two sea-level events were switched, so that the major sea-level event corresponds to the 67 Ma sequence boundary. Few papers (e.g. Flexer & Reyment 1989; Schmitz *et al.* 1992) have provided independent (palaeontological) data suggesting sea-level rise or transgression across the K/Pg boundary.

At El Kef both the relative abundance of land-derived sporomorphs and the proportion of nearshore groups of dinoflagellates (*Spiniferites* and *Cyclonephelium*) suggest a marked sea-level fall starting some 50 cm ($\approx$12 ka) below the boundary and culminating in the K/Pg boundary interval (Brinkhuis & Zachariasse 1988).

Highest relative numbers of sporomorphs and nearshore dinoflagellates are found between 20 cm above and below the boundary (and also between 50 and 85 cm above the K/Pg boundary). Based on these frequency patterns Brinkhuis & Zachariasse (1988) erected a relative sea-level curve for El Kef with shallowest conditions in the K/Pg boundary interval, followed by a gradually rising, though fluctuating, sea-level in the earliest Palaeocene.

Our analysis of benthic foraminiferal assemblages from the upper 30 m of upper Maastrichtian of El Kef reveals no particular trend that can be related to either shallowing or deepening. Even in a relatively deep water site as El Kef eustatic sea-level variations in the order of 50–100 m as suggested by Donovan et al. (1988) would result in a distinct pattern of taxonomic replacement and frequency changes. Therefore, we can only conclude that such sea-level variations did not occur in El Kef within the studied part of the upper Maastrichtian. Moreover, both the rather stable benthic assemblages as well as the continuously high P/B ratios in the uppermost 4.5 m of the Maastrichtian also argue against a major sea-level fall at the end of the Cretaceous. A refined foraminifera-based palaeobathymetrical distribution model for the late Maastrichtian southern Tethys is required to trace minor sea-level events as suggested by Haq et al. (1987, 1988) in El Kef. Without such refinement we cannot be conclusive with respect to possible low amplitude sea-level variations. Also, since Brinkhuis & Zachariasse (1988) do not indicate the relative magnitude of sea-level change across the K/Pg boundary, we are unable to test their hypothesis by means of alternative, foraminifera-based, palaeobathymetrical analyses.

Independently of Brinkhuis & Zachariasse (1988), Keller (1988b) argued for a significant sea-level fall at the K/Pg boundary because of the disappearance of many upper bathyal to outer neritic benthic foraminifera. In a later paper the results of El Kef are discussed in a wider, Tethyan context (Keller 1992). In that paper much credit is given to the sea-level curve of Brinkhuis & Zachariasse (1988), which as outlined above, indicates falling sea-level at least up to the boundary (i.e. to the base of the boundary clay). The turnover in Tethyan benthic foraminifera, however, is related to a pre-boundary sea-level rise and resulting OMZ expansion (Keller 1992). Since at the same time the upper Maastrichtian assemblages are considered to indicate greater depths than the lowermost Palaeocene assemblages (Keller 1992), this scenario appears rather contradictory.

We agree with Keller (1988b) that the faunal change at the K/Pg boundary may suggest a significant shallowing. However, according to our palaeodepth estimates a very rapid shallowing of at least 150 m up to 400 m would be required to explain these faunal changes independently of any other parameters. Furthermore, this apparent lowstand of sea-level would have persisted well into the early Palaeocene i.e. until outer neritic and upper bathyal taxa reappeared. Such prolonged sea-level lowstand is incompatible with the sea-level curve of either Brinkhuis & Zachariasse (1988) or Haq et al. (1987, 1988), nor does the sedimentary record show any signs of erosion or redeposited shallow water sediments which could be expected with a sea-level fall of that magnitude.

Quasi-palaeobathymetrical faunal change

In previous discussions it was shown that the benthic faunal changes reflect fundamental ecosystem alterations. In our view these are not necessarily related to sea-level change. The tight coincidence between faunal change and proxy records for oxygenation and productivity suggests a more complex situation. Food and oxygen supply are generally considered to be the fundamental parameters that determine the overall composition of benthic foraminiferal assemblages (e.g. Harman 1964; Van der Zwaan 1982; Lutze & Coulbourn 1984; Jorissen 1987; Corliss & Chen 1988; Sjoerdsma & Van der Zwaan 1992; Sen Gupta & Machain-Castillo 1993). Palaeoecological model simulations predict that a significant change in either one of these parameters could result in considerable community alteration, e.g. reflecting a quasi-bathymetrical change (Sjoerdsma & Van der Zwaan 1992). The 'delta effect' in the Gulf of Mexico (Pflum & Frerichs 1976) might be the expression of such a change. It was shown that in the Mississippi delta front the bathymetrical distribution of many species differs significantly from the distribution in other parts of the Gulf of Mexico: enhanced food supply and variation in other environmental parameters generated these modifications.

Examples from the fossil record of deep water community turnover and temporal replacement by a shallow water community, are documented for the deep-sea extinction event at the end of the Palaeocene (Thomas 1990a, b; Speijer & Van der Zwaan 1994). At Maud Rise (Southern Ocean; 1500–2000 m palaeodepth) a low diversity assemblage dominated by T. selmensis temporarily replaced the highly diverse late Palaeocene deep-sea assemblage (Thomas

1990b). *Tappanina selmensis* occurs more often in bathyal deposits, but it is typically a neritic species (Van Morkhoven *et al.* 1986; Gibson & Bybell 1995). This faunal change has been related to deep-sea oxygen deficiency in the latest Palaeocene, favouring the more tolerant and opportunistic species *T. selmensis* (Thomas 1990b). A similar faunal response was observed in Wadi Nukhl; at the end of the Palaeocene this site had deepened to 500–700 m and a highly diverse upper–middle bathyal benthic assemblage flourished (Speijer 1995). Within a sapropelitic bed this fauna was temporarily but abruptly replaced by a low diversity assemblage dominated by *Anomalinoides aegyptiacus*, a species that typifies middle neritic deposits in southern and eastern Egypt (Speijer & Van der Zwaan 1994; Speijer *et al.* 1996). As at Maud Rise, oxygen deficiency played an important role here. A similar faunal response to changing trophic and redox conditions, unrelated to a significant sea-level fluctuation, may have occurred as well at El Kef during the K/Pg boundary transition. We realise, however, that proposing a similar scenario for El Kef is rather speculative and more detailed information from shallower deposits in Tunisia is required to test this hypothesis.

In a similar way we interpret the apparent deepening during the early Palaeocene primarily as reflecting gradual ecosystem restoration, closely following ameliorating redox and trophic conditions. Restoration started with the remigration of many Lazarus taxa (probably from deeper waters) in an otherwise rather empty ecospace. Subsequent immigration of other taxa eventually led to the development of a normal Palaeocene upper bathyal fauna that is present in many places along the north African continental margin.

Although sea-level variations (Haq *et al.* 1987, 1988; Brinkhuis & Zachariasse 1988) may have played a role within the studied interval as well, we believe that these were not instrumental in the prominent palaeoenvironmental changes during the earliest–early Palaeocene. Thus, rather than depth controlled faunal changes, we suggest that major perturbations in oxygen and food supply determined a quasi-palaeobathymetrical faunal sequence.

Tethyan continental margin productivity and oxygen deficiency

Basically, the results from El Kef can be reduced into signals of two main palaeoenvironmental events: at the K/Pg boundary a sudden drop in productivity (surface and bottom) and reduced

bottom oxygenation, followed by gradual recovery during the early Palaeocene. The surface productivity failure appears to have been an ocean-wide phenomenon (Perch-Nielsen *et al.* 1982; Hsü & McKenzie 1985; Arthur *et al.* 1987; Meyers & Simoneit 1989; Stott & Kennett 1989; Zachos *et al.* 1989), although at several high latitude sites this seems not to be the case (Hollis 1993; Keller 1993; Barrera & Keller 1994). Restoration of ocean productivity up to pre-boundary levels is thought to have taken place over a time-span of 0.3–1.5 Ma (Arthur *et al.* 1987; Stott & Kennett 1989; Zachos *et al.* 1989). This time-span probably stretches beyond our record of El Kef; several productivity proxies indeed indicate that pre-boundary levels were not reached within the studied time interval. Earliest Palaeocene oxygen deficiency also seems to have been a widespread phenomenon, in particular in Tethyan marginal basins (Magaritz *et al.* 1985; Lahodynsky 1988; Schmitz 1988; Wiedmann 1988; Kaijwara & Kaiho 1992; Sarkar *et al.* 1992; Kuhnt & Kaminski 1993; Coccioni & Galeotti 1994), although quite variable in duration.

Oxygen concentrations at the sea-floor depend on the balance between oxygen supply and consumption (e.g. Demaison & Moore 1980). Below the mixed layer oxygen is supplied through turbulent mixing and advection, whereas it is consumed by degradation of organic matter and respiration (e.g. Southam *et al.* 1982). Furthermore, oxygen solubility decreases with increasing temperature and aging of the water, i.e. longer exposure to organic breakdown, results in decreasing oxygen contents. There is considerable evidence that supply of organic matter from Tethyan surface waters was greatly reduced during the earliest Palaeocene. In spite of this, oxygen concentration at the sea-floor was depressed; therefore we can only conclude that the oxygen supply to intermediate waters also strongly diminished. We speculate that the production of Tethyan intermediate water was curtailed at the K/Pg boundary, perhaps as a result of climatic perturbations at low–middle latitudes, triggered by the impact of an extraterrestrial object and amplified by various feedback mechanisms.

Extinction and survivorship selectivity

In El Kef, the extinction of many species in the earliest Palaeocene is related to both oxygen deficiency and nutrient limitations. As conditions ameliorated in the early Palaeocene, half the number of taxa that became locally extinct at the K/Pg boundary, were able to re-occupy a niche. Meanwhile, many other species had become extinct, although at least some (e.g. *C. incrassata gigantea* and *C. beaumontianus*) persisted a while longer in the early Palaeocene.

Although it may be argued that widespread oxygen deficiency was the prime cause for benthic extinctions in Tethyan margin localities as El Kef and Caravaca (see Coccioni & Galeotti 1994), it cannot account for extinctions in the deep-sea. There, oxygen concentration of bottom waters appears to have remained normal (Thomas 1990*a*, 1992; Nomura 1991; Widmark & Malmgren 1992*a*). Another mechanism must account for these extinctions.

Our data suggest extinction selectivity between various morphogroups: elongate bi-/triserial taxa appear more severely affected than trochospiral taxa. In modern environments abundance of the bi-/triserial morphogroup (in particular, *Bulimina*, *Bolivina* and *Stainforthia*) is often related to depressed oxygen levels, enhanced food levels, or to a combination of both (e.g. Van der Zwaan 1982; Lutze & Coulbourn 1984; Corliss & Chen 1988; Sen Gupta & Machain-Castillo 1993; Rathburn & Corliss 1994). In a general sense, there also appears to be a relationship between test-shape and microhabitats: low trochospiral forms are usually found to live on the sediment surface, whereas elongate bi-/triserial forms often occupy a habitat within the sediment (Corliss 1985, 1991; Corliss & Chen 1988; but see also Barmawidjaja *et al.* 1992; Linke & Lutze 1993; Rathburn & Corliss 1994). Combining these morphogroup–environment relationships has led to the suggestion that the amount of organic carbon within the sediment may determine the abundance of endobenthic (and thus often bi-/triserial) taxa (Rathburn & Corliss 1994).

Three aspects of benthic extinction selectivity in the earliest Palaeocene suggest that the long-term surface productivity crisis, starting at the K/Pg boundary was the prime controlling parameter. (1) Shallower more eutrophic environments show a more distinct turnover and a greater loss in number of species than oligotrophic deeper waters (Thomas 1990*b*; Kaiho 1992). (2) Many of the species that became extinct were rather common in these more eutrophic environments. This is particularly clear in our data from El Kef and Gebel Duwi. (3) Species that are supposed to have lived within organic carbon-rich sediments were most prone to extinction. Elongate endobenthic taxa (e.g. *Coryphostoma*, *Sitella*, *Bolivinoides*, *Orthokarstenia*, *Pyramidina* and *Eouvigerina*) were much more severely affected than trochospiral taxa. This pattern observed earlier by Keller

(1992), is very consistent in upper bathyal as well as neritic assemblages: despite the much higher number of trochospiral taxa in our dataset, we record only four extinctions in this morphogroup, against ten in the bi-/triserial morphogroup.

The pelagic productivity crisis not only caused a decrease in the amount of food supply to the sea-floor, but also a reduction in nutrient heterogeneity, as evidenced by widely recognized oligotaxic algal blooms of, for example, *Thoracosphaera* and *Braarudosphaera*, that characterize the earliest Palaeocene (Romein 1979*a, b*; Perch-Nielsen *et al.* 1982; Alcala-Herrera *et al.* 1992; Pospichal 1994). Since some recent foraminifera are considered to be selective feeders (e.g. Lee 1974; Murray 1991), it seems likely that at least some extinctions may be due to this reduction of nutrient heterogeneity.

With the present data we can only speculate on where the Lazarus taxa (SA 2) of El Kef survived, but most likely the majority survived in deep-sea environments unaffected by oxygen deficiency. Taxa such as *G. pyramidata, P. plummerae, V. depressa* and *Pullenia* spp. are indeed widely documented from Upper Cretaceous and Palaeocene deep-sea deposits (e.g. Sliter 1977*a, b*; Dailey 1983; Thomas 1990*b*; Nomura 1991; Widmark & Malmgren 1992*b*). These taxa were unable to withstand dysoxic conditions at El Kef, but were probably less dependent on the amount and heterogeneity of the food supply than the species that eventually became extinct.

Conclusions

The K/Pg boundary succession of El Kef provides a distinct sequence of sudden ecosystem collapse followed by gradual recovery. A severe reduction of oxygenation and trophic resources on the sea-floor invoked temporal replacement of a stable upper bathyal palaeocommunity by one primarily consisting of more tolerant and opportunistic shallow water taxa. Gradual amelioration of the environment was followed by the settlement of a normal Palaeocene upper bathyal fauna, very different from the Maastrichtian one.

In contrast to the deep-sea, shallow palaeocommunities were severely affected by the K/Pg event. Many taxa that are thought to have been dependent on high and heterogeneous nutrient resources (in particular endobenthic deposit feeders) suffered extinction most severely during the earliest Palaeocene. The differential effects in various bathymetrical compartments and ecological groups result from a prolonged period of reduced surface fertility, that started at the K/Pg boundary.

We thank Alexandra Nederbragt and Jan Smit for providing samples, Anders Henriksson for calcareous nannoplankton determinations, and Peter Luger for enabling us to study his sample collection from Egypt. RPS thanks his late friend Albert Langejans for all computer hardware and software support and advice. Constructive criticism by Henk Brinkhuis and Norman MacLeod improved the quality of the manuscript. An earlier version of this paper was part of the first author's doctoral thesis.

Appendix 1.

Discussions on taxonomy and synonymies are beyond the general scope of this volume. For more information on these topics we refer the reader to the taxonomy section of an earlier version of this paper in the first author's doctoral thesis (Speijer 1994, p. 44–64). Here we confine ourselves to listing the primary references of the species discussed in this paper.

Alabamina wilcoxensis Toulmin, 1941, p. 603, pl. 81, figs 10–14.

Ammomarginulina aubertae Gradstein & Kaminski, 1989, p. 74, pl. 3, figs 1–8; pl. 4, figs 1–3; text-fig. 2.

Anomalinoides cf. *acutus* (Plummer) = cf. *Anomalina ammonoides* (Reuss) var. *acuta* Plummer, 1926, p. 149, pl. 10, fig. 2.

Anomalinoides affinis (Hantken) = *Pulvinulina affinis* Hantken, 1875, p. 78, pl. 10, fig. 6.

Anomalinoides ekblomi (Brotzen) = *Cibicides ekblomi* Brotzen, 1948, p. 82, pl. 13, fig. 2.

Anomalinoides praeacutus (Vasilenko) = *Anomalina praeacuta* Vasilenko, 1950, p. 208, pl. 5, figs 2, 3.

Anomalinoides rubiginosus (Cushman) = *Anomalina rubiginosa* Cushman, 1926, p. 607, pl. 21, fig. 6.

Anomalinoides simplex (Brotzen) = *Cibicides simplex* Brotzen, 1948, p. 83, pl. 13, figs 4, 5.

Anomalinoides umboniferus (Schwager) = *Discorbis umbonifer* Schwager, 1883, p. 126, pl. 27, fig. 14.

Anomalinoides susanaensis (Browning) = *Cibicides susanaensis* Browning (in Mallory), 1959, p. 271, pl. 32, figs 11, 12.

Bolivinoides decoratus (Jones) = *Bolivina decorata* Jones, 1886, p. 330, pl. 27, figs 7, 8.

Bolivinoides draco draco (Marsson) = *Bolivina draco* Marson, 1878, p. 157, pl. 3, fig. 25.

Bolivinopsis clotho (Grzybowski) = *Bolivinopsis? clotho* (Grzybowski) in: Cushman, 1946, p. 103, pl. 44, figs 10–13.

Bulimina strobila Marie, 1941, p. 265, pl. 32, fig. 302.

Cibicides beaumontianus (D'Orbigny) = *Truncatulina beaumontiana* D'Orbigny, 1840, p. 35, pl. 3, figs 17–19.

Cibicidoides abudurbensis (Nakkady) = *Cibicides abudurbensis* Nakkady, 1950, p. 691, pl. 90, figs 35–38.

Cibicidoides cf. *hyphalus* (Fisher) = cf. *Anomalinoides hyphalus* Fisher, 1969, p. 197, pl. 3. (pc) = planoconvex morphology; (bc) = biconvex morphology.

Cibicidoides pseudoacutus (Nakkady) = *Anomalina pseudoacuta* Nakkady, 1950, p. 691, pl. 90, figs 29–32.

Cibicidoides suzakensis (Bykova) = *Cibicides suzakensis* Bykova in: Salaj et al. 1976, p. 158, pl. 16, figs 7, 8.

Coryphostoma incrassata gigantea (Wicher) = *Bolivina incrassata* Reuss var. *gigantea* Wicher, 1956, p. 120, pl. 12, figs 2–3.

Coryphostoma plaitum (Carsey) = *Bolivina plaitum* Carsey, 1926, p. 26, pl. 4, fig. 2.

Eouvigerina subsculptura McNeil & Caldwell, 1981, p. 231, pl. 18, figs 20, 21.

Elhasaella cf. *allanwoodi* Hamam = cf. *Elhasaella allanwoodi* Hamam, 1976, p. 454, pl. 1, figs 1–7; pl. 2, figs 1–8.

Gaudryina pyramidata Cushman = *Gaudryina laevigata* Franke var. *pyramidata* Cushman, 1926, p. 587, pl. 16, fig. 8.

Gavelinella beccariiformis (White) = *Rotalia beccariiformis* White, 1928, p. 287, pl. 39, figs 2–4.

Gavelinella martini (Sliter) = *Gyroidinoides quadratus martini* Sliter, 1968, p. 121, pl. 22, fig. 9.

Gyroidinoides girardanus (Reuss) = *Rotalina girardana* Reuss, 1851, p. 73, pl. 5, fig. 34.

Gyroidinoides octocameratus (Cushman & Hanna) = *Gyroidina soldanii* D'Orbigny *octocamerata* Cushman & Hanna, 1927, p. 223, pl. 14, figs 16–18.

Gyroidinoides tellburmaensis Futyan, 1976, p. 532, pl. 81, figs 10–12.

Loxostomoides applinae (Plummer) = *Bolivina applini* Plummer, 1926, p. 69, pl. 4, fig. 1.

Neobulimina canadensis Cushman & Wickenden, 1928, p. 13, pl. 1, figs 1–2.

Oridorsalis plummerae (Cushman) = *Eponides plummerae* Cushman, 1948, p. 44, pl. 8, fig. 9.

Orthokarstenia oveyi (Nakkady) = *Siphogenerinoides oveyi* Nakkady, 1950, p. 686, pl. 89, fig. 20.

Osangularia plummerae Brotzen, 1940, p. 30, text-fig. 8.

Praebulimina reussi Morrow = *Bulimina reussi* Morrow, 1934, p. 195, pl. 29, fig. 12.

Pseudoeponides cf. *elevatus* (Plummer) = cf. *Truncatulina elevatus* Plummer, 1926, p. 142, pl. 11, fig. 1.

Pseudouvigerina plummerae Cushman, 1927, p. 115, pl. 23, fig. 8.

Pulsiphonina prima (Plummer) = *Siphonina prima* Plummer, 1926, p. 148, pl. 12, fig. 4.

Pyramidina aegyptiaca (Nakkady) = *Reussella aegyptiaca* Nakkady, 1950, p. 687, pl. 90, fig. 1.

Sitella colonensis (Cushman & Hedberg) = *Buliminella colonensis* Cushman & Hedberg, 1930, p. 65, pl. 9, figs 6, 7.

Sitella cushmani (Sandidge) = *Buliminella cushmani* Sandidge, 1932, p. 280, pl. 42, figs 18, 19.

Sitella fabilis (Cushman & Parker) = *Buliminella fabilis* Cushman & Parker, 1936, p. 7, pl. 2, fig. 5.

Tappanina selmensis (Cushman) = *Bolivinita selmensis* Cushman, 1933, p. 58, pl. 7, figs 3–4.

Valvalabamina depressa (Alth) = *Gyroidina depressa* (Alth) in: Cushman, 1946, p. 139, pl. 58, figs 1–4.

Stensioeina pommerana Brotzen, 1936, p. 165.

References

ALEALA-HERRERA, J. A., GROSSMAN, E. L. & GARTNER, S. 1992. Nannofossil diversity and equitability and fine-fraction $\delta^{13}C$ across the Cretaceous–Tertiary boundary at Walvis Ridge Leg 74, South Atlantic. *Marine Micropaleontology*, **20**, 77–88.

ALVAREZ, L. W., ALVAREZ, W., ASARO, F. & MICHEL, H. V. 1960. Extraterrestrial cause for the Cretaceous–Tertiary extinction. *Science*, **208**, 1095–1108.

ARCHIBALD, J. D. & BRYANT, L. J. 1990. Differential Cretaceous/Tertiary extinctions of nonmarine vertebrates; evidence from northeastern Montana. *In:* SHARPTON, V. L. & WARD, P. D. (eds) *Global Catastrophes in Earth History. An Interdisciplinary Conference on Impacts, Volcanism, and Mass Mortality.* Geological Society of America Special Paper, **247**, 549–562.

ARTHUR, M. A., ZACHOS, J. C. & JONES, D. S. 1987. Primary productivity and the Cretaceous/Tertiary boundary event in the oceans. *Cretaceous Research*, **8**, 43–54.

AUBERT, J. & BERGGREN, W. A. 1976. Paleocene benthic foraminiferal biostratigraphy and paleoecology of Tunisia. *Bulletin de Centre de Recherche Pau-SNPA*, **10**, 379–469.

BARMADWIDJAJA, D. M., JORISSEN, F. J., PUŠKARIĆ, S. & VAN DER ZWAAN, G. J. 1992. Microhabitat selection by benthic foraminifera in the northern Adriatic Sea. *Journal of Foraminiferal Research*, **22**, 297–317.

BARRERA, E. & KELLER, G. 1994. Productivity across the Cretaceous/Tertiary boundary in high latitudes. *Geological Society of America Bulletin*, **106**, 1254–1266.

BENJAMINI, C., KELLER, G. & PERELIS-GROSSOVICZ, L. 1992. On benthic foraminiferal paleoenvironments across the K/T boundary, Negev, Israel.

In: SARNTHEIN, M., THIEDE, J. & ZAHN, R. (eds) *Fourth International Conference on Paleoceanography, ICP IV; Short- and Long-Term Global Change: Records and Modelling. 21–25 September 1992, Kiel/Germany, Program and Abstracts.* GEOMAR Report 15, GEOMAR, Kiel, Germany, 61.

BERGER, W. H. & DIESTER-HAASS, L. 1988. Paleoproductivity; the benthic/planktonic ratio in foraminifera as a productivity index. *Marine Geology*, **81**, 15–25.

BERGGREN, W. A. & AUBERT, J. 1975. Paleocene benthonic foraminiferal biostratigraphy, paleobiogeography and paleoecology of Atlantic-Tethyan regions; Midway-type fauna. *Palaeogeography, Palaeoclimatology, Palaeoecology*, **18**, 73–192.

——, KENT, D. V., SWISHER, C. C. III & AUBRY, M. P. 1995. A revised Cenozoic geochronology and chronostratigraphy. *In:* BERGGREN, W. A., KENT, D. V., AUBRY, M. P. & HARDENBOL, J. (eds) *Geochronology, Time Scales and Stratigraphic Correlations: Framework for an Historical Geology.* Society for Economic Paleontologists and Mineralogists, Special Publication, **54**, 129–212.

BRAMLETTE, M. M. & MARTINI, E. 1964. The great change in calcareous nannoplankton fossils between the Maestrichtian and Danian. *Micropaleontology*, **10**, 291–322.

BRINKHUIS, H. & ZACHARIASSE, W. J. 1988. Dinoflagellate cysts, sea level changes and planktonic foraminifers across the Cretaceous–Tertiary boundary at El Haria, Northwest Tunisia. *Marine Micropaleontology*, **13**, 153–191.

——, ROMEIN, A. J. T., SMIT, J. & ZACHARIASSE, J.-W. 1994. Danian–Selandian dinoflagellate cysts from lower latitudes with special reference to the El Kef section, NW Tunisia. *GFF*, **116**, 46–48.

BROTZEN, F. 1936. Foraminiferen aus dem schwedischen, untersten Senon von Eriksdal in Schonen. *Sveriges Geologiska Undersökning Serie C*, **30**.

—— 1940. Flintrännans och Trindelrännans geologi (Öresund)–Die Geologie der Flint- und Trindelrinne (Öresund). *Sveriges Geologiska Undersökning Serie C*, **34**, 1–33.

—— 1948. The Swedish Paleocene and its foraminiferal fauna. *Sveriges Geologiska Undersökning Serie C*, **42**.

BUZAS, M. A., CULVER, S. J. & JORISSEN, F. J. 1993. A statistical evaluation of the microhabitats of living (stained) infaunal benthic foraminifera. *In:* LANGER, M. R. (ed.) *Foraminiferal Microhabitats*, Marine Micropaleontology, **20**, 311–320.

CARSEY, D. O. 1926. Foraminifera of the Cretaceous of central Texas. *Texas University, Bulletin*, **2612**.

COCCIONI, R. & GALEOTTI, S. 1994. K–T boundary extinction: Geologically instantaneous or gradual event? Evidence from deep-sea benthic foraminifera. *Geology*, **22**, 779–782.

CORLISS, B. H. 1985. Microhabitats of benthic foraminifera within deep-sea sediments. *Nature*, **314**, 435–438.

—— 1991. Morphology and microhabitat preferences of benthic foraminifera from the Northwest Atlantic Ocean. *Marine Micropaleontology*, **17**, 195–236.

—— & CHEN, C. 1988. Morphotype patterns of Norwegian Sea deep-sea benthic foraminifera and ecological implications. *Geology*, **16**, 716–719.

CUSHMAN, J. A. 1926. The Foraminifera of the Velasco shale of the Tampico embayment San Luis Potosi, Mexico. *AAPG Bulletin*, **10**, 581–612.

—— 1927. New and interesting foraminifera from Mexico and Texas. *Contributions from the Cushman Laboratory for Foraminiferal Research*, **3**, 111–117.

—— 1933. New American Cretaceous foraminifera. *Contributions from the Cushman Laboratory for Foraminiferal Research*, **9**, 49–64.

—— 1946. *Upper Cretaceous foraminifera of the Gulf coastal region of the United States and adjacent areas*. United States Geological Survey, Professional Paper, **206**.

—— 1948. Additional new foraminifera from the American Paleocene. *Contributions from the Cushman Laboratory for Foraminiferal Research*, **24**, 43–45.

—— & HANNA, G. D. 1927. Foraminifera from the Eocene near Coalinga, California. *California Academy of Science, Proceedings, 4th Series*, **16**, 205–228.

—— & HEDBERG, H. D. 1930. Notes on some foraminifera from Venezuela and Colombia. *Contributions from the Cushman Laboratory for Foraminiferal Research*, **6**, 64–69.

—— & PARKER, F. L. 1936. Notes on some Cretaceous species of *Buliminella* and *Neobulimina*. *Contributions from the Cushman Laboratory for Foraminiferal Research*, **12**, 5–10.

—— & WICKENDEN, R. T. D. 1928. A new foraminiferal genus from the Upper Cretaceous. *Contributions from the Cushman Laboratory for Foraminiferal Research*, **4**, 12–13.

DAILEY, D. H. 1983. Late Cretaceous and Paleocene benthic foraminifers from Deep Sea Drilling Project Site 516, Rio Grande Rise, western South Atlantic Ocean. *In:* BARKER, P. J., CARLSON, R. L., JOHNSON, D. A. *et al.* (eds) *Initial reports of the Deep Sea Drilling Project*. **72**, U.S. Government Printing Office, Washington, DC, 757–782.

DEMAISON, G. J. & MOORE, G. T. 1980. Anoxic environments and oil source bed genesis. *AAPG Bulletin*, **64**, 1179–1209.

D'HONDT, S. & KELLER, G. 1991. Some patterns of planktic foraminiferal assemblage turnover at the Cretaceous–Tertiary boundary. *Marine Micropaleontology*, **17**, 77–118.

DONOVAN, A. D., BAUM, G. R., BLECHSCHMIDT, G. L., LOUTIT, T. S., PFLUM, C. E. & VAIL, P. R. 1988. Sequence stratigraphic setting of the Cretaceous–Tertiary boundary in central Alabama. *In:* WILGUS, C. K. *et al.* (eds) *Sea-Level Changes: An Integrated Approach*. Society for Economic Paleontologists and Mineralogists, Special Publication, **42**, 299–307.

DONZE, P., COLIN, J. P., DAMOTTE, R., OERTLI, H. J.,

PEYPOUQUET, J. & SAID, R. 1982. Les ostracodes du Campanien terminal a l'Eocene inférieur de la coupe du Kef, Tunisie Nord-Occidentale. *Bulletin des Centres de Recherches Exploration–Production Elf Aquitaine*, **6**, 273–307.

D'ORBIGNY, A. 1840. Mémoire sur les foraminifères de la craie blanche du bassin de Paris. *Société de Géologie de France, Mémoire*, **4**, 1–51.

EKDALE, A. A. & BROMLEY, R. G. 1984. Sedimentology and ichnology of the Cretaceous–Tertiary boundary in Denmark; implications for the causes of the terminal Cretaceous extinction. *Journal of Sedimentary Petrology*, **54**, 681–703.

FARIS, M. 1982. *Micropaléontologie et Biostratigraphie du Crétacé Supérieur a l'Eocene Inférieur de l'Egypte Centrale (Région de Duwi, Vallée du Nil, Oasis de Kharga et de Dakhla)*. Doctorate Thesis, Université de Paris.

FISHER, M. J. 1969. Benthonic foraminifera from the Maestrichtian chalk of Galicia bank, west of Spain. *Palaeontology*, **12**, 189–200.

FLEXER, A. & REYMENT, R. A. 1989. Note on Cretaceous transgressive peaks and their relation to geodynamic events for the Arabo-Nubian and the northern African shields. *Journal of African Earth Sciences*, **8**, 65–73.

FUTYAN, A. I. 1976. Late Mesozoic and early Cainozoic benthic foraminifera from Jordan. *Palaeontology*, **19**, 517–537.

GALLAGHER, W. B. 1991. Selective extinction and survival across the Cretaceous/Tertiary boundary in the Northern Atlantic Coastal Plain. *Geology*, **19**, 967–970.

GERSTEL, J., THUNNELL, R. & EHRLICH, R. 1987. Danian faunal succession; planktonic foraminiferal response to a changing marine environment. *Geology*, **15**, 665–668.

GIBSON, T. G. & BYBELL, L. M. 1995. Sedimentary patterns across the Paleocene–Eocene boundary in the Atlantic and Gulf Coastal Plains of the United States. *In:* LAGA, P. (ed.) *Paleocene–Eocene Boundary Events*. Bulletin de la Société belge de Géologie, **103** for 1994, 237–265.

——, —— & OWENS, J. P. 1993. Latest Paleocene lithologic and biotic events in neritic deposits of southwestern New Jersey. *Paleoceano-graphy*, **8**, 495–514.

GRADSTEIN, F. M. & KAMINSKI, M. A. 1989. Taxonomy and biostratigraphy of new and emended species of Cenozoic deep-water agglutinated foraminifera from the Labrador and North seas. *Micropaleontology*, **35**, 72–92.

HABIB, D., MOSHKOVITZ, S. & KRAMER, C. 1992. Dinoflagellate and calcareous nannofossil response to sea-level change in Cretaceous–Tertiary boundary sections. *Geology*, **20**, 165–168.

HANCOCK, J. M. & KAUFFMAN, E. G. 1979. The great transgressions of the late Cretaceous. *Journal of the Geological Society, London*, **136**, 175–186.

HANSEN, T. A., FARRELL, B. R. & UPSHAW, B. III 1993. The first 2 million years after the Cretaceous–Tertiary boundary in East Texas; rate and paleoecology of the molluscan recovery. *Paleobiology*, **19**, 251–265.

HANTKEN, M. 1875. Die Fauna der *Clavulina* Szaboi–Schichten; Theil I – Foraminiferen. *Königlich-Ungarische Geologische Anstalt, Mitteilungen Jahrbuch, Budapest,* **4** 1–93.

HAQ, B. U., HARDENBOL, J. & VAIL, P. R. 1987. Chronology of fluctuating sea levels since the Triassic. *Science,* **235**, 1156–1167.

——, —— & —— 1988. Mesozoic and Cenozoic chronostratigraphy and cycles of sea level change. *In:* WILGUS, C. K. *et al.* (eds) *Sea-Level Changes: An Integrated Approach.* Society for Economic Paleontologists and Mineralogists, Special Publication, **42**, 71–108.

HARMAN, R. A. 1964. Distribution of foraminifera in the Santa Barbara Basin, California. *Micropaleontology,* **10**, 81–96.

HENDRIKS, F., LUGER, P., BOWITZ, J. & KALLENBACH, H. 1987. Evolution of depositional environments of SE-Egypt during the Cretaceous and lower Tertiary. *Berliner Geowissenschaftliche Abhandlungen, Reihe A: Geologie und Paläontologie,* **75**, 49–82.

HERGUERA, J. C. & BERGER, W. H. 1991. Paleoproductivity from benthic foraminifera abundance; glacial to postglacial change in the west-equatorial Pacific. *Geology,* **19**, 1173–1176.

HOLLIS, C. J. 1993. Latest Cretaceous to late Paleocene radiolarian biostratigraphy; a new zonation from the New Zealand region. *In:* LAZARUS, D. B. & DE WEVER, P. D. (eds) *Interrad VI.* Marine Micropaleontology, **21**, 295–327.

HSÜ, K. J. & MCKENZIE, J. A. 1985. A 'strangelove' ocean in the earliest Tertiary. *In:* SUNDQUIST, E. T. & BROECKER, W. S. (eds) *The Carbon Cycle and Atmospheric CO₂; Natural Variations Archaean to Present.* Geophysical Monograph, **32**, AGU, Washington, DC, 487–492.

HUBER, B. T. 1991. Maestrichtian planktonic foraminifer biostratigraphy and the Cretaceous/Tertiary boundary at Hole 738C (Kerguelen Plateau, southern Indian Ocean). *In:* BARRON, J. A., LARSEN, B. *et al.* (eds) *Proceedings of the Ocean Drilling Program, Scientific Results.* **119**, ODP, College Station, TX, 451–465.

HULTBERG, S. U. & MALMGREN, B. A. 1986. Dinoflagellate and planktonic foraminiferal paleobathymetrical indices in the Boreal uppermost Cretaceous. *Micropaleontology,* **32**, 316–323.

JÄGER, M. 1993. Danian serpulidae and spirorbidae from NE Belgium and SE Netherlands: K/T boundary extinction, survival and origination patterns. *Contributions to Tertiary and Quaternary Geology,* **29**, 73–117.

JONES, T. R. 1886. A list of the Cretaceous foraminifera of Keady Hill, County Derry. *Proceedings of the Belfast Nature Field Club,* **330**.

JONES, D. S., MÜLLER, P. A., BRYAN, J. R., DOBSON, J. P., CHANNEL, J. E. T., ZACHOS, J. C. & ARTHUR, M. A. 1987. Biotic, geochemical, and paleomagnetic changes across the Cretaceous/Tertiary boundary at Braggs, Alabama. *Geology,* **15**, 311–315.

JORISSEN, F. J. 1987. The distribution of benthic foraminifera in the Adriatic Sea. *Marine*

Micropaleontology, **12**, 21–48.

——, BUZAS, M. A., CULVER, S. J. & KUEHL, S. A. 1994. Vertical distribution of living benthic foraminifera in submarine canyons off New Jersey. *Journal of Foraminiferal Research,* **24**, 28–36.

KAIHO, K. 1992. A low extinction rate of intermediate water benthic foraminifera at the Cretaceous/Tertiary boundary. *Marine Micropaleontology,* **18**, 229–259.

KAJIWARA, Y. & KAIHO, K. 1992. Oceanic anoxia at the Cretaceous/Tertiary boundary supported by the sulfur isotopic record. *Palaeogeography, Palaeoclimatology, Palaeoecology,* **99**, 151–162.

KELLER, G. 1988a. Biotic turnover in benthic foraminifera across the Cretaceous/Tertiary boundary at El Kef, Tunisia. *Palaeogeography, Palaeoclimatology, Palaeoecology,* **66**, 153–171.

—— 1988b. Extinction, survivorship and evolution of planktic foraminifera across the Cretaceous/Tertiary boundary at El Kef, Tunisia. *Marine Micropaleontology,* **13**, 239–263.

—— 1989a. Extended Cretaceous/Tertiary boundary extinctions and delayed population change in planktonic foraminifera from Brazos River, Texas. *Paleoceanography,* **4**, 287–332.

—— 1989b. Extended period of extinctions across the Cretaceous/Tertiary boundary in planktonic foraminifera of continental-shelf sections; implications for impact and volcanism theories. *Geological Society of America Bulletin,* **101** 1403–1419.

—— 1992. Paleoecologic response of Tethyan benthic foraminifera to the Cretaceous–Tertiary boundary transition. *In:* TAKAYANAGI, Y. & SAITO, T. (eds) *Studies in Benthic Foraminifera.* Tokai University Press, Tokyo, 77–91.

—— 1993. The Cretaceous–Tertiary boundary transition in the Antarctic Ocean and its global implications. *Marine Micropaleontology,* **21**, 1-45.

—— & LINDINGER, M. 1989. Stable isotope, TOC and CaCO₃ record across the Cretaceous–Tertiary boundary at El Kef, Tunisia. *Palaeogeography, Palaeoclimatology, Palaeoecology,* **73**, 243–265.

——, BARRERA, E., SCHMITZ, B. & MATTSON, E. 1993. Gradual mass extinction, species survivorship, and long-term environmental changes across the Cretaceous–Tertiary boundary in high latitudes. *Geological Society of America Bulletin,* **105**, 979–997.

——, BENJAMINI, C., MAGARITZ, M. & MOSHKOVITZ, S. 1990. Faunal, erosional and CaCO₃ events in the early Tertiary eastern Tethys. *In:* SHARPTON, V. L. & WARD, P. D. (eds) *Global Catastrophes in Earth History. An Interdisciplinary Conference on Impacts, Volcanism, and Mass Mortality.* GSA Special Paper, **247**, 471–480.

KROON, D. & NEDERBRAGT, A. J. 1990. Ecology and paleoecology of triserial planktic foraminifera. *Marine Micropaleontology,* **16**, 25–38.

KUHN, W. 1992. Paleozäne und untereozäne Benthos-Foraminiferen des bayerischen und salzburgischen Helvetikums – Systematik, Stratigraphie und Palökologie. *Münchner Geowissenshaftliche*

Abhandlungen, Reihe A: Geologie und Paläontologie, **24**.

KUHNT, W. & KAMINSKI, M. A. 1993. Changes in the community structure of deep water agglutinated foraminifers across the K/T Boundary in the Basque basin (Northern Spain). *Revista Espanola de Micropaleontologia*, **25**, 57–92.

LAHODYNSKY, R. 1988. Lithostratigraphy and sedimentology across the Cretaceous–Tertiary boundary in the Flyschgosau, Eastern Alps, Austria. *In:* LAMOLDA, M. A., KAUFFMAN, E. G. & WALLISER, O. H. (eds) *Palaeontology and Evolution: Extinction Events*. Revista Espanola de Paleontologia, Numero extraordinario, 73–82.

LEE, J. J. 1974. Towards understanding the niche of foraminifera. *In:* HEDLEY, R. H. *et al.* (eds) *Foraminifera, 1*. Academic, New York, 207–260.

LEROY, L. W. 1953. *Biostratigraphy of the Maqfi section, Egypt*. Geological Society of America, Memoir, **54**.

LINKE, P. & LUTZE, G. F. 1993. Microhabitat preferences of benthic foraminifera; a static concept or a dynamic adaptation to optimize food aquisition? *In:* LANGER, M. R. (ed.) *Foraminiferal Microhabitats*. Marine Micropaleontology, **20**, 215–234.

LOEBLICH, A. R. J. & TAPPAN, H. 1988. *Foraminiferal Genera and Their Classification*. Van Nostrand Reinhold, New York.

LUGER, P. 1985. Stratigraphie der marinen Oberkreide und des Alttertiärs im südwestlichen Obernil–Becken (SW-Ägypten) unter besonderer Berücksichtigung der Mikropaläontologie, Paläoekologie und Paläogeographie. *Berliner Geowissenshaftliche Abhandlungen, Reihe A: Geologie und Paläontologie*, **63**.

—— 1988. Maestrichtian to Paleocene facies evolution and Cretaceous/Tertiary boundary in middle and southern Egypt. *In:* LAMOLDA, M. A., KAUFFMAN, E. G. & WALLISER, O. H. (eds) *Palaeontology and Evolution: Extinction Events*. Revista Espanola de Paleontologia, Numero extraordinario, 83–90.

LUTERBACHER, H. P. & PREMOLI-SILVA, I. 1964. Biostratigrafia del limite cretaceo-terziario nell'Appennino centrale. *Rivista Italiana di Paleontologia*, **70**, 67–117.

LUTZE, G. F. & COULBOURN, W. T. 1984. Recent benthic foraminifera from the continental margin of Northwest Africa; community structure and distribution. *Marine Micropaleontology*, **8**, 361–401.

MACLEOD, N. & KELLER, G. 1991*a*. Hiatus distributions and mass extinctions at the Cretaceous/Tertiary boundary. *Geology*, **19**, 497–501.

—— 1991*b*. How complete are Cretaceous/Tertiary boundary sections? A chronostratigraphic estimate based on graphic correlation. *Geological Society of America, Bulletin*, **103** 1439–1457.

MCNEIL, D. H. & CALDWELL, W. G. E. 1981. *Cretaceous rocks and their foraminifera in the Manitoba Escarpment*. Geological Association of Canada, Special Paper, **21**.

MAGARITZ, M., BENJAMINI, C., KELLER, G. &

MOSHKOVITZ, S. 1992. Early diagenetic isotopic signal at the Cretaceous/Tertiary boundary, Israel. *Palaeogeography, Palaeoclimatology, Palaeoecology*, **91**, 291–304.

——, MOSHKOVITZ, S., BENJAMINI, C., HANSEN, H. J., HAKANSSON, E. & RASMUSSEN, K. L. 1985. Carbon isotope-, bio- and magnetostratigraphy across the Cretaceous–Tertiary boundary in the Zin Valley, Negev, Israel. *Newsletters on Stratigraphy*, **15**, 100–113.

MALLORY, V. S. 1959. *Lower Tertiary Biostratigraphy of the California Coast Ranges*. American Association of Petroleum Geologists, Tulsa, OK.

MARIE, P. 1941. *Les foraminifères de la craie à Belemnitella mucronata du bassin de Paris*. Muséum nationale d'Histoire naturelle, Paris, Mémoir, **12**.

MARSSON, T. H. 1878. Die Foraminiferen der weissen Schreibkreide der Insel Rügen. *Naturwissenschaftlicher Verein für Neu-Vorpommern und Rügen in Greifswald*, **10**, 115–196.

MASTERS, B. A. 1993. Re-evaluation of the species and subspecies of the genus *Plummerita* Bronnimann and a new species of *Rugoglobigerina* Bronnimann (Foraminiferida). *Journal of Foraminiferal Research*, **23**, 267–274.

MATSUMOTO, T. 1980. Inter-regional correlation of transgressions and regressions in the Cretaceous period. *Cretaceous Research*, **1**, 359–373.

MEYERS, P. A. & SIMONEIT, B. R. T. 1989. Global comparisons of organic matter in sediments across the Cretaceous/Tertiary boundary. *Organic Geochemistry*, **16**, 641–648.

MORROW, A. L. 1934. Foraminifera and ostracoda from the Upper Cretaceous of Kansas. *Journal of Paleontology*, **8**, 186–205.

MOSHKOVITZ, S. & HABIB, D. 1993. Calcareous nannofossil and dinoflagellate stratigraphy of the Cretaceous–Tertiary boundary, Alabama and Georgia. *Micropaleontology*, **39**, 167–191.

MURRAY, J. W. 1991. *Ecology and Palaeoecology of Benthic Foraminifera*. Elsevier, Amsterdam.

NAKKADY, S. E. 1950. A new foraminiferal fauna from the Esna shales and upper Cretaceous Chalk of Egypt. *Journal of Paleontology*, **24**, 675–692.

NEDERBRAGT, A. J. 1991. Late Cretaceous biostratigraphy and development of Heterohelicidae (planktic foraminifera). *Micropaleontology*, **37**, 329–372.

—— 1992. Paleoecology of late Maastrichtian Heterohelicidae (planktic foraminifera) from the Atlantic region. *In:* MALMGREN, B. A. & BENGTSON, P. (eds) *Biogeographic Patterns in the Cretaceous Ocean*. Palaeogeography, Palaeoclimatology, Palaeoecology, **92**, 361–374.

NOMURA, R. 1991. Paleoceanography of upper Maestrichtian to Eocene benthic foraminiferal assemblages at sites 752, 753, and 754, eastern Indian Ocean. *In:* WEISSEL, J., PEIRCE, J. *et al.* (eds) *Proceedings of the Ocean Drilling Program, Scientific Results*. **121**, ODP, College Station, TX, 3–29.

NYONG, E. E. & OLSSON, R. K. 1984. A paleoslope model of Campanian to lower Maestrichtian foraminifera in the North American Basin and adjacent continental margin. *Marine Micropaleontology*, **80**, 437–477.

OLSSON, R. K. 1977. Mesozoic foraminifera; western Atlantic. *In:* SWAIN, F. M. (ed.) *Stratigraphic Micropaleontology of Atlantic Basin and Borderlands*. Elsevier, Amsterdam, 205–230.

—— & LIU, C. 1993. Controversies on the placement of Cretaceous–Paleogene boundary and the K/P mass extinction of planktonic foraminifera. *Palaios*, **8**, 127–139.

—— & NYONG, E. E. 1984. A paleoslope model for Campanian–lower Maestrichtian foraminifera of New Jersey and Delaware. *Journal of Foraminiferal Research*, **14**, 50–68.

—— & WISE, S. W. JR. 1987. Upper Paleocene to middle Eocene depositional sequences and hiatuses in the New Jersey Atlantic Margin. *In:* ROSS, C. A. & HAMAN, D. (eds) *Timing and Depositional History of Eustatic Sequences; Constraints of Seismic Stratigraphy*. Cushman Foundation for Foraminiferal Research, Special Publication, **24**, 99–112.

PARRISH, J. T. & CURTIS, R. L. 1982. Atmospheric circulation, upwelling, and organic-rich rocks in the Mesozoic and Cenozoic eras. *Palaeogeography, Palaeoclimatology, Palaeoecology*, **40**, 31–66.

PERCH-NIELSEN, K. 1981. Nouvelles observations sur les nannofossiles calcaires à la limite Crétacé–Tertiaire, près de El Kef, Tunisie. *Cahiers de Micropaléontologie*, **3** 25–36.

——, MCKENZIE, J. A. & HE, Q. 1982. Biostratigraphy and isotope stratigraphy and the 'catastrophic' extinction of calcareous nannoplankton at the Cretaceous/Tertiary boundary. *In:* SILVER, L. T. & SCHULTZ, P. H. (eds) *Geological Implications of Impacts of Large Asteroids and Comets on the Earth*. Geological Society of America Special Paper, **190**, 353–371.

PETTERS, S. W. 1982. Central West African Cretaceous–Tertiary benthic foraminifera and stratigraphy. *Palaeontographica, Abteilung A: Palaeozoologie, Stratigraphie*, **179**.

PEYPOUQUET, J. P., GROUSSET, F. & MOURGUIART, P. 1986. Paleoceanography of the Mesogean Sea based on ostracods of the northern Tunisian continental shelf between the Late Cretaceous and Early Paleogene. *Geologische Rundschau*, **75**, 159–174.

PFLUM, C. E. & FRERICHS, W. E. 1976. *Gulf of Mexico deep-water foraminifers*. Cushman Foundation for Foraminiferal Research, Special Publication, **14**.

PLUMMER, H. J. 1926. Foraminifera of the Midway formation in Texas. *Texas University, Bulletin*, **2644**.

POSPICHAL, J. J. 1994. Cretaceous nannofossils at the K–T boundary, El Kef; no evidence for stepwise, gradual, or sequential extinctions. *Geology*, **22**, 99–102.

RATHBURN, A. E. & CORLISS, B. H. 1994. The ecology of the living (stained) deep-sea benthic foraminifera from the Sulu Sea. *Paleoceanography*, **9**, 87–150.

RAUP, D. M. & JABLONSKI, D. 1993. Geography of end-Cretaceous marine bivalve extinctions. *Science*, **260**, 971–973.

REISS, Z. 1988. Assemblages from a Senonian high-productivity sea *Revue de Paléobiologie, Special Issue*, **2**, 323–332.

REUSS, A. E. 1851. Die Foraminiferen und Entomostraceen des Kreidemergels von Lemberg. *Naturwissenschaftliche Abhandlungen, Wien*, **4**, 17–52.

RHODES, M. C. & THAYER, C. W. 1991. Mass extinctions: ecological selectivity and primary production. *Geology*, **19**, 877–880.

ROHLING, E. J., ZACHARIASSE, W. J. & BRINKHUIS, H. 1991. A terrestrial scenario for the Cretaceous–Tertiary boundary collapse of the marine pelagic ecosystem. *Terra Nova*, **3**, 41–48.

ROMEIN, A. J. T. 1979a. Cretaceous nannofossils from the Cretaceous/Tertiary boundary interval in the Nahal Avdat Section, Negev, Israel. *In:* CHRISTENSEN, W. K. & BIRKELUND, T. (eds) *Symposium on Cretaceous–Tertiary Boundary Events, Proceedings*, 2. University of Copenhagen, Copenhagen, Denmark, 202–206.

—— 1979b. Lineages in early Paleogene calcareous nannoplankton. *Utrecht Micropaleontological Bulletins*, **22**.

—— & SMIT, 1981 The Cretaceous/Tertiary boundary: calcareous nannofossils and stable isotopes. *Proceedings of the Koninklijke Nederlandse Akademie van Wetenschappen, Series B*, **84**, 295–314.

ROSENFELD, A., FLEXER, A., HONIGSTEIN, A., ALMOGI-LABIN, A. & DVORACHEK, M. 1989. First report on a Cretaceous/Tertiary boundary section at Mikhtesh Gadol, southern Israel. *Neues Jahrbuch für Geologie und Paläontologie, Monatshefte*, **1989**, 474–486.

SAID, R. 1978. *Étude Stratigraphique et Micropaléontologique du Passage Crétacé–Tertiaire du Synclinal d'Elles (Region Siliana–Sers), Tunisie Centrale*, Doctorat 3e Cycle Thesis, Université de Paris.

—— 1990a. Cenozoic. *In:* SAID, R. (ed.) *The Geology of Egypt*. Balkema, Rotterdam, 451–486.

—— 1990b. Cretaceous paleogeographic maps. *In:* SAID, R. (ed.) *The Geology of Egypt*. Balkema, Rotterdam, 439–449.

—— & KENAWY, A. 1956. Upper Cretaceous and lower Tertiary foraminifera from northern Sinai, Egypt. *Micropaleontology*, **2**, 105–173.

SAINT-MARC, P. 1993. Biogeographic and bathymetric distribution of benthic foraminifera in Paleocene El Haria Formation of Tunisia. *Journal of African Earth Sciences*, **15**, 473–487.

—— & BERGGREN, W. A. 1988. A quantitative analysis of Paleocene benthic foraminiferal assemblages in central Tunisia. *Journal of Foraminiferal Research*, **18**, 97–113.

SALAJ, J. 1978. Contribution à la microbiostratigraphie des hypostratotypes tunisiens du Crétacé supérieur du Danien et du Paléocène. *Annales des Mines et de Géologie*, **28**, 119–145.

—— 1980. *Microbiostratigraphie du Crétacé et du Paléogène de la Tunisie Septentrionale et Orientale (Hypostratotypes Tunisiens)*. Institut Géologique de Dionyz Stúr, Bratislava, Slovakia.

POZARYSKA, K. & SZCZECHURA, J. 1976. Foraminiferida, zonation and subzonation of the Paleocene of Tunisia. *Acta Palaeontologica Polonica*, **21**, 127–190.

SANDIDGE, J. R. 1932. Foraminifera from the Ripley formation of western Alabama. *Journal of Paleontology*, **6**, 265–287.

SARKAR, A., BHATTACHARYA, S. K., SHUKLA, P. N., BHANDARI, N. & NAIDIN, D. P. 1992. High-resolution profile of stable isotopes and iridium across a K/T boundary section from Koshak Hill, Mangyshlak, Kazakhstan. *Terra Nova*, **4**, 585–591.

SCHMITZ, B. 1988. Origin of microlayering in world-wide distributed Ir-rich marine Cretaceous/Tertiary boundary clays. *Geology*, **16**, 1068–1072.

——, KELLER, G. & STENVALL, O. 1992. Stable isotope and foraminiferal changes across the Cretaceous–Tertiary boundary at Stevns Klint, Denmark: arguments for long-term oceanic stability before and after bolide-impact event. *Palaeogeography, Palaeoclimatology, Palaeoecology*, **96**, 233–260.

SCHWAGER, C. 1883. Die Foraminiferen aus dem Eocänablagerungen der libyschen Wüste und Ägyptens. *Palaeontographica*, **30**, 81–153.

SEN GUPTA, B. K. & MACHAIN-CASTILLO, M. L. 1993. Benthic foraminifera in oxygen-poor habitats. *In:* LANGER, M. R. (ed.) *Foraminiferal Microhabitats*. Marine Micropaleontology, **20**, 183–201.

SHAHIN, A. 1992. Contribution to foraminiferal biostratigraphy and paleobathymetry of the Late Cretaceous and early Tertiary in the western central Sinai, Egypt. *Revue de Micropaléontologie*, **35**, 157–175.

SHARPTON, V. L. & WARD, P. D. 1990. *Global Catastrophes in Earth History. An Interdisciplinary Conference on Impacts, Volcanism, and Mass Mortality*. Geological Society of America Special Paper, **247**.

SHEEHAN, P. M. & FASTOVSKY, D. E. 1992. Major extinctions of land-dwelling vertebrates at the Cretaceous–Tertiary boundary, eastern Montana. *Geology*, **20**, 556–560.

SJOERDSMA, P. G. & VAN DER ZWAAN, G. J. 1992. Simulating the effect of changing organic flux and oxygen content on the distribution of benthic foraminifers. *In:* VAN DER ZWAAN, G. J., JORISSEN, F. J. & ZACHARIASSE, W. J. (eds) *Approaches to Paleoproductivity Reconstructions*. Marine Micropaleontology, **19**, 163–180.

SLITER, W. V. 1968. *Upper Cretaceous Foraminifera from southern California and northwestern Baja California, Mexico*. Kansas University Paleontological Contributions, **49**.

—— 1977a. Cretaceous benthic foraminifers from the western South Atlantic Leg 39, Deep Sea Drilling Project. *In:* SUPKO, P. R., PERCH-NIELSEN, K. *et al.* (eds) *Initial Reports of the Deep Sea Drilling Project*. **39**, U.S. Government Printing Office, Washington, DC, 657–697.

—— 1977b. Cretaceous foraminifers from the southwestern Atlantic Ocean, Leg 36, Deep Sea Drilling Project. *In:* BARKER, P., DALZIEL, I. W.

D. *et al.* (eds) *Initial Reports of the Deep Sea Drilling Project*. **36**, U.S. Government Printing Office, Washington, DC, 519–573.

—— & BAKER, R. A. 1972. Cretaceous bathymetric distribution of benthic foraminifers. *Journal of Foraminiferal Research*, **2**, 167–181.

SMIT, J. 1982. Extinction and evolution of planktonic foraminifera after a major impact at the Cretaceous/Tertiary boundary. *In:* SILVER, L. T. & SCHULTZ, P. H. (eds) *Geological Implications of Impacts of Large Asteroids and Comets on the Earth*. Geological Society of America Special Paper, **190**, 329–352.

—— 1990. Meteorite impact, extinctions and the Cretaceous–Tertiary boundary. *Geologie en Mijnbouw*, **69**, 187–204.

SOUTHAM, J. R., PETERSON, W. H. & BRASS, G. W. 1982. Dynamics of anoxia. *In:* BARRON, E. J. (ed.) *Paleogeography and Climate*. Palaeogeography, Palaeoclimatology, Palaeoecology, **40**, 183–198.

SPEIJER, R. P. 1992. Benthic foraminiferal response to environmental changes following the K/T boundary. *In:* SARNTHEIN, M., THIEDE, J. & ZAHN, R. (eds) *Fourth International Conference on Paleoceanography, ICP IV; Short- and Long-Term Global Change: Records and Modelling. 21–25 September 1992, Kiel/Germany, Abstracts and Programs*. GEOMAR Report 15, GEOMAR, Kiel, Germany, 267–268.

—— 1994. *Extinction and Recovery Patterns in Benthic Foraminiferal Paleocommunities across the Cretaceous/Paleogene and Paleocene/Eocene Boundaries*. Geologica Ultraiectina, 124.

—— 1995. The late Paleocene benthic foraminiferal extinction event as observed in the Middle East. *In:* LAGA, P. (ed.) *Paleocene–Eocene Boundary Events*. Bulletin de la Société belge de Géologie, **103** for 1994, 267–280.

—— & VAN DER ZWAAN, G. J. 1994. The differential effect of the Paleocene/Eocene boundary event: extinction and survivorship in shallow to deep water Egyptian benthic foraminiferal assemblages. *In:* SPEIJER, R. P. (ed.) *Extinction and Recovery Patterns in Benthic Foraminiferal Paleocommunities across the Cretaceous/Paleogene and Paleocene/Eocene Boundaries*. Geologica Ultraiectina, **124**, 121–168.

——, —— & SCHMITZ, B. 1996. The impact of Paleocene/Eocene boundary events on middle neritic benthic foraminiferal assemblages from Egypt. *Marine Micropaleontology*, in press.

STOTT, L. D. & KENNETT, J. P. 1989. New constraints on early Tertiary palaeoproductivity from carbon isotopes in foraminifera. *Nature*, **342**, 526–529.

SUTHERLAND, F. L. 1994. Volcanism around K/T boundary time – its role in an impact scenario for the K/T extinction events. *Earth Science Reviews*, **36**, 1–26.

THOMAS, E. 1990a. Late Cretaceous–early Eocene mass extinctions in the deep sea. *In:* SHARPTON, V. L. & WARD, P. D. (eds) *Global Catastrophes in Earth History. An Interdisciplinary Conference on Impacts, Volcanism, and Mass Mortality*. GSA Special Paper, **247**, 481–496.

—— 1990*b*. Late Cretaceous through Neogene deep-sea benthic foraminifers (Maud Rise, Weddell Sea, Antarctica). *In:* BARKER, P. F., KENNETT, J. P. *et al.* (eds) *Proceedings of the Ocean Drilling Program, Scientific Results.* 113, ODP, College Station, TX, 571–594.

—— 1992. Cenozoic deep-sea circulation: evidence from deep-sea benthic foraminifera. *AGU Antarctic Research Series*, 56, 141–165.

TJALSMA, R. C. & LOHMANN, G. P. 1983. *Paleocene–Eocene bathyal and abyssal benthic foraminifera from the Atlantic Ocean.* Micropaleontology, Special Publication, 4.

TOULMIN, L. D. JR. 1941. Eocene smaller Foraminifera from the Salt Mountain limestone of Alabama. *Journal of Paleontology*, 15, 567–611.

VAIL, P. R., MITCHUM, R. M. & THOMPSON, S. III 1977. Global cycles of relative changes of sea level. *In:* PAYTON, C. E. (ed.) *Seismic Stratigraphy – Applications to Hydrocarbon Exploration.* American Association of Petroleum Geologists, Memoir, 26, 83–97.

VAN DER ZWAAN, G. J. 1982. *Paleoecology of late Miocene Mediterranean foraminifera.* Utrecht Micropaleontological Bulletins, 25.

——, JORISSEN, F. J. & DE STIGTER, H. C. 1990. The depth dependency of planktonic/benthic foraminiferal ratios: Constraints and applications. *Marine Geology*, 95, 1–16.

——, —— & ZACHARIASSE, W. J. 1992. Approaches to Paleoproductivity Reconstructions. *Marine Micropaleontology*, 19. Elsevier, Amsterdam.

VAN MORKHOVEN, F. P. C. M., BERGGREN, W. A. & EDWARDS, A. S. 1986. *Cenozoic cosmopolitan deep-water benthic foraminifera.* Bulletin des Centres de Recherches Exploration–Production Elf Aquitaine, Memoir, 11.

VASILENKO, V. P. 1950. Paleocene foraminifera of the central part of the Dnieper-Donets basin. *In:* *Microfauna of the U.S.S.R.* 4, Trudy, VNIGRI, Kavkaz, 177–224 [in Russian].

VERBEEK, J. W. 1977. *Calcareous nannoplankton*

biostratigraphy of Middle and Upper Cretaceous deposits in Tunisia, southern Spain and France. Utrecht Micropaleontological Bulletins, 16.

WHITE, M. P. 1928. Some index Foraminifera of the Tampico Embayment area of Mexico (part 1, 2). *Journal of Paleontology*, 2, 177–215, 280–316.

WICHER, C. A. 1956. Die Gosau-Schichten in Becken von Gams (Österreich) und die Foraminiferengliederung der hoheren Oberkreide in der Tethys. *Paläontologisches Zeitschrift, Sonderhefte*, 30, 87–136.

WIDMARK, J. G. V. 1995. Multiple deep-water sources and trophic regimes in the latest Cretaceous deep sea: evidence from benthic foraminifera. *Marine Micropaleontology*, 26, 361–384.

—— & MALMGREN, B. A. 1992*a*. Benthic foraminiferal changes across the Cretaceous/Tertiary boundary in the deep sea; DSDP sites 525, 527, and 465. *Journal of Foraminiferal Research*, 22, 81–113.

—— 1992*b*. Biogeography of terminal Cretaceous deep-sea benthic foraminifera in the Atlantic and Pacific oceans. *In:* MALMGREN, B. A. & BENGTSON, P. (eds) *Biogeographic Patterns in the Cretaceous Ocean.* Palaeogeography, Palaeoclimatology, Palaeoecology, 92, 375–405.

WIEDMANN, J. 1988. The Basque coastal sections of the K/T boundary; a key to understanding mass extinction in the fossil record. *In:* LAMOLDA, M. A., KAUFFMAN, E. G. & WALLISER, O. H. (eds) *Palaeontology and Evolution: Extinction Events.* Revista Espanola de Paleontologia, Numero extraordinario, 127–140.

WONDERS, A. A. H. 1980. *Middle and Late Cretaceous planktonic foraminifera of the western Mediterranean area.* Utrecht Micropaleontological Bulletins, 24.

ZACHOS, J. C., ARTHUR, M. A. & DEAN, W. E. 1989. Geochemical evidence for suppression of pelagic marine productivity at the Cretaceous/Tertiary boundary. *Nature*, 337, 61–64.

Recovery of the naticid gastropod predator–prey system from the Cretaceous–Tertiary and Eocene–Oligocene extinctions

PATRICIA H. KELLEY[1] & THOR A. HANSEN[2]

[1]*Department of Geology and Geological Engineering, University of North Dakota, Grand Forks ND 58202-8358, USA*

[2]*Department of Geology, Western Washington University, Bellingham, WA 98225, USA*

Abstract: Naticid gastropods have been important shell-drilling predators of molluscs since the Cretaceous. Preliminary compilations of drilling frequencies by Vermeij (1987) were used to support his hypothesis of escalation, involving temporal increase in the hazard of predation and in adaptation to this hazard.

A comprehensive survey of naticid predation on molluscs in the Cretaceous and Palaeogene of the North American Coastal Plain revealed a complex pattern of escalation influenced by mass extinctions and recoveries. Data from more than 46 000 specimens from 17 formations show that dynamics of the naticid predator–prey system were affected significantly by the Cretaceous–Tertiary and Eocene–Oligocene mass extinctions. In the Cretaceous, drilling was low for gastropod prey (4–6%) and moderate for bivalves (13–19%). Drilling frequencies significantly increased above the K/T boundary in an initial recovery phase that lasted no more than 1–3 Ma. Recovery from the Eocene–Oligocene extinction also involved a significant increase in drilling, especially on bivalves (from 8% to 24%).

Based on these data (and similar patterns reported for fossil ostracode prey), we propose a model of cyclic escalation of the naticid predator–prey system, in which mass extinctions perturb the system and initiate cycles. Drilling frequencies in recovery faunas are greatly increased, possibly owing to preferential extinction of highly escalated prey. As the recovery proceeds, escalation of prey defences causes drilling frequencies to stabilize or decline. Reorganization of the predator–prey system also involves reduction in predator behavioural stereotypy after mass extinctions, followed by gradual recovery.

Mass extinctions, although generally attributed to changes in the physical environment, exact their toll on a fauna by disturbing biological systems. Well developed ecological systems may be disrupted, for instance, by elimination of key members of the food web. During recovery from mass extinctions, ecosystems are reconfigured as new taxa assume niches occupied by predecessors eliminated from the fauna.

One such system that apparently was affected by mass extinctions was the naticid gastropod predator–prey system. Naticids are shell-drilling gastropods that have been important predators of molluscs since the Cretaceous (Sohl 1969). Juvenile naticids also attack ostracodes (Reyment 1963, 1966, 1967) and foraminifera (Reyment 1966, 1967; Livan 1937, although see objections by Sohl 1969 to Livan's conclusions). Detailed descriptions of naticid predatory behaviour are provided by Ziegelmeier (1954), Reyment (1967), Carriker (1981) and Kabat (1990). Naticids normally are infaunal predators; however, in the laboratory *Naticarius intricatoides* has been observed to attack, without burrowing, specimens of the infaunal bivalve *Chamalea*

gallina that had been placed on the sediment surface (Guerrero & Reyment 1988). Savazzi & Reyment (1989) also reported subaerial foraging, at low tide, by *Natica gualteriana*.

Naticid predator–prey interactions can be reconstructed readily from the fossil record because drilled specimens provide not only information about the prey but also about the predator and the predation event (for instance, whether or not the attack was successful, predator size and selectivity of the predator with respect to borehole site on the shell of the prey). Holes drilled by naticids can be distinguished from those drilled by epifaunal muricid gastropods based on drillhole morphology and size (Reyment 1963; Carriker 1981, among others).

Vermeij (1987) suggested that the history of drilling gastropods, including the naticid gastropod predator–prey system, has been characterized by 'escalation'. According to Vermeij (1994, p. 220), the hypothesis of escalation 'states that enemies – predators, competitors, and dangerous prey – are the most important agents of selection among individual organisms, and that enemy-related adaptation brought about long-

From Hart, M. B. (ed.), 1996, *Biotic Recovery from Mass Extinction Events*, Geological Society Special Publication No. 102, pp. 373–386

term evolutionary trends in the morphology, ecology, and behavior of organisms over the course of the Phanerozoic'. Thus hazards such as predation are claimed to have intensified and prey to have developed antipredatory aptations during geological time.

Vermeij's (1987) hypothesis that escalation has characterized drilling gastropod predation was based on evidence, from four (for gastropods) or five (for bivalves) local assemblages, that drilling frequencies were low in the Cretaceous and achieved modern levels by sometime in the Eocene. Vermeij (1987) also reported that the occurrence of antipredatory aptations among prey increased through time. Systematic collection of a larger database was needed, however, to test the hypothesis rigorously.

Kelley & Hansen (1993) provided such a test by documenting the development of the naticid gastropod predator–prey system in the North American Coastal Plain. We performed a comprehensive survey of naticid predation on mollusc assemblages from seventeen Upper Cretaceous through Oligocene formations. Data on naticid drilling were reported for more than 40 000 bivalve and gastropod prey specimens. We studied spatial patterns in drilling for specific intervals (Hansen & Kelley 1995), as well as temporal trends in drilling patterns (Kelley & Hansen 1993).

The present paper reviews the results of the temporal study (expanding the database now to include results for more than 46 000 specimens) and analyses in greater detail predation patterns across the Cretaceous–Tertiary and Eocene–Oligocene boundaries. Kelley & Hansen (1993) found a pattern of moderate predation mortality in the Cretaceous, a decrease in drilling across the K/T boundary, a substantial increase in the early Palaeocene, a decline again in the late Eocene and a moderate Oligocene recovery. These results are consistent in part with Vermeij's (1987) suggestion that escalation may have occurred episodically: environmental deterioration and mass extinctions interrupt escalation, whereas climatic warming and increases in primary productivity foster escalation.

This paper also presents an empirically based, refined model of escalation of the naticid system with more specific expectations than presented by Vermeij (1987), although aspects of the model were anticipated by Vermeij. Our model recognizes an overall increase in prey effectiveness through time with a superimposed pattern of 'escalation cycles.' Such cycles are initiated by mass extinctions, after which drilling frequencies rise abruptly during the recovery of the fauna (possibly owing to extinction of escalated prey)

and then stabilize or decrease until the next extinction. Future work is designed to test this model; this paper presents results of a preliminary analysis of the level of escalation before and after extinction events. We also review data on ostracode predation from Maddocks (1988), which support our model.

Methodology used in the drilling survey

Kelley & Hansen (1993) provided a full description of the materials studied and the methods used in surveying predation patterns in the North American Coastal Plain Cretaceous and Palaeogene. That information is reviewed briefly here.

Molluscan assemblages were examined from 14 stratigraphic levels (17 formations) of Late Cretaceous through early Oligocene age from Texas to Virginia. The samples represent normal marine, inner to middle shelf, clastic sedimentary environments, except for those from the outer-shelf Yazoo Formation. (The Yazoo was examined in order to obtain data on drilling from a broader range of palaeoenvironments. However, the Yazoo molluscan assemblage differs markedly from the roughly correlative Moodys Branch assemblages, particularly in the proportion of carnivorous gastropods, and drilling frequencies were significantly greater in the Yazoo than in the Moodys Branch (see Hansen & Kelley 1995). For completeness, Yazoo drilling frequencies are reported in this paper, but should not be compared directly with those from the inner to middle shelf environments.) With the exception of the Cretaceous Ripley and Providence collections, all assemblages studied were from bulk samples. The samples were wet sieved and all shells were picked, identified and counted.

All complete or nearly complete specimens were surveyed for naticid and muricid drillholes. Except in very thin-shelled prey, naticid drillholes can be recognized by their parabolic shape in cross section (Carriker 1981). Holes made by naticids also tend to be larger than those drilled by muricids (Reyment 1963; Kowalewski 1993). Muricid drillholes were less frequent in our samples, probably because muricid predation is restricted to epifauna and our samples were dominated by infaunal specimens (about 90%). Therefore we have focused our analyses on the naticid predator–prey system. By concentrating on naticid predation, we also avoid confounding patterns for naticids with potentially divergent patterns for muricids.

Drilling frequency for gastropod species was calculated as the frequency of individuals with

Results of drilling survey

Drilling frequencies, mass extinctions, and recoveries

one or more complete naticid drillholes. For bivalves, drilling frequency was calculated by doubling the frequency of valves with one or more complete naticid drillholes (since drilling of only one valve was sufficient to kill a bivalved individual). Drillhole position on the prey shell was recorded with respect to a nine-sector grid for bivalves (Kelley 1988, fig. 1); for gastropods, drillhole position was recorded as near the apex, near the middle or near the aperture. (The magnitude of the survey precluded more detailed measurements of drillhole position on gastropods.)

Data were tabulated on the occurrence of incomplete drillholes, which fail to penetrate to the interior of the shell and therefore represent failed predation attempts. We also determined the occurrence of multiply drilled specimens. Several explanations have been offered for multiple drillholes. (1) Laboratory experiments by Kitchell *et al.* (1986) demonstrated that multiple drillholes can result from successful prey escape tactics. Arnold *et al.* (1985) also inferred predator-avoidance behaviour in the foraminiferid *Siphouvigerina* with multiple, presumably naticid, boreholes. They suggested that retreat of the protozoan into earlier chambers caused the predator to drill multiple exploratory holes. (2) Reyment *et al.* (1987), citing Fischer-Piette (1935) and Carriker (1981), claimed that muricids may drill empty shells. However, Carriker (pers. comm.) stated that empty shells in the laboratory are bored only if the predator is 'teased' by placing the flesh of a prey animal inside an empty shell. (3) Reyment (pers. comm.) stated that 'there are numerous cases of more than one predator mounting a bivalve', although Carriker (pers. comm.) knows of no cases of multiple naticids attacking the same prey individual simultaneously. According to Carriker (1981, pers. comm.), the naticid predator extends its proboscis through a fleshy tube formed by the folding of the predator's propodium around the prey; this mode of predation renders unlikely simultaneous attacks by multiple predators on one prey individual. (4) Gastropod prey with inflated whorls may show two holes in adjacent whorls, caused by one attack, because rounding of the shell interferes with rasping by the radula (Reyment 1966, 1967). Boring of highly curved shell surfaces may thus account for some multiple drilling (Reyment *et al.* 1987). We believe that prey escape tactics are the most likely explanation for multiple drilling in our samples, and that multiple holes thus indicate predator inefficiency. (Explanation 2, which we consider less likely, also represents inefficient predator behaviour.)

Drilling frequencies, especially for bivalves, exhibited a cyclical pattern apparently related to mass extinctions and recoveries from such events (Table 1, Fig. 1). Drilling frequencies in the Upper Cretaceous Ripley and Providence Formations were relatively high for bivalves (0.13 and 0.19, respectively). The Ripley and Providence samples, from the US Geological Survey, Reston, Virginia, were collected over many years by Norman Sohl and others. Because these were not bulk samples, relative species abundances were not maintained and overall drilling frequencies may not be completely comparable to those of other samples. However, drilling frequencies on individual species should be accurate. In that case, the percentage of abundant species with drilling frequencies greater than 0.10 (the measure used by Vermeij 1987) may be a better indicator of the level of drilling in the fauna. The Ripley and Providence samples included 13 and 6 'abundant' bivalve species respectively (those with more than 20 valves); 62% and 83% of those abundant species had drilling frequencies greater than 0.10. Thus the level of drilling of bivalves appears to be relatively high in those pre-extinction assemblages.

Preservational problems complicate the estimation of drilling frequencies crossing the Cretaceous–Tertiary boundary. Preservation is relatively poor in the Kincaid and Corsicana Formations, which occur immediately above and below the boundary in the Texas Coastal Plain. Because wet sieving was not possible, specimens instead were cleaned with needles and left exposed on matrix pedestals. Only bivalve exteriors could be examined for drillholes, because incomplete holes could not be detected by examining interiors of valves. No data were collected for gastropods because of the preservational problems.

Very low drilling frequencies on bivalves were observed in the samples immediately below and above the K/T boundary (0.055 in the Corsicana and 0.034 in the Kincaid). Only two Corsicana species had more than 20 valves present; neither was drilled at a frequency exceeding 0.10. Of the 13 'abundant' Kincaid species, only one (7.7%) had a drilling frequency greater than 0.10. Despite the preservational problems, the two lines of evidence suggest significantly lower levels of drilling crossing the K/T boundary. (Providence and Corsicana drilling frequencies

Table 1. *Drilling on bivalves through time*

Formation	Age	# Spec.	Dr. freq.	# Ab. sp.	% Dr.
Ripley	L. Cret.	2629	0.132	13	61.6
Providence	L. Cret.	639	0.194	6	83.3
Corsicana	L. Cret.	291	0.055	2	0
Kincaid	E. Palaeo.	1054	0.034	13	7.7
Brightseat	E. Palaeo.	1743	0.327	11	81.8
Mt. Landing	M. Palaeo.	235	0.111	3	33.3
Bells Landing	L. Palaeo.	62	0	0	–
Bashi	E. Eo.	1124	0.415	5	100.0
Cook Mountain	L. m. Eo.	5035	0.287	31	74.2
Gosport	L. m. Eo.	468	0.073	5	20.0
Moodys Branch	L. Eo.	15950	0.087	30	36.7
Yazoo*	L. Eo.	928	0.186	4	75.0
Red Bluff	E. Oligo.	2197	0.235	8	50.0
Mint Spring	E. Oligo.	1596	0.204	13	76.9
Byram	E. Oligo.	662	0.139	3	100.0

Abbreviations: # Spec., number of specimens; Dr. freq., drilling frequency; # Ab. sp., number of abundant species (those with 20 or more valves); % Dr., percent of abundant species with drilling frequencies > 0.10; Mt. Landing, Matthews Landing.
*Data for Yazoo Formation included for completeness, but drilling frequencies are not comparable owing to environmental differences.

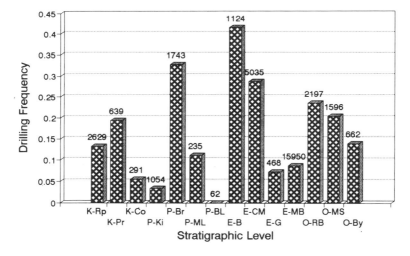

Fig. 1. Drilling frequencies through time for bivalves. Drilling frequency calculated as $2D/N$ where D = number of valves with one or more complete naticid drillholes and N = total number of valves. Number above each bar is total number of complete specimens examined for that stratigraphic level. Abbreviations for stratigraphic levels as follows. Cretaceous levels (designated K): Rp, Ripley Formation of Georgia; Pr, Providence Formation of Georgia; Co, Corsicana Formation of Texas. Palaeocene levels (designated P): Ki, Kincaid Formation of Texas; Br, Brightseat Formation of Maryland; ML, Matthews Landing Member of the Naheola Formation of Alabama; BL, Bells Landing Member of the Tuscahoma Formation of Alabama. Eocene levels (designated E): B, Bashi Member of the Hatchetigbee Formation of Alabama; CM, Cook Mountain level (Cook Mountain Formation of Texas and Louisiana and the correlative Piney Point Formation of Virginia and Upper Lisbon Formation of Alabama); G, Gosport Formation of Alabama; MB, Moodys Branch Formation of Louisiana and Mississippi. Oligocene levels (designated O): RB, Red Bluff Formation of Mississippi; MS, Mint Spring Formation of Mississippi; By, Byram Formation of Mississippi.

Table 2. *Drilling on gastropods through time*

Formation	Age	# Spec.	Dr. freq.	# Ab. sp.	% Dr.
Ripley	L. Cret.	2426	0.037	23	8.7
Providence	L. Cret.	516	0.060	5	20.0
Brightseat	E. Palaeo.	414	0.377	6	50.0
Mt. Landing	M. Palaeo.	297	0.465	7	100.0
Bells Landing	L. Palaeo.	742	0.352	10	60.0
Bashi	E. Eo.	271	0.203	6	66.7
Cook Mountain	L. m. Eo.	2761	0.162	22	45.5
Gosport	L. m. Eo.	200	0.075	3	33.3
Moodys Branch	L. Eo.	1291	0.098	16	37.5
Yazoo*	L. Eo.	1513	0.220	22	63.6
Red Bluff	E. Oligo.	329	0.119	5	20.0
Mint Spring	E. Oligo.	737	0.145	15	66.7
Byram	E. Oligo.	178	0.174	6	66.7

Abbreviations as for Table 1. Abundant species are those with 10 or more individuals.

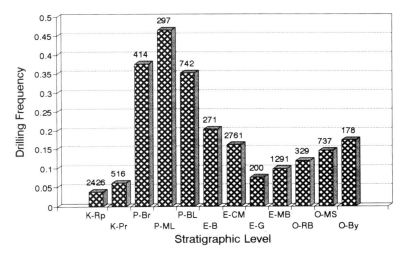

Fig. 2. Drilling frequencies through time for gastropods. Drilling frequency calculated as the ratio of number of individuals with one or more complete naticid drillholes to total number of individuals. Number above each bar is total number of complete specimens examined for that stratigraphic level. Abbreviations as in Fig. 1.

are significantly different: $\chi^2 = 13.89$, $p < 0.001$.)

The Brightseat Formation, dating from about 1–3 Ma after the end-Cretaceous extinction (Hazel et al. 1984), exhibited a very sharp increase in naticid drilling. Drilling frequency on bivalves rose to 0.33, and 82% of the 11 abundant species had drilling frequencies greater than 0.10. The increase in drilling frequency between the Kincaid and Brightseat is highly significant ($\chi^2 = 145.80$, $p \ll 0.001$). Drilling frequencies remained high (for good samples) into the middle Eocene (0.42 and 0.29 for Bashi and Cook Mountain samples respectively; 100% and 74% of the abundant species in those two units had drilling frequencies greater than 0.10).

Drilling then declined significantly ($\chi^2 = 42.12$, $p \ll 0.001$) through the Gosport and Moodys Branch Formations, to 7–9%. Only 20% and 36% of the abundant species in those two formations, respectively, had drilling frequencies greater than 0.10. (Drilling in the deep-water equivalent of the Moodys Branch, the Yazoo Formation, was significantly greater, but differences in the physical environment prohibit direct comparison of drilling in the two formations; see Hansen & Kelley 1995.)

Following the Eocene–Oligocene extinction, drilling frequencies on bivalves rose significantly ($\chi^2 = 206.44$, $p \ll 0.001$). Samples taken from just above the E/O boundary in the Red Bluff

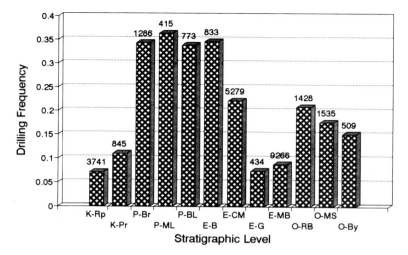

Fig. 3. Drilling frequencies through time for the entire fauna (bivalves and gastropods). Drilling frequency calculated as the ratio of number of individuals with one or more complete naticid drillholes to total number of individuals. Number above each bar is total number of individuals examined for that stratigraphic level. Abbreviations as in Fig. 1.

Formation of Mississippi showed drilling frequencies on bivalves of 0.235. Drilling frequencies then declined somewhat through the early Oligocene (0.204 in the Mint Springs and 0.139 in the Byram Formation).

Patterns of drilling on gastropods partially reinforce those observed for bivalves (Table 2, Fig. 2). Unfortunately, drilling frequencies could not be calculated for the formations immediately above and below the K/T boundary because of poor preservation of the gastropod fauna. Nevertheless, drilling frequencies exhibit the same pattern of dramatic increase above the K/T boundary. Drilling frequencies in the Cretaceous Ripley and Providence Formations were 0.037 and 0.060, with only 9% and 20% of the abundant species (those with more than 10 individuals) drilled greater than 0.10. In contrast, the Palaeocene Brightseat, Matthews Landing and Bells Landing had drilling frequencies of 0.377, 0.465 and 0.352 respectively. The increase in drilling frequencies is statistically significant ($\chi^2 = 143.44$, $p \ll 0.001$, Providence v. Brightseat). Drilling declined through the Eocene (with the exception of the deeper-water Yazoo Formation, for which results are not directly comparable) and then rose again in the early Oligocene. The increase above the E/O boundary was not as marked as for bivalves, however.

In general, then, patterns in drilling were characterized by the following features: (1) significant increases in drilling levels in recovery faunas just above mass extinction boundaries; (2) stable or declining levels of predation as the recovery from extinction continued (Fig. 3).

Stereotypy of naticid drilling behaviour

If escalation occurred in the naticid predator-prey system, the dynamics of the system may have changed through time. Drilling by extant naticids is highly stereotyped with respect to drillhole site and selectivity of prey size and species (Kitchell et al. 1981; Kitchell 1986). (Reyment 1966 reported lack of size selectivity for drilling predation in some samples from the modern Niger Delta; however, his analysis of size selectivity appears to have combined all prey species, a procedure that may have masked size selectivity within individual prey species.) Selectivity of prey size, species and drillhole site ensures maximum benefit per drilling effort by the predator. Based on the hypothesis of escalation, Kelley & Hansen (1993) suggested that such selectivity of drillhole site and prey item may have increased through time.

Kelley & Hansen (1993) reported results of drillhole site selectivity analyses for all Cretaceous-through-Oligocene bivalve species with sufficient drillholes for statistical analysis. No general pattern of increase in selectivity of drillhole site was found; evidence of selectivity occurred at all stratigraphic levels. There is some

GASTROPOD PREDATOR–PREY RECOVERY FROM EXTINCTION

Table 3. *Positions of drillholes on gastropod shells*

Formation	Taxon	Drillhole position			χ^2	p
		Apex	Mid	Apert.		
Cretaceous						
Ripley	*Urceolabrum* sp.	0	13	1	22.4	< 0.005
	Laxispira monilifera	0	13	2	19.6	< 0.005
	Turritella trilira	0	10	1	16.5	< 0.005
	Turritella cf. *T. tippana*	0	8	1	12.7	< 0.005
	Avellana sp.	0	21	0	42.0	≪ 0.005
Prov.	*Turritella bilira*	0	21	0	42.0	≪ 0.005
Palaeocene						
Brightseat	*Haustator ?gnoma*	0	115	4	214.8	≪ 0.005
Mt. Landing	*Rissoina alabamensis*	0	19	11	18.2	< 0.005
	Turritella aldrichi	4	77	12	103.4	≪ 0.005
	Microdrillia sp.	0	11	0	22.0	< 0.005
Bells L.	*Mesalia alabamiensis*	8	208	9	353.8	≪ 0.005
	Turritella eurynome	0	10	1	16.5	< 0.005
	Turritella sp.	0	3	5	4.75	n.s.
Eocene						
Bashi	*Turritella gilberti*	2	23	3	30.1	≪ 0.005
Cook Mt.	*Mesalia claibornensis*	0	12	0	24.0	≪ 0.005
	Mesalia vetusta	0	9	4	9.4	< 0.01
	Turritella nasuta	1	9	4	6.6	< 0.05
	Polinices aratus	0	46	2	84.5	≪ 0.005
	Buccitriton texanum	110	100	2	100.8	≪ 0.005
Gosport	*Mesalia vetusta*	0	10	1	16.5	< 0.005
Moodys Br.	*Turritella perdita*	1	9	0	14.6	< 0.005
	Turritella alveata	2	36	1	61.1	≪ 0.005
	Hipponix pygmaea (all)	34	21	23	3.8	n.s.
	H. pygmaea (complete)	5	17	22	10.4	< 0.01
	H. pygmaea (incomplete)	28	3	2	39.4	≪ 0.005
Oligocene						
Red Bluff	*Turritella* aff. *T. premimetes*	16	7	2	12.1	< 0.005
	Syntomodrillia collarubra	2	5	0	5.4	n.s.
Mint Sp.	*Turritella premimetes*	0	8	0	16.0	< 0.005
	Sinum mississippiensis	3	15	1	18.1	< 0.005
	Terebra alaba	1	10	4	18	< 0.005
	Pyramidella leafensis	2	11	1	13	< 0.005

Drillhole positions designated as near apex, near middle of shell, or near aperture (apert.). Other abbreviations: Prov., Providence Formation; Mt. Landing, Matthews Landing Member of the Naheola Formation; Bells L., Bells Landing Member of the Tuscahoma Formation; Cook Mt., Cook Mountain Formation or correlatives; Moodys Br., Moodys Branch Formation; Mint Sp., Mint Spring Formation; n.s., not significant at p < 0.05

indication, though, that selectivity was strongest prior to mass extinction events and less prominent following such events. Based on chi-squared tests, all six Cretaceous bivalve species examined showed selectivity of drillhole site. In contrast, shortly above the K/T boundary in the Brightseat Formation, only 43% (three of seven species) exhibited such site selectivity. Drillhole site selectivity was more common in the Eocene (characterizing nine of 13 species, or 69%) and then declined again above the Eocene–Oligocene boundary. Only two of five cases examined for the Oligocene showed selectivity of drillhole site (40%).

In the present study, we analysed drillhole positions on gastropod prey. Holes were classified in a very general way as occurring near the apex, middle, or aperture of the shell. Chi-squared tests determined whether a significant preference occurred for drilling a particular area. Perhaps because of the broad categories used (owing to the magnitude of the study), nearly all species showed selectivity of drillhole site (Table 3). All six Cretaceous gastropod species showed a strong preference for mid-shell drillhole positions. In the Palaeocene, six of seven species (86%) were preferentially drilled in the middle area of the shell. The seventh species, *Turritella* sp. from the Bells Landing Member of the Tuscahoma Formation, showed no significant

Table 4. *Temporal occurrence of incomplete and multiple drilling*

	Incomplete drillholes		Multiple drillholes	
	Bivalves	Gastropods	Bivalves	Gastropods
Cretaceous	5%	0.8%	0	3%
Palaeocene	10%	0.5%	3%	9%
Eocene	11%	9%	6%	11%
Oligocene	17%	6%	15%	18%

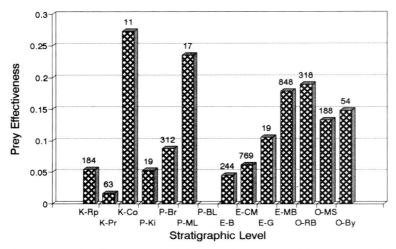

Fig. 4. Prey effectiveness (ratio of incomplete drillholes to total attempted drillholes) through time for bivalves. Number above each bar is the total number of attempted drillholes at that stratigraphic level. Abbreviations as in Fig. 1.

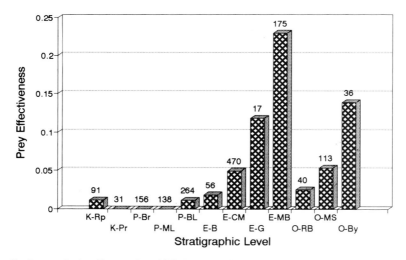

Fig. 5. Prey effectiveness (ratio of incomplete drillholes to total attempted drillholes) through time for gastropods. Number above each bar is the total number of attempted drillholes at that stratigraphic level. Abbreviations as in Fig. 1.

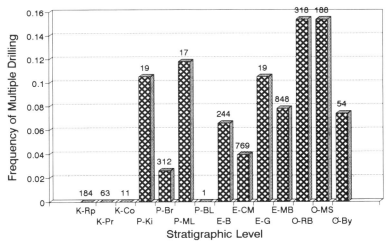

Fig. 6. Frequency of multiple drilling (ratio of number of drillholes in multiply bored valves to total attempted drillholes) through time for bivalves. Number above each bar is the total number of attempted drillholes at that stratigraphic level. Abbreviations as in Fig. 1.

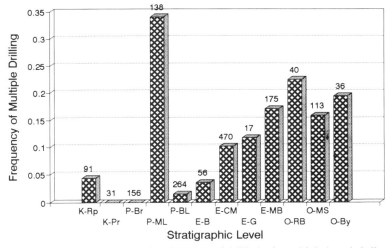

Fig. 7. Frequency of multiple drilling (ratio of number of drillholes in multiply bored shells to total attempted drillholes) through time for gastropods. Number above each bar is the total number of attempted drillholes at that stratigraphic level. Abbreviations as in Fig. 1.

preference for drillhole position (though only 8 drillholes occurred).

Ten Eocene gastropod species had sufficient drillholes for study. Nine species had significant chi-squared values; with the exception of *Buccitriton texanum* (Cook Mountain level), all nine showed a strong preference for mid-shell drilling. *Hipponix pygmaea* of the Moodys Branch showed no significant preference for drillhole position when all drillholes were considered. However, complete drillholes showed a significantly different distribution on the shell than did incomplete drillholes. Chi-squared values for complete and incomplete drillholes calculated separately were significant.

Five of six Oligocene gastropod species had significant chi-squared values; most were drilled at mid-shell. *Syntomodrillia collarubra*, of the

Red Bluff Formation, had a nonsignificant distribution of drillholes (but only seven drill-holes).

In general, no clear temporal patterns of drillhole site selectivity are evident for gastropod prey. It may be noted, however, that the cases of non-stereotyped drilling occurred for prey species from intervals above mass extinction boundaries.

Changing effectiveness of predator and prey

Vermeij (1987) proposed that prey have become more effective against drilling since the Cretaceous. Our data indicate a significant increase in relative effectiveness of prey through time (Kelley & Hansen 1993). Incomplete and multiple drillholes were used to identify unsuccessful drilling attempts. Such drillholes were nearly absent from Cretaceous samples, but became more common through time (Table 4, Figs 4–7). The only Cretaceous or Palaeocene samples with relatively high frequencies of incomplete or multiple drillholes were those with very small sample sizes (excluding the Matthews Landing, in which a single prey species, *Turritella aldrichi*, was responsible for the high frequency of multiple drilling).

Individual prey taxa, such as corbulids and turritellids, also exhibited increases in the frequency of incomplete and multiple holes through time. This increase in prey effectiveness was interpreted by Kelley & Hansen (1993) to be consistent with the hypothesis of escalation.

A refined model of escalation for the naticid predator–prey system

Features of the model

The history of the naticid gastropod predator/molluscan prey system through the Cretaceous and Palaeogene suggests certain generalizations regarding escalation in the system. We propose the following model as an empirically derived refinement of Vermeij's escalation hypothesis applicable to the naticid predator–prey system.

(1) Over the long term, there has been an overall trend of increasing prey effectiveness. However, superimposed on this trend is a more complex pattern of escalation cycles, punctuated by mass extinctions and involving recovery of the fauna from the extinctions.

(2) Mass extinctions cause significant perturbation to the system and initiate the cycles. Predator behavioural stereotypy (for instance, with respect to selectivity of drillhole site) is reduced immediately after mass extinction events, followed by eventual recovery.

(3) After mass extinctions, naticid drilling frequencies rise rapidly in the recovery fauna, perhaps because, as suggested by Vermeij (1987), highly escalated prey have been eliminated from the fauna.

(4) The rapid rise in drilling is followed by stabilization or decline in drilling frequencies, as the prey escalate defences.

(5) The next mass extinction interrupts this trend by eliminating highly escalated prey and causing drilling frequencies to rise rapidly again. This change in the ecosystem initiates a new escalation cycle.

Testing the escalation model

This model proposes that mass extinctions play a pivotal role in the escalation of the naticid predator–prey system, interrupting and resetting the process. Why should mass extinctions initiate escalation cycles? Vermeij (1987) suggested that highly escalated species with antipredatory aptations would be selectively lost at extinctions (particularly those involving cooling or a decline in primary productivity) due to the higher energy requirements for maintaining those aptations. Although causes of the Cretaceous–Tertiary and Eocene–Oligocene extinctions are still being debated, evidence suggests a loss in primary productivity at the K–T boundary (Hsü 1986; Zachos *et al.* 1989; Hansen *et al.* 1993) and cooling at the E–O boundary (Berggren & Prothero 1992; Hansen 1992). If Vermeij's suggestion is correct, the incidence of antipredatory aptations should be relatively high in faunas immediately prior to those mass extinctions. In contrast, post-extinction recovery faunas should exhibit significantly reduced antipredatory aptations, yielding high drilling frequencies.

To test this hypothesis, we are examining the incidence of antipredatory aptations before and after the Cretaceous–Tertiary and Eocene–Oligocene mass extinctions to see if differential survival of morphotypes occurred. The purpose of this analysis is to determine whether extinctions were random with respect to degree of escalation, or whether highly escalated taxa were eliminated preferentially, as suggested by Vermeij (1987).

Previous studies have suggested a number of morphological aptations that may deter drilling predation. Large prey size may pose a manipulation problem for naticid predators (Taylor 1970; Adegoke & Tevesz 1974; Stump 1975; Thomas 1976; Kitchell *et al.* 1981). Increased thickness prolongs drilling time and may effectively deter predation (Carriker 1951; Carter 1968; Kitchell *et al*. 1981; Vermeij 1978, 1987; Kelley 1989). Tight valve closure in pelecypods (for instance, involving crenulated margins) may prevent chemical cues from attracting predators (Vermeij 1978, 1987; see also Carter 1968). Ornamentation may also be a deterrent to predators of molluscs and ostracodes (Reyment 1967; Thomas 1978; Reyment *et al*. 1987; Maddocks 1988). The same or additional traits may protect molluscs from shell-breaking predation; for instance, gastropods with evolute, uncoiled or umbilicate shells are more vulnerable than are high-spired small-apertured shells to shell-breaking predation (Vermeij 1987).

As a first step in determining the relative escalation of faunas before and after extinction events, we examined the morphology of bivalve taxa reported from assemblages below and above the Cretaceous–Tertiary and Eocene–Oligocene boundaries. Epifaunal species were excluded from the analysis because they are less likely to be attacked by normally infaunal naticids. This preliminary analysis has been based on the published literature. We compiled species lists for the Nacatoch Sand and its equivalents (Navarro Group, Upper Cretaceous of Texas; Stephenson 1941), the lower Palaeocene of the Gulf Coastal Plain (Palmer & Brann 1965, 1966), the upper Eocene Moodys Branch Formation of Mississippi and Louisiana (Palmer & Brann 1965, 1966; Dockery 1977), and the lower Oligocene Red Bluff Formation of Mississippi (Dockery 1982). Following Kelley (1982), for each infaunal species reported, we assigned scores indicating level of escalation to several characters related to predation.

Four characters were examined for each species: shell size, thickness, ornamentation and nature of margin. Scores ranging from 1 to 3 were assigned to each character, in order to weight all characters equally in the analysis (Table 5). Assignments were based on figured specimens illustrated in Stephenson (1941), Dockery (1977, 1982), or the references cited in Palmer & Brann (1965, 1966). Shell thickness could not be determined from published illustrations, but was coded based on the species descriptions or our own familiarity with the faunas. Species size was estimated from the maximum dimension of the largest figured specimen shown in the above monographs. In addition, a total score was determined for each species by summing the values for the four characters. Total scores for a species potentially ranged from 4 (minimally escalated) to 12 (very highly escalated). Scores for all species in a fauna were then averaged for each character

(including the 'total score' character) in order to indicate the relative level of escalation of the fauna.

Table 5. *Scores assigned to antipredatory characters*

Character	Score	Explanation
Shell size	1	< 10 mm
	1.5	10–20 mm
	2	20–30 mm
	2.5	30–40 mm
	3	> 40 mm
Thickness	1	very thin
	1.5	moderately thin
	2	average thickness
	2.5	moderately thick
	3	very thick
Ornamentation	1	smooth
	1.5	fine striations
	2	ribbed
	2.5	coarse ribs
	3	very coarse ribs
Margin	1	gape present
	2	no gape; margin smooth
	3	no gape; crenulated margin

Preliminary results based on the published faunas are ambiguous. Vermeij's (1987) suggestion was not upheld by the Cretaceous–Tertiary analysis, but is partly consistent with the Eocene–Oligocene results. Mean scores for all characters within each fauna are given in Table 6.

The Cretaceous fauna yielded lower mean scores than did the lower Palaeocene fauna for all characters; total scores were 7.32 and 8.51 for the Cretaceous and Palaeocene respectively. The differences between faunas are significant for all characters ($p < 0.001$ for size and total score; $p < 0.01$ for ornamentation and margin; $p < 0.02$ for thickness; Mann-Whitney U-test). Thus, these preliminary results suggest that Cretaceous species in general were smaller, thinner, less ornamented, and with less tight valve closure than were lower Palaeocene species.

Mean scores for the upper Eocene Moodys Branch Formation were greater than the lower Oligocene Red Bluff scores for size, thickness and total score, in support of Vermeij's hypothesis. However, only the difference in size between the two faunas is statistically significant (Mann-Whitney U-test, $p < 0.01$). The ornamentation and margin characters yielded greater mean scores in the Oligocene than the Eocene, but the difference between the two populations is not significant at the 0.05 level (Mann-Whitney U-test).

These literature-based results thus fail to

Table 6. *Mean scores of antipredatory characters for four faunas*

Fauna	# of species	Size	Mean score			
			Thickness	Ornament	Margin	Total
Cretaceous	79	1.76	1.89	1.50	2.16	7.32
Palaeocene	56	2.22	2.12	1.82	2.35	8.51
Eocene	62	1.79	2.12	1.63	2.28	7.83
Oligocene	30	1.42	2.08	1.82	2.38	7.70

support Vermeij's (1987) suggestion that mass extinctions eliminated escalated species in the Cretaceous, and provide only partial support for the hypothesis with respect to the E/O extinction. These results suggest that either escalated species were not eliminated preferentially, at least by the Cretaceous–Tertiary event, or that our test of Vermeij's hypothesis is inadequate. At present, we prefer the latter explanation for the following reasons.

(1) Thus far we have worked only with species lists from the published literature. These lists do not indicate the relative importance of any of the species to the faunas in which they occur; abundant species are weighted equally with rare species. Work in progress uses only the taxa we have found in our bulk sampling of the faunas and weights species according to their abundance within the fauna. This approach will provide a better estimate of the degree of escalation of the important species in each of the faunas.

(2) The preliminary analysis relies on armour as an indicator of escalation of a fauna, because Vermeij (1987) and others have stressed the role of armour as an antipredatory aptation. However, multiple strategies for escaping predation exist. Focusing on armour as a measure of escalation may obfuscate other equally valid criteria for determining escalation levels. For instance, non-morphological antipredatory traits of bivalves may include either rapid or deep burrowing. Many rapid burrowers have smooth, thin shells; many deep burrowers are characterized by gapes. Both groups would be assigned low escalation scores based on the criteria used in this study, despite their success in escaping naticid predation. Future work will consider such non-morphological antipredatory aptations in determining levels of escalation. Such an approach is consistent with Vermeij's (1983) suggestion that bivalves, in contrast to gastropods, have tended to emphasize avoidance and escape rather than armour.

Comparison with results for ostracodes

Maddocks (1988) presented data on drilling on ostracode assemblages of Cretaceous, Palaeo-gene, and Holocene age from Texas (80 000 specimens). Mortality was calculated as the percent of adult individuals that were drilled. No naticid drillholes were recorded from the Albian, but by the Cenomanian, mortality due to naticid predation was approximately 1%. Naticid-caused mortality reached 4% in the Navarro Group and increased abruptly above the Cretaceous/Tertiary boundary to 8%. Mortality due to naticid predation apparently stabilized at about 7–8% for the remainder of the Midway Group (Palaeocene) and the Claiborne Group (middle Eocene).

Maddocks' results indicate that naticid predation on ostracodes followed a pattern resembling that on molluscs. In particular, the increase in predation on ostracodes above the Cretaceous–Tertiary boundary parallels that observed for molluscs in the early Palaeocene and is consistent with our model of escalation cycles initiated by mass extinctions. Although Maddocks (1988) provided no information on overall levels of escalation of the ostracode fauna, she noted that naticid predation was particularly intense on *Cytherella*, *Bairdoppilata* and *Brachycythere*, all of which have smooth exteriors. (For instance, naticid predation on adult *Brachycythere plena* was 26% during the Palaeocene.) This observation is also consistent with our model.

Work by Reyment and colleagues (Reyment 1966, 1967; Reyment et al. 1987) also supported a preference for drilling on ostracodes with a smooth surface. In both the Santonian of Israel and the Palaeocene of Nigeria, the level of predation on smooth shells exceeded that on moderately ornate shells; drilling was least on strongly ornamented shells. Reyment et al. (1987), however, recorded lower drilling frequencies in the Palaeocene (0.0205) than in the Cretaceous (0.0775), in contrast to Maddocks' and our results. We suggest that the difference in drilling of the Israeli and Nigerian assemblages may be environmentally based; Hansen & Kelley (1995) found that drilling frequencies may vary with latitude. In addition, Reyment et al. (1987) noted that the Israeli samples contained virtually no juveniles, while the Nigerian samples included adults and juveniles. Because naticids apparently prefer adult ostracodes as prey (Reyment 1963; Reyment *et al.* 1987), the nature of the samples may have affected the drilling frequencies. Our results are more appropriately compared to those of Maddocks, because both studies focus on the North American Coastal Plain.

Conclusions

The naticid gastropod predator–prey system in the North American Coastal Plain during the Cretaceous and Palaeogene exhibited a complex pattern of escalation linked to mass extinctions and recoveries. Naticid drilling frequencies on molluscs were moderate in the Cretaceous, declined crossing the K/T boundary, and increased dramatically within the recovery fauna. A similar pattern of drilling frequencies was associated with the Eocene–Oligocene mass extinction and recovery. Predator behavioural stereotypy also was reduced at mass extinctions and recovered more gradually.

The empirical observations yield a model of cyclic escalation for the naticid predator–prey system. Mass extinctions perturb the system and initiate cycles; predator behavioural stereotypy is reduced after extinctions, but drilling frequencies are elevated. The rapid rise in drilling in the recovery fauna is followed by eventual stabilization or decline in drilling frequencies until the next mass extinction disrupts the system.

Naticid drilling frequencies on ostracode assemblages from Texas showed a pattern in the Cretaceous and Palaeogene that supports our escalation model. Nevertheless, further testing of the model is required. The generalizations about drilling frequencies that are incorporated in the proposed escalation model are based only on observations of Late Cretaceous through early Oligocene naticid predation. This interval includes the end of one 'escalation cycle' (the Late Cretaceous), one complete cycle (the Palaeocene–Eocene), and the beginning of the next (the early Oligocene). The database needs to be extended stratigraphically to determine if the pattern of escalation cycles continues in the Neogene.

Work in progress will test the proposed model by: (1) continuing the comprehensive survey of drilling through the Neogene to determine if drilling frequencies rise quickly after Miocene and Pliocene extinction episodes and then stabilize or slowly decline; and (2) performing a more detailed analysis of the mass extinctions and recoveries for both the Palaeogene and Neogene (Cretaceous–Tertiary, Eocene–Oligocene, Miocene, and end-Pliocene extinctions). In particular, we are testing the validity of Vermeij's (1987) hypothesis that mass extinctions preferentially eliminate highly escalated species.

We thank David Dockery for guiding us to collecting localities, David Haasl for assistance in the field, and Elizabeth Akins, Ben Farrell, and David Haasl for help in data collection. Warren Blow, Norman Sohl and Lauck Ward provided access to additional samples. The paper benefited from a review by Richard Reyment and discussion with Melbourne Carriker. This research has been supported by National Science Foundation grant EAR 8915725 (to Kelley and Hansen), and collaborative grants 9405104 (to Kelley) and 9406479 (to Hansen).

References

ADEGOKE, O. S. & TEVESZ, M. J. S. 1974. Gastropod predation patterns in the Eocene of Nigeria. *Lethaia*, **7**, 17–24.

ARNOLD, A. J., D'ESCRIVAN, F. & PARKER, W. C. 1985. Predation and avoidance responses in the foraminifera of the Galapagos hydrothermal mounds. *Journal of Foraminiferal Research*, **15**, 38–42.

BERGGREN, W. A. & PROTHERO, D. R. 1992. Eocene–Oligocene climatic and biotic evolution: an overview. *In:* PROTHERO, D. R. & BERGGREN, W. A. (eds) *Eocene–Oligocene Climatic and Biotic Evolution.* Princeton University Press, 1–28.

CARRIKER, M. R. 1951. Observations on the penetration of tightly closing bivalves by *Busycon* and other predators. *Ecology*, **32**, 73–83.

—— 1981. Shell penetration and feeding by naticacean and muricacean predatory gastropods: A synthesis. *Malacologia*, **20**, 403–422.

CARTER, R. M. 1968. On some predators of bivalved Mollusca. *Palaeogeography, Palaeoclimatology, Palaeoecology*, **4**, 29–65.

DOCKERY, D. T. III. 1977. Mollusca of the Moodys Branch Formation, Mississippi. *Mississippi Geological Survey Bulletin*, **120**.

—— 1982. Lower Oligocene Bivalvia of the Vicksburg Group in Mississippi. *Mississippi Bureau of Geology Bulletin*, **123**.

FISCHER-PIETTE, E. 1935. Histoire d'une moulière. Observations sur une phase déséquilibre faunique. *Bulletin Biologique France et Belgique*, **69**, 153–177.

GUERRERO, S. & REYMENT, R. A. 1988. Predation and feeding in the naticid gastropod *Naticarius intricatoides* (Hidalgo). *Palaeogeography, Palaeoclimatology, Palaeoecology*, **68**, 49–52.

HANSEN, T. A. 1992. The patterns and causes of molluscan extinction across the Eocene/Oligocene boundary. *In:* PROTHERO, D. R. & BERGGREN, W. A. (eds) *Eocene–Oligocene Climatic and Biotic Evolution.* Princeton University Press, 341–348.

—— & KELLEY, P. H. 1995. Spatial variation of naticid gastropod predation in the Eocene of North America. *Palaios*, **10**, 268–278.

——, FARRELL, B. R. & UPSHAW, B. III. 1993. The first two million years after the Cretaceous–Tertiary boundary in east Texas: rate and paleoecology of the molluscan recovery. *Paleobiology*, **19**, 251–265.

HAZEL, J. E., EDWARDS, L. E. & BYBELL, L. M. 1984. Significant unconformities and the hiatuses represented by them in the Paleogene of the Atlantic and Gulf Coastal Province. *In:* SCHLEE, J. S. (ed.) *Interregional Unconformities and Hydrocarbon Accumulation.* American Association of Petroleum Geologists Memoir, **36**, 59–66.

HSÜ, K. J. 1986. Environmental changes in time of biotic crisis. *In:* RAUP, D. M. & JABLONSKI, D. (eds) *Patterns and Processes in the History of Life.* Dahlem Konferenzen. Springer, Berlin, 297–312.

KABAT, A. R. 1990. Predatory ecology of naticid gastropods with a review of shell boring predation. *Malacologia*, **32**, 155–193.

KELLEY, P. H. 1982. Predator-resistant adaptations of shallow water Tertiary bivalves. *Geological Society of America Abstracts with Programs*, **14**, 30.

—— 1988. Predation by Miocene gastropods of the Chesapeake Group: Stereotyped and predictable. *Palaios*, **3**, 436–448.

—— 1989. Evolutionary trends within bivalve prey of Chesapeake Group naticid gastropods. *Historical Biology*, **2**, 139–156.

—— & HANSEN, T. A. 1993. Evolution of the naticid gastropod predator–prey system: An evaluation of the hypothesis of escalation. *Palaios*, **8**, 358–375.

KITCHELL, J. A. 1986. The evolution of predator–prey behavior: Naticid gastropods and their molluscan prey. *In:* NITECKI, M. & KITCHELL, J. A. (eds) *Evolution of Animal Behavior: Paleontological and Field Approaches.* Oxford University Press, 88–110.

——, BOGGS, C. H., KITCHELL, J. F. & RICE, J. A. 1981. Prey selection by naticid gastropods: Experimental tests and application to the fossil record. *Paleobiology*, **7**, 533–552.

——, ——, RICE, J. A., KITCHELL, J. F., HOFFMAN, A. & MARTINELL, J. 1986. Anomalies in naticid predatory behavior: A critique and experimental observations. *Malacologia*, **27**, 291–298.

KOWALEWSKI, M. 1993. Morphometric analysis of predatory drillholes. *Palaeogeography, Palaeoclimatology, Palaeoecology*, **102**, 69–88.

LIVAN, M. 1937. Über Bohr-Löcher an rezenten und fossilen Invertebraten. *Senckenbergiana*, **19**, 138–150.

MADDOCKS, R. F. 1988. One hundred million years of predation on ostracodes: the fossil record in Texas. *In:* HANAI, T. *et al.* (eds) *Evolutionary Biology of Ostracoda. Developments in Palaeontology and Stratigraphy.* Elsevier, Amsterdam, **11**, 637–657.

PALMER, K. V. W. & BRANN, D. C. 1965, 1966. Catalogue of the Paleocene and Eocene Mollusca of the Southern and Eastern United States. *Bulletin of American Paleontology*, **48**, no. 218.

REYMENT, R. A. 1963. Bohrlöcher bei Ostrakoden. *Paläontologische Zeitschrift*, **37**, 283–291.

—— 1966. Preliminary observations on gastropod predation in the western Niger Delta. *Palaeogeography, Palaeoclimatology, Palaeoecology*, **2**, 81–102.

—— 1967. Paleoethology and fossil drilling gastropods. *Transactions of the Kansas Academy of Sciences*, **70**, 33–50.

——, REYMENT, E. R. & HONIGSTEIN, A. 1987. Predation by boring gastropods on Late Cretaceous and early Palaeocene ostracods. *Cretaceous Research*, **8**, 189–209.

SAVAZZI, E. & REYMENT, R. A. 1989. Subaerial hunting behaviour in *Natica gualteriana* (naticid gastropod). *Palaeogeography, Palaeoclimatology, Palaeoecology*, **74**, 355–364.

SOHL, N. F. 1969. The fossil record of shell boring by snails. *American Zoologist*, **9**, 725–734.

STEPHENSON, L. W. 1941. *The larger invertebrate fossils of the Navarro Group of Texas.* University of Texas Publication, No. **4101**.

STUMP, T. E. 1975. Pleistocene molluscan paleoecology and community structure of the Puerto Libertad region, Sonora, Mexico. *Palaeogeography, Palaeoclimatology, Palaeoecology*, **17**, 177–226.

TAYLOR, J. D. 1970. Feeding habits of predatory gastropods in a Tertiary (Eocene) molluscan assemblage from the Paris basin. *Palaeontology*, **13**, 254–260.

THOMAS, R. D. K. 1976. Gastropod predation on sympatric Neogene species of *Glycymeris* (Bivalvia) from the Eastern United States. *Journal of Paleontology*, **50**, 488–499.

—— 1978. Shell form and the ecological range of living and extinct Arcoida. *Paleobiology*, **4**, 181–194.

VERMEIJ, G. J. 1978. *Biogeography and Adaptation: Patterns of Marine Life.* Harvard University Press, Cambridge, Massachusetts.

—— 1983. Traces and trends of predation, with special reference to bivalved animals. *Palaeontology*, **26**(3), 455–465.

—— 1987. *Evolution and Escalation: An Ecological History of Life.* Princeton University Press.

—— 1994. The evolutionary interaction among species: Selection, escalation, and coevolution. *Annual Review of Ecology and Systematics*, **25**, 219–236.

ZACHOS, J. C., ARTHUR, M. A. & DEAN, W. E. 1989. Geochemical evidence for suppression of pelagic marine productivity at the Cretaceous/Tertiary boundary. *Nature*, **337**, 61–64.

ZIEGELMEIER, E. 1954. Beobachtungen über den Nahrungserwerb bei der Naticide *Lunatia nitida* Donovan (Gastropoda Prosobranchia). *Helgoländer Wissenschaftliche Meeresforschung*, **5**, 1–33.

Index

acritarchs 131
Aeronian, recovery patterns 127–33
Alaska, graptolites 119–26
Albian, Late, Albian–Cenomanian oceanic anoxic
  events (OAEs) 240–1
ammonites 231
  Campanian 299–308
  desmoceratacean, E Russia, Sakhalin 299–308
  Santonian–Maastrichtian, stratigraphy 300–3
  see also goniatites
ammonoids
  early stages 164–8
  goniatite survival 163–85
  Japan 306
  juvenile ornament 169
  morphological sequence 169
  protoconch size 166, 181
  recovery, Sakhalin 304, 306
  taxonomic diversity, dynamics 305
Amphipora-bearing limestone 135–61
angiosperms, origination, extinction and diversity 73
Anisian Stage 223, 224
  Lazarus taxa 227
Annulata Event, Devonian 178
Apterygota 65
archaeocyaths 82, 86
Ashgill, correlation of biotic events 129
Atavograptus atavus Zone 124
Avalonian plate 125

bacterial chemosymbioses 49–50
benthic communities
  Phanerozoic trends 2, 3
  tiering 3, 4
bioturbation, ichnofabrics 5–6
Botomian, Early–Middle Botomian, Sinsk Event 80–1
brachiopods
  first appearances, Europe 233
  Ordovician mass extinction events 102
Brazil, Cretaceous–Tertiary boundary (K/T) event,
  Poty section 318–36
Brotzenella, evolutionary lineage 341
bryozoans 61

calcimicrobes 82–5, 87
Cambrian
  cyanobacteria, interactions with reef-building
    organisms 82
  Early Cambrian mass extinction
    reef ecosystem recovery 79–96
    reef-building organisms 82–8
    Sinsk Event 80–1
  explosion of phyla, status 9–10
Caninia 187
Cantabrian Mountains, rugose corals 187–99
Capitan Limestone, Texas Permian Basin 224–5
Capitanian Stage 223

Carbonate Dagestan 259
carbonate ramps, Frasnian–Famennian 135
Carboniferous
  'lesser mass extinction event'
    rugose corals
      extinction 188–9
      recovery 192–7
      survival interval 189–91
catastrophic mass extinction 54
Caucasus, N
  foraminifera, Danian extinctions 337–42
  locations 337
Caucasus, NE
  foraminifera, Cenomanian–Turonian Boundary
    Event 259–64
  location map 260
Cauvery Basin, oceanic anoxic events (OAEs) 238
Cenomanian–Turonian Boundary Event
  dinoflagellate cyst assemblages recovery, England,
    oceanic anoxic events 279–97
  England, S 267
  food chain recovery 265–77
  foraminifera 237–44, 259–64
  Milankovitch rhythms 246
  oceanic anoxic events (OAEs)
    India, SE, Cauvery Basin 237–44
    NE Caucasus 259–64
    Spain, Menoyo section 245–58
  Turonian lithological logs, England, SE 280–2
Changxingian Stage 223, 224
Chotec Event, Eifelian 171
Clementsian ecosystem theories 216–17
climate and ocean-state, Ordovician, Late, mass
  extinction event, conodonts 105–18
climax cut-off model 61–2
clymeniid ammonoids 163
coal-balls 202–4, 215–16
community structure, mass extinction recovery 61–3
Condroz Event, Devonian 178
conodonts
  F–F facies 136–7
  Middle *varcus* Zone 174
  Ordovician, Late, mass extinction event
    appearance of new taxa 112–14
    climate and ocean-state models 105–18
    per taxon rates of extinction and origination
      114–15
coralomorphs 82
corals 127–8
  see also reef-building organisms
corals, rugose
  Carboniferous 'lesser mass extinction event' 187–99
  morphoecotypes 189
  morphotypes 187
Coronograptus gregarius Zone 124
Corsicana Formation, naticid drilling frequencies
  375–7
Cretaceous
  marine food chain model 266

388 INDEX

mollusc species 'fabric', US Atlantic and Gulf coastal plain 309–18
oceanic anoxic events (OAEs) 238
Cretaceous–Palaeogene (K/Pg) see Cretaceous–Tertiary boundary (K/T) event
Cretaceous–Tertiary boundary (K/T) event
  extinction and survivorship selectivity 363–4
  foraminifera
    Danian, Early 319–35
    Tethyan, S 343–71
  gastropod predator–prey system 373–86
  insects 74
  North American Coastal Plain 373
  recoveries, mixed and graded communities 62–3
crinoids 231
crisis progenitor taxa
  mass extinction recovery 15–39
  Phanerozoic examples 20–1, 23–6
cyanobacteria
  interactions with reef-building organisms
    Cambrian 82
    Lower–Middle Ordovician 89
cycadophytes, recoveries 62
Czech Republic, Frasnian–Famennian events 135–61
Czekanowskia, recovery 62

Dakhla Formation, Tunisia 346
Daleje Event, Emsian 170–1
Danian extinctions, foraminifera
  Caucasus, N 337–42
  K/T boundary event 319–35
Decorah Shale 97, 102
Deicke metabentonite 97–104
  Deicke-K horizon 102
dendrolites, Cambrian–Lower Ordovician 87
desmoceratacean ammonites, Russia, E, Sakhalin, Santonian, Campanian boundary event 299–308
Devonian
  Annulata Event 178
  Condroz Event 178
  Enkeberg Event 178
  Hangenberg Event 179–81
  Kellwasser Events 135–61, 163, 177
Devonian, Late
  goniatite survival 163–85
  Nehden Event 177–8
  reef-building organisms 135–61
dinoflagellate cyst assemblages recovery
  England, S
    Cenomanian–Turonian Boundary Event 279–97
    list 284
dinosaurs
  gigantism tendency 63
  recoveries 62–3
disaster species, trophic/life habit 45, 47
Djulfian Stage 223, 225
dormancy 49
Dubuque–Maquoketa boundary, US 102

ecosystem reconstructions, England, S 268–73
Edenian, Stewartville Formation 102
Egypt, K/T foraminifera 346, 348

Eifelian, Chotec Event 171
El Kef, Tunisia, K/T foraminifera 343–71
Elburgan Formation 337
Elvis-taxa 79
Emsian
  Daleje Event 170–1
  Zlichov Event 168–70
England, S
  ecosystem reconstructions 268–73
  location maps 279
  oceanic anoxic events, dinoflagellate cyst assemblages recovery, Cenomanian–Turonian Boundary Event 279–97
  see also UK
Enkeberg Event, Devonian 178
Eocene–Oligocene extinction, Cretaceous–Tertiary extinction, gastropod predator–prey system, naticids 373–86
Estonia, and Ordovician mass extinction 127–33
Europe
  bivalve molluscs 234
  brachiopods 233–4
    first appearances 233
  (end)-Triassic mass extinctions, marine fauna 231–6
  Frasnian–Famennian–Kellwasser events, Mokrá Section, Moravia 135–61
evolutionary palaeoecology
  comparative approach 6–10
  Permian–Triassic mass extinction 7–9

ferns, opportunists 216
food chain recovery, Cenomanian–Turonian Boundary Event 265–77
foraminifera 231, 251, 252
  accumulation rates (PFARs) 348–9
  benthic K/T boundary, El Kef 349–54
  Cenomanian–Turonian Boundary Event
    Britain 265–77
    Caucasus, N 337–42
    Caucasus, NE 259–64
    India 237–44
    Spain, Menoyo section 245–58
  Cretaceous–Tertiary boundary (K/T) event
    Danian, Early 319–35
    Tethyan, southern 343–71
  Danian extinctions, Caucasus, N 337–42
  dynamics of diversification 262–4
  oceanic anoxic events 259–64
    India, SE, Cauvery Basin, Cenomanian–Turonian Boundary Event 237–44
  productivity, Maastrichtian P/B ratios 348
  zones, rugose corals 189
Frasnian
  Frasnes Event 175–6
  Middlesex Event 176
  Rhinestreet Event 176–7
Frasnian–Famennian–Kellwasser events, Mokrá Section, Moravia, Central Europe, extinction recovery gradients 135–61

gastropods, naticids, predator–prey system, Cretaceous–Tertiary extinction and Eocene–

INDEX

Oligocene extinction 373–86
Genundewa Event, Frasnian 175
gigantism tendency 63
Stephanian (Late Pennsylvanian) 213–15
Givetian
Kačák Event 171–3
Pumilio Events 173–4
Gleasonian ecosystem theories 216–17
Glenwood formation 97, 102
Globigerina eugubina Zone 337
Globigerinida 323
Globorotalida 323
goniatites, juvenile, survival strategies, Devonian
extinction events 163–85
graptolites
Ordovician, Late, mass extinction event, Alaska
119–26
Silurian, Early, recovery patterns 128–31
Guadaloupian, gastropods 224
guilds, Phanerozoic trends 2, 3

habitats
protected or buffered 44
refugia species 21–2, 44–5
Hangenberg Event, Devonian 163, 179–81
Haustator bilira Assemblage Zone 309
Hawke Bay Event 79–96, 81–2
herbivores, gigantism tendency 63
Hettangian, Lower Jurassic 231–2
hexapods *see* insects
high-resolution stratigraphy 55–6
Hirnantian 130
Holometabola 65
Homoceras Zone 187

ichnofabrics 5–6
India, SE, Cauvery Basin
Cenomanian–Turonian Boundary Event, oceanic
anoxic events (OAEs), foraminifera 237–44
Cretaceous rocks distribution 238
Indian Ocean, development 237
Induan stage 224
Triassic recovery 226
Inoceramidae, *Mytiloides* 26–34
Inoceramus, displacing *Mytiloides* 33
Inoceramus labiatus Zone 260
insects
Appendix 1, dating update 76–7
apterygotes, origination, extinction and diversity 69,
70
Cretaceous–Tertiary extinction 74
diversity trends 65–78
ordinal diversity 74
early evolution 69–75
oligoneopteran vs polyneopteran life-cycles 70
origination, extinction and diversity
families 67, 68
genera 68
oligoneopterans 72
orders 66
palaeopterans 71
paraneopterans 71

Phanerozoic 65–78
polyneopterans 71
Israel, K/T foraminifera 346, 348

Jurassic, Lower, Britain, species diversity 232
Jurassic–Cretaceous boundary, recoveries, mixed and
graded communities 62–3

Kačák Event, Givetian 171–3
Karai–Kulakkainattam, India, SE, Cauvery Basin
237–44
Karanaiskaya Formation 260
Kellwasser events 163
Central Europe 135–61
Frasnian–Famennian 135–61, 163, 177
Kincaid Formation, naticid drilling frequencies 375–7

Lagarograptus acinaces Zone 124
Laurentian craton 107, 119
Lazarus taxa 21, 41, 55, 223–4
Liassic, Britain, species diversity 231–4
Llandovery
acritarchs 131
correlation of biotic events and env/parameters 129
recovery patterns 127–33
lycopsids, *Chaloneria* 201

Maastrichtian, Upper, benthic foraminifera, Middle
East 355–9
Maastrichtian–Danian boundary *see* Cretaceous–
Tertiary boundary (K/T) event
Manticoceras Event 175
marine fauna, Europe, (end)-Triassic mass extinctions
231–6
mass extinctions
mass extinction–survival–recovery intervals 15–16
stepwise or graded 52, 54
see also recoveries; survivors and survival
mechanisms; specific events
metabentonite horizons, Ordovician 97–104
microbial structures 6
Middle East
benthic foraminifera 355–60
benthic foraminiferal assemblages 355–7
Middlesex Event, Frasnian 176
models for biotic survival, mass extinction recovery
41–60
Mokrá Section, Moravia, Central Europe
Frasnian–Famennian–Kellwasser events
extinction recovery gradients 135–61
interpretation 158–60
stratigraphy, data analysis 137–58
molluscs
antipredatory characters, four faunas 384
bivalve molluscs 231
first appearances, Europe and S America 234
gastropod predator–prey system, Cretaceous–
Tertiary boundary (K/T) event and Eocene–
Oligocene extinction 373–86
gastropods, Permo-Triassic mass extinction 223–9

390 INDEX

proportions by habitat, USA 313, 315
species 'fabric', Cretaceous, US Atlantic and Gulf
coastal plain 309–18
Monte Carlo simulations 135
Moodys Branch, molluscan assemblages 374–85
Mytiloides, crisis progenitor bivalve 26–34

naticids
drillhole morphology and size 373
drillholes through time
bivalves 376
gastropods 376
drilling behaviour, stereotypy 378–82
gastropods predator–prey system
Cretaceous–Tertiary extinction, Eocene–Oligocene
extinction 373–86
model of escalation 382–5
nomismogenesis 305–6
North America
Alaska 122–3
Atlantic and Gulf coastal plain, mollusc species
'fabric', Cretaceous 309–18
Canadian Cordillera 121–2
Coastal Plain, Cretaceous and Palaeogene 373–86
Great Basin 120–1
metabentonite horizons, Ordovician 97–104
molluscan assemblages 374
north central USA
Ordovician extinction events, Ostracode
speciation 97–104
ostracode speciation 97–104
Ordovician, Late, mass extinction event, graptolites
119–26
Ordovician, Middle and Upper, mass extinction
event, summary correlation chart 98
peat-forming environments, Pennsylvanian,
Middle–Late, transition 201–21
Pennsylvanian Transition 201–21
Novaya Zemija, rugose corals, distribution 189, 190,
191

ocean-state model
biotic changes, summary 110
composite standard reference section (CSRS) range
charts 111–15
Ordovician mass extinction events 107–12
oceanic anoxic events (OAEs)
Cenomanian–Turonian Boundary Event, India, SE,
Cauvery Basin 237–44
and continental productivity, Tethyan, S 362–3
dinoflagellate cyst assemblages recovery, England,
S, Cenomanian–Turonian Boundary Event 279–
97
foraminifera 259–64
Olenekian, Triassic recovery 226–7
Ordovician, cyanobacteria, interactions with reef-
building organisms 89
Ordovician mass extinction events
conodonts
climate and ocean-state 105–18
IGCP Project 335 model 106

Estonia 127–33
graptolites, North America 119–26
Ordovician–Silurian, global sea-levels and ocean-
state 106–12
Ostracoda, speciation 97–104, 231
summary correlation chart 98
Ostracoda, K/T boundary 359
Ostracoda, speciation 231
Ordovician (Middle) extinction events, United
States, north central 97–104
Upper Mississippi Valley, stratigraphic distribution
99–101

paedomorphosis 44
Palaeocene, Early, S Tethyan 359–60
palaeoecology, assessment 1–13
Palaeosmilia 187
Parakidograptus acuminatus Zone 120, 123–4
Paraorthograptus pacificus Zone 121–5
peat-forming environments, North America, Pennsylvanian, Middle–Late, transition 201–21
Pennsylvanian, Middle–Late, transition, peat-forming
environments, North America 201–21
Permian, percent extinctions 226
Permian Basin, Texas, Capitan Limestone 224–5
Permian–Triassic mass extinction 7–9
gastropods 223–9
PFARs (planktonic foraminifera accumulation rates)
348–9
Phanerozoic
insect origination and extinction 65–78
palaeoecological trends 1–13
patterns of recovery and radiation 22–3
trends, benthic communities
onshore–offshore patterns 4
tiering 3, 4
plants *see* vascular plants; *named groups*
Platteville Limestone 97, 102
Plenus Marl Formation 279
Pliensbachian Stage 231
Lower Jurassic 232
population dynamics, widespread dispersion 48–9
Poty section, Brazil, foraminifera, K/T boundary event
319–35
Praeglobotruncana imbricata Zone 260
preadaptation 43, 44
Prionoceras
differentiation from *Mimimitoceras* 179
Stockum Limestone 180–1
Prosser Formation 102
Pterygota 65
Pumilio Events, Givetian 173–4

recoveries
community structure 61–3
crisis progenitor taxa 15–39
defined 265
dinoflagellate cyst assemblages, Cenomanian–
Turonian Boundary Event 279–97
foraminifera, oceanic anoxic events 237–44
gradients, Frasnian–Famennian–Kellwasser events,
Central Europe 135–61
graptolites, Silurian, Early 128–31

INDEX

391

Jurassic–Cretaceous boundary, mixed and graded
  communities 62–3
models for biotic survival 41–60
reef ecosystem, Cambrian, Early, 79–96
  Silurian, Early 127–33
reef-building organisms
  Early Cambrian extinction
    ecosystem recovery 79–96
    generic diversity 80
  end-Triassic mass extinctions 234
  Hawke Bay Event 81–2
  Late Devonian 135–61
  Lower–Middle Ordovician 88–9
  organism interactions
    Late Cambrian reefs 88
    Middle Cambrian reefs 86–8
  see also corals
  Sinsk Event 80–1
  terminology 160
refugia species 21–2
  habitats 44–5
  temporal and abundance patterns 46
Rhinestreet Event, Frasnian 176–7
Rhuddanian Stage 127
  recovery patterns 127–33
rhynchonellid brachiopods 231
Rocklandian Stage ash fall 97
Rotalipora, disappearance 260
Rotalipora cushmani Zone 245–58, 259–64
rudist bivalves 61
Russia, *see also* Caucasus, N and NE
Russia, E, Sakhalin, desmoceratacean ammonites,
  Santonian, Campanian boundary event 299–308

Santonian, Campanian boundary event, desmoceratacean ammonites, Russia, E, Sakhalin 299–308
Santonian–Maastrichtian, stratigraphy 300, 301
sea-levels
  and benthic changes 361–2
  Cretaceous–Tertiary boundary (K/T) event 360–1
  Ordovician mass extinction events 106
  Silurian 120
Selaginella, recoveries 62
Shannon–Weaver heterogeneity index H(S) 247
shell beds 6
Shermanian, Prosser Formation 102
Silurian, Early
  diachronous recovery patterns
    acritarchs 131
    corals 127–8
    graptolites 128–31
  Llandovery palaeogeography map 120
  recovery patterns 127–33
Sinemurian Stage, Lower Jurassic 231, 232
Sinsk Event 80–1
Skolithos, in ichnofabrics 5–6
Spain, Menoyo section
  Cenomanian–Turonian Boundary Event
    foraminiferal 245–58
    lithostratigraphy 246
  location 245
Spain, Zumaya section, Maastrichtian 314
Spathian Stage 223

Stephanian (Late Pennsylvanian) 201
  gigantism 213–15
  post-extinction system 210–12
  Westphalian–Stephanian transition 206–10
stepwise or graded mass extinction 52, 54
Stewartville Formation 102
Stockum Limestone, *Prionoceras* 180–1
Strangelove Ocean period 319
stromatolites
  Cambrian–Lower Ordovician 87
  Late Cambrian 88
  Lower–Middle Ordovician 88–9
  trends 6, 8
survivors and survival mechanisms 16–18
  adaptive mechanisms, summary 51–5
  common groups 18–22
  crisis progenitors 20–1
  disaster species 19–20
  ecological generalists 18, 47–8
  ecological opportunists 20
  habitats, protected or buffered 44
  opportunists 47
  paedomorphosis 44
  persistent trophic resource exploiters 48
  physiological traits
    bacterial chemosynbioses 49–50
    dormancy 49
    reproductive adaptation 50–1
    skeletonization requirements 50
  population dynamics 48–9
  preadaptation 43, 44
  preadapted survivors 20
  rapid evolution 43
  refugia species 21–2
  stranded populations 21
  survivorship selectivity, K/T event 363–4

Taghanic Event, Devonian 174
Tatarian stage 224
Telychian, recovery patterns 127–33
terminology, reef-dwelling fauna 160
Tethyan, S
  benthic communities, stages 357–60
  continental productivity, and oxygen deficiency
    362–3
  foraminifera, Cretaceous–Tertiary boundary (K/T)
    343–71
  Palane, Early 359–60
Texas Coastal plain, naticid drilling frequencies 375–7
Texas Permian Basin, Capitan Limestone 224–5
thrombolites
  Late Cambrian 88
  Lower–Middle Ordovician 88–9
Timan, N, rugose corals, distribution 192
Toarcian, mass extinctions 231–6
Tommotian Stage 79
tree ferns 201
(end)-Triassic mass extinctions, Europe, marine fauna
  231–6
Triassic recovery, phases 1–3 226–7
trophic/life habit
  disaster species 45, 47
  ecological generalists 47–8

opportunists 47
persistent trophic resource exploiters 48
tropical peat-forming environments 201–21
Tunisia
K/T foraminifera 343–71
benthic, common taxa 349–54
faunal turnover 354
productivity 348–9
Turonian *see* Cenomanian–Turonian Boundary Event

UK, Ordovician, Middle and Upper, mass extinction
event, summary correlation chart 98
USA *see* North America

vascular plants
origination, extinction and diversity
angiosperms 73
families 73

Vendian–Cambrian *see* Cambrian explosion

Wenlock Stage 127
Estonia 130–1
recovery patterns 127–33
Westphalian D mires 204–6
Westphalian Stage extinction 201
Westphalian–Stephanian transition 206–10
pre-extinction–postextinction landscapes 212–15
Whiteinella archaeocretacea Zone 245–58
Wordian Stage 223–5
Lazarus taxa 224

Yazoo Formation, molluscan assemblages 374

Zlichov Event, Emsian 168–70